T0191673

IFIP Advances in Information and Communication Technology 436

IFIP – The International Federation for Information Processing

IFIP was founded in 1960 under the auspices of UNESCO, following the First World Computer Congress held in Paris the previous year. An umbrella organization for societies working in information processing, IFIP's aim is two-fold: to support information processing within its member countries and to encourage technology transfer to developing nations. As its mission statement clearly states,

> *IFIP's mission is to be the leading, truly international, apolitical organization which encourages and assists in the development, exploitation and application of information technology for the bene t of all people.*

IFIP is a non-profitmaking organization, run almost solely by 2500 volunteers. It operates through a number of technical committees, which organize events and publications. IFIP's events range from an international congress to local seminars, but the most important are:

- The IFIP World Computer Congress, held every second year;
- Open conferences;
- Working conferences.

The flagship event is the IFIP World Computer Congress, at which both invited and contributed papers are presented. Contributed papers are rigorously refereed and the rejection rate is high.

As with the Congress, participation in the open conferences is open to all and papers may be invited or submitted. Again, submitted papers are stringently refereed.

The working conferences are structured differently. They are usually run by a working group and attendance is small and by invitation only. Their purpose is to create an atmosphere conducive to innovation and development. Refereeing is also rigorous and papers are subjected to extensive group discussion.

Publications arising from IFIP events vary. The papers presented at the IFIP World Computer Congress and at open conferences are published as conference proceedings, while the results of the working conferences are often published as collections of selected and edited papers.

Any national society whose primary activity is about information processing may apply to become a full member of IFIP, although full membership is restricted to one society per country. Full members are entitled to vote at the annual General Assembly, National societies preferring a less committed involvement may apply for associate or corresponding membership. Associate members enjoy the same benefits as full members, but without voting rights. Corresponding members are not represented in IFIP bodies. Affiliated membership is open to non-national societies, and individual and honorary membership schemes are also offered.

Lazaros Iliadis   Ilias Maglogiannis
Harris Papadopoulos (Eds.)

Artificial Intelligence Applications and Innovations

10th IFIP WG 12.5 International Conference, AIAI 2014
Rhodes, Greece, September 19-21, 2014
Proceedings

 Springer

Volume Editors

Lazaros Iliadis
Democritus University of Thrace
Department of Forestry and Management of the Environment
Pandazidou 193, 68200 Orestiada, Greece
E-mail: liliadis@fmenr.duth.gr

Ilias Maglogiannis
University of Piraeus
Department of Digital Systems
80, Karaoli and Dimitriou Str., 18534 Piraeus, Greece
E-mail: imaglo@unipi.gr

Harris Papadopoulos
Frederick University
Department of Computer Science and Engineering
7, Yianni Frederickou Str., Pallouriotissa, 1036 Nicosia, Cyprus
E-mail: h.papadopoulos@frederick.ac.cy

ISSN 1868-4238   ISSN 1868-422X (eBook)
ISBN 978-3-662-52599-9   ISBN 978-3-662-44654-6 (eBook)
DOI 10.1007/978-3-662-44654-6
Springer Heidelberg New York Dordrecht London

Typesetting: Camera-ready by author, data conversion by Scientific Publishing Services, Chennai, India

Printed on acid-free paper

Springer is part of Springer Science+Business Media (www.springer.com)

Preface

It has been 58 years since the term artificial intelligence (AI) was coined in 1956 by John McCarthy at the Massachusetts Institute of Technology USA. Since then, after huge efforts of the international scientific community, sophisticated and advanced approaches— e.g., games playing, (computers capable of playing games against human opponents) natural languages, computers able to see, hear, and react to sensory stimuli— that would appear only as science fiction in the past are gradually becoming a reality. Multi-agent systems and autonomous agents, image processing, and biologically inspired neural networks (Spiking ANN) are already a reality. Moreover, AI has offered the international scientific community many mature tools that are easily used, well documented, and applied. These efforts have been continuously technically supported by various scientific organizations like the IFIP.

The International Federation for Information Processing (IFIP) was founded in 1960 under the auspices of UNESCO, following the first historical World Computer Congress held in Paris in 1959. The first AIAI conference (Artificial Intelligence Applications and Innovations) was organized in Toulouse, France, in 2004 by the IFIP. Since then, it has always been technically supported by the Working Group 12.5 "Artificial Intelligence Applications." After 10 years of continuous presence, it has become a well-known and recognized mature event, offering AI scientists from all over the globe the chance to present their research achievements. The 10th AIAI was held in Rhodes, Greece, during September 19–21, 2014.

Following a long-standing tradition, this Springer volume belongs to the IFIP AICT series and it contains the accepted papers that were presented orally at the AIAI 2014 main conference. An additional volume comprises the papers that were accepted and presented to the workshops and were held as parallel events. Four workshops were organized, by invitation to prominent and distinguished colleagues, namely:

- The Third CoPA (Conformal Prediction and Its Applications)
- The Third MHDW (Mining Humanistic Data Workshop)
- The Third IIVC (Intelligent Innovative Ways for Video-to-Video Communications in Modern Smart Cities)
- The First MT4BD (New Methods and Tools for Big Data)

It is interesting that three of the above workshops were organized for the third time in the row, which means that they are well established in the AI community.

As the title of the conference denotes, there are two core orientations of interest, basic research AI approaches and also applications in real-world cases. The diverse nature of papers presented demonstrates the vitality of AI computing methods and proves the wide range of AI applications.

All papers went through a peer-review process by at least two independent academic reviewers. Where needed, a third and a fourth reviewer was consulted to resolve any potential conflicts. In the 10th AIAI conference, 41.3% of the submitted manuscripts (62) were accepted for oral presentation. From these, only 33 (22%) were accepted as full papers, whereas 29 (19.3%) were accepted as short ones. The authors of accepted papers of the main event come from 20 countries, namely: Brazil, Bulgaria, Canada, Cyprus, China, Czech Republic, Denmark, Finland, Germany, Great Britain, Greece, Iran, Italy, Pakistan, Poland, Russia, Spain, Switzerland, Tunisia, and Turkey.

Three distinguished keynote speakers were invited to present a lecture at the 10th AIAI conference.

1. Professor Hojjat Adeli, Ohio State University, USA.

Title: "Multi-Paradigm Computational Intelligence Models for EEG-Based Diagnosis of Neurological and Psychiatric Disorders"

- Professor in the Departments of Biomedical Informatics, Civil and Environmental Engineering and Geodetic Science and Neuroscience and Centers of Biomedical Engineering and Cognitive Science
- Director of Knowledge Engineering Lab at the Ohio State University
- Author of nearly 400 research and scientific publications in various fields of computer science, engineering, and applied mathematics since 1976 when he received his Ph.D. from Stanford University
- Author of nine books
- Founder and Editor-in-Chief of the international research journals *Computer-Aided Civil and Infrastructure Engineering* and *Integrated Computer-Aided Engineering*
- Over 100 academic, research, and leadership awards, honors, and recognition
- Keynote/plenary lecturer at 43 national and international computing conferences held in 28 different countries

2. Professor Plamen Angelov, Lancaster University, UK.

Title: "Autonomous Learning Systems: Association-Based Learning"

- Chair in Intelligent Systems and leads the Intelligent Systems Research within the School of Computing and Communications, Lancaster University, UK
- Founding Chair of the Technical Committee on Evolving Intelligent Systems with Systems, Man and Cybernetics Society, IEEE
- Co-recipient of several best paper awards at IEEE conferences (2006 and 2009, 2012, 2013)
- Co-recipient of two prestigious Engineer 2008 Technology + Innovation awards for Aerospace and Defense
- Co-recipient of the Special Award as well as the Outstanding Contributions Award by IEEE and INNS (2013)

- Editor-in-Chief of the Springer journal *Evolving Systems*, Associate Editor of the prestigious *IEEE Transactions on Fuzzy Systems* and of Elsevier's *Fuzzy Sets and Systems*

3. Professor Tharam Dillon, La Trobe University, Australia

Title: "Conjoint Mining of Data and Content with Applications in Business, Bio-medicine, Transport Logistics and Electrical Power Systems"

- Life Fellow IEEE, FACS, FIE
- Editor-in-Chief of the *International Journal of Computer Systems Science & Engineering* (UK) 1986–1991 Butterworths, 1992–1996 CRL Publishing
- Editor-in-Chief of the *International Journal of Engineering Intelligent Systems* (UK) 1993–1996
- Chief Co-editor of the *International Journal of Electric Power and Energy Systems* (UK) 1978–1991, Butterworths 1992–1996 Elsevier
- Associate Editor of *IEEE Transactions on Neural Networks* (USA) 1994–2004

The accepted papers of the 10th AIAI conference are related to the following thematic topics:

- Artificial neural networks
- Bioinformatics
- Feature extraction
- Clustering
- Control systems
- Data mining
- Engineering applications of AI
- Face recognition, pattern recognition
- Filtering
- Fuzzy logic
- Genetic algorithms, evolutionary computing

- Hybrid clustering systems
- Image and video processing
- Multi-agent systems
- Environmental applications
- Multi-attribute DSS
- Ontology, intelligent tutoring systems
- Optimization, genetic algorithms
- Recommendation systems
- Support vector machines, classification
- Text mining

- We wish to thank Professors Harris Papadopoulos (Frederick University, Cyprus), Alex Gammerman and Vladimir Vovk (Royal Holloway University of London, UK) for their common efforts toward the organization of the third CoPA workshop.
- We are also grateful to Professors Spyros Sioutas, Katia Lida Kermanidis (Ionian University, Greece), Christos Makris (University of Patras, Greece), and Giannis Tzimas (TEI of Western Greece). Thanks to their invaluable contribution and hard work, the third MHDW workshop was held successfully once more and it has already become a well-accepted event running in parallel with AIAI.

- The third IIVC workshop was an important part of the AIAI 2014 event and it was driven by the hard work of Drs. Ioannis P. Chochliouros and Ioannis M. Stephanakis (Hellenic Telecommunications Organization— OTE, Greece) and Professors Vishanth Weerakkody (Brunel University, UK) and Nancy Alonistioti (National and Kapodistrian University of Athens, Greece).
- It was a pleasure to host the MT4BD 2014 in the framework of the AIAI conference. We wish to sincerely thank its organizers for their great efforts. More specifically we wish to thank Professors Spiros Likothanassis (University of Patras, Greece), Dimitrios Tzovaras (CERTH/ITI, Greece), Eero Hyvönen (Aalto University, Finland), and Jörn Kohlhammer (Fraunhofer-Institut für Graphische Datenverarbeitung IGD, Germany).

A Keynote lecture was given by Dr. Dimitrios Tzovaras in the framework of the MT4BD 2014 workshop.
Title: Visual Analytics Technologies for the Efficient Processing and Analysis of Big Data

The AIAI 2014 conference had a high attendance from scientists from all parts of the globe and we would like to thank all participants for this. The organization of the 10th AIAI was truly a milestone. After 10 years, it has been established as a mature event with loyal followers and it has plenty of new and qualitative research results to offer the international scientific community. We hope that the readers of these proceedings will be highly motivated and stimulated for further research in the domain of AI in general.

September 2014   Lazaros Iliadis
                                                                Ilias Maglogiannis
                                                               Harris Papadopoulos

Organization

Executive Committee

General Chair

Tharam Dillon — Latrobe University, Melbourne, Australia

Program Committee Co-chairs

Lazaros Iliadis — Democritus University of Thrace, Greece
Ilias Maglogiannis — University of Piraeus, Greece
Harris Papadopoulos — Frederick University, Cyprus

Workshop Co-chairs

Spyros Sioutas — Ionian University, Greece
Christos Makris — University of Patras, Greece

Organizing Co-chairs

Yannis Manolopoulos — Aristotle University of Thessaloniki, Greece
Andreas Andreou — Cyprus University of Technology, Cyprus

Advisory Co-chairs

Elias Pimenidis — University of East London, UK
Chrisina Jayne — Coventry University, UK
Haralambos Mouratidis — University of Brighton, UK

Website and Advertising Chair

Ioannis Karydis — Ionian University, Greece

Honorary Co-chairs

Nikola Kasabov — KEDRI Auckland University of Technology, New Zealand
Vera Kurkova — Czech Academy of Sciences, Czech Republic
Hojjat Adeli — The Ohio State University, USA

Program Committee

El-Houssaine Aghezzaf — Ghent University, Belgium
Michel Aldanondo — Toulouse University, Mines Albi, France
George Anastassopoulos — Democritus University of Thrace, Greece
Ioannis Andreadis — Democritus University of Thrace, Greece
Andreas Andreou — Cyprus University of Technology, Cyprus

Demosthenes Vouyioukas	University of the Aegean, Greece
Vladimir Vovk	Royal Holloway, University, UK
Arlette Van Wissen	VU University Amsterdam, The Netherlands
Michalis Xenos	Hellenic Open University, Greece
Xin-she Yang	University of Cambridge, UK
Engin Yesil	Technical University of Istanbul, Turkey
Contantine Yialouris	Agricultural University of Athens, Greece
Peter Yuen	Cranfield University, UK
Drago Zagar	University of Osijek, Croatia

Table of Contents

Hybrid - Changing Environments

Agents (AGE)

Classification Pattern Recognition

Genetic Algorithms

Image - Video Processing

Feature Extraction

Environmental AI

Simulations and Fuzzy Modeling

Data Mining-Forecasting

AI Applications - Mobile Applications

Image Video Processing 4

Conjoint Mining of Data and Content with Applications in Business, Bio-medicine, Transport Logistics and Electrical Power Systems

Tharam S. Dillon[1,2], Yi-Ping Phoebe Chen[1], Elizabeth Chang[2], and Mukesh Mohania[3]

[1] Department of Computer Science and Computer Engineering,
La Trobe University, Melbourne, Victoria 3086, Australia
[2] School of Business, Australian Defence Force Academy,
University of New South Wales, Canberra, Australia
[3] IBM India Research Lab.
Tharam.dillon7@gmail.com, Phoebe.Chen@latrobe.edu.au,
Elizabeth.chang@unsw.edu.au, mkmukesh@in.ibm.com

1   Introduction

Digital information within an enterprise consists of (1) *structured data* and (2) *unstructured content*. The structured data includes enterprise and business data like sales, customers, products, accounts, inventory and enterprise assets, etc. while the content includes contracts, reports, emails, customer opinions, transcribed calls, on-line inquires, complements and complaints. Further, cutting edge businesses also using GPS tracking or surveillance monitors as well as sensor technologies for productivity, performance and efficiency measures, and these are provided by outsourcers etc. Similarly in the Biomedical area, resources can be structured data say in Swiss-Prot or unstructured text information in journal articles stored in content repositories such as PubMed. The structured data and the unstructured content generally reside in entirely separate repositories with the former being managed by a DBMS and the latter by a content manager frequently provided by an outsourcer or vendor [76]. This separation is undesirable since the information content of these sources is complementary. Further, each outsourcer or vendor keep the data on their own Cloud, and data are not sharable between the vendor systems, and most vendor system were not integrated with the enterprise systems, and leaves the organization to consolidate the data and information manually for data analytics. Effective knowledge and information use requires seamless access and intelligent analysis of information in its totality to allow enterprises to gain enhanced critical insights. This is becoming even more important, as the proportion of structured to unstructured information has shifted from 50-50 in the 1960s to 5-95 today [1]. Unless we can effectively utilize the unstructured content conjointly with the structured data, we will only obtain very limited and shallow knowledge discovery from an increasingly narrow slice of information. The techniques developed in our research will then be used to address significant issues in three application areas, but potential applications with significant impact are much more extensive.

L. Iliadis et al. (Eds.): AIAI 2014, IFIP AICT 436, pp. 1–19, 2014.

2   Aims and Underlying Issues

We develop a methodology and techniques for deriving deep knowledge from structured and unstructured information conjointly. This methodology would be useful in
several Business Intelligence and Advanced Analytics Applications utilizing Data and
Content (AADC), information integration applications, such as managing customer
attrition, targeted marketing, fraud detection and prevention, compliance, and
customer relationship management as well as a number of fields which involve information of high complexity such as Bioinformatics, Transport Logistics, Business
and Electrical Power Systems (EPS). The broad aims of this research are to:

a) use structured data to disambiguate text segments and link them to records for
   AADC investigations
b) use ontologies to disambiguate and annotate content segments to permit AADC
   investigations
c) develop methods of AADC for conjoint analysis of linked structured and content
   elements
d) apply these techniques for investigation of business problems, biomedical problems, electrical power system problems and logistics and transportation problems.

To achieve the aims (a) to (c), we have a number of sub-aims that develop
techniques for:

1. cleansing and filtering the noisy unstructured data with respect to structured data;
2. text annotation to enrich the unstructured data semantically;
3. fuzzy matching and searching over structured data based on annotated values for
   deriving the correlation between data and content;
4. discovering the linkages between data and content;
5. representations of the conjoint information that is suitable for AADC;
6. allowing new insights in the form of business intelligence and deep knowledge
   from the integrated data and content together.

The applications of the techniques developed (aim (d)) help to make better decisions
and to better understand behaviors in the field of business, biomedicine, electrical
power systems and logistics. In regards to aim (d), we study the following:

1. in business applications: customer experience and focused cross/up-selling and
   personalized marketing using the data and 'the voice of customer';
2. in biomedical area: Use of structured information from Protein Databases such as
   Swiss-Prot and epidemiological medical databases together with information from
   journal articles in PubMed to determine their relationship to specific conditions
   and obtain diagnostic knowledge for specific diseases;
3. In transport logistics area: movement of people, goods and services tracking for
   productivity, security and safety; trust, reputation and quality of service (QoS);
   track and trace of sub-contractor's performance (against SLA); track and trace fuel
   and vehicle performance, track and trace carbon emission, etc.

4. in Electric Power Systems: (i) Determination of risk events to profitability related to contractual terms for supply through bi-level contracts, in conjunction with structured data such as inflows, maintenance, failure and spot prices; (ii) Refinement of contractual terms and conditions and (iii) Remedial actions to manage and mitigate potential risk conditions.

Advanced Analytics (AADC) is an automated or semi-automated process for eliciting novel embedded knowledge, patterns and associations from data and information repositories, that is useful and understandable [2]. Thus information could be structured data and unstructured information such as freeform text. Furthermore the distribution of data both in an output class or category could be relatively flat or peaky with a multi-modal distribution.

Structured discrete data works well with (i) Association Rule Mining Techniques [3, 4], (ii) Decision Tree Methods [5, 6], and (iii) Rule Search Methods [7]. Effective methods for continuous structured data include Neural Net Based Rule extraction methods [8, 9]. Association Rules are of the form $X => Y$ (s, c) where X and Y are sets of items and s, c are support and confidence respectively [10]. Association Rule Mining has since been extended to include efficient Apriori-like mining methods [10], query mining [11], constraint based rule mining [12, 13], mining correlations and causal structures[14] and interesting associations [15] and mining associations from semi-structured data [16-18]. An XML document possesses a hierarchical document structure, and this is frequently modelled using a labelled ordered tree.

Our research group [17] initiated an XML-enabled association rule framework. It extends the notion of associated items to XML fragments or subtrees so that one finds associations among ordered trees rather than simple items as in classical association rules. Work has been proposed for mining XML documents [16-19], including a collection of semi-structured objects [17]. To help with mining XML documents and other structures, frequent sub tree mining algorithms are being developed [20-22] which focus on extraction of different types of tree patterns for different applications. Unstructured information has no schema to describe its structure, unlike structured data. We will confine our attention to textual information, and particularly to text mining which has concentrated on two issues, namely (1) categorization of textual documents, and (2) information extraction from a document to elicit a collection of facts or named entities from text documents. The most common approach to text categorization involves three phases:

1. text pre-processing;
2. encoding key information about the document using a feature vector, which is used with the knowledge discovery technique in (3) below;
3. a classification technique or a cluster technique to categorize the document.

Encoding represents the document by a vector of features such as word or phrase or clause, weighted by an importance factor. Classification techniques used include nearest neighbor classification, decision trees, support vector machines, naive Bayes Classifier and clustering methods include SOM, K-means or statistical measures such as regression. Information Extraction essentially involves extracting factual information or named entities from textual data of a certain type. Powerful entity extraction methods include support vector machines, hidden Markov Models [23], Random

Fields Methods [24] and Maximum Entropy Models. An application in bioinformatics [25], found rather poor results showing there remain challenging issues for some domains.

3   Significance of Work

Currently when deriving business intelligence by knowledge discovery from structured data sources one can often answer the questions related to patterns of what is happening. To answer questions related to why it is happening it will be necessary to derive conjoint mining of content (unstructured/) together with structured data. Thus, for example, a bank examining its database for patterns of customers who have decided to cancel their credit cards could through Data Mining techniques determine some of the features of the customer who is cancelling their credit card. e.g. co-branded cards, branches they belong to, their addresses, length of holding card etc. They would not be able to obtain information from the database of the number and nature of complaints, requests the customer may have made which would be contained in unstructured content repositories holding emails, transcribed phone call information, etc. In order to carry out this conjoint mining it is necessary to link segments of text from the content repository with individual records of interest in the structured data base. This is the problem of semantic integration between structured data and content (unstructured). Considerable work has been done for semantic integration between different sources of structured data. Some work has been done on semantic integration of structured data sources and XML data (semi structured) for the purpose of querying [26, 27]. Very limited work has been done on semantic integration of unstructured data and structured data sources [28, 29], and all of these concentrate on semantic integration for querying these diverse data sources (whether they involve XML data or Unstructured Content). Almost no work to the best of the authors' knowledge has been carried out on semantic integration of unstructured content and structured data in a form suitable for deep knowledge discovery through AADC. The information integration approaches above also assume that the schema of the different data sources is available but this is not always possible as enterprises frequently outsource supporting processes to different vendors. For example, Customer Contact Centers are generally outsourced to third parties, who maintain the unstructured information in isolation from the enterprise structured data creating information silos where the schema is not always known. Therefore, we need to develop efficient techniques for correlating the data (structured) and (unstructured) content conjointly, and this task has many technical challenges.

The first issue is that the unstructured data is typically noisy in nature. For example, unstructured data received from different contact channels, like calls, SMS, emails, is very noisy. The problems of cleaning this are somewhat different to that of cleaning structured data. We need to clean this data before we can correlate it with the structured data. The domain of noisy text correction is comparatively new, and we use new techniques for cleansing drawing on the field of automatic spelling corrections [30] and use of structured data and various ontologies to provide reference information to clarify terms in the text and fill in missing values and terms. The second issue is we need to discover the semantic knowledge/features from the text data. Typical information extraction tools or annotators take plain text as input and identify named entities and simple relations (e.g. works-for) and other text mining annotations based on a data dictionary or a gazetteer approach. For example, given a dictionary of product names, one can identify the product name in the text document. However, the current annotation systems, like UIMA, cannot annotate the documents based on concept, ontology and hierarchy. Therefore, we will extend the UIMA type techniques for annotating the documents based on domain ontology/concept. The third issue is to carryout semantic integration of content and structured data. This requires discovery of the entities in the text data based on the semantic knowledge, which can be matched with the structured entities for data correlation. One of the challenges is that no explicit identifiers of the entity, such as a unique transaction number, may be available in the document. Additionally, the document is noisy, so that a term in the text does not exactly match the corresponding attribute of the entity in the structured data. For instance a customer may mention a different transaction amount in her email or spell her name differently from that in the database. This naturally affects recall. It also affects precision when the noisy and partial information in a document leads to an incorrect entity being identified. We propose the use of a new fuzzy matching algorithm that uses an information theoretic basis. This fuzzy matching algorithm helps overcome the problem of the lack of precise information that permits exact matching. The fourth issue is to develop an intermediate representation suitable for Data Mining. This intermediate representation should be capable of capturing the embedded structure in content as well as the values. It should be in a form suitable to enable the use of conjoint Data Mining techniques. We propose an XML based approach for doing this as it enables one to represent domain information in a more meaningful and specialized way. The fifth issue is the development of algorithms to carry out Data mining conjointly from the content and structured data.

There are also frequently complex and important relationships between information that is stored in the form of structured data and content. To date the ability to find patterns, knowledge and relationships in unstructured text mining is aimed at document classification or entity extraction or simple associations between entities. The ability to find patterns or knowledge relationships between entities that might exist within unstructured textual documents is severely limited. For instance, in the biomedical area one may wish to find the chains involved in metabolic pathways. Furthermore existing techniques for document classification or entity extraction flatten the information structure using feature vectors, losing any structure other than the immediately preceding or following words that might be implicitly embedded or emergent within the document. These implicit or emergent structures may represent how the text is arranged under headings (which are semantically meaningful) or reflect chains of argumentation. Thus when grading an essay, the grader does not only look for the presence or absence of certain words or phrases but also the logical validity of the organization and the structure and clarity of the presentation; otherwise inserting the required words in a semi random fashion could give a good grade. More sophisticated AADC techniques should be capable of detecting such implicit or emergent structures. The approach adopted here, which converts the unstructured text into a semi structured intermediate form such as XML or RDF, will allow better representation of implicit structures. The extension of XML Mining and RDF Mining techniques [26] previously developed by the present authors allows one to investigate the presence of patterns and associations between tree structured items (sub trees within the embedded structures), graph structured items (sub graphs) and sequences within the conjoint data and content instead of just the values of attributes or terms. Correlation of data and content conjointly enables the discovery of interesting relationships and analytics that involve predicates and groupings and their arrangements in combinations. To obtain valuable insights, it is important to find useful associations among concepts which could be dimensions from content and from structured data. The applications in the fields of business, biomedicine and electric power systems and logistics allow one to discover innovative new insights that would be difficult to obtain without the use of conjoint AADC from content and data. The problems tackled in this proposal are very hard and complex problems that are becoming critical with the large proportion of content as well as data that is currently being generated. We develop novel and ground breaking techniques to address each of these issues, which are highly innovative that have the potential to produce a paradigm shift in business intelligence and deep knowledge discovery.

4 Approach and Methodology

Data and content are stored in repositories that are isolated from each other. Hence, the problem of semantic integration of data and content must be addressed in a form suitable for AADC. One also needs representations and techniques for AADC conjointly from data and content. To resolve these issues, we use two base ontologies: (i) a static concept relation ontology and (ii) an event, transformation ontology which captures the key types of transformations and events allowed. These provide a conceptual framework that enforces an agreement on the organization of information, without losing any of the flexibility of allowing people to express and view parts in their own familiar expression language. An ontology, which is a shared conceptualization of some domain [31,73], captures and represents the key concepts, relationships and constraints which permits coherent understanding of the meaning of shared information.

The base ontologies will ensure a common ground for understanding content. One way to restrict the scope of disambiguation of particular content within the base ontology is by creating sub-ontology or specialized ontology (also known as *Ontology Commitment* [32], *Ontology Version* [33], *Materialized Ontology View* [34,74] appropriate for the category of information being considered. Next, we consider the approaches used for the sub aims.

4.1   Unstructured Data Cleansing

The unstructured content is generally noisy, the extent of which is different for each problem e.g. for biomedicine, the unstructured content in journal papers is cleaner. Content like transcribed calls or emails in customer contact centers is very noisy. We need to clean this information. Processing SMS and email requires different data cleansing, e.g., removing spam messages, disclaimers, promotional material, and previous historical exchanges by the customer can be removed using heuristics for the domain. Even the body of the message is very noisy, using incomplete product name, spelling mistakes, added binary characters, etc. We will use two different approaches to this, (1) which borrows techniques from automatic spelling and grammar checkers [30] and (2) text cleansing methods which use structured data, and various ontologies (eg. Word Net) to provide reference data to clarify terms in the text and fill in missing values and terms, and deal with term variation arising from synonyms and acronyms. Spelling error correction is related to exact and approximate pattern matching respectively. Spell checking techniques involve non-word error detection and spell correction involves isolated-word or context-dependent error correction. The task involves three steps: (i) morphological analysis to identify a word-stem from a full word-form; (ii) isolating the misspelled words using techniques such as dictionary lookup and n-gram analysis; and (iii) offering a list of suggested correct spellings using one or more of six techniques, such as minimum edit distance, similarity key techniques, rule-based techniques, n-gram-based techniques, probabilistic techniques, and neural networks [30]. When doing this, we compare it to structured terms in the database or synonyms provided from WordNet.

4.2   Unstructured Data Annotation

In a document, we distinguish between:

1. Entity Extraction;
2. Identification of a relationship between two entities;
3. A network of relationships between entities;
4. Associations between several entities;
5. Associations between groups of entities such as ones arranged in subtexts and/or subsections.

Most work is on Named Entity Extraction (NER) which finds terms (words or phrases) for a specific named entity. Measures used to judge the efficacy of the algorithms are precision P (classification accuracy), recall R (coverage) and the F-score (harmonic mean of precision and recall). NER methods include probabilistic methods such as Hidden Markov [23, 35] or Conditional random fields [24], rule based methods [36] or Lexicon methods. Problems in NER are due to the same word or phrase referring to different entities, or many synonyms and/or abbreviations referring to an entity. To resolve these, we use an ontology for the domain of interest which maps these terms to concepts and relationships, recasting it as named concept recognition (NCR). Unlike a lexicon, which defines various items, an ontology has definitions for concepts and their relationships. There are several text to XML Annotators such as UIMA [37], GLOSS [38], and XI [39]. We will use the UIMA (Unstructured Information Management Architecture) [37] back-end that uses both statistical and rule-based annotators for text. Typically such annotators use a data dictionary or a gazette for information extraction from text of named entities (persons, organizations) and simple relations between terms (works-for). To enhance their capability, we will extend the UIMA techniques for annotating the documents based on domain ontology/concept. This permits disambiguation of terms through concept relationships in the same segment of text. We use Annotators to extract relevant tokens from a document and map them to a small subset of the attributes for determining matches in structured data. E.g. by using NER in an annotator one can extract names from a document, and match them against the customer and product name attributes of the transaction table. One could also extract chunked text such as noun clauses by using a part of speech tagger for matching. This allows us to determine a score for an entity in a document. The highest scoring entity or best matching one can be found without computing explicit scores for all entities. Performing fuzzy match on each extracted token results in a ranked list of possible entities. Entities and relationships determined can be used to define the XML profiles for those documents. The Extraction Methods can be used to determine the values for each tag for an instance document. We use a combination approaches namely (1) ontology-based [40, 41] ones that extract terms from text and map them to concepts in an ontology to give semantics and (2) ontology-driven [42-44] ones that make active use of an ontology to guide or constrain the analysis.

4.3   Fuzzy Matching and Data and Content Correlation

Discovering the business insights conjointly requires correlation between data and content. How can we link information from a text document (TD) with structured data in a database (DB), i.e. find the best matching entities from the DB for the given TD. We filter the annotated TD to retain only the relevant terms while the DB is considered as a set of entity instances and their associated related information. These DB entities are represented for matching as a collection of entity microschematas. We consider a single-type entity identification problem. We define the microschemata, as a rooted tree with the base table as the root and the related tables as the non-root nodes. If any tables have a foreign key relationship in the schema, their nodes are linked by edges. Each row in the base table is identified as an entity, having its own attribute values $e.A_j$. Here, an entity is an instance (a row in the base table) rather than a class level abstraction. The entity row is connected to the appropriate rows in the related table through foreign keys. We also have a collection $\{di\}$ of TDs (the content) that have references to the entities. Each TD d has a set of terms $\{ti\}$. The TD consists of sentences. One or more sentences taken together will be referred to as a segment s. Let e be the central entity for this TD, then each term ti may correspond to some attribute $e.A_j$ of entity e. For instance, a TD about a transaction entity refers to the customer name, shop name, date attributes of a specific transaction. Given the terms in a TD, our goal is to identify the central entity (e.g. the Customer A/C#) from the structured table. No explicit identifiers of the entity, e.g. a customer number, may be available in the TD. Also noise in the TD means that t_i does not exactly match the corresponding attribute of e. This may lead to (i) not identifying the entity associated with the piece of content – poor recall or (ii) incorrectly identifying a wrong entity with the piece of content-poor precision. We want to link a given segment of the given document with an entity in the DB. There may also be information related to multiple entities in the given segment and we will need to identify these. First we define the microschemata for the structured DB, which is a rooted tree with the base table as the root and the related tables as the non-root nodes. The *TD* is filtered to retain noun phrases using a part of speech parser and annotated using the annotation techniques referred to above. If necessary "semantic integration within text document" techniques [45] can be used to identify terms which refer to the same concept. One next annotates the term using the annotation techniques (say with UIMA) or alternatively using database look ups to identify the column it occurs in. The key idea for matching a term is to determine the information content contained in a term in predicting the entity it refers to and we use an information theoretic formulation for this purpose. From Information theory [46], for a finite probability distribution p_i ($i = 1,\ldots, m$), the entropy is given by

$$H\left(p_1, p_2,\ldots, p_m\right)= -\sum_{i=1}^{m} p_i \log_2 p_i$$

This measure of information content can be considered to be the uncertainty of the occurrence of a term corresponding to the particular entity. Let us assume that the term is contained in the contexts of $n(t)$ distinct entities and there are N entities in total. Hence the probability of occurrence of a given entity e if t_i is present is $p(t_i) = n(t_i)/N$. Hence the associated information content

$$I\left(e, t_i\right)= -\left(\frac{n\left(t_i\right)}{N}\right) * \log_2\left(\frac{n\left(t_i\right)}{N}\right)$$

Given that t_i occurs $f(t_i)$ times in a particular segment d, the information content associated with t_i occurring in the segment d linked to entity e is

$$I\left(e, d\right)= \sum_{\forall t_i \in d} f\left(t_i\right) * \left\{-\left(\frac{n\left(t_i\right)}{N}\right) * \log_2\left(\frac{n\left(t_i\right)}{N}\right)\right\}$$

A larger value for I indicates a greater predictive capability. A matching cache that contains a collection of pairs *(e,t)* (i.e term *t* is contained in entity *e*) is populated using two queries, one which returns the set of entities containing the term t and another which returns the set of terms contained in the contexts of the entity e. This avoids repetition of the queries for the same term in a new segment. This cache can be used with the expression above to produce a ranked list of entities that match a segment of text. An alternative to the above approach is to use the Zhou and Dillon Symmetrical Tau [6] to produce the ranked list. The unstructured text is annotated and represented as an XML document which is matched and merged with the structured data into XML documents using mapping between the concepts. These will be matched using a combination of semantic concept matching, online dictionaries, thesauri, schemas, and structure/related information, and extensions using an ontology. Then both documents will be adjusted to use common concept labels. We then find the common knowledge segments using our U3 mining algorithm [47], as in [48]. The additional information from unstructured content in XML corresponds to the unmatched knowledge segments and is used to augment the XML repository.

4.4   Techniques for Conjoint Data and Content Mining

To obtain valuable insight, it is important to find useful associations among concepts, from content (unstructured) and from structured data conjointly. It is useful to pre-identify valuable relationships. E.g. identify the top five products that experienced a sharp increase in complaints, common features of the customers and the nature of their complaints, products receiving the most inquiries and the profile of the inquirer and their reasons for the inquiry. Answering such questions give the enterprise timely insights about the customers' concerns. We note that structured data and content (unstructured documents) each have a very different representation. It is important for Integrated Data Mining to get a common intermediate form and we have chosen XML, as it has been successfully used for exchanging data between heterogeneous data sources, can capture the essentials of unstructured textual information, and it allows for mining of values and structures. We have the information held in a structured database with entities, i.e. $E = \{e_1,..., e_n\}$ which we represent in its XML form as:

$\{Ex = (e_{x1},..., e_{xn})$, and an unstructured document repository $U = \{u_1,..., u_m\}$.

The corresponding XML representation of the unstructured repository is

$Ux = \{ux1,..., uxn)$; An extended XML representation transaction is defined which consists of $EX = \{<exi, ux11> ... <exn, uxn>\}$

where the tuple $<e_{xi}, u_{xi}>$ consists of a concatenation of the XML representations of the two categories of information (E_x and U_x).

We used this approach for data records to obtain inter-transactional association rules [4]. As all data is conjointly represented in an XML document, the problem now becomes one of using the powerful XML mining algorithms to tackle mining of collections of these augmented XML documents. Our recent work has demonstrated the feasibility of conjoint mining of structured databases and XML repositories [49]. An XML-enabled framework for mining of association rules in XML repositories was first presented in [17] where the rules extracted are more powerful than traditional ones in expressing association relationships at both a structural and semantic level. To extract such rules the most difficult task is to find all the frequent sub trees from an XML database. This is known as the frequent sub tree mining (FSM) problem and is defined as: Given a tree database T_{db} a minimum support threshold (σ) find all subtrees that occur at least σ times in T_{db} [17]. Being able to mine all different subtree types using different support definitions is particularly important when we work on an XML representation of textual information, since these concepts can be repeated within many fragments of text and there exist different relationships among the concepts in the text, given the flexibility in its representation and expressiveness. The present authors have developed amongst the most powerful algorithms for mining XML document repositories and tree structured data.

The current work builds on this considerable body of expertise. The performance bottlenecks in FSM algorithms are candidate generation and counting, and this is often affected by the ability to effectively represent the document structure. Our work in the FSM field is characterized by a Tree Model Guided (TMG) [47, 50] candidate generation approach. This non-redundant systematic enumeration model uses the underlying tree structure of the data to generate only valid candidates which conform to the underlying tree structure of the data. We proposed the so called *Dictionary*, *Embedding List* (EL) [60] and the *Recursive List* (RL) structures [47] whose purpose is to capture the structural aspects of a document and allow for efficient access to necessary information. The RL is a more compact representation of the EL that reduces the memory and serves as a global lookup list and also encodes the embedding relationships of the subtrees to be mined. To enable efficient counting we use the *Vertical Occurrence List* (VOL) [50, 51] which stores a representation of a subtree encoding together with coordinates. Using the *TMG* framework with the above representation structures we have presented FSM algorithms for mining of following subtrees (under any support definitions): ordered induced [52] and embedded [53,77], ordered [54] and unordered [51] distance-constrained embedded, unordered induced [55] and embedded [47]. We have also extended *TMG* for sequence mining [4]. An important aspect of this process is Trust[72,75].

5   Applications to the Electric Power Industry

The deregulated market allows power consumers to purchase power from different generation companies (gencos) who price the power based on the system Marginal Price [56]. To maintain a competitive position, gencos form bilateral contracts with their clients (particularly large ones). These provide the clients with a guarantee of their required energy at a defined cost over a long period (say 5 years). These contracts are in textual form with possibly different terms for each client and are stored in a content repository. The bilateral contracts are legally binding and the genco has to ensure that it can meet the required demand from all the different clients with contracts and other customers who purchase power from the genco as needed. It may also have contracts for supply to it from other gencos. Thus, the genco has to deliver the expected Ensured Energy (EE) to meet its obligations. Otherwise there are many quite severe penalties which are normally staged according to proportion of power not delivered. Uncertainty, in a hydro thermal system being able to meet this EE, is caused by scheduled maintenance, unplanned outages arising from equipment breakdown, power from hydro plants being uncertain due to uncertain inflows, uncertainty in non-contract demand. Historical information related to these factors is stored in structured data bases together with historical spot prices for electrical energy. These will be used to produce forecasts for several of these factors and develop schedules for others such as hydro scheduling [57] and scheduled maintenance which are stored in structured DBs. This has led to approaches using the structured data to assess the risk especially the loss of load probability and expected unserved energy in the case of no complex contractual terms [58].

However what is needed is a risk assessment and management approach including textual contract terms. The profitability of a company will be impacted by any inability to meet the ensured energy. In this event the genco would try either (1) to purchase the extra energy at spot prices which are generally much higher than from its own supplies (and may exceed the contract price) or (2) to not fulfil the required demand under some contracts, or (3) try to put in place different contracts with other suppliers for the duration of any energy shortfalls, or (4) establish potential new contracts with new clients, or (5) renegotiate existing terms in existing contracts. To make such decisions, requires the genco to extract business intelligence conjointly from (i) the content repository containing information on the different contracts with its clients and suppliers (perhaps with notes on the feasibility of term variations) and (ii) the structured information in the different databases on the different factors. By linking and analyzing this information one can find associations which could result in risky situations and also determine potential remedial actions. Examples could be maintenance patterns, inflow levels (say in a dry year and failure rates which result in energy deficits leading to non-fulfilment of contractual obligations). This would alert the genco early of the need to purchase power from other gencos using supply contracts and refrain from certain client contracts.

6   Applications to Business Problems

We propose a new approach for consumer expectation-based market segmentation through conjoint mining of content and data. Consumer expectation has long been considered as an important satisfaction determinant that represents market demand and shapes the consumer behaviors [59, 60]. Hence, customer segmentation based on their expectations is of great significance for firms to predict dynamics of targeted markets.

However, consumer expectations data are often unobservable and prohibitive to collect using traditional market research methods such as interviews, self-reported surveys, experiments, etc. Conjoint mining provides a new means by which marketers are able to understand the 'minds' of millions of consumers on a daily basis without having to physically interact with consumers (e.g. shadowing, field experiment.) or explicitly soliciting opinions (e.g. interviews, surveys) from them. Consequently, data reliability and model validity bears more substantial rigor than previous approaches. Using conjoint mining techniques, consumer expectation can be inferred from customer satisfaction data gleaned from online customer reviews which include both structured and unstructured data, e.g. the reviews on Epinions.com hold numerical ratings towards each product attribute (e.g. 'battery life' for digital cameras, 'memory capacity' for MP3 players) as well as free text comments from consumers on their opinions and experiences.

Conjoint mining allows the use of structured data to reveal latent customer expectations based on unobservable concomitant variables such as consumer preferences, taste, values and the use of content to mine consumer opinions in free text, to (1) extract product features from the reviews, and (2) obtain consumer affects and sentiments towards these. We develop algorithms which augment opinion mining methods used in the interactive data analysis from [61] and the overall processes of opinion tracking from [62]. To discover heterogeneous expectations towards different brands, comparison-based algorithms [63] will be leveraged and redeveloped. The Web usage mining [64], sentiment-based algorithms [65] and opinion holder identification algorithms [66] will be integrated in order to make markets segments 'actionable' for managers and produce a custom score function to classify the sentiment [66].

Product extraction will augment the method in [67] and borrow some ideas from entity-based search engines [68] to mark items from free texts. Both product and feature extraction will allow business analysts to annotate text with an ontology and allow the unstructured customer opinions/ complaints to be linked with structured customer data in internal DBs or Customer Relationship Management systems.

7 Application in Transport Logistics Industry

It is important to understand that today's transport logistics providers spent 50% of their time on managing the physical mess and 50% on managing the related information mess [78]. Here, intelligent transportation has enabled vehicle to driver, vehicle to vehicle and vehicle to infrastructure communications and emerging intelligent infrastructure that provides embedded un-manned situation awareness 24/7 that enables greater mobility, security and safety.

It is also important to realise that over the years, the Transport Logistics sector has generated and accumulated much more valuable economic information than Facebook. This informs us on Big data impacts on global financial movement and Financial forecast including financial forecasts. Logistics professionals around the world know that they are no longer just transport and logistics operators, they are required to be "Data Experts" or at least to have Data Experts in their organization. Our ARC (Australia Research Council) Logistics Industry Partners in New South Wales and Queensland have been pushing their data to the Cloud since 2009 with vendor support. However, this Big Data has not been fully utilised, due to the lack of availability of co-joint data and content mining technology.

Further, many manufactured items, goods or assets today utilizing the Internet of Things are already Internet enabled, they have capability to talk to Internet, talk to each other, talk to logistics providers and talk to logistics infrastructure. This has sped up the automated people, goods and asset movement in logistics, transportation, warehouse and distribution [78] sector.

Intelligent Tracking powered by co-joint data and content mining is the core technology that is needed in transport logistics industry today. Tracking movement of people, goods and services in the entire logistics network, tracking quality of services, service providers performance, through entire life cycle of supply chain and asset management, track and trace of data and information shared over the logistics alliances, coalition partners and joint forces, situation awareness and ambient intelligent, for productivity, security and safety. Intelligent tracking powered by conjoint predictive analytics with real time data and in real time environment is a major challenge for all modern transport logistics providers. We have been working with our industry partner to adopt conjoint predictive analytics and co-joint mining for monitoring, visualisation, sharing, control and management of physical mess (goods and assets) and information mess (data) as well as business processes for their Business Intelligence including maximisation of human, transport and infrastructure performance and minimisation of the costs and security risk.

The co-joint data and content mining on Big Data in transport logistics sector including the combined RFID and wireless sensors data on the goods and assets handling, warehousing and transportation, GPS, GPRS and position location system for transport vehicle and shipment tracking, Surveillance Systems for Operator Performance and situation awareness, provenance of Goods and Asset tracking. The co-joint data and content mining are also needed for Inter- and intra-logistics partners transactions data monitoring, customers based tracking of trades data, smart phone, bluetooth, and black-box (on heavy vehicles and ships vessels) communication and even logistics social networks to support auto and semi-automated physical flow and information flow which enables business intelligent.

We use transport logistics ontology to help manage the Big data by defining the meaning of data through adding context that gives information on the data. Our works include Ontologies, RDF annotations and contexts. We carry out mining and visualization of big data both relational data (warehouse data or 3PL data) and complex data includes tree structured data (Geo-data), XML documents (procedures and workflows), unstructured textual data (smart phone notification and web data), image data (positions and locations), multimedia data (surveillance data), graphical data (Asset tracking data).

One of our biggest challenges in the co-joint data mining has been the assurance that the Big data are from trusted sources, the data services for Big data are trusted such as Clouds, and the Quality of Data, especially in the automated environment utilising Internet of Things and Cyber-Physical Systems. If the wrong decision is made based on the poor data set, it could result in major financial losses, high casualties and possible terrorist attack through the use of transport.

8   Applications to Biomedical Applications

Existing biomedical information is distributed across a large number of information resources and is heterogeneous in its content, format and structure. This hinders effective information retrieval. Targeted searches are very difficult with current search engines as they look for the specific string of letters within the text rather than its meaning. Use of highly expressive knowledge models such as ontologies enables the machines to view the text as meaningful expressions. This increases the semantics and forms the basis of a more efficient approach to finding the right information. An ontology can be used for creating metadata by semantic annotation of text through three steps: tokenization (splitting the sentences into tokens), matching the tokens against the ontology terms and matching the tokens against the ontology relationships until the best fit is found. New web pages created can be annotated automatically during their creation process. This semantic annotation allows machines to access web content, understand it, retrieve and process the data automatically rather than only display it. In our research work , we have developed a number of ontologies, such as Protein Ontology [69], Human Disease Ontology [70] and Mental Health Ontology [71]. The ontologies can be used to annotate target information in content sources and enable intelligent retrieval of specific information, analysis of it and linking with the existing pool of knowledge. E.g., protein and bibliographical reference data available via Swiss-Prot can be linked with the related publications from PubMed and with the epidemiological data in medical databases. Conjoint content and data mining of the linked content/data provides quality knowledge that can help build effective prevention and intervention strategies. Thus the presence of the protein, PSA, at a given level or given form (free or complex) at a certain age or ethnicity or lifestyle, might be indicative of a certain probability of the existence of cancer. Conjoint mining of the two structured databases and the textual information in PubMed will help with the discovery of such knowledge. There may be situations which coincide with some ambiguity or inconsistency. This will help researchers identify what requires further investigation.

9   Conclusions

The paper presents a methodology for conjoint mining of structured and unstructured information. The potential impact in industry of use of BI and AADC can be inferred from a 2003 IDC study of 40 US and European companies that use predictive analytics KDD who achieved a median Return of Investment of 145%, achieved higher investment levels and yielded higher overall returns over five years. These improvements occurred in just effectively utilizing the 5% of information available as structured data. This effect would be considerably amplified if one could in an integrated fashion exploit the remaining 95% of content as well as the 5% of structured data.

This research, by developing an integrated approach for BI and AADC of data and content, will provide a competitive edge in handling such information. This integrated knowledge discovery techniques could improve policy formation and effectiveness by Government and non-Government in such areas as compliance by companies, reduction of aberrant behavior in areas such as health benefits allocation, pension entitlements etc. It will provide an intellectually rigorous approach to underpin the trend in electronic document handling.

References

1. Brodie, M.L.: Computer science 2.0: A new world of data management. In: Proc. 33rd VLDB Conf., p. 1161 (2007)
2. Fayyad, U.M., Piatetsky-Shapiro, G., Smyth, P.: From data mining to knowledge discovery: An overview. In: Advances in knowledge discovery and data mining: American Assoc. for Artificial Intel., pp. 1–34 (1996)
3. Feng, L., Dillon, T.S., Liu, J.: Inter-transactional association rules for multi-dimensional contexts for prediction and their application to studying meteorological data. Data Knowl. Eng. 37(1), 85–115 (2001)
4. Tan, H., Dillon, T.S., Hadzic, F., Chang, E.: Sequest: Mining frequent subsequences using DMA strips. In: Proc. 7th Intl. Conf. on Data Mining and Inf. Engineering, Prague, Czech Republic, pp. 315–328 (2006)

5. Zhou, X.J., Dillon, T.S.: Theoretical and practical considerations of uncertainty and complexity in automated knowledge acquisition. IEEE Trans. Knowl. Data Eng. 7(5), 699–712 (1995)
6. Zhou, X.-J.M., Dillon, T.S.: A statistical-heuristic feature selection criterion for decision tree induction. IEEE Transactions on Pattern Analysis and Machine Intelligence 13(8), 834–841 (1991)
7. Hadzic, F., Dillon, T.: Using competitive learning between symbolic rules as a knowledge learning method. In: Bramer, M. (ed.) Proc. IFIP 20th World Computer Congress, Milan, Italy. IFIP, vol. 276, pp. 351–360. Springer, Heidelberg (2008)
8. Sestito, S., Dillon, T.S.: Knowledge acquisition of conjunctive rules using multi-layered neural networks. Int. J. Intell. Syst. 8(7), 779–806 (1993)
9. Sestito, S., Dillon, T.S.: Automated knowledge acquisition. Prentice Hall, Sydney (1994)
10. Agrawal, R., Srikant, R.: Fast algorithms for mining association rules. In: Proc. 20th Intl. Conf. Very Large Data Bases (VLDB), Chile, pp. 487–499 (1994)
11. Tsur, D., Ullman, J.D., et al.: Query flocks: A generalization of association-rule mining. In: ACM Intl. Conf. on Management of Data, Seattle, USA (1998)
12. Srikant, R., Vu, Q., Agrawal, R.: Mining association rules with item constraints. In: Proc. 3rd Intl. Conf. on Knowledge Discovery and Data Mining, USA, pp. 67–73 (1997)
13. Lakshmanan, L.V.S., Ng, R., Han, J., Pang, A.: Optimization of constrained frequent set queries with 2-variable constraints. In: ACM SIGMOD Intl. Conf. on Management of Data, USA, pp. 157–168 (1999)
14. Silverstein, C., Brin, S., Motwani, R., Ullman, J.: Scalable techniques for mining causal structures. Data Min. Knowl. Disc. 4(2), 163–192 (2000)
15. Ramaswamy, S., Mahajan, S., Silberschatz, A.: On the discovery of interesting patterns in association rules. In: Proc. 24rd Intl. Conf. on Very Large Data Bases (VLDB), pp. 368–379 (1998)
16. Wang, K., Liu, H.: Discovering structural associations of semistructured data. IEEE Trans. Knowl. Data Eng. 12(3), 353–371 (2000)
17. Feng, L., Dillon, T.S., Weigand, H., Chang, E.: An XML-enabled association rule framework. In: Mařík, V., Štěpánková, O., Retschitzegger, W. (eds.) DEXA 2003. LNCS, vol. 2736, pp. 88–97. Springer, Heidelberg (2003)
18. Zaki, M.J., Aggarwal, C.C.: XRules: An effective structural classifier for XML data. In: Proc. 9th ACM SIGKDD Intl. Conf. on Knowledge Discovery and Data Mining, Washington D.C., USA, pp. 316–325 (2003)
19. Yang, L.H., Lee, M.L., Hsu, W.: Efficient mining of XML query patterns for caching. In: Proc. 29th Intl. Conf. on Very Large Data Bases (VLDB), Berlin, Germany, pp. 69–80 (2003)
20. Asai, T., Arimura, H., Uno, T., Nakano, S.-I.: Discovering frequent substructures in large unordered trees. In: Grieser, G., Tanaka, Y., Yamamoto, A. (eds.) DS 2003. LNCS (LNAI), vol. 2843, pp. 47–61. Springer, Heidelberg (2003)
21. Zaki, M.J.: Efficiently mining frequent trees in a forest: Algorithms and applications. IEEE Trans. Knowl. Data Eng. 17(8), 1021–1035 (2005)
22. Termier, A., Rousset, M.-C., Sebag, M.: Treefinder: A first step towards XML data mining. In: Proc. 2nd IEEE Intl. Conf. on Data Mining (ICDM), Maebashi City, Japan, pp. 450–458 (2002)
23. Bikel, D.M., Schwartz, R., Weischedel, R.M.: An algorithm that learns what's in a name. Mach. Learn. 34(1) (1999)

24. Lafferty, J.D., McCallum, A., Pereira, F.C.N.: Conditional random fields: Probabilistic models for segmenting and labeling sequence data. In: 18th Intl. Conf. on Machine Learning (ICML), pp. 282–289 (2001)
25. Kim, J.-D., Ohta, T., Tsuruoka, Y., Tateisi, Y., Collier, N.: Introduction to the bio-entity recognition task at JNLPBA. In: Proc. Workshop on Natural Language Processing in Biomedicine and its Apps., pp. 70–75 (2004)
26. Bergamaschi, S., Castano, S., Vincini, M., Beneventano, D.: Semantic integration of heterogeneous information sources. Data Knowl. Eng. 36(3), 215–249 (2001)
27. Mçbrien, P., Poulovassilis, A.: A semantic approach to integrating XML and structured data sources. In: Dittrich, K.R., Geppert, A., Norrie, M. (eds.) CAiSE 2001. LNCS, vol. 2068, p. 330. Springer, Heidelberg (2001)
28. Chakaravarthy, V.T., Gupta, H., Roy, P., Mohania, M.: Efficiently linking text documents with relevant structured information. In: 32nd Intl. Conf. on Very Large Data Bases (VLDB), Seoul, Korea, pp. 667–678 (2006)
29. Bhide, M.A., Gupta, A., et al.: LIPTUS: Associating structured and unstructured information in a banking environment. In: Proc. ACMSIGMOD Intl. Conf. on Management of Data, Beijing, China, pp. 915–924 (2007)
30. Kukich, K.: Techniques for automatically correcting words in text. ACM Comp. Surv. 24(4), 377–439 (1992)
31. Gruber, T.R.: A translation approach to portable ontology specifications. Knowl. Acquis. 5(2) (1993)
32. Jarrar, M., Meersman, R.: Formal ontology engineering in the DOGMA approach. In: Confederated Intl. Conf. CoopIS, DOA, and ODBASE, California, USA, pp. 1238–1254 (2002)
33. Klein, M., Fensel, D., Kiryakov, A., Ognyanov, D.: Ontology versioning and change detection on the web. In: Proc. 13th Intl. Conf. on Knowledge Eng., Spain, pp. 247–259 (2002)
34. Wouters, C., Dillon, T.S., Rahayu, J.W., Chang, E., Meersman, R.: A practical approach to the derivation of a materialized ontology view. In: Taniar, D., Rahayu, W. (eds.) Web information systems (2004)
35. Zhou, G., Zhang, J., Su, J., Shen, D., Tan, C.: Recognizing names in biomedical texts: A machine learning approach. Bioinformatics 20(7), 1178–1190 (2004)
36. Tanabe, L., Wilbur, W.J.: Tagging gene and protein names in biomedical text. Bioinformatics 18(8), 1124–1132 (2002)
37. Ferrucci, D., Lally, A.: UIMA: An architectural approach to unstructured information processing in the corporate research environment. Nat. Lang. Eng. 10(3-4), 327–348 (2004)
38. Kaye, R.: The gloss system for trans. from plain text to XML. In: Proc. MathUI (2006), http://www.activemath.org/~paul/MathUI06/
39. Marchal, B.: XI: Open-source conver. of legacy text files to XML, http://www.ananas.org/xi/index.html (Ac.: November 20, 2008)
40. Embley, D.W., Campbell, D.M., Smith, R.D., Liddle, S.W.: Ontology-based extraction and structuring of information from data- rich unstructured documents. In: Proc. 7th Intl. Conf. on Inform. & Knowl. Mgmt. (CIKM), USA, pp. 52–59 (1998)
41. Alani, H., Kim, S., Millard, D.E., Weal, M.J., Hall, W., et al.: Automatic ontology-based knowledge extraction from web documents. IEEE Intel. Syst. 18(1), 14–21 (2003)
42. Handschuh, S., Staab, S., Maedche, A.: Cream: Creating relational metadata with a component-based, ontology-driven annotation framework. In: 1st Intl. Conf. on Knowledge Capture, Canada, pp. 76–83 (2001)

43. Vargas-Vera, M., Motta, E., et al.: Mnm: Ontology driven semi-automatic and automatic support for semantic markup. In: Proc. 13th Intl. Conf. on Knowl. Eng. and Know. Mgmt., Spain, pp. 213–221 (2002)
44. Gaizauskas, R., Demetriou, G., Artymiuk, P., Willett, P.: Protein structures and information extraction from biological texts: The pasta system. Bioinformatics 19(1), 135–143 (2003)
45. Li, X., Morie, P., et al.: Semantic integration in text: from ambiguous names to identifiable entities. AI. Mag. 26(1) (2005)
46. Reza, F.M.: An introduction to information theory. Dover Publications, New York (1994)
47. Hadzic, F., Tan, H., Dillon, T.S.: U3 – mining unordered embedded subtrees using TMG candidate generation. In: Proc. IEEE/WIC/ACM Intl. Conf. on Web Intelligence, Sydney, Australia (2008)
48. Hadzic, F., Dillon, T.S., Chang, E.: Tree mining application to matching of heterogeneous knowledge representations. In: Proc. IEEE Intl. Conf. on Granular Computing (GRC), California, USA, pp. 351–351 (2007)
49. Pan, Q.H., Hadzic, F., Dillon, T.S.: Conjoint data mining of structured and semi-structured data. In: Proc. 4th Intl. Conf. on the Semantics, Knowledge and Grid (SKG), Beijing, China, pp. 87–94 (2008)
50. Tan, H., Hadzic, F., Dillon, T.S., Feng, L., Chang, E.: Tree model guided candidate generation for mining frequent subtrees from XML. ACM Trans. Knowl. Discov. Data 2(2) (July 2008)
51. Hadzic, F., Tan, H., Dillon, T.S.: Mining unordered distance-constrained embedded subtrees. In: Proc. 11th Intl. Conf. on Discovery Science (DS), Budapest, Hungary (2008)
52. Tan, H., Dillon, T.S., Hadzic, F., Chang, E., Feng, L.: IMB3-miner: Mining induced/Embedded subtrees by constraining the level of embedding. In: Ng, W.-K., Kitsuregawa, M., Li, J., Chang, K. (eds.) PAKDD 2006. LNCS (LNAI), vol. 3918, pp. 450–461. Springer, Heidelberg (2006)
53. Tan, H., Hadzic, F., Feng, L., Chang, E.: MB3-Miner: Mining embedded subtrees using tree model guided candidate generation. In: 1st Intl. W'shop on Mining Complex Data in Conj. with ICDM 2005, USA, (2005)
54. Tan, H., Dillon, T.S., Hadzic, F., Chang, E.: Razor: Mining distance-constrained embedded subtrees. In: Proc. Workshop on Ontology Mining and Knowledge Discovery from Semistructured documents (MSD) in Conjunction with 2006 Intl. Conf. on Data Mining, Hong Kong, pp. 8–13 (2006)
55. Hadzic, F., Tan, H., Dillon, T.S.: UNI3 - efficient algorithm for mining unordered induced subtrees using TMG candidate generation. In: IEEE Sym. on Comp. Intel. and Data Mining (CIDM), USA, pp. 568–575 (2007)
56. Szkuta, B.R., Sanabria, L.A., Dillon, T.S.: Electricity price short-term forecasting using artificial neural networks. IEEE Transactions on Power Systems (PES) 14(3), 851–857 (1999)
57. Sjelvgren, D., Andersson, S., Andersson, T., Nyberg, U., Dillon, T.S.: Optimal operations planning in a large hydro-thermal power system. IEEE Trans. Power App. Syst., PAS 102(11), 3644–3651 (1983)
58. Dillon, T.S., Martin, R.W., Sjelvgren, D.: Stochastic optimization and modelling of large hydro-thermal systems for long term regu- lation. Intl Journal of Electrical Power and Energy Systems 2(1), 2–20 (1980)
59. Oliver, R.L.: A cogni. model of the antecedents and conseq. of satisfaction decisions. J. Marketing Rsch. 17 (1980)

60. Rust, R.T., Inman, J.J., Jia, J., Zahorik, A.: What you don't know about customer-perceived quality: The role of customer expectation distributions. Marketing Science 18(1), 77–92 (1999)

61. Ku, L.-W., Chen, H.-H.: Mining opinions from the web: Beyond relevance retrieval. J. Amer. Soc. Inf. Sci. Technol. 58(12), 1532–2882 (2007)

62. Glance, N., Hurst, M., Nigam, K., Siegler, M., et al.: Deriving marketing intelligence from online discussion. In: Proc. 11th ACM SIGKDD Intl. Conf. on Knowl. Discov. in Data Mining (KDD), USA, pp. 419–428 (2005)

63. Jindal, N., Liu, B.: Identifying comparative sentences in text documents. In: Proc. 29th ACM SIGIR Intl. Conf. on Research and Development in Information Retrieval, Seattle, USA, pp. 244–251 (2006)

64. Büchner, A.G., Mulvenna, M.D.: Discovering internet marketing intelligence through online analytical web usage mining. ACM SIGMOD Rec. 27(4), 54–61 (1998)

65. Turney, P.: Thumbs up or thumbs down? Semantic orientation applied to unsupervised classification of reviews. In: Proc. 40th Ann. Meeting on Assoc. for Comp. Lingui., USA, pp. 417–422 (2001)

66. Kim, S.-M., Hovy, E.: Determ. the sentiment of opinions. In: 20th Intl. Conf. on Comp. Lingustics, Switzerland (2004)

67. Drenner, S., Harper, M., Frankowski, D., et al.: Insert movie refer. here: A system to bridge conversation and item-oriented web sites. In: SIGCHI Conf. on Human Fact. in Comp. Syst (CHI), Canada, pp. 951–954 (2006)

68. Cheng, T., Yan, X., Chang, K.C.-C.: Entityrank: Searching entities directly and holistically. In: Proc. 33rd Intl. Conf. on Very Large Data Bases, Vienna, Austria, pp. 387–398 (2007)

69. Sidhu, A.S., Dillon, T.S., Chang, E., Sidhu, B.S.: Protein ontology: vocabulary for protein data 2005. In: Third.Int. Conf. Information Technology and Application, ICITA 2005 (2005)

70. Hadzic, M., Chang, E.: Medical ontologies to support human disease research and control. J. Web Grid Serv. 1(2) (2005)

71. Hadzic, M., Chen, M., Dillon, T.S.: Towards the mental health ontology. In: Proc. IEEE Intl. Conf. on Bioinformatics and Biomedicine (BIBM), USA, pp. 284–288 (2008)

72. Alhamad, M., Dillon, T., Chang, E.: SLA-based trust model for cloud computing. In: 2010 13th International Conference Network-Based Information Systems (NBiS) (2010)

73. Aberer, K., et al.: Emergent semantics systems. In: Bouzeghoub, M., Goble, C.A., Kashyap, V., Spaccapietra, S. (eds.) ICSNW 2004. LNCS, vol. 3226, pp. 14–43. Springer, Heidelberg (2004)

74. Wouters, C., Dillon, T., Rahayu, W., Chang, E., Meersman, R.: Ontologies on the MOVE Database Systems for Advanced Applications, pp. 812–823

75. Chang, E., Dillon, T.S., Hussain, F.K.: Trust and reputation relationships in service-oriented environments 2005. In: Third Int. Conf. Information Technology and Applications, ICITA 2005 (2005)

76. Wu, C., Chang, E.: Searching services on the web: A public web services discovery approach. In: 2007 Conf. Signal-Image Technologies and Internet-Based System, SITIS 2007 (2007)

77. Tan, H., Dillon, T.S., Hadzic, F., Chang, E.: L Feng MB3-Miner: efficiently mining eM-Bedded subTREEs using Tree Model Guided candidate generation. Department of Mathematics and Computing Science Saint Marys University

78. Chang, E.: Transport Logistics, the Grand Challenges, Australian Defence Force Academy (2014)

Breaking Ties of Plurality Voting in Ensembles of Distributed Neural Network Classifiers Using Soft Max Accumulations

Yiannis Kokkinos and Konstantinos G. Margaritis[*]

Parallel and Distributed Processing Laboratory, Department of Applied Informatics,
University of Macedonia, 156 Egnatia str., P.O. Box 1591, 54006, Thessaloniki, Greece

Abstract. An ensemble of distributed neural network classifiers is composed when several different individual neural networks are trained based on their local training data. These classifiers can provide either a single class label prediction, or the normalized via the soft max real value class-outputs that represent posterior probabilities which give the confidence levels. To form the ensemble decision the individual classifier decisions can be combined via the well known majority (or plurality) voting that sums the votes for each class and selects the class that receives most of the votes. While the majority voting is the most popular combination rule many ties in votes can occur, especially in multi-class problems. Ties are usually broken either randomly where the unknown instance is assigned randomly to one of the tied classes or using the class proportions where the tied class with the largest proportion wins. We present a tie breaking strategy that uses soft max confidence accumulations. Every class accumulates a vote and a confidence for this vote. If a tie occurs then the tied class with the maximum confidence sum wins. The proposed tie breaking in the voting process performs very well in all cases of different data distributions on various benchmark datasets.

Keywords: Neural Networks, distributed computing, ensemble classifiers, majority voting, plurality voting.

1 Introduction

Ensembles of neural network classifiers [1][2][3][4] are very popular tools in pattern recognition. Ensembles methods [5][6][7] are well known for their high accuracy and robustness since the combined predictions of several classifiers outperforms their individual predictions. In addition to the generalization performance, for which they are preferred, ensemble methods are usually the only choice available for other reasons like: (i) reduction of computational complexity by partitioning the data set into several smaller sub-sets, training different estimators on these sub-sets, and combining their predictions, (ii) modularity, by building modular systems that work on different input spaces, (iii) distributed learning from physically distributed data

---

[*] Corresponding author.

L. Iliadis et al. (Eds.): AIAI 2014, IFIP AICT 436, pp. 20–28, 2014.

repositories as well as Peer-to-Peer systems where gathering large volumes of data to a single location is unfeasible, (iv) their native parallelism and data scaling via the divide-and-conquer approach.

Distributed data collections can usually lay either in physically distributed database systems, in which case a small number of locations hold large volumes of data, or in Peer-to-Peer systems, where a large number of locations have typically small volumes of data. In any case the individual neural network classifiers are independently constructed in parallel, based on their local training data. After this training phase is finished, the classification predictions are combined via several schemes [8] in which the most popular one is majority voting [9][10][11][12] or its most frequently used version the plurality voting combining scheme. In plurality voting (for reviews see [5] [8] [10][11] [12]) a classification of an unlabeled instance is performed according to the class that obtains the highest number of votes (most of the votes). The strict version of (majority) voting needs the agreement of more than half of the participants to reach a decision, although this scheme can work for binary class problems only. Thus in reality, to cover the multi-class problems, the commonly used voting process is plurality voting, which selects the candidate class that receives the most votes. That is why usually one do not distinguish between plurality voting and majority voting, and the term "majority voting" is used even if the underlining criterion is plurality voting. This method is also known as the basic ensemble method [6].

However there may be more than one class that receives the largest number of classifications (or votes) by the ensemble members. Hence a tie can occur. In real life cases a tie-breaking chairperson can resolve this problem [13]. In classification problems such ties are usually broken arbitrary by randomly selecting one of the tied classes [5]. This random tie-breaking receives justification by the fact that all possible ways that other voters can vote are equi-probable. In the case of strictly binary class datasets one can just use an odd number of classifiers. In the case of k-nearest neighbour classifiers a nearest neighbour tie-break can be used. Another common strategy is choosing the class that is selected most often among the tied classes or tied classifiers [14]. This strategy predicts the class based on the largest class proportion among the ties classes. Another similar strategy can use the frequencies of predicted classes that occur during the training phase.

In this work we utilize confidence (soft max) accumulation as a method of breaking the ties in majority voting. Our motivation was the fact that a neural network classifier outputs, by default, the predicted class label accompanied by the confidence of this prediction given by the soft max function. Hence, we have tried to exploit further this extra information to improve the tie-breaking process and we have find out that it usually performs much better. Thus, the proposed tie-breaking method requires no additional computing resources other than those of simple majority voting. It works with classifiers, like neural networks, than in principle can accompany their class-prediction with a level of confidence, by using the soft max function. Every neural network classifier simply sends the predicted class *label* and its confidence *conf* for this class-prediction. Thus every prediction constitutes a pair {*label*, *conf*}. When a tie occurs in the voting process and two or more classes get the same number of votes then the proposed tie-breaking method sums up the received confidences for each of these tied classes. The class that accumulates the maximum confidence sum wins.

2   The Ensemble of Distributed Regularization Networks

We employ Regularization Neural Networks [15] [16] [17] for implementing the Neural Network classifiers. Regularization Networks are well known for using the real training data points as centers for their hidden neuron kernels. This is valuable when data features have discrete values, like in cases of image processing, computer vision and data mining. Using the real training data as kernel centers is a common strategy in kernel methods that can capture the data closeness approximate of any underlined problem distribution. These qualities elevate such type of stable kernel based learning methods to state of the art in modern machine learning [18].

We assume that each data location holds N_p training examples and there are L different data locations. Thus location p can train a neural network classifier $f_p()$ based on its local training set$\{\mathbf{x}_n, y_n\}_{n=1}^{N_p}$. Such an ensemble composed of Regularization Networks is illustrated in fig. 1 for the three-class case.

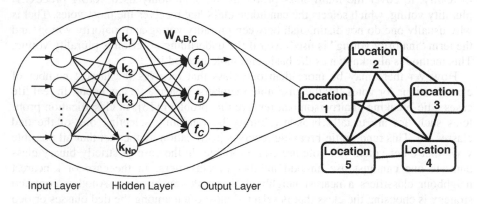

Input Layer   Hidden Layer   Output Layer

Fig. 1. An ensemble of Regularization Neural Networks distributed in several locations

The Regularization Network input neurons are as many as the number of data features. The hidden neurons are as many as the number of the training instances. A location p has N_p training instances. The output neurons are as many as the classes. The hidden-to-output layer weights **w** are computed for each class-output by solving a regularized risk functional. Assume three classes A, B and C then for the paradigm in fig. 1 the class output estimations of classifier $f_p()$ for an unknown **x** will be given by:

$$f_{p,A}(\mathbf{x}) = \sum\nolimits_{n=1}^{N_p} w_{A,n} \cdot k(\mathbf{x}_n, \mathbf{x}) \qquad (1a)$$

$$f_{p,B}(\mathbf{x}) = \sum\nolimits_{n=1}^{N_p} w_{B,n} \cdot k(\mathbf{x}_n, \mathbf{x}) \qquad (1b)$$

$$f_{p,C}(\mathbf{x}) = \sum\nolimits_{n=1}^{N_p} w_{C,n} \cdot k(\mathbf{x}_n, \mathbf{x}) \qquad (1c)$$

The class-output with the maximum value is chosen for the predicted class. The kernel function $k(\cdot,\cdot)$ can be any symmetric, positive semi-definite function, and the most commonly used is the Gaussian kernel $k(\mathbf{x}_n, \mathbf{x}) = \exp(-\|\mathbf{x}_n - \mathbf{x}\|^2/\sigma^2)$ centered at a particular $\mathbf{x}_n$ point. The local kernel matrix $\mathbf{K}$ with entries $K_{ij} = k(x_i, x_j)$ and size $N_p \times N_p$ contains the information regarding the high density regions inside classes and the separating planes in between classes. To find the weights $\mathbf{w}$ in the Regularization Network (RN) [15] [16] [17] the learning problem is stated as the minimization on a Reproducing kernel Hilbert space (RKHS) of a regularized risk functional which has the usual data error term plus a regularization term that embodies the stabilizer:

$$\arg \min_{f \in H_K} \left\{ \frac{1}{N_p} \sum_{n=1}^{Np} (y_n - f_p(\mathbf{x}_n))^2 + \gamma \|f_p\|_K^2 \right\} \tag{2}$$

The data term is scaled proportionally to the number of data points, and $\gamma > 0$ is the regularization parameter. The minimization solution set is unique and the weight vectors $\mathbf{w}_A$, $\mathbf{w}_B$, and $\mathbf{w}_C$ for the three class paradigm that correspond to a regularization network classifier p are given by solving the linear systems [15] [1] [17] in eq. 3:

$$\mathbf{w}_A = (\mathbf{K} + N_p \gamma \mathbf{I})^{-1} \mathbf{y}_A \tag{3a}$$

$$\mathbf{w}_B = (\mathbf{K} + N_p \gamma \mathbf{I})^{-1} \mathbf{y}_B \tag{3b}$$

$$\mathbf{w}_C = (\mathbf{K} + N_p \gamma \mathbf{I})^{-1} \mathbf{y}_C \tag{3c}$$

where $\mathbf{I}$ is the identity matrix, $\mathbf{K}$ is the local kernel matrix (of size $N_p \times N_p$) and $\mathbf{y}_A$, $\mathbf{y}_B$ and $\mathbf{y}_C$ are the vectors (of size N_p) that hold the desired labels for each class A, B and C respectively, which in the hot encoding are 1 for labels of the same class points and 0 for the others. The rescaled regularization parameter $N_p\gamma$ is usually small and can be found during training via cross validation. In the experimental runs we use a typical value $N_p\gamma = 0.1$ for all the local Regularization Networks.

For an unknown $\mathbf{x}$ the class-output responses $f_{p,A}(\mathbf{x})\, f_{p,B}(\mathbf{x})$ and $f_{p,C}(\mathbf{x})$ of each neural network f_p are normalized, so that they fall in the unity interval [0,1] and thus to reflect posterior probabilities. These posteriors produced by the softmax function are interpreted as confidence [19]. For instance the softmax confidence for the A class is:

$$h_{p,A}(\mathbf{x}) = \exp(f_{p,A}(\mathbf{x})) / (\exp(f_{p,A}(\mathbf{x})) + \exp(f_{p,B}(\mathbf{x})) + \exp(f_{p,C}(\mathbf{x}))) \tag{4}$$

Eq. 4 gives theoretically the Bayesian posterior probability (see also [19] for proof) as approximated by the regularization network. Thus each classifier outputs the predicted class together with the confidence for this class.

3   Combining Predictions Using Majority Voting

In a distributed parallel environment the combination of the decisions is straightforward. The majority voting is used in classification tasks where the individual classifiers produce a single class label, where in this case each classifier "votes" for a particular class, and the class with the majority vote on the ensemble wins. This means that for an unknown **x** each classifier $f_p(\mathbf{x})$ outputs one integer number, given by the max argument of its class-outputs $f_{p,m}(\mathbf{x})$, that designates the most probable class label. With M classes and L classifiers the correct class is the kth class that collects the largest number of votes, given by:

$$\text{class}(\mathbf{x}) = \arg\max_{k=1}^{M} \left(\sum_{p}^{L} \{ \text{ if } k = \arg\max_{m=1}^{M} \left(f_{p,m}(\mathbf{x}) \right) \text{ then 1 else 0} \} \right) \qquad (5)$$

The simplest method to handle ties relays on a *First Labelled* basis which in case of a tie favours the class that was labelled earlier (or first). The drawback is that in a tie the class 1 (the first) will never lose, while the class M (the last) will never win.

The most commonly used tie-breaking method during the voting process is to break the ties arbitrary by *Randomly Selecting* one of the tied classes [5]. If a tie occurs this method choose a random class among the tied ones.

Differently from the random selection another way is to check which of the tied classes has higher prior probability and accordingly make the decision. Usually the class proportions given by the number of examples in every class (for the three class problem these are N_A/N, N_B/N and N_C/N) can representing the priors. Hence this method uses *Class Proportions*, and the ties are going in favour of the largest proportions.

Another way for counting priors is to monitor how many times a class occurs during the training process. Thus if class A and class B receive the same number of votes but the class B has higher frequency of occurrence, meaning that is predicted more often that the class A during the training of all the classifiers, then B is selected. Hence this method uses *Class Frequencies* during training.

A variant of the previous method is to count the correct predictions during the training of all classifiers and use these *Class Correctly Frequencies* as priors.

Using neural networks each classifier can afford to send a pair {*label, conf*}, where *label* is the predicted class m of the unknown **x** and *conf* is the soft max confidence of this prediction. The proposed method in this paper when a tie occurs is to sum the confidences for the received votes of the tied classes and select the one class that accumulates the *Maximum Confidence sum*. Therefore, with L classifiers which output their prediction m and their soft max confidence $h_{p,m}(\mathbf{x})$ when classifying **x** in class m, the correct class is the kth class that collects the largest confidence sum (eq. 6):

$$\text{class}(\mathbf{x}) = \arg\max_{k=1}^{M} \left(\sum_{p}^{L} \{ \text{if } k = \arg\max_{m=1}^{M} \left(f_{p,m}(\mathbf{x}) \right) \text{ then } h_{p,m}(\mathbf{x}) \text{ else 0} \} \right) \qquad (6)$$

To provide an example consider the running paradigm in fig. 1, which illustrates a three-class problem with 5 neural network classifiers where each one can produce a pair {*label, conf*}. Assuming that for an unknown **x** the classifier $f_1()$ votes for class A with confidence 0.7 which gives a pair {A, 0.7} while $f_2()$ gives {A, 0.8}, $f_3()$ gives {B, 0.5}, $f_4()$ gives {B, 0.6} and $f_5()$ gives {C, 0.7}. A tie occurs for the classes A and B with 2 votes each. In this case the class A wins with sum of confidences 1.5.

4   Experimental Simulations

The classification performance of every tie-breaking method for majority voting are measured on a number of publicly available real-world benchmark problems which are downloaded from the UCI machine learning data repository [20] (http://archive.ics.uci.edu/ml). The specific details of these datasets are illustrated in table 1, where they are ordered by the number of their classes.

Table 1. Benchmark dataset details (ordered by class population)

Dataset Name	instances	Features	Classes
Sonar	208	60	2
Wisconsin (Diagnostic)	683	9	2
Diabetes Pima Indians	768	8	2
Spambase	4601	57	2
Phoneme	5404	5	2
Wine	178	13	3
Vehicle Silhouettes	846	18	4
Ecoli	336	7	5
Page Blocks	5473	10	5
Glass	212	9	6
Dermatology	358	34	6
Satellite Image	6435	36	6
Yeast	1479	8	9
Optical Digits	5620	64	10
Vowel Context	990	11	11
Spectrometer	531	100	24
Letter	20000	16	26

The groups of experiments aim at comparing the error rates of the proposed tie-breaking method and to make direct comparisons against the others. The experimental design is as follows:

(i)   A dataset is randomly split into a training set (80%) and a test set (20%) with stratification

(ii)   The training set of size N, is divided equally into L disjoint partitions $(N_p=N_1=N_2=\ldots=N_L)$ which are distributed to the different locations,

(iii)   every location p holds N_p data and builds a Regularization Neural Network classifier $f_p()$ based on its own local examples,

(iv)   The classifiers are combined and the ensemble is tested on the test set.

(v)   The procedure is repeated 30 times for each benchmark dataset and each ensemble population.

We measure the classification error rate which counts the number of incorrectly classified examples and divides by the test set population number.

The comparison results of the error rates and the standard deviations for each dataset and each tie-breaking method are illustrated in table 2.

Table 2. Error rates for the majority voting tie-breaking methods

Dataset	Ensemble size	First labeled	Random selection	Class Proportions	Class Frequencies	Class Correctly	Use max Confidence
Sonar	10	25.58	26.35	27.36	27.36	27.36	**23.25**
		±6.33	±6.47	±6.34	±6.34	±6.34	±5.50
Wisconsin	20	4.50	**4.28**	4.50	4.50	4.50	**4.28**
		±1.62	±1.57	±1.62	±1.62	±1.62	±1.55
Diabetes	20	24.95	25.08	24.95	24.95	24.95	**24.27**
		±2.42	±2.32	±2.43	±2.43	±2.43	2.54
Spambase	40	9.64	9.66	9.65	9.65	9.65	**9.54**
		±0.83	±0.91	±0.95	±0.95	±0.95	±0.90
Phoneme	40	**18.42**	18.54	**18.42**	**18.42**	**18.42**	18.56
		±1.15	±1.29	±1.15	±1.15	±1.15	±1.25
Wine	10	2.61	2.52	2.43	**2.34**	2.43	**2.34**
		±2.25	±2.50	±2.45	±2.29	±2.45	±2.39
Vehicle	20	31.03	31.55	31.50	31.48	31.77	**30.86**
		±2.59	±2.48	±2.37	±2.47	±2.30	±2.63
Ecoli	10	14.25	14.49	14.58	14.58	14.58	**13.94**
		3.26	3.21	3.36	3.36	3.36	3.47
Page Blocks	40	6.37	6.33	6.37	6.37	6.37	**6.28**
		±0.35	±0.32	±0.35	±0.35	±0.35	±0.37
Glass	10	39.55	39.55	39.70	39.70	40.00	**39.40**
		±5.33	±5.72	±5.73	±5.87	±5.62	±5.03
Dermatology	10	**3.12**	3.56	**3.12**	3.60	**3.12**	3.51
		±1.88	±2.27	±1.88	±2.18	±1.88	±2.35
Satellite Image	40	12.85	12.89	12.89	12.85	12.85	**12.81**
		±0.54	±0.55	±0.58	±0.56	±0.56	±0.86
Yeast	20	39.79	39.73	39.79	39.87	39.87	**39.65**
		±1.95	±2.02	±1.95	±1.90	±1.90	±1.97
Optical Digits	40	2.87	2.83	2.88	2.88	2.88	**2.76**
		±0.41	±0.40	±0.40	±0.41	±0.39	±0.40
Vowel Context	10	24.17	23.95	24.17	24.66	23.85	**19.12**
		±4.16	±4.45	±4.16	±4.90	±3.85	±4.32
Spectrometer	10	55.66	56.16	56.16	56.19	56.31	**54.95**
		±4.37	±4.02	±4.64	±4.72	±4.35	±4.49
Letter	40	11.51	11.44	11.47	11.60	11.45	**11.04**
		±0.59	±0.61	±0.57	±0.59	±0.60	±0.62

In table 2 the dataset name is in the first column. The second column has the number of classifiers in the ensemble. The third column illustrates the tie-breaking using the simplest "first labelled" method. The fourth column shows the results for the widely used method of breaking ties arbitrary by randomly selecting one of the ties classes in the voting. The fifth column is the tie-breaking by selecting among the tied classes the one with the largest "class proportion", namely the largest number of training examples.

The sixth column uses "class frequencies" and selects among the tied classes the one class that had the maximum occurrence during training process of the classifiers (the one that appears most frequently). The seventh column corresponds to the tie-breaking that uses "class correctly frequencies" which selects the class among the tied classes that had the maximum number of correctly occurrences during the training process. The last column is the proposed method that sums also the received confidence of those votes and selects the tied class with the *maximum confidence sum*. Note that in many cases there are very small differences of the proposed method with the others. The differences depend only on the number of ties since the voting algorithm is otherwise the same. This simply means that not many ties where encountered during the testing.

We run several experiments and in most of the cases the proposed method of using confidence produces better results than the other tie-breaking strategies. In general, one can notice from table 2 that the differences of the methods in the comparisons are not substantial when not many ties occur. However when several ties occur during the voting process then the difference of the proposed method as compared with the others becomes considerable. In some multi-class datasets like the Vowel the proposed tie-breaking method has 4 units less error rate than that of the other methods.

5   Conclusions

Majority or plurality voting is the most popular decision combination rule in neural network ensembles. In many cases, regardless of number of classes and neural networks in the ensemble a tie in voting occurs. Usually such ties are broken arbitrary. We present a tie-breaking technique that uses confidence. Since each neural network classifier can afford to send by default the predicted label and the confidence for this prediction (via the soft max function), the proposed method receives pairs of {predicted class, confidence} for every unknown **x**. Thus every class accumulates a voting sum and a confidence sum. If a tie occurs then the tied class with the maximum confidence sum wins. When compared with other tie-breaking methods, like random class selection, or class selection using prior class probabilities, the proposed tie breaking performs very well in all cases of different data distributions on various benchmark datasets.

References

1. Hansen, L.K., Salamon, P.: Neural network ensembles. IEEE Transactions on Pattern Analysis and Machine Intelligence 12, 993–1001 (1990)
2. Perrone, M.P., Cooper, L.N.: When networks disagree: ensemble method for neural networks. In: Mammone, R.J. (ed.) Neural Networks for Speech and Image Processing. Chapman & Hall, Boca Raton (1993)
3. Hashem, S.: Optimal linear combinations of Neural Networks. Neural Networks 10(4), 599–614 (1997)
4. Zhou, Z.H., Wu, J., Tang, W.: Ensembling neural networks: many could be better than all. Artificial Intelligence 137, 239–263 (2002)

5. Kuncheva, L.I.: Combining Pattern Classifiers: Methods and Algorithms. Wiley Interscience (2004)
6. Rokach, L.: Ensemble-based classifiers. Artificial Intelligence Review 33, 1–39 (2010)
7. Seni, G., Elder, J.: Ensemble Methods in Data Mining. Morgan & Claypool Publishers (2010)
8. Kittler, J., Hatef, M., Duin, R.P.W., Matas, J.: On Combining Classifiers. IEEE Transactions on Pattern Analysis and Machine Intelligence 20(3), 226–239 (1998)
9. Xu, L., Krzyzak, A., Suen, C.Y.: Methods of combining multiple classifiers and their applications to handwriting recognition. IEEE Transactions on Systems, Man, and Cybernetics 22(3), 418–435 (1992)
10. Lam, L., Suen, C.Y.: A theoretical analysis of the application of majority voting to pattern recognition. In: 12th International Conf. on Pattern Recognition II, pp. 418–420 (1994)
11. Ho, T.K., Hull, J.J., Srihari, S.N.: Decision combination in multiple classifier systems. IEEE Transactions on Pattern Analysis and Machine Intelligence 16(1), 66–75 (1994)
12. Lam, L., Suen, C.Y.: Application of majority voting to pattern recognition: An analysis of the behavior and performance. IEEE Transactions on Systems, Man, and Cybernetics 27(5), 553–567 (1997)
13. Brams, S.J., Fishburn, P.C.: Voting Procedures. In: Arrow, K.J., Sen, A.K., Suzumura, K. (eds.) Handbook of Social Choice and Welfare, vol. 1. Elsevier (2002)
14. Woods, K., Kegelmeyer, W.P., Bowyer, K.: Combination of Multiple Classifiers Using Local Accuracy Estimates. IEEE Transactions on Pattern Analysis and Machine Intelligence 19(4), 405–410 (1997)
15. Poggio, T., Girosi, F.: Regularization algorithms for learning that are equivalent to multilayer networks. Science 247, 978–982 (1990)
16. Girosi, F., Jones, M., Poggio, T.: Regularization theory and neural networks architectures. Neural Computation 7, 219–269 (1995)
17. Evgeniou, T., Pontil, M., Poggio, T.: Regularization Networks and Support Vector Machines. Advances in Computational Mathematics (2000)
18. Kashima, H., Ide, T., Kato, T., Sugiyama, M.: Recent Advances and Trends in Large-scale Kernel Methods. IEICE Transactions on Information and systems E92-D (7), 1338–1353 (2009)
19. Bishop, C.M.: Neural Network for Pattern Recognition. Oxford University Press Inc., New York (1995)
20. Frank, A., Asuncion, A.: UCI Machine Learning Data Repository. University of California, CA (2014), http://archive.ics.uci.edu/ml

Predicting Firms' Credit Ratings Using Ensembles of Artificial Immune Systems and Machine Learning – An Over-Sampling Approach

Petr Hájek and Vladimír Olej

Institute of System Engineering and Informatics
Faculty of Economics and Administration, University of Pardubice, Studentská 84
532 10 Pardubice, Czech Republic
{petr.hajek,vladimir.olej}@upce.cz

Abstract. This paper examines the classification performance of artificial immune systems on the one hand and machine learning and neural networks on the other hand on the problem of forecasting credit ratings of firms. The problem is realized as a two-class problem, for investment and non-investment rating grades. The dataset is usually imbalanced in credit rating predictions. We address the issue by over-sampling the minority class in the training dataset. The experimental results show that this approach leads to significantly higher classification accuracy. Additionally, the use of the ensembles of classifiers makes the prediction even more accurate.

Keywords: Credit rating, artificial immune systems, machine learning, neural networks, classification performance, balanced and imbalanced dataset, SMOTE, AdaBoost.

1   Introduction

Credit rating addresses an issuer's overall capacity and willingness to meet its financial obligations, thus reducing the information asymmetry between issuers and investors. However, credit ratings are based on costly analysis performed by professionals, which is why credit rating forecasting has attracted considerable recent interest. Various artificial intelligence (AI) methods have been applied to model the complex non-linear relations between input variables and target classes (see [7] for an example of a review). Recent efforts have shown that approaches integrating feature selection process and an appropriate AI method provide the best classification performance [6]. However, although the ensembles of classifiers have shown promising results in related fields such as credit risk and bankruptcy forecasting [15], significantly insufficient attention has been paid to them in credit rating forecasting. In addition, no research has been found that examines the effect of over-sampling the minority class in credit rating data. Therefore, the aim of this paper is to examine the effect of: (1) over-sampling the minority class in credit rating data; and (2) ensembling base classifiers. The difficulty in predicting credit ratings also stems from the fact that a multitude of sources is used in the credit rating analysis,

L. Iliadis et al. (Eds.): AIAI 2014, IFIP AICT 436, pp. 29–38, 2014.

involving both quantitative data (accounting and financial data drawn from financial statements) and qualitative assessments. The sentiment analysis of corporate annual reports has recently been used to address the issue of qualitative assessment [8]. We follow this approach and use the chosen sentiment categories in addition to corporate financial indicators as input variables.

We employ two categories of AI methods to examine the given aims, (1) artificial immune systems (AISs) and (2) machine learning (ML). AISs mimic the processes and mechanisms of biological immune systems and we specifically use Artificial Immune Recognition Systems (AIRSs) [16] and the Clonal Selection Classification Algorithm (CSCA) [1]. Out of the second category, we use C4.5 decision trees (DTs) [12], Support Vector Machines (SVMs) [13], Radial Basis Function Neural Networks (RBFs) [9] and Multilayer Perceptrons (MLPs) [9]. The research to date has tended to focus on the ensembles of classifiers of the latter category while little attention has been paid to the ensembles of AISs [5].

This paper is organized as follows. In the next section, we provide the description of the dataset. Given the fact that the classes are imbalanced in the dataset, we use the Synthetic Minority Oversampling Technique (SMOTE) algorithm [10] to modify the training dataset. The third section presents the results of experiments. First, the effect of over-sampling is tested and then we employ the AdaBoost algorithm [4] to generate the ensembles of the base classifiers (AIRS1, AIRS2, AIRS2-p, CSCA; and DT, SVM, RBF, MLP). For both categories, the classification performance is examined on (imbalanced) testing dataset depending on the proportion of training dataset generated. The last section discusses the results and concludes the paper.

2   Problem Formulation and Dataset

The prediction of firms' credit ratings was realized as a two-class problem. The classes were represented by investment grade (IG, low default risk) and non-investment grade (NG, high default risk), assigned by a highly regarded Standard & Poor's rating agency in 2011. The investment grade position is critical to many investors due to the restrictions imposed on investment instruments.

We used two main groups of input variables in this study, financial indicators and sentiment indicators (see Table 1). More specifically, we used several subgroups of financial indicators such as size, profitability ratios, liquidity ratios, leverage ratios and market value ratios. Sentiment indicators refer to the business position of a company (business risk, character (reputation), organizational problems, management evaluation, accounting quality, etc.). The financial indicators were drawn from the Value Line database, while sentiment indicators were drawn from annual reports available at the U.S. Securities and Exchange Commission EDGAR System. Both groups of input variables were collected for 520 U.S. companies (selected from the Standard & Poor's database) in the year 2010, 195 classified as IG and 325 as NG. Mining and financial companies were excluded from the dataset since they require specific input variables.

The sentiment indicators required linguistic pre-processing (tokenization and lemmatization) and subsequent comparison with the financial dictionary provided by [11]. Then, the tf.idf term weighting scheme was applied to obtain the importance of terms and an average weight was calculated for each sentiment category (negative, positive, uncertainty, litigious, modal strong and modal weak). See [8] for the detailed information on the collection of data.

Table 1. Input and output variables describing the dataset

	Variable		Variable
x_1	Enterprise value	x_{11}	Dividend yield
x_2	Cash	x_{12}	Payout ratio
x_3	Revenues	x_{13}	Standard deviation of stock price
x_4	Earnings per share	x_{14}	Frequency of negative terms
x_5	Return on equity	x_{15}	Frequency of positive terms
x_6	Price to book value	x_{16}	Frequency of uncertainty terms
x_7	Enterprise value/earnings	x_{17}	Frequency of litigious terms
x_8	Price to earnings per share	x_{18}	Frequency of strong modal terms
x_9	Market debt / total capital	x_{19}	Frequency of weak modal terms
x_{10}	High to low stock price	class	{IG, NG}

The given dataset was randomly divided into training and testing dataset (2:1). This procedure was repeated five times. Thus, the training dataset O_{train} contained 347 companies, 217 classified as IG and 130 as NG. The testing dataset O_{test} covered 173 companies, 108 as IG and 65 as NG. Both datasets were imbalanced, with the less frequent NG category (minority class). There are several approaches to handle this issue. The under-sampling of the majority class may represent a good way to reduce the sensitivity of classifiers, but in our case this approach resulted in a decrease in classification performance. This may be related to both the small size of the dataset and important decision information stored in most of the objects in the majority class. Another way is to apply the over-sampling of the minority class so that all classes are represented equally in the training dataset. We used the SMOTE procedure [10] to generate additional objects (firms) for the NG class to make the training dataset balanced. Thus, O_{train} contained 433 (O_{train}^{100}) companies, 217 classified as IG and 216 as NG. These training datasets O_{train}^{100} were considered to be the base balanced training datasets with 100 % of companies. We further examined the effect of generating additional training datasets, increasing the number by 25 % up to 300 % (O_{train}^{125}, O_{train}^{150}, ... ,O_{train}^{300}). The testing set remained imbalanced and fixed for all training sets (O_{train}^{100}, O_{train}^{125}, O_{train}^{150}, ... ,O_{train}^{300}).

3 Modelling Financial and Sentiment Indicators

In this section we employed commonly used artificial immune classification algorithms, AIRS1, AIRS2, AIRS2-p and CSCA. Further, the AISs were compared with DTs, SVMs, RBFs and MLPs. We used 10-fold cross-validation on training data to find the optimum settings of the classifiers' parameters.

The measures of classification performance are represented by the averages of standard statistics applied in classification tasks [14]: true positives (TP rate), false positives (FP rate), precision (Pre) and recall (Re), F-measure (F-m), the area under the receiver operating characteristic (ROC) curve and misclassification cost (MC). F-m is the weighted harmonic mean of Pre and Re, or the Matthews correlation coefficient, which is a geometric mean of the chance-corrected variants. A ROC is a graphical plot which illustrates the performance of a binary classifier system, which represents a standard technique for summarization classifier performance over a range of tradeoffs between TP and FP error rates. In particular, FP rate is important in this study owing to its possible serious financial consequences. This is due to the significant difference between default rates of IG and NG firms. For example, Standard & Poor's reported 0.03 % for IG and 1.71 % for NG in 2011. Therefore, we designed an MC matrix, where MC = 0.03 was assigned to each false classified IG firm and MC = 1.71 to each false classified NG firm, respectively.

3.1   Artificial Immune Classification Algorithms

3.1.1   Methods

The group of AIRS algorithms represents AISs using populations [2,3]. The algorithms are based on the principle of Recognition Ball (RB) or Artificial Recognition Ball (ARB), which can be described as recognition areas or artificial recognition areas that combine feature vector (antibody) and vector class. The principle solves the issue of the completeness of AISs. Each antibody is surrounded (in the sense of antibody representation in the state space) by a small area called the RB, in which the antibody recognizes all antigens (training dataset). Further, the AIRS algorithms use the principle of a limited resource. Each ARB area competes for limited resource according to its stimulation level. The classification performance of the AIRS algorithms proposed by [16] depends on several user-defined parameters. We used the AIRS1, AIRS2 and AIRS2-p algorithms to predict the firms' credit ratings. The following parameters of the AIRSs were examined to obtain the best classification performance: affinity threshold scalar = {0.1,0.2, ... ,0.9}, clonal rate = {1,2,4, ... ,32}, hypermutation rate = {1,2, ... ,10}, number of k nearest neighbors = {1,2, ... ,10}, initial memory cell pool size = 50, number of instances to compute the affinity threshold = all, stimulation threshold = 0.9 and total resources = 150.

The CSCA [1] uses a fitness function to maximize classification accuracy (minimize misclassification accuracy). The performance of the CSCA depends on the following user-defined parameters: clonal scale factor = 1.0, initial population size = {10,20, ... ,100}, number of nearest neighbors = {1,2, ... ,10}, minimum fitness threshold = 1, number of partitions = 1 and total generations = {1,5, 10, ... ,100}.

3.1.2   Results

The best classification results for the AIRSs and CSCA simulations on testing dataset are shown in Table 2.

Here, we present the average performance measures over the five datasets. The measures are represented by the averages of standard statistics applied in classification tasks [14]. The results in Table 2 show that the AIRS2 performed best while the

CSCA performed significantly worse (using paired t-test on $P < 0.01$). Further, the sizes of training datasets suggest that the AIRSs require larger datasets to achieve good classification performance (see Fig. 1).

Table 2. Best results for AIRS1, AIRS2, AIRS2-p and CSCA on testing dataset

	AIRS1		AIRS2		AIRS2-p		CSCA	
O_{train}	O_{train}^{175}		O_{train}^{225}		O_{train}^{225}		O_{train}^{150}	
Accuracy [%]	83.24		**85.24**		83.27		80.92	
MC	41.15		35.95		37.89		**22.76**	
Class	IG	NG	IG	NG	IG	NG	IG	NG
TP rate	0.923	0.778	0.938	0.806	0.892	0.796	0.692	0.880
FP rate	0.222	0.077	0.194	0.062	0.204	0.108	0.120	0.308
Precision	0.714	0.944	0.744	0.956	0.725	0.925	0.776	0.826
Recall	0.923	0.778	0.938	0.806	0.892	0.796	0.692	0.880
F-m	0.805	0.853	0.830	0.874	0.800	0.856	0.732	0.852
ROC	0.850	0.850	0.872	0.872	0.844	0.844	0.786	0.786

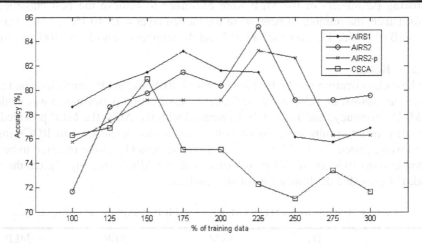

Fig. 1. Classification accuracy of AIRS1, AIRS2, AIRS2-p and CSCA on O_{test} depending on O_{train} size (Source: own)

3.2 Machine Learning and Neural Networks

3.2.1 Methods

A DT is a tree representation assigning a class to an object based on its attributes (variables), which can be continuous or discrete. An attribute with the best value of a splitting criterion is assigned to each root and intermediate node. Classes are assigned to leaf nodes in a DT. The DT with pruning was employed in this study. The classification performance of the DT depends on the following parameters: the confidence factor used for pruning (= 0.25 in this study) and the minimum number of instances per leaf = 2.

SVMs represent an essential kernel-based method with many modifications proposed recently. SVMs use kernel functions to separate the hyperplane between two classes by maximizing the margin between the closest data points. This is done in a higher-dimensional space where the data become linearly separable. We used the SVMs trained by the sequential minimal optimization (SMO) algorithm [13]. The classification performance of the SVM was tested for the following user-defined parameters: kernel functions = {polynomial, RBF}, $\gamma = 0.01$, the level of polynomial function = 2, complexity parameter $C = \{1,2,4, \ldots ,256\}$, round-off error $\varepsilon = 1.0E\text{-}12$ and tolerance parameter = 0.001. The RBF was trained with the Broyden-Fletcher-Goldfarb-Shanno method. The initial centres for the Gaussian RBFs were found using a k-means algorithm. The initial sigma values were set to the maximum distance between any centre and its nearest neighbor in the set of centres. The classification performance of the RBF depends on the following user-defined parameters: maximum number iteration = not limited, minimum standard deviation for the clusters = {0.1,0.2,0.3}, the number of clusters for k-means = {1,2,4, ... ,64} and ridge factor = 1.0E-8.

The MLP was trained using the backpropagation algorithm with momentum. The following parameters of the MLP were examined to achieve the best classification performance: the number of neurons in the hidden layer = {5,10,15, ... ,30}, learning rate = {0.01,0.05,0.1}, momentum = 0.2 and the number of epochs = {100,500,1000}.

3.2.2   Results

The best classification results for the DT, SVM, RBF and MLP simulations on testing dataset are shown in Table 3. Here, the best classification performance was achieved by MLP (Accuracy) and DT (MC). Compared with the AISs, the MLP provided significantly better results in terms of both classification accuracy and ROCs (again, tested using paired t-test at $P < 0.01$). Furthermore, less O_{train} were required to be generated in order to learn the MLP compared with the AIRS2 (see Fig. 2). On the other hand, DT provided the lowest MC of all classifiers.

Table 3. Best results for DT, SVM, RBF and MLP algorithms on testing dataset

	DT		SVM		RBF		MLP	
O_{train}	$O_{train}{}^{125}$		$O_{train}{}^{150}$		$O_{train}{}^{150}$		$O_{train}{}^{125}$	
Accuracy [%]	86.13		87.28		84.39		**88.44**	
MC	**17.59**		30.96		36.01		24.19	
Class	IG	NG	IG	NG	IG	NG	IG	NG
TP rate	0.785	0.907	0.938	0.833	0.908	0.806	0.908	0.870
FP rate	0.093	0.215	0.167	0.062	0.194	0.092	0.130	0.092
Precision	0.836	0.875	0.772	0.957	0.738	0.935	0.808	0.940
Recall	0.785	0.907	0.938	0.833	0.908	0.806	0.908	0.870
F-m	0.810	0.891	0.847	0.891	0.814	0.866	0.855	0.904
ROC	0.871	0.871	0.886	0.886	0.884	0.881	0.947	0.947

3.3   Comparison across Classifiers

In Fig. 3 we compare the ROC achieved for the balanced and imbalanced datasets. Except for the AIRS2-p and RBF, the performance of the classifiers improved significantly (at $P < 0.05$) when the O_{train} were balanced using the SMOTE ($O_{train}{}^{100}$). Depending upon the amount of over-sampling required, neighbors from k nearest neighbors are randomly chosen. In our experiments we used $k = 5$ nearest neighbors. The results of the classification of O_{test} for the original imbalanced and balanced O_{train} are shown in Table 4.

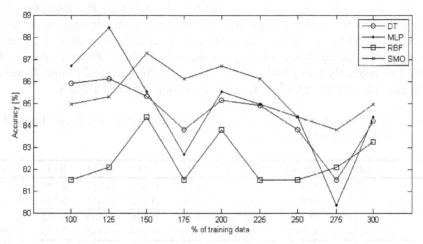

Fig. 2. Classification accuracy of DT, SVM, RBF and MLP on O_{test} depending on O_{train} size (Source: own)

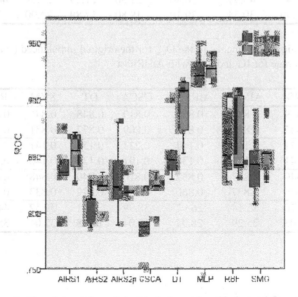

Fig. 3. ROC on testing dataset for original imbalanced and balanced O_{train} (Source: own)

We further examined the effect of the ensembles of classifiers. We employed the AdaBoost method [4] that combines many 'weak' classifiers to obtain an accurate learning algorithm. More precisely, AdaBoost is a meta-algorithm that can be used in conjunction with many other learning algorithms to improve their performance. This meta-algorithm is adaptive in the sense that subsequent classifiers built are tweaked in favor of those instances misclassified by previous classifiers. The number of iterations to be performed was set to 10. Again, we tested the previously described classification algorithms as base learners.

The results are depicted in Fig. 4. Here, the balanced training dataset was used from prior experiments. The ROC was significantly higher (at $P < 0.05$) for all the classifiers except for the MLP. AdaBoost provided noteworthy improvements for the DTs, AISs and SVMs especially. The results of the classification of the O_{test} for the original imbalanced and balanced O_{train} are shown in Table 5. DT with balanced data provided significantly lower MC (at $P < 0.05$) for both individual (Table 4) and ensemble classifiers (Table 5). For details, see confusion matrix in Table 6.

Table 4. Classification performance on O_{test} for the original imbalanced (i) and balanced (b) O_{train} (weighted average for IG and NG)

		AIRS1	AIRS2	AIRS2-p	CSCA	DT	SVM	RBF	MLP
TP	i	0.821	0.814	0.834	0.814	0.861	0.851	0.799	0.845
	b	0.853	0.801	0.798	0.801	0.875	0.853	0.786	0.834
FP	i	0.143	0.201	0.174	0.241	0.138	0.163	0.202	0.161
	b	0.148	0.160	0.155	0.160	0.133	0.148	0.196	0.148
F-m	i	0.822	0.814	0.834	0.810	0.862	0.851	0.800	0.846
	b	0.854	0.804	0.801	0.804	0.876	0.854	0.788	0.835
MC	i	42.42	45.99	50.80	21.56	21.29	27.99	44.80	36.14
	b	26.96	30.54	29.10	19.80	**17.89**	24.00	37.33	27.33

Table 5. Classification performance of the O_{test} for the original imbalanced (i) and balanced (b) O_{train} (weighted average for IG and NG) with AdaBoost

		AIRS1	AIRS2	AIRS2-p	CSCA	DT	SVM	RBF	MLP
TP	i	0.839	0.845	0.850	0.819	0.848	0.846	0.828	0.852
	b	0.845	0.836	0.858	0.809	0.872	0.831	0.798	0.836
FP	i	0.171	0.166	0.144	0.233	0.155	0.141	0.180	0.150
	b	0.141	0.147	0.138	0.191	0.134	0.140	0.179	0.138
F-m	i	0.840	0.857	0.852	0.816	0.849	0.848	0.829	0.853
	b	0.846	0.837	0.860	0.803	0.872	0.833	0.800	0.838
MC	i	32.70	35.14	26.97	38.99	22.51	39.52	44.09	37.83
	b	27.03	25.29	28.97	19.04	**17.28**	32.04	35.26	26.93

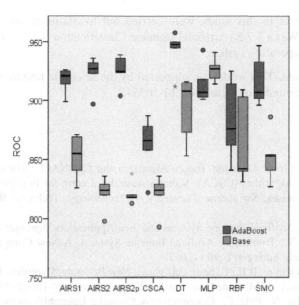

Fig. 4. ROC for base classifiers and AdaBoost (Source: own)

Table 6. Confusion matrix of the O_{test} for the balanced O_{train} trained using DT with AdaBoost

	IG	NG
IG	56.8±6.3	9.2±5.2
NG	13.0±6.1	95.0±6.1

4   Conclusion

The aim of this paper was to examine the performance of AISs and ML on a complex classification task where financial indicators are combined with sentiment analysis. We attempted to address an important issue of imbalanced dataset. On the one hand, the over-sampling of the minority class showed promising classification improvement, yet on the other hand, the under-sampling of the majority class resulted in weaker classification performance. We performed many experiments to find the best setting of the learning parameters of individual classifiers to achieve the best classification performance. The results also confirmed that an ensemble of weaker base classifiers may perform better than the individual classifiers, especially in terms of ROC measure. More importantly, the decrease in MC may lead to the substantial financial savings of investors. For example, the ensemble of DTs trained on balanced data reduced the MC from 21.29 to 17.28 (18.8% savings). Thus, this study represents a good basis for further experiments in the field of financial distress forecasting, where the problem of imbalanced datasets is usually to be addressed. Additionally, we encourage future research in multi-class datasets.

The experiments in this study were carried out in Statistica 10 (linguistic pre-processing) and Weka 3.7.5 (Artificial Immune Classification Algorithms and ML) in MS Windows 7 operating system.

Acknowledgments. This work was supported by the scientific research project of the Czech Sciences Foundation Grant No: 13-10331S.

References

1. Brownlee, J.: Clonal Selection Theory Algorithm and CLONALG: The Clonal Selection Classification Algorithm (CSCA). Victoria Australia, Centre for Intelligent Systems and Complex Processes, Swinburne University of Technology, Technical Report ID: 2-01 (2005)
2. Dasgupta, D.: Artificial Immune Systems and their Applications. Springer, Berlin (1999)
3. De Castro, L.N., Timmis, J.: Artificial Immune Systems: A New Computational Intelligence Approach. Springer, Berlin (2002)
4. Freund, Y., Schapire, R.E.: Experiments with a New Boosting Algorithm. In: 13th International Conference on Machine Learning, San Francisco, pp. 148–156 (1996)
5. García-Pedrajas, N., Fyfe, C.: Construction of Classifier Ensembles by means of Artificial Immune Systems. Journal of Heuristics 14(3), 285–310 (2008)
6. Hajek, P.: Municipal Credit Rating Modelling by Neural Networks. Decision Support Systems 51(1), 108–118 (2011)
7. Hajek, P., Olej, V.: Credit Rating Modelling by Kernel-Based Approaches with Supervised and Semi-Supervised Learning. Neural Computing and Applications 20(6), 761–773 (2011)
8. Hájek, P., Olej, V.: Evaluating Sentiment in Annual Reports for Financial Distress Prediction Using Neural Networks and Support Vector Machines. In: van Zee, G.A., van de Vorst, J.G.G. (eds.) Shell Conference 1988. LNCS, vol. 384, pp. 1–10. Springer, Heidelberg (1989)
9. Haykin, S.: Neural Networks: A Comprehensive Foundation, 2nd edn. Prentice-Hall Inc., New Jersey (1999)
10. Chawla, N.V., Bowyer, K.W., et al.: SMOTE: Synthetic Minority Over-sampling Technique. Journal of Artificial Intelligence Research 16, 321–357 (2002)
11. Loughran, T., McDonald, B.: When Is a Liability Not a Liability? Textual Analysis, Dictionaries, and 10-Ks. The Journal of Finance 66(1), 35–65 (2011)
12. Quinlan, J.R.: C4. 5: Programs for Machine Learning. Morgan Kaufmann (1993)
13. Platt, J.C.: Fast Training of Support Vector Machines using Sequential Minimal Optimization. In: Schoelkopf, B., Burges, C., Smola, A. (eds.) Advances in Kernel Methods - Support Vector Learning. MIT Press (1998)
14. Powers, D.M.W.: Evaluation: from Precision, Recall and F-measure to ROC, Informedness, Markedness and Correlation. Journal of Machine Learning Technologies 1(2), 37–63 (2011)
15. Ravi Kumar, P., Ravi, V.: Bankruptcy Prediction in Banks and Firms via Statistical and Intelligent Techniques - A Review. European Journal of Operational Research 180(1), 1–28 (2007)
16. Watkins, B.A., Timmis, J., Boggess, L.: Artificial Immune Recognition System (AIRS): An Immune Inspired Supervised Learning Algorithm. Genetic Programming and Evolvable Machines 5, 291–317 (2004)

Automating Transition Functions:
A Way To Improve Trading Profits
with Recurrent Reinforcement Learning

Jin Zhang

Faculty of Economics and Business Administration, University of Basel,
Peter Merian-Weg 6, Switzerland
jin.zhang@unibas.ch

Abstract. This paper presents an application of the logistic smooth transition function and recurrent reinforcement learning for designing financial trading systems. We propose a trading system which is an upgraded version of the regime-switching recurrent reinforcement learning (RS-RRL) trading system referred to in the literature. In our proposed system (RS-RRL 2.0), we use an automated transition function to model the regime switches in equity returns. Unlike the original RS-RRL trading system, the dynamic of the transition function in our trading system is driven by utility maximization, which is in line with the trading purpose. Volume, relative strength index, price-to-earnings ratio, moving average prices from technical analysis, and the conditional volatility from a GARCH model are considered as possible options for the transition variable in RS-RRL type trading systems. The significance of Sharpe ratios, the choice of transition variables, and the stability of the trading system are examined by using the daily data of 20 Swiss SPI stocks for the period April 2009 to September 2013. The results from our experiment show that our proposed trading system outperforms the original RS-RRL and RRL trading systems suggested in the literature in terms of better Sharpe ratios recorded in three consecutive out-of-sample periods.

Keywords: Daily equity trading, Recurrent reinforcement learning, Transition variable selection, Automated transition functions.

1   Introduction

Recent developments in algorithmic trading have shown that people never stop their effort of searching and developing trading strategies. In the computer science literature, artificial intelligence (AI) techniques have been increasingly applied in the field of technical analysis for equity trading, e.g. the application of a genetic algorithm (GA) in technical trading rules optimization [1], the adaptive reinforcement learning system which uses 14 commonly-used technical indicators as a part of the system inputs [5], and a Logitboost method to select combinations of technical trading rules for stocks and futures trading [4].

Although AI has been used widely in research on financial trading platform design, these studies have concentrated on the optimization of trading signals

L. Iliadis et al. (Eds.): AIAI 2014, IFIP AICT 436, pp. 39–49, 2014.

in technical analysis, in isolation from financial fundamentals and other available tools. Various approaches based on technical analysis [13] and fundamental analysis [8] have been developed to forecast future trends in stock prices. Additionally, tools from financial engineering have been used to facilitate price movement analysis. In real world, financial news, quotes, company earnings, pre-market and after-hours data, market research, company analysis are commonly used for price movement analysis. For example, researchers suggest to use trading systems which use fundamental indicators to select equity and technical indicators to produce trading signals [14].

Recurrent reinforcement learning (RRL), an online learning technique which finds approximate solutions to stochastic dynamic programming problems, was used by researchers to tune financial trading systems for the purpose of utility maximization [12]. In the literature, most discussions of RRL trading have been in the context of high-frequency FX and equity indices trading (see [6,5,2,7]). Due to the effectiveness of RRL in training financial trading systems, the technique has been adopted by researchers to build more sophisticated financial trading systems. For example, the multi-layer trading system integrated a risk management module with a RRL trading system for utility maximization [5].

Over the last few decades, models such as regime switching (RS) have become popular in econometrics. For example, the threshold autoregressive (TAR) model [15] and the smooth transition autoregressive (STAR) model allowing for smooth transition [3] draw people's attention. Although interest in RS models has grown, most papers on RS have focused on model development, with only a few applications of RS models concerned with artificial intelligence being found in the financial trading field.

In the literature, it has been found that transitions between economic regimes are often signaled by price patterns. Researchers proposed a regime-switching RRL trading system to cope with the price changes which is non-linear due to regime switches [9,10]. The basic RS-RRL trading system consists of two independent RRL systems which produce two trading signals for two scenarios, i.e. for low and high volatility regimes. The trading signal is made on the basis of the weighted total value of the trading signals from the two RRL systems. The weight is an output from a logistic smooth transition autoregressive (LSTAR) function, in which the conditional variance from a GARCH model is used as the transition variable.

In this paper, we aim to add to the existing literature on RS-RRL. We do so by studying the transition variable selection problem in the RS-RRL trading system. Five indicator variables (i.e. volume, relative strength index, price to earnings ratio, conditional volatility and moving average prices) are considered as possible transition variables in the RS-RRL trading system. We also introduce an upgraded version of the original RS-RRL trading system for the purpose of enhancing trading profits. Our upgraded version, the RS-RRL 2.0 trading system uses an automated transition function which is different from the constant transition function in the original RS-RRL trading system.

The paper is organized as follows. Section 2 provides a brief overview of RRL-type trading systems and introduces the our proposed trading system: RS-RRL 2.0. Section 3 provides the results of the experiment. Section 4 concludes the paper.

2 RRL-Type Trading Systems

2.1 Recurrent Reinforcement Learning

Recurrent reinforcement learning (RRL) was a technique to tune financial trading systems for the purpose of utility maximization [12]. The RRL technique is a stochastic gradient ascent algorithm which continuously optimizes a utility measure by using new market information. Although, in most discussions on RRL, the market information usually comprises a series of lagged price returns. The basic RRL trading system in [12] is designed to trade a single-asset with a two-position action (long/short), which is produced by using linear combinations of returns and a tanh function (see Figure 1). $\boldsymbol{w}$ denotes a series parameters of the input signals. F_{t-1} refers to the current holding position; and F_t denotes the asset's holding position on the following day.

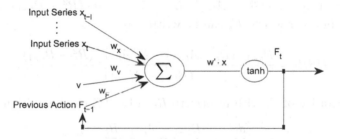

Fig. 1. Recurrent reinforcement learning

The goal of the RRL trading system is to maximize a wealth measure U_t by adjusting the parameter set $\boldsymbol{w}_t$ in a continuous manner:

$$\max U_t(R_t; \boldsymbol{w}_t), \tag{1}$$

$$R_t = \nu \cdot (\text{sgn}(F_{t-1}) \cdot r_t - \delta \cdot |\,\text{sgn}(F_t) - \text{sgn}(F_{t-1})|), \tag{2}$$

and

$$F_t = \tanh(\boldsymbol{w}_t \times \mathbf{I}_t), \tag{3}$$

where ν is the number of shares and δ is the transactions cost rate. $\mathbf{I}_t$ denotes a series of inputs, including F_{t-1}, a constant v with a value of 1, and a set of lagged returns $r_t, r_{t-1}, r_{t-2}, \ldots, r_{t-l+1}$, $t = 1, \ldots, T$, and l is an integer number representing the length of the lags.

The signal parameters $\boldsymbol{w}_t$ are updated by RRL based on stochastic gradients. The gradients of U_t with respect to the signal parameter set $\boldsymbol{w}_t$ can be written as:

$$\frac{dU_t(\boldsymbol{w}_t)}{d\boldsymbol{w}_t} = \frac{dU_t}{dR_t} \left\{ \frac{dR_t}{dF_t} \frac{dF_t}{d\boldsymbol{w}_t} + \frac{dR_t}{dF_{t-1}} \frac{dF_{t-1}}{d\boldsymbol{w}_{t-1}} \right\}, \tag{4}$$

$$\frac{dF_t}{d\boldsymbol{w}_t} \approx \frac{\partial F_t}{\partial \boldsymbol{w}_t} + \frac{\partial F_t}{\partial F_{t-1}} \frac{dF_{t-1}}{d\boldsymbol{w}_{t-1}}, \tag{5}$$

$$\frac{dR_t}{dF_{t-1}} = \nu \cdot (r_t + \delta \cdot \mathrm{sgn}(F_t - F_{t-1})), \tag{6}$$

$$\frac{dR_t}{dF_t} = -\nu \cdot \delta \cdot \mathrm{sgn}(F_t - F_{t-1}). \tag{7}$$

In the literature, the utility measure used in RRL trading systems is the differential Sharpe ratio (DSR), which is the first-order term after taking the Taylor series expansion of a performance measure, namely the exponential moving average Sharpe ratio (EMSR) at $\eta \to 0$:

$$EMSR_t = \frac{A_t}{K_\eta \cdot (B_t - A_t^2)^{1/2}}, \tag{8}$$

where $A_t = A_{t-1} + \eta \cdot (R_t - A_{t-1})$, $B_t = B_{t-1} + \eta \cdot (R_t^2 - B_{t-1})$, and $K_\eta = (\frac{1-\eta/2}{1-\eta})^{1/2}$. In other words, U_t can be written as:

$$DSR_t = \frac{B_{t-1} \cdot (R_t - A_{t-1}) - \frac{1}{2} \cdot A_{t-1} \cdot (R_t^2 - B_{t-1})}{(B_{t-1} - A_{t-1}^2)^{\frac{3}{2}}}, \tag{9}$$

and the derivative of U_t with respect to R_t in Eq. (4) can be written as:

$$\frac{dU_t}{dR_t} = \frac{B_{t-1} - A_{t-1} \cdot R_t}{(B_{t-1} - A_{t-1}^2)^{3/2}}. \tag{10}$$

At time t, the signal parameter set $\boldsymbol{w}_t$ is updated by using $\boldsymbol{w}_t = \boldsymbol{w}_{t-1} + \rho \cdot \frac{dU_t(\boldsymbol{w}_t)}{d\boldsymbol{w}_t}$, where ρ is the learning step.

2.2   Regime-Switching RRL

Although the basic RRL trading system uses linear combinations to generate trading signals, nonlinear behavior in financial time series has been documented in the literature. To model the nonlinearities in price changes caused by regime switches, researchers proposed a RS-RRL trading system [9], in which the trading signal F_t is defined as:

$$F_t = F_t^A \cdot (1 - G_t) + F_t^B \cdot G_t, \tag{11}$$

$$F_t^A = \tanh\left(\boldsymbol{w}_{t-1}^A \times \mathbf{I}_{t-1}\right), \tag{12}$$

$$F_t^B = \tanh\left(\boldsymbol{w}_{t-1}^B \times \mathbf{I}_{t-1}\right). \tag{13}$$

In other words, the final output F_t of the system is a weighted sum of F_t^A and F_t^B from two individual RRL systems. The weight is a value generated by using a logistic function:

$$G_t(s_t; \gamma, c) = \frac{1}{1 + \exp(-\gamma \cdot (s_t - c))}, \tag{14}$$

where s_t, γ and c refer to the transition variable, the transition rate and the threshold value respectively.

In the econometric literature, G_t refers to the transition function in a LSTAR model. The LSTAR models a univariate return series r_t as:

$$r_t = (\phi_{0,1} + \phi_{1,1} r_{t-1} + \ldots + \phi_{l,1} r_{t-l}) \cdot (1 - G(s_t; \gamma, c))$$
$$+ (\phi_{0,2} + \phi_{1,2} r_{t-1} + \ldots + \phi_{l,2} r_{t-l}) \cdot G(s_t; \gamma, c) + \varepsilon_t. \tag{15}$$

It is found that the conditional volatility variable from a GARCH model is a suitable transition variable for RS-RRL trading; and the transition rate and the threshold value in Eq. (14) are derived by using a quasi-maximum likelihood estimation (for further details, see [11]).

2.3   Regime-Switching RRL 2.0

There are two main concerns with the original RS-RRL trading system. The first is that, if users do not manually update the parameters in G_t, the transition function will be constant. Constant transition models may not work well in the real world as the transition rate and the threshold value can be time-varying. The second concern is that, the likelihood based inference for the transition function does not serve the purpose of RRL in the trading context. In other words, the derivation of LSTAR parameters with quasi-maximum likelihood estimation (i.e. minimizing error terms), does not fit well with the purpose of trading, which is utility maximization.

We therefore propose a new RS-RRL trading system which we call RS-RRL 2.0 (see Figure 2), in order to improve the trading performance of the RS-RRL trading system. In our proposed trading system, in addition to the two RRL trading systems, we use a control unit which consists of a summation function and a sigmoid function to mimic the transition function G_t in the LSTAR model. Inputs of the control unit include the transition variable S_t, the threshold value c, and the current holding position F_{t-1}. To maximize the utility function, RRL updates the signal parameters in the trading units and the control unit. The design of RS-RRL 2.0 allows the transition function G_t and the utility measure to interact in an automated fashion in real time.

In the proposed RS-RRL 2.0 trading system, F_t is the weighted sum of the outputs F_t^A and F_t^B from the two individual RRL trading systems, which are given by Eq. (11), Eq. (12) and Eq. (13). G_t is a function of the transition variable, the threshold value, and the signal parameters $\boldsymbol{w}^G$:

$$G_t = \mathrm{sigmoid}(\boldsymbol{w}_{t-1}^G \times \mathbf{I}_{t-1}) = \frac{1}{1 + \exp(-\boldsymbol{w}_{t-1}^G \times \mathbf{I}_{t-1})}, \tag{16}$$

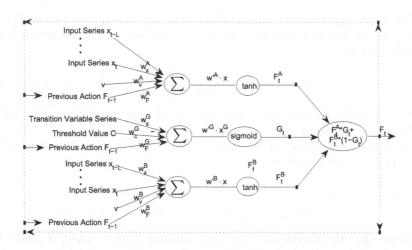

Fig. 2. The regime-switching recurrent reinforcement learning 2.0

which is similar to the logistic transition function in LSTAR models. We consider only one transition function in this paper; other transition functions from the econometric literature, such as exponential STAR (ESTAR) functions, also can be modified and used in the system.

Gradients of U_t with respect to the signal parameter sets $\boldsymbol{w}_t^A$, $\boldsymbol{w}_t^B$ and $\boldsymbol{w}_t^G$ can be written as:

$$\frac{dU_t(\boldsymbol{w}_t^A)}{d\boldsymbol{w}_t^A} = \frac{dU_t}{dR_t}\left\{\frac{dR_t}{dF_t}\frac{dF_t}{d\boldsymbol{w}_t^A} + \frac{dR_t}{dF_{t-1}}\frac{dF_{t-1}}{d\boldsymbol{w}_{t-1}^A}\right\}, \tag{17}$$

$$\frac{dU_t(\boldsymbol{w}_t^B)}{d\boldsymbol{w}_t^B} = \frac{dU_t}{dR_t}\left\{\frac{dR_t}{dF_t}\frac{dF_t}{d\boldsymbol{w}_t^B} + \frac{dR_t}{dF_{t-1}}\frac{dF_{t-1}}{d\boldsymbol{w}_{t-1}^B}\right\}, \tag{18}$$

and

$$\frac{dU_t(\boldsymbol{w}_t^G)}{d\boldsymbol{w}_t^G} = \frac{dU_t}{dR_t}\left\{\frac{dR_t}{dF_t}\frac{dF_t}{d\boldsymbol{w}_t^G} + \frac{dR_t}{dF_{t-1}}\frac{dF_{t-1}}{d\boldsymbol{w}_{t-1}^G}\right\}. \tag{19}$$

We use the reward measure and the utility function given in Eq. (2) and Eq. (9), so that the derivatives $\frac{dU_t}{dR_t}$, $\frac{dR_t}{dF_t}$ and $\frac{dR_t}{dF_{t-1}}$ are the same as Eq. (10), Eq. (6) and Eq. (7). The other components in Eq. (17), Eq. (18) and Eq. (19) can be found in the following:

$$\frac{dF_t}{d\boldsymbol{w}_t^A} \approx \frac{\partial F_t^A}{\partial \boldsymbol{w}_t^A} + \frac{\partial F_t}{\partial F_{t-1}}\frac{dF_{t-1}}{d\boldsymbol{w}_{t-1}^A}, \tag{20}$$

$$\frac{dF_t}{d\boldsymbol{w}_t^B} \approx \frac{\partial F_t^B}{\partial \boldsymbol{w}_t^B} + \frac{\partial F_t}{\partial F_{t-1}}\frac{dF_{t-1}}{d\boldsymbol{w}_{t-1}^B}, \tag{21}$$

and

$$\frac{dF_t}{d\boldsymbol{w}_t^G} \approx \frac{\partial G_t}{\partial \boldsymbol{w}_t^G} + \frac{\partial F_t}{\partial F_{t-1}}\frac{dF_{t-1}}{d\boldsymbol{w}_{t-1}^G}, \tag{22}$$

where

$$\frac{\partial F_t}{\partial F_{t-1}} = (G_t \frac{\partial F_t^A}{\partial F_{t-1}} + F_t^A \frac{\partial G_t}{\partial F_{t-1}}) - (G_t \frac{\partial F_t^B}{\partial F_{t-1}} + F_t^B \frac{\partial G_t}{\partial F_{t-1}}) + \frac{\partial F_t^B}{\partial F_{t-1}},$$

and all the other required components can be derived using the chain rule of derivatives.

3 Results of the Experiment

3.1 The Context of the Experiment

The companies selected for this experiment are Swiss SPI stocks (1 April 2009 – 26 September 2013, each series containing 1126 observations). We consider volume, relative strength index (RSI), price-to-earnings (P/E) ratio and moving average price changes from technical analysis, and conditional volatility from a GARCH model as options for selecting the transition variable. Of over 100 SPI stocks, 20 have full historical information (downloaded from Bloomberg) of the above variables. The 1126 observations of each stock are partitioned into an initial training set consisting of the first 750 samples. We consider three consecutive out-of-sample periods, with each out-of-sample period consisting of 125 observations.

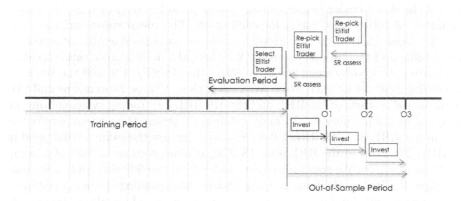

Fig. 3. Training and trading

The RS-RRL 2.0 trading system is designed to trade a single asset, there are 20 trading systems to trade the 20 SPI stocks. Every trading system consists of a group of 500 simulated traders (sim-traders). In the real world, as different traders may have different levels of information asymmetry, we initialize the sim-traders by using different numbers (i.e. random numbers from a Gaussian distribution with a mean of 0 and a standard deviation of 0.05). The sim-traders are then trained with the first 750 samples. At the end of the training period, we select the best-performing trader (referred to as the elitist trader hereafter) from the 500 sim-traders according to their Sharpe ratio rankings in an evaluation period (i.e. the last 125 trades before out-of-sample trading). The elitist trader is reselected at the beginning of each out-of-sample period.

Based on preliminary tests, the following parameters were found to be suitable settings for the daily equity trading problem: the number of shares traded $\nu = 1$; the learning rate $\rho = 0.15$; the adaption rate $\eta = 0.05$. We expect that new information will be reflected in stock prices in a maximum period of two weeks, therefore we use a value of $l = 10$. The transaction cost δ has a value of 3 bps. The parameters of the transition variable and the threshold value have an initial value of 1. We use an optimization approach, differential evolution to estimate the threshold value (see [11]), other parameters of the system are initialized using random numbers.

3.2   Out-of-Sample Trading Performance

Profitability and stability are two particularly important factors in a financial trading system. In this study, we use the Sharpe ratio to measure the profitability and we calculate the Sharpe ratio using daily returns. It should be noted that the trading performance of RRL-type trading systems relates directly to the initialization of signal parameters. Therefore, stability refers to the consistency of the Sharpe ratios recorded from independent restarts of the trading system.

We restart the RS-RRL 2.0 trading system 100 times, saving the daily Sharpe ratio of the elitist trader in each out-of-sample period from each trial. We collect a number of 100 Sharpe ratios for each stock. The mean value and the standard deviation (SD) of the 100 Sharpe ratios are considered as the measures for profitability and stability respectively. We check the daily Sharpe ratio statistics which are produced by using the basic RRL, the RS-RRL 1.0, and the RS-RRL 2.0 trading systems for each stock. We collect these Sharpe ratio means and SDs based on 100 restarts of the RRL and the RS-RRL 1.0 trading systems from the three out-of-sample periods. To make a fair comparison, the system parameters used in the RRL and RS-RRL 1.0 trading systems are the same as that used in RS-RRL 2.0. Both of the RRL and RS-RRL trading systems consist of a group of 500 sim-traders. We select an elitist trader from the 500 sim-traders for out-of-sample trading on the basis of their Sharpe ratio rankings from evaluation.

The means and standard deviations which are computed based the 100 restarts from the three out-of-sample periods are reported in Table 1 and Table 2. As these tables show that, the RS-RRL 2.0 trading system tends to perform at least as well as the RRL and the RS-RRL 1.0 trading system, and often better with respect to higher means and lower standard deviations. We perform the one-way analysis of variance (ANOVA) to compare the means of the Sharpe ratios produced by the RS-RRL 1.0 and the RS-RRL 2.0 trading systems. In Table 1, the Star symbols indicate the cases where the two Sharpe ratio means are significantly different from each other at a significance level of 5%, given a same transition variable. The results of the experiment show that most of the SR means from the RS-RRL 2.0 trading system are statistically greater than those

Table 1. Sharpe ratio means

		Volume		Con.Vol		RSI 9		PE		MV 30	
	RRL	RSRRL1	RSRRL2	RSRRL1	RSRRL2	RSRRL1	RSRRL2	RSRRL1	RSRRL2	RSRRL1	RSRRL2
1	0.033	0.004	0.098*	0.026	0.096*	0.028	0.094*	0.026	0.095*	0.040	0.101*
2	0.020	0.027	0.031	0.005	0.029*	0.016	0.029	0.024	0.025*	0.046	0.025
3	0.056	0.076	0.046*	0.026	0.025	0.060	0.055	0.033	0.027	0.013	0.036*
4	0.040	0.041	0.033	0.038	0.039	0.041	0.033	0.043	0.040	0.006	0.023*
5	0.008	-0.007	0.070*	0.007	0.070*	0.006	0.059*	0.018	0.072*	0.038	0.076*
6	0.057	0.023	0.113*	0.060	0.112*	0.061	0.107*	0.097	0.112*	0.009	0.110*
7	0.033	0.016	0.020	-0.003	0.042*	0.039	0.055*	0.009	0.021	0.009	0.014
8	0.014	0.020	0.079*	0.021	0.076*	0.011	0.059*	0.023	0.080*	0.044	0.077*
9	0.027	0.052	0.028*	0.035	0.031	0.024	0.023	0.067	0.041*	-0.006	0.019*
10	0.032	0.029	0.047*	0.062	0.074	0.025	0.019	0.016	0.034*	-0.002	0.054*
11	0.026	0.028	0.024	0.007	0.037*	0.017	0.031*	0.003	0.042*	0.023	0.036*
12	0.050	0.069	0.078	0.075	0.071	0.051	0.058	0.027	0.073*	0.048	0.080*
13	0.034	0.030	0.023	0.013	0.014	0.030	0.025	0.014	0.024*	0.029	0.015*
14	0.064	0.072	0.082	0.060	0.066	0.077	0.094*	0.056	0.073*	0.069	0.087*
15	0.032	0.043	0.068*	0.033	0.072*	0.036	0.061*	0.063	0.065	0.031	0.051*
16	0.020	0.014	-0.015*	-0.026	-0.022	0.023	0.001*	-0.003	0.045*	-0.011	0.008*
17	0.035	0.026	0.104*	0.029	0.087*	0.045	0.053	0.062	0.111*	0.004	0.106*
18	0.019	0.019	0.006*	0.001	0.011	0.025	0.002*	0.028	-0.009*	0.024	0.004*
19	0.035	0.042	0.067*	0.049	0.089*	0.043	0.087*	0.044	0.084*	0.036	0.081*
20	0.066	0.005	0.049*	0.048	0.055	0.062	0.090*	0.082	0.048*	0.040	0.055*

Table 2. Sharpe ratio standard deviations

		Volume		Con.Vol		RSI 9		PE		MV 30	
	RRL	RSRRL1	RSRRL2	RSRRL1	RSRRL2	RSRRL1	RSRRL2	RSRRL1	RSRRL2	RSRRL1	RSRRL2
1	0.081	0.077	0.044	0.077	0.043	0.078	0.052	0.089	0.037	0.075	0.039
2	0.070	0.072	0.060	0.069	0.053	0.067	0.059	0.065	0.047	0.081	0.049
3	0.072	0.075	0.051	0.075	0.050	0.068	0.065	0.063	0.050	0.083	0.052
4	0.071	0.070	0.042	0.069	0.043	0.069	0.051	0.060	0.043	0.059	0.055
5	0.078	0.077	0.052	0.069	0.049	0.077	0.078	0.065	0.054	0.083	0.054
6	0.073	0.072	0.037	0.077	0.039	0.075	0.050	0.064	0.036	0.095	0.038
7	0.091	0.085	0.068	0.083	0.077	0.095	0.087	0.092	0.075	0.077	0.062
8	0.099	0.092	0.105	0.094	0.112	0.099	0.107	0.086	0.111	0.104	0.112
9	0.079	0.071	0.057	0.075	0.063	0.076	0.070	0.070	0.061	0.077	0.069
10	0.081	0.081	0.126	0.099	0.154	0.085	0.110	0.076	0.131	0.094	0.145
11	0.079	0.091	0.056	0.078	0.065	0.080	0.055	0.065	0.073	0.077	0.068
12	0.068	0.081	0.060	0.072	0.063	0.074	0.054	0.080	0.076	0.075	0.080
13	0.071	0.076	0.047	0.072	0.055	0.073	0.044	0.056	0.045	0.076	0.057
14	0.072	0.073	0.064	0.078	0.067	0.076	0.058	0.069	0.069	0.076	0.079
15	0.080	0.073	0.050	0.070	0.051	0.074	0.055	0.065	0.045	0.084	0.056
16	0.082	0.095	0.085	0.077	0.080	0.077	0.082	0.086	0.080	0.079	0.096
17	0.075	0.087	0.047	0.076	0.059	0.078	0.066	0.085	0.048	0.077	0.055
18	0.078	0.073	0.083	0.086	0.072	0.072	0.090	0.071	0.083	0.090	0.068
19	0.074	0.068	0.077	0.074	0.078	0.076	0.093	0.083	0.078	0.067	0.077
20	0.095	0.085	0.092	0.082	0.073	0.094	0.097	0.093	0.079	0.076	0.087

from the RS-RRL 1.0 trading system at a significance level of 5%. Although most Sharpe ratio means are not statistically greater than zero, we can still produce a positive return after netting the profits and losses from trading the 20 stocks by using the RS-RRL 2.0, which confirms the same findings in [11].

4   Conclusions

This paper presents an upgraded version of the RS-RRL trading system referred to in the literature. In the proposed RS-RRL 2.0 trading system, we use an automated transition function to model regime switches in equity returns. Unlike the original RS-RRL trading system, the dynamic of the transition function in our system is driven by utility maximization. Although the proposed system looks similar to an artificial neural network (ANN), we use a performance measure, i.e. the differential Sharpe ratio, which is different from the error function in traditional ANNs to implement the online learning for equity trading. We are able to reveal the time-varying feature of the threshold value by studying the parameter w_C^G (as shown in Figure 2), which provides a deeper insight into the impact of regime switches on stock returns comparing to the constant transition function suggested in the literature.

The results of our experiment show that regime-switching trading system brings value-added to trading performance. The RS-RRL 1.0 trading system referred to in the literature, produces higher Sharpe ratio means than the RRL trading system without impairing the stability of trading results. We also show that most the Sharpe ratios from the proposed RS-RRL 2.0 trading system are statistically greater than that from the RS-RRL 1.0 trading system at a significance level of 5%, given a same transition variable. With respect to transition variable selection, it seems that the price-to-earnings ratio is apt to produce higher Sharpe ratio means comparing the other four indicators; however, it is difficult to tell a 'one size fits all' indicator which improves the Sharpe ratio in general based on such a small sample of stocks.

Acknowledgements. This work has been partially supported by the Swiss National Science Foundation project SNSF-138095.

References

1. Allen, F., Karjalainen, R.: Using genetic algorithms to find technical trading rules. Journal of Financial Economics 51, 245–271 (1999)
2. Bertoluzzo, F., Corazza, M.: Making Financial Trading by Recurrent Reinforcement Learning. In: Apolloni, B., Howlett, R.J., Jain, L. (eds.) KES 2007, Part II. LNCS (LNAI), vol. 4693, pp. 619–626. Springer, Heidelberg (2007)
3. Chan, K.S., Tong, H.: On estimating thresholds in autoregressive models. Journal of time series analysis 7, 179–190 (1986)
4. Creamer, G.: Model calibration and automated trading agent for euro futures. Quantitative Finance 12, 531–545 (2012)
5. Dempster, M.A.H., Leemans, V.: An automated FX trading system using adaptive reinforcement learning. Expert Systems with Applications 30, 543–552 (2006)
6. Gold, C.: FX trading via recurrent reinforcement learning. In: Proceedings of IEEE International Conference on Computational Intelligence in Financial Engineering, pp. 363–370. IEEE Computer Society Press, Los Alamitos (2003)
7. Gorse, D.: Application of stochastic recurrent reinforcement learning to index trading. In: European Symposium on Artificial Neural Networks (2011)
8. Graham, B., Zweig, J.: The Intelligent Investor: The Definitive Book on Value Investing. HarperCollins (2003)
9. Maringer, D., Ramtohul, T.: Regime-switching recurrent reinforcement learning for investment decision making. Computational Management Science 9, 89–107 (2012)

10. Maringer, D., Ramtohul, T.: Regime-switching recurrent reinforcement learning in automated trading. In: Brabazon, A., O'Neill, M., Maringer, D.G. (eds.) Natural Computing in Computational Finance, vol. 4, pp. 93–121. Springer (2012)
11. Maringer, D., Zhang, J.: Transition variable selection for regime switching recurrent reinforcement learning. In: Proceedings of the 2014 IEEE Conference on Computational Intelligence for Financial Engineering and Economics, pp. 407–412 (2014)
12. Moody, J., Wu, L., Liao, Y., Saffell, M.: Performance functions and reinforcement learning for trading systems and portfolios. Journal of Forecasting 17, 441–470 (1998)
13. Murphy, J.: Technical analysis of the financial markets: a comprehensive guide to trading methods and applications. New York Institute of Finance (1999)
14. Silva, A., Neves, R., Horta, N.: Portfolio optimization using fundamental indicators based on multi-objective ea. In: Proceedings of the 2014 IEEE Conference on Computational Intelligence for Financial Engineering and Economics, pp. 158–165 (2014)
15. Tong, H.: On a threshold model. In: Chen, C. (ed.) Pattern Recognition and Signal Processing. NATO ASI Series E: Applied Sc. (29), Sijthoff & Noordhoff, Netherlands (1978)

Utilising Tree-Based Ensemble Learning for Speaker Segmentation

Mohamed Abou-Zleikha[1,*], Zheng-Hua Tan[1], Mads Græsbøll Christensen[2], and Søren Holdt Jensen[1]

[1] Department of Electronic Systems, Aalborg University, Denmark
[2] Audio Analysis Lab, ad:mt, Aalborg University, Denmark
{moa,zt,shj}@es.aau.dk,mgc@create.aau.dk

Abstract. In audio and speech processing, accurate detection of the changing points between multiple speakers in speech segments is an important stage for several applications such as speaker identification and tracking. Bayesian Information Criteria (BIC)-based approaches are the most traditionally used ones as they proved to be very effective for such task. The main criticism levelled against BIC-based approaches is the use of a penalty parameter in the BIC function. The use of this parameters consequently means that a fine tuning is required for each variation of the acoustic conditions. When tuned for a certain condition, the model becomes biased to the data used for training limiting the model's generalisation ability.

In this paper, we propose a BIC-based tuning-free approach for speaker segmentation through the use of ensemble-based learning. A forest of segmentation trees is constructed in which each tree is trained using a sampled version of the speech segment. During the tree construction process, a set of randomly selected points in the input sequence is examined as potential segmentation points. The point that yields the highest ΔBIC is chosen and the same process is repeated for the resultant left and right segments. The tree is constructed where each node corresponds to the highest ΔBIC with the associated point index. After building the forest and using all trees, the accumulated ΔBIC for each point is calculated and the positions of the local maximums are considered as speaker changing points. The proposed approach is tested on artificially created conversations from the TIMIT database. The approach proposed show very accurate results comparable to those achieved by the-state-of-the-art methods with a 9% (absolute) higher F_1 compared with the standard ΔBIC with optimally tuned penalty parameter.

1 Introduction

Speaker segmentation is the process of determining the speaker switching points in a speech signal leading to an accurate separation of the signal into speaker homogeneous subsegments. This is an essential initial stage in several speech and audio applications such as speaker tracking [5], audio classification [14] and speaker diarization

---

* This work was supported in part by the Danish Council for Strategic Research of the Danish Agency for Science Technology and Innovation under the CoSound project, case number 11-115328. This publication only reflects the authors' views.

systems [3]. Segmentation is usually performed without any prior knowledge about who or how many speakers are present in the speech segment, with an unknown maximum and minimum length of the speakers' segments and with no prior information about the acoustic conditions and the noise level and type. Having no or very limited prior information about these conditions makes speaker segmentation one of the very challenging task in the speech processing domain.

Several approaches have been proposed for speaker segmentation [7,13,4,10,8,15] (an extensive review can be find in [3]). Among these methods, ΔBIC metric is the most widely used for this task [7,13,4,10,8]. The main drawback of BIC-based approaches is that they require fine tuning of a penalty factor that highly affects the quality of segmentation. This penalty factor is related to the number of parameters those are used to model the speech segment. Traditionally, the penalty factor is tuned for each specific acoustic conditions which limits the generalisation capabilities of the system [16,21]. An alternative BIC function has been proposed which aims at eliminating the effect of the number of parameters in the BIC calculation [2]. This is accomplished by estimating the speech segment by two mixtures of the Gaussian mixture model and estimating the left and right segments around the segmentation point by one mixtures of the Gaussian mixture model.

In this paper, we introduce a novel approach based on the ensemble-based mechanism to estimate the speakers changing points in multi-speakers segments. The main advantage of the proposed approach is that it eliminates the need to tune the penalty factor originally employed in the BIC-based segmentation function. Through using the proposed method we were able to increase the robustness of the traditional BIC-based techniques when used in various acoustic conditions.

The approach works by building a set of segmentation trees called segmentation forest; each tree is built using an approach similar to the one proposed in [8]. Each node in each segmentation tree examines a set of randomly selected points as potential segmentation points. The gain function in each tree is the value of the BIC and the goal is to select the point that gives the highest BIC. The value of the penalty factor is randomly assigned for each BIC calculation. Each segmentation tree assigns a BIC value for the examined points. The final resultant BIC values for the sequence points are the accumulated values using all segmentation trees in the forest.

The paper is organised as follows: In section 3, the standard BIC-based speaker changing point detection is presented, followed by a BIC-based speaker segmentation using divide-and-conquer strategy in Section 3. In Section 4, the proposed approach is explained. The experiments and evaluations conducted to validate the proposed approach are presented in Section 5 and finally a conclusion is presented in section 6.

2   Detecting Speaker Changing Points Using BIC

The BIC is a asymptotically optimal Bayesian-based model-selection criterion traditionally used to determine the parametric model that best fits a set of data samples $X = x_1, ..., x_N$, where $x_i \in \mathbb{R}^d$, where d is the dimension of the feature space [7].

According to the BIC approach, the model M_j that best fits a set of data samples is the model that maximises the function:

$$BIC_j = logP(X|M_j) - \lambda\frac{1}{2}k_j logN \tag{1}$$

where $logP(X|M_j)$ is the log likelihood of the data X for M_j, λ is the weight of the second term that relates to the number of parameters k_j in the model M_j and the number of samples N in the data.

Suppose we have two models representing the data X namely M_1 and M_2, the model M_1 is chosen over M_2 to fit the data X if $\Delta BIC = BIC_1 - BIC_2$ is positive.

For the speaker segmentation task, suppose a segment of speech X. To check if the speaker changes at a point index i, M_1 represents the model drawn from two full-covariance Gaussians using the separated segments at point i and M_2 is the model drawn from one full-covariance Gaussian. We calculate the ΔBIC between M_1 and M_2: the number of parameters k_1 in M_1 is twice the number of parameters in M_2, where $k_1 = d + \frac{d(d+1)}{2}$. ΔBIC is calculated as:

$$\Delta BIC = N*log|\Sigma| - i*log|\Sigma_{left}| - (N-i+1)*log|\Sigma_{right}| - \frac{1}{2}\lambda(d+\frac{d(d+1)}{2})logN \tag{2}$$

where $|.|$ is the determinant of the covariance matrix, λ is the penalty factor that tunes the segmentation sensitivity [9], ideally this parameter is equal to 1. Point i is considered a speaker switching point, if $\Delta BIC > 0$.

Applying this approach on a speech segment detects a single changing point. In order to be able to detect multiple-switching points, a window growing mechanism is usually employed [20]. This approach starts by processing a small window N_{init} and tries to detect a changing point within this window. It then extends this window by a N_g until a changing point is detected or the window size reaches a maximum size N_{max}. If the search reaches N_{max} with no detection of a changing point, the window is shifted by N_s. Otherwise, the search process is repeated starting from the newly discovered switching point.

3   BIC-Based Speaker Segmentation Using Divide-and-Conquer

This approach uses the ΔBIC function to perform a hierarchical splitting of a speech sequence into its most two dissimilar parts [8]. This is done by scanning the whole segment and choosing the point that gives the highest ΔBIC. The resultant point is then considered as a potential speaker switching point (called i). The same approach is then repeated on the left and right segments of the point i. This process is applied recursively until the size of the segment becomes smaller than a threshold.

After processing the left and right segments, ΔBIC is checked, if it is positive, point i is considered as a switching point; otherwise, the leftmost sub-segment from the right segment and the rightmost sub-segment from the left segment are considered as one segment and the highest ΔBIC is calculated. If ΔBIC is positive, the new point is added to the set of changing points.

One can note that the process described can be seen as generating a tree-like structure, where each internal node in that tree is a segmentation point. It is also worth noticing that this approach still uses the standard ΔBIC where the tuning process of λ is still required.

The approach proposed in this paper employs a similar mechanism for generating a tree structure for the segmentation. These resultant trees are used to detect the changing points. In the next section, the approach proposed for speaker segmentation using a segmentation forest is explained.

4   BIC-Based Speaker Segmentation Using Segmentation Forest

The ensemble-based learning approaches such as random forest and density forest [6] have been successfully employed for several tasks in the audio domain such as emotion recognition [19,18], paralinguistic event detection [1] and audio event detection [11]. In this work, a segmentation forest is utilised as a special case of the random forest approach. Random forest is a tree-based non-parametric classification and regression approach. The principle is to grow an assemble of trees on a random selection of samples in a training set. Each tree is a classification and regression tree (CART). While constructing the trees and at each tree node, a randomly selected set of features is considered and the features are investigated as potential predictors that decide the split of the data. The splitting robustness is calculated as the information gain resulted from the splitting.

The proposed approach applies a similar mechanism to the one used by the random forest. However, instead of building decision trees, we build a set of segmentation trees, where the randomly selected features to be examined at each node are the potential segmentation positions, the gain function in our case is the ΔBIC.

Formally, a segmentation forest SF is a set of segmentation trees

$$SF = \{t_i\} : i = [1...T] \tag{3}$$

where t_i is the ith individual tree and T is the total number of trees. Each tree t_i, in the forest is trained independently in a similar manner to the one used for building the tree in the divide-and-conquer approach explained in Section 3. The main differences between the two processes are:

- Instead of scanning the full set of positions in the sequence, a randomly selected set of points is examined.
- The value of λ in the ΔBIC function is randomly chosen between a min and a max values.
- We do not recheck internal points if the value of ΔBIC is negative, instead we assign the negative value obtained for ΔBIC to the corresponding point.
- Each tree is trained using a sampling from the original sequence without replacement.

A detailed description of the tree building process is presented in Algorithm 1.

The result of the pervious process is a set of segmentation trees where each node in each tree is associated with a position and a ΔBIC value. The final ΔBIC for

---

Algorithm 1. Building a segmentation tree $n \leftarrow train(X)$

- X : the speech segment
- n : the node that represents the segmentation of the input segment

---

if ($length(X) < threshold$) **then**
    return empty
end if
- $pp \leftarrow$ get round of $\sqrt{(length(X))}$ randomly chosen points from X as potential segmentation points to examine
- Find the point i from pp that gives the highest ΔBIC (called $mBIC$), where λ is randomly chosen for each point
- Split X at position i into X_l and X_r
- call the same process on X_l to get $n.child_l$
- call the same process on X_r to get $n.child_r$
- store the position i in $n.position$
- - store the value $mBIC$ in $n.\Delta BIC$
return n

---

each point is the accumulated ΔBIC from all trees. As a result, positive accumulated ΔBIC values are produced on and around the speaker segmentation points. To detect the exact changing point positions, the resultant ΔBIC value sequence is grouped forming a set of local regions according to the distance between the points. The position of the maxima of each local region represents a speaker changing point as illustrated in Figure 1.

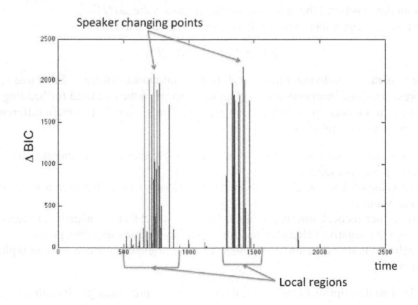

Fig. 1. An illustration of a accumulated ΔBIC and the local regions with their maxima

5 Experiment and Evaluation

The purpose of the experiments conducted is to validate the proposed approach and to check its efficiency for the speaker segmentation task. For this purpose, The performance and execution time of the proposed method is compared with several other BIC-based approaches reported in the literature.

5.1 Experiment Setup

In order to evaluate the proposed approach, a comparison between its performance and several other BIC-based approaches is performed. A artificially created conversations by concatenating speech from the TIMIT database are used for evaluation [12]. TIMIT is an acoustic-phonetic databased which consists of 6300 utterances for 630 english speakers. Two conversation sets (A and B) are generated, the first set is for penalty parameter tuning in the standard and divide-and-conquer BIC-based approaches. It consists of 20 conversation, each contains 2 to 6 speakers and the length of each speaker segment varies between 2 and 6 seconds. The testing dataset, B, consists of 100 conversation. The number of speakers per conversation changes from 2 to 6 and the length of each speaker segment is between 2 and 6 seconds. The testing set contains 338 switching points. The feature vectors used were 23-dimensional mel-frequency cepstral coefficients (MFCC) and log energy extracted every 10 ms, with a window size of 25 ms.

In the process of detecting the changing points in the speaker segmentation modules, two types of errors occur: the first is due to missing a true speaker changing point. This type of errors can be measured using the precision (PRC), which is calculated as:

$$PRC = \frac{number\ of\ correctly\ found\ changes}{total\ number\ of\ changes\ found} \qquad (4)$$

and the other measurement is the Missed Detection Rate (MDR), which is calculated as:

$$MDR = 100 * \frac{MD}{RC} \qquad (5)$$

where MD is the number of missed changing points and RC is the number of true changing points. The second error type occurs when a false changing point is detected. This type can be measured using the recall (RCL), which is defined as:

$$RCL = \frac{number\ of\ correctly\ found\ changes}{total\ number\ of\ correct\ changes} \qquad (6)$$

and the other measurement is False Alarm Rate (FAR), which is calculated as:

$$FAR = 100 * (1 - RCL) \qquad (7)$$

Another measurement that combines the PRC and RCL in one value is the F_1-measure. This measurement is defined as:

$$F_1 = 2 * \frac{PRC * RCL}{PRC + RCL} \qquad (8)$$

Table 1. Average and standard deviation values for the results obtained from each approach using the five error measurements

	BIC	$DICBIC$	$GMMBIC$	$SF - BIC$
$PRC(mean)$	0.77	0.89	0.59	0.83
$PRC(std)$	0.30	0.27	0.28	0.24
$RCL(mean)$	0.74	0.75	0.75	0.84
$RCL(std)$	0.31	0.29	0.31	0.25
$F_1(mean)$	0.74	0.79	0.65	0.83
$F_1(std)$	0.29	0.27	0.23	0.23
$FAR(mean)$	13.31%	3.97%	50.67%	10.78 %
$FAR(std)$	16.94	11.00	23.88	14.22
$MDR(mean)$	26.03%	25.33%	35.03%	16.47%
$MDR(std)$	30.48	28.82	30.61	24.85

The value of F_1 is between 0 and 1, the closer the value to 1 is, the better the system. For MDR and FAR measurements, the values are between 0 and 100, the smaller the value is, the better the system.

Four approaches are evaluated using the conversation set. The first approach is the standard window-growing approach (referred to as BIC) as described in Section 2. The λ value is tuned using the set A and the result value is set to $\lambda = 2.8$. The second approach is the divide-and-conquer approach (referred to as DACBIC) as described in Section 3 and the λ value is also tuned using the set A and it gives $\lambda = 3.2$. The third one is the window-growing approach using one mixture per GMM for separated segments and two mixture per GMM for combined segments to estimate the covariance matrices [2]. This removes the effect of the penalty parameter (referred to as GMM-BIC). Diagonal covariance GMM is used for GMMBIC implementation. The fourth approach is the proposed approach (referred to as SF-BIC). The number of trees in our approach is set to 50 and the stopping criterion threshold (minimum segment length) is 500. The number of points to examine at each node is $\sqrt{\frac{length(segment)}{N_{min}}}$ where N_{min} is the shifting factor. λ value was generated randomly between $[0..6]$. The tolerance value for the detected points is 0.5 second, i.e. if a detected point is positioned within 0.5 distance around a reference point, it is classified as a correctly detected point.

5.2   Evaluation

The five accuracy measures discussed are calculated for each examined approach. We used the same evaluation protocol proposed in [17]. According to this protocol, the previously discussed error metrics are first calculated followed by applying ANOVA and Tukey test.

Table 1 presents the mean and standard divination values of each measure for each of the examined system.

The results show that a better performance is obtained by the proposed method compared with the other approaches with respect to the overall performance of the system as depicted by the F_1 measure. The results also indicate that our approach is able to more accurately predict the true changing points according to the RCL and MDR measures. The DICBIC detects less false alarm points compared with the other approaches as the results of PRC and FAR show.

To check if the performance obtained is significantly different among the compared approaches, a One-way ANOVA is applied for a 95% confidence level. The null hypothesis tested is that the groups means are equal, i.e. the approaches are not significantly different. The alternative hypothesis states that the groups means are unequal, which consequently means that at least one of the systems differs from the rest with respect to the examined measure. The $F - statistic$ values and the $p - values$ of all approaches are calculated and presented in Table 2. The results show that the performance is significantly different among all five measures.

Table 2. $F - statistic$ and $p - value$ results obtained from ANOVA for each approach using the five measurements

	$F - statistic$	$p - value$
PRC	66.64	1.42e-34
RCL	6.59	2.36e-04
F_1	44.85	8.45e-25
FAR	142.39	1.22e-61
MDR	6.59	2.36e-04

Since ANOVA test does not provide any information about which system is different from the other, the Tukey range test, or honestly significant differences method, is employed. Tukeys method provides a pair-wise comparison of the means while maintaining the confidence level at a predefined value. If the confidence interval includes zero, the differences are not significant, otherwise, the differences are significant.

The Tukey test is applied between the proposed approach and the three other approaches and the results obtained are presented in Table 3.

Table 3. The Tukey test results between the proposed approach and the other examined approaches

	$SF - BIC$ vs $GMMBIC$	$SF - BIC$ vs $DICBIC$	$SF - BIC$ vs BIC
PRC	0.374,0.448	-0.126,-0.0540	-0.0130 , -0.0635
RCL	0.105 , 0.1856	0.01201, 0.0886	0.01655 , 0.09561
F_1	0.3151 , 0.3805	-0.0401, 0.0302	0.0077 , 0.0820
FAR	-45.477 , -39.888	3.197 , 6.8120	-6.977 , -2.530
MDR	-26.489 , -18.561	-16.512, -8.860	-17.468 , -9.561

The results show that the differences in the performance between all approaches are significant except for the F_1 measure between the proposed approach and the DICBIC approach.

Table 4. Average and standard deviation values for the execution time of each approach

	BIC	$DICBIC$	$GMMBIC$	$SF - BIC$
$time(mean)(s)$	50.16	19.82	91.81	124.77
$time(std)$	25.76	12.26	50.24	54.18

The results obtained from the statistical analysis performed indicate that the performance of our approach is comparable to those achieved by the-state-of-the-art method as demonstrated by the insignificant difference between the accuracies obtained.

The execution time for each approach (shown in Table 4) shows that the proposed approach has the highest execution time. This is due to the time required to build a set of segmentation models instead of one. However, since each tree is built independently, this time consumption issue can be solved by constructing the trees in parallel.

6   Conclusion

In this paper, a tree-based ensemble method for speaker changing point detection is proposed. The approach trains a set of trees using a sampled version of the speech segment. To build a node in the tree, a randomly selected points are examined as potential segmentation points. The point that gives the highest ΔBIC is chosen. This process is recursively applied on the left and right subsegments until a stopping criterion is reached. As a result, a tree is constructed where each node stores the highest ΔBIC with the associated point index. Once the model is built, the accumulated ΔBIC for each point is calculated using the all trees in the forest. The final changing points are then calculated as the positions of the local maximums after grouping the points those have a positive ΔBIC into groups according to the distance between them.

We conduct a set of experiments to test the proposed approach and analyse its performance. For this purpose, a comparison is performed with three other state-of-the-art methods. The comparison shows that the proposed approach achieves better average results an insignificant performance difference from the best models reported in the literature. Our model, however, has the advantage of eliminating the need of parameter tuning and thereafter demonstrates more robustness and generalisation capability over changes in the acoustic conditions.

The future work includes building an interactive speaker changing point detection, where the user can modify, add or delete a set of changing points and the model uses this information to adapt itself.

References

1. Abou-Zleikha, M., , Tan, Z.H., Christensen, M.G., Jensen, S.H.: Non-linguistic vocal event detection and localisation using online random forest. In: Proceedings of 37th International Convention of Information and Communication Technology (MIPRO). IEEE (2014)
2. Ajmera, J., McCowan, I., Bourlard, H.: Robust speaker change detection. IEEE Signal Processing Letters 11(8), 649–651 (2004)

3. Anguera Miro, X., Bozonnet, S., Evans, N., Fredouille, C., Friedland, G., Vinyals, O.: Speaker diarization: A review of recent research. IEEE Transactions on Audio, Speech, and Language Processing 20(2), 356–370 (2012)
4. Ben, M., Betser, M., Bimbot, F., Gravier, G.: Speaker diarization using bottom-up clustering based on a parameter-derived distance between adapted gmms. In: Proceedings of ICSLP (2004)
5. Bonastre, J.F., Delacourt, P., Fredouille, C., Merlin, T., Wellekens, C.: A speaker tracking system based on speaker turn detection for nist evaluation. In: Proceedings of IEEE International Conference on Acoustics, Speech, and Signal Processing, vol. 2, pp. II1177–II1180. IEEE (2000)
6. Breiman, L.: Random forests. Machine learning 45(1), 5–32 (2001)
7. Chen, S., Gopalakrishnan, P.: Speaker, environment and channel change detection and clustering via the bayesian information criterion. In: Proceedings of DARPA Broadcast News Transcription and Understanding Workshop (1998)
8. Cheng, S.S., Wang, H.M., Fu, H.C.: Bic-based speaker segmentation using divide-and-conquer strategies with application to speaker diarization. IEEE Transactions on Audio, Speech, and Language Processing 18(1), 141–157 (2010)
9. Grašič, M., Kos, M., Kačič, Z.: Online speaker segmentation and clustering using cross-likelihood ratio calculation with reference criterion selection. IET signal processing 4(6), 673–685 (2010)
10. Kotti, M., Benetos, E., Kotropoulos, C.: Automatic speaker change detection with the bayesian information criterion using mpeg-7 features and a fusion scheme. In: IEEE International Symposium on Circuits and Systems, p. 4. IEEE (2006)
11. Kumar, A., Dighe, P., Singh, R., Chaudhuri, S., Raj, B.: Audio event detection from acoustic unit occurrence patterns. In: Proceedings of IEEE International Conference on Acoustics, Speech, and Signal Processing, pp. 489–492 (2012)
12. Lamel, L.F., Kassel, R.H., Seneff, S.: Speech database development: Design and analysis of the acoustic-phonetic corpus. In: Speech Input/Output Assessment and Speech Databases (1989)
13. Li, R., Schultz, T., Jin, Q.: Improving speaker segmentation via speaker identification and text segmentation. In: Proceedings of INTERSPEECH 2009 (2009)
14. Meinedo, H., Neto, J.: Audio segmentation, classification and clustering in a broadcast news task. In: Proceedings of IEEE International Conference on Acoustics, Speech, and Signal Processing, vol. 2, pp. II–5. IEEE (2003)
15. Mohammadi, S.H., Sameti, H., Langarani, M.S.E., Tavanaei, A.: Knndist: A non-parametric distance measure for speaker segmentation. In: Proceedings of INTERSPEECH (2012)
16. Mori, K., Nakagawa, S.: Speaker change detection and speaker clustering using vq distortion for broadcast news speech recognition. In: Proceedings of IEEE International Conference on Acoustics, Speech, and Signal Processing, vol. 1, pp. 413–416. IEEE (2001)
17. Moschou, V., Kotti, M., Benetos, E., Kotropoulos, C.: Systematic comparison of bic-based speaker segmentation systems. In: Proceedings of IEEE 9th Workshop on Multimedia Signal Processing, pp. 66–69. IEEE (2007)
18. Rong, J., Li, G., Chen, Y.P.P.: Acoustic feature selection for automatic emotion recognition from speech. Information processing & management 45(3), 315–328 (2009)
19. Schuller, B., Batliner, A., Seppi, D., Steidl, S., Vogt, T., Wagner, J., Devillers, L., Vidrascu, L., Amir, N., Kessous, L., et al.: The relevance of feature type for the automatic classification of emotional user states: low level descriptors and functionals. In: Proceedings of INTERSPEECH, vol. 2007, pp. 1–4 (2007)
20. Tritschler, A., Gopinath, R.A.: Improved speaker segmentation and segments clustering using the bayesian information criterion. In: Proceedings of Eurospeech, vol. 99, pp. 679–682 (1999)
21. Vandecatseye, A., Martens, J.P., Neto, J.P., Meinedo, H., Garcia-Mateo, C., Dieguez-Tirado, J., Mihelic, F., Zibert, J., Nouza, J., David, P., et al.: The cost278 pan-european broadcast news database. In: Proceedings of LREC (2004)

Speakers' Language Characteristics
Analysis of Online Educational Videos

Dimitrios Kravvaris and Katia Lida Kermanidis

Department of Informatics, Ionian University, Corfu, Greece
jkravv@gmail.com, kerman@ionio.gr

Abstract. Research in the field of educational videos and the contribution of data mining to education can affect the instructors' approach to learning. This particular study focuses on online educational videos and more specifically on their speakers. Initially a survey is conducted related to the popularity of educational videos on the YouTube which are then divided into two categories the more popular and the less popular. Then the characteristics related to language are extracted from the transcript of the speakers and after a clustering procedure the differences between the two categories are stated. The characteristics related to the language of the speakers of the popular videos present very interesting results. That is, the pace of speaking is faster and the complexity off the sentences is higher than the ones in the less popular videos.

Keywords: Educational video, transcript, popularity, clustering, k-means.

1 Introduction

The advancement of social media adds a large amount of data on the web on a daily basis and especially in content-based communities such as YouTube and Daily Motion. A very large number of videos in social media concern education, and in many cases, constitute part of the traditional online courses [1] and the upcoming massive open online courses [2]. They are usually created by universities, companies, organizations or even individual users. In many cases transcripts of the video lectures are available.

The present research focuses on the study of educational videos from social media, oriented both to verbal content and to metadata of the pages that contain them. The present study thus, examines questions concerning issues such as why some educational videos are more popular than others and what are the characteristics that make a video popular. The issues arising are both interesting and complex. Our research innovation is that we examine them based on the audio language used in the educational videos. Through a qualitative study of the transcripts of the videos we extract the characteristics of the language used by the speakers (i.e. pace of speech, sentence length, commas, range of vocabulary etc.), which will be utilized in order to designate the speakers into to two basic types of speakers. These two types are based on whether the videos they take part in are popular or not. At this point another interesting question arises which we had to answer as well, i.e. what are these characteristics that define the popularity of a video after all, and how we can measure it? Using the

L. Iliadis et al. (Eds.): AIAI 2014, IFIP AICT 436, pp. 60–69, 2014.

metadata of the web pages that contain the videos we moved to an analysis related to the issue and propose a formula for defining video popularity. Finding the language characteristics of the speakers of the popular educational videos is very important both for the educational organizations and the individuals as creators of educational videos and for the scientific community since this study contributes to the research of linguistic data in the social media.

In the first part of the study we present relevant studies and point out how our research differs. In the second part we analyze the concept of popularity of online videos and we propose a formula for its estimation. In the third part we present the methodology used and include a thorough analysis of the characteristics used in our research. In the fourth part we present the experiments conducted as well as a commentary on the findings. Finally in the last part we present the findings of our research and how these can be utilized.

2   Related Work

As far as the videos are concerned, a lot of studies have been conducted in various field studies concerning video classification [3], [15] in order for the videos to fall automatically in certain categories using video and text data. Studies that concern the searching of videos and more specifically studies focusing on information retrieval browsing very large document collections [4] and video retrieval on the web utilizing the integration of multiple features [5], [7]. Finally, there have been studies that focus on video comparison [6] in order to estimate the percentage of visually similar frames.

The special characteristic of our research concerns the transcript of what the speakers say in each video. This has been used in other relevant studies concerning text mining such as text classification [13,14] and text clustering [8], [12], as well as studies concerning natural language processing [9,10]. Our research was inspired by the research conducted by Jin and Murakami [11] who studied the authors' characteristic writing styles as seen through their use of commas.

When it comes to social media, and more specifically YouTube, it has been shown that the introduction of videos in higher education has opened new horizons both to the educators who want to contribute to education and to learners who want to learn [16,17]. Thus, a new effort is being made in order for success in learning to be maximized. At this point our study comes in order to examine the educational videos in relation to their popularity on the YouTube. More specifically, we use the clustering method on metadata and on content data of the transcript of the video. Our purpose is to divide the videos in two categories: the most popular video category and the least popular videos category, and then to study which are qualitative speech characteristics of each category, regardless of the subject content of the videos.

3   Video Popularity

YouTube contains quite a few characteristics that could be utilized in order to define the popularity of a video, such as the number of views, of likes, of dislikes, the users' comments, the number of those who have chosen it as favorite and finally the number of video responses [18,19]. The favorites and responses are the least used characteristics by the users. The views characteristic refers to the number of times the video has been viewed, while the likes, dislikes and comments can be used by registered users only. Especially for comments we should mention that we face two problems: the first concerns the complicated and time consuming procedure required in order to characterize the users' opinion [20,21], and the second concerns the ability to comment the video lecture, which could be deactivated by the creator and, thus, we would have no relevant comments.

Thus, in order to be fair concerning the videos in focus we chose to keep the characteristics that are definitely present and that attribute a positive value to the video. We ended up, therefore, using the views and the likes, in order to estimate the popularity of the videos. These two characteristics are based on human actions that show how many times a video has been viewed and how many people liked it.

We define as popularity P of a video i, which belongs to a certain category c, the normalized value of likes L and the number of views V according to formula

$$P_{i,c} = \frac{L_i}{maxL_c} + \frac{V_i}{maxV_c}$$

where $maxL$ and $maxV$ are the maximum values of likes and views correspondingly, that were observed in the particular video category. Since designating the value of P as *high* or *low* is subjective, we used the mathematical method of median [31]. The median is a measure of central tendency. In our case it represents the value for which half of videos' popularity are higher and the other half are lower. In that way splitting in half the videos of high and low popularity we can use machine learning methods in order to extract knowledge concerning what makes a video more popular than another.

We chose not to use the lifetime of a video on YouTube as a parameter in estimating the popularity of a video, because there seemed to be a problem: The new videos with few likes and views seemed to be more popular which was wrong because older videos had more likes and views.

4   Methodology

4.1   Data

Our data were collected from YouTube, which is the third most visited social media site worldwide [22] and the largest provider of videos [23]. YouTube provides its users with a specific space to upload videos that fall into the category of educational videos[1]. Searching through the category of Education of YouTube by inserting keywords from different scientific fields such as computer science, physics, medicine, art, health, philosophy, energy and others, 20830 videos were collected among which 1108 (5.3%) had English transcripts. The total duration of the 1108 videos used in our research is 473 hours and have over 242 million views in total. From each video metadata attributes were collected using the YouTube API v2 [24] as well as qualitative

---

[1] http://www.youtube.com/education

attributes of speech (after processing the transcripts of videos). Grzybek's et al. [32] and Mahowald's et al. [33] research shows the importance of words, Hill's and Murray's research [34] note the value of commas and Palmer's study [35] highlights the importance of sentence segmentation of a natural language text. Thus, we used these important structural elements for the definition of our qualitative attributes. Below we refer to these attributes and their description in categories.

- Metadata
  In this category there are two attributes. The first one is the *Duration* attribute which refers to the second of the total appearance of the online video. The second attribute is the *AuthorUri* which concerns the unique identity of the owner of the video on YouTube, which may refer to a University, an educational organization or an individual. Both attributes come from metadata of the YouTube page, which contain the video in focus.

- Words
  The words category contains the qualitative characteristics of the transcript of the educational video. More analytically, the attribute *NumOfWords* concerns the number of words used by the speakers of the video. This attribute shows the real duration of speech, since we count neither the duration of speech, which contains times pauses, nor the duration of a video which contains other elements such as ads or short introductions before the educational video begins. The second attribute *AvgWordLength* concerns the average word length. This attribute helps us form a complete view of the net length of speech we referred to earlier, since videos differ also in the length of words used, besides the number of words.

- Transcript Sentences
  This category contains four attributes which concern: the number of the transcript sentences (*NumOfSent*), the minimum sentence length in characters (*MinSentLength*), the average sentence length in characters (*AvgSentLength*) and the maximum sentence length in characters (*MaxSentLength*). All these four attributes describe the number and the length of the transcript sentences. Thus, through the transcripts we can extract qualitative information concerning the sentence length used by the speakers, supposing that longer sentences are more likely to contain more information for the listener that shorter ones.

- Sentences complexity
  This category contains two attributes concerning the commas contained in the transcripts. The *NumOfCommas* attribute refers to the total number of commas contained in the transcript while the *AvgNumOfCommasPerSent* attribute shows the average number of commas per sentence. Commas are used in order to avoid ambiguity. They are mainly used in lists, for separation causes, to set off certain adverbs at the beginning of a sentence and in parenthetical phrases. All the above indicate that a sentence with commas is more complicated in structure and in meaning that one without commas [30].

- Vocabulary
  This category contains the *NumOfUniqueWords* attribute which shows the number of unique words in the transcript. The more unique words a transcript contains the wider the vocabulary used by the speaker, without it necessarily being more advanced since the videos come from different scientific fields and contain the domain-specific terminology of the corresponding fields.

- Flow of words
  This category contains two attributes: the *MicroRhythm* attribute that refers to the micro flow of words and the *MacroRhythm* attribute which refers to the macro flow of words. More analytically, the *MicroRhythm* attribute measures the average flow of words in the time (measured in seconds) the corresponding transcript text is displayed on the screen, and the *MacroRhythm* attribute measures the flow of words in the total time the transcript texts are displayed on the screen.
- Evaluation
  This category contains the evaluation attribute of our study named *Popularity*. This attribute is used for the binary classification of the educational videos. It has two values *high* and *low* as it was described in the previous section of the present paper.

4.2   Experimental Procedure

In the beginning we conduct a statistical analysis of our data. At this point we study any extreme cases and we suggest solutions to deal with them. The purpose is to pre-process our data so as to avoid problems during the experimental procedure, such as missing values in the data of our datasets.

In the experimental procedure we used the Weka version 3.6.10 software [25]. We employed unsupervised learning methods for the clustering experiments. More specifically, a centroid-based clustering algorithm [26] using SimpleKMeans, with the Euclidian distance function [27] has been used. The SimpleKMeans method is quite suitable for our experiments since it is easy to understand and to explain its clustering outcome [28]. Two clusters were chosen for the value of K (in SimpleKmeans), since we have two class values: high/low popularity. Moreover, we chose to use the clustering mode classes-to-clusters evaluation [29], which assigns classes to the clusters based on majority and computes the classification error of the videos that have different value from the class value of the cluster they belong to. With the above procedure, on the one hand, we can study the differences between the qualitative characteristics of the videos (that come from the transcript), and, on the other hand, to evaluate how these characteristics can define the videos' popularity.

5   Experimental Results

5.1   Data Analysis

While analyzing our data we found out that there is a great difference in the duration of videos and for this reason we have discretized their duration in 10-minute intervals. The results are presented in figure 1 below, which shows the number of videos in each time category they belong to. We find that the greatest number of educational videos fall into the 1 to 10 minute category (47.5% of videos), while the 41 to 50 minute and 51 to 60 minute categories contain 25% of videos in total. In the first case, there are short videos concerning the time duration, while in the second case long ones, for this reason, thus, we divided the initial dataset into two new ones based on their duration. In this way we can conduct our study on data that have similar characteristics.

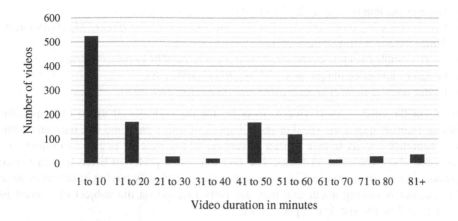

Fig. 1. Distribution of videos based on time duration

5.2   Short Videos

After conducting our experiment on the dataset that contains short (in terms of duration) videos we extract the following results shown on table 1. The videos that belong
to *Cluster-0* are in majority of high popularity, while the videos that belong to *Clus-
ter-1* are of low popularity. It should be mentioned that 66.54% of the videos have
been correctly clustered. Considering that we have to do with data that are based on
the human activity of speech the percentage can be characterized as highly positive.

Table 1. Clustering results for short videos

Attribute	Cluster-0 (209 videos)	Cluster-1 (317 videos)
Duration	398,0909	179,7476
AuthorUri	TEDEducation	Udacity
MicroRhythm	3,2269	3,0254
MacroRhythm	2,9277	2,6793
NumOfWords	1064,9809	417,3817
AvgWordLength	4,5849	4,6857
NumOfSent	64,8852	27,1735
MaxSentLength	479,3923	323,6498
MinSentLength	19,3541	35,2114
AvgSentLength	141,7913	124,3602
NumOfCommas	56,3636	19,4795
AvgNumOfCommasPerSent	0,9938	0,7671
NumOfUniqueWords	383,6268	200,6467

Based on the qualitative characteristics of the transcripts of the video speakers, as they
were described above, we can describe the types of the speakers of each cluster. More
specifically, the speakers of the popular videos (*Cluster-0*) present the following language characteristics, compared to speakers of the less popular videos (*Cluster-1*).

- Greater net length of speech (*NumOfWords, AvgWordLength*).
- Sentences with more information for the listener (*NumOfSent, MaxSentLength, MinSentLength, AvgSentLength*).
- More complex sentences (*NumOfCommas, AvgNumOfCommasPerSent*).
- Greater number of unique words (*NumOfUniqueWords*).
- Faster pace of flow of speech (*MicroRhythm, MacroRhythm*).

Thus, on the case of short videos, in order for the educational video to be popular among internet users, the speaker should speak at a fast pace and use long and complex sentences, so as to take greater advantage of the time he is given to inform the audience about the issue in focus. The fact that users prefer to listen to a fast pace speaker is very interesting which means that they are closely paying attention to what the speaker is talking about and they are fully focused on the subject of interest in order to follow the speaker's speech pace.

5.3   Long Videos

From the experiment conducted on the dataset that contains long (in terms of duration) videos we extract the following results shown on table 2. In this case, also, as was also shown in the previous experiment, videos that belong to *Cluster-0* are in majority of high popularity, while videos that belong to *Cluster-1* are of low popularity. The correctly clustered videos reach 86.36%, which is extremely positive for the classification of the videos.

Table 2. Clustering results for long videos.

Attribute	Cluster-0 (167 videos)	Cluster-1 (119 videos)
Duration	2965,9641	2974,3529
AuthorUri	MIT	YaleCourses
MicroRhythm	3,0471	2,6756
MacroRhythm	2,3240	2,4421
NumOfWords	6779,0719	6684,5630
AvgWordLength	4,2704	4,6612
NumOfSent	465,0659	364,9244
MaxSentLength	392,9521	818,7227
MinSentLength	1,9162	10,3193
AvgSentLength	77,5774	130,3118
NumOfCommas	461,1617	322,4202
AvgNumOfCommasPerSent	1,0453	0,9979
NumOfUniqueWords	974,8323	1343,6555

Following the same logic, as in the previous case of short videos, we can describe the types of speakers of every cluster. In this way, we can record comparatively the language characteristics of the speakers of the popular videos (*Cluster-0*) compared to the speakers of the less popular videos (*Cluster-1*). We find out that the speakers of the popular videos have:

- Practical the same length of speech as the speakers of the less popular videos. (*NumOfWords, AvgWordLength*).
- Sentences containing less information for the listener (*NumOfSent, MaxSentLength, MinSentLength, AvgSentLength*).
- More complex sentences (*NumOfCommas, AvgNumOfCommasPerSent*).
- Lower number of unique words (*NumOfUniqueWords*).
- Faster pace of micro flow of words (*MicroRhythm*), and almost the same pace of macro flow of words (*MacroRhythm*) as the speakers of less popular videos. This means that on average the speaker in a popular video uses more words at a given period of time.

To sum up, in order for a long video to be frequently viewed and positively reviewed, the speaker has to speak at a fast pace, to limit his vocabulary to the issue in question and to use complex sentences, which, however, do not carry too much information. In that way the user stays focused on the speaker's words and does not get confused or bored while watching the video.

5.4   Similarities

Based on our findings, there are similarities between the characteristics of the speakers of the popular short videos and the popular long videos as these are shown in column *Cluster-0* of tables 1 and 2. The similarities concern: a) the pace of speech, where we see that the speakers use almost the same number of words at a given period of time and b) the complexity of the sentences, where it is shown that speakers prefer to use more complicated in structure and in meaning sentences for their listeners. Thus, we conclude that, in order to create a popular educational video, regardless of its duration, the main speaker has to have a good command of the audio language and to be fully aware of the lecture subject so as to be able to express complex issues at a fast pace of speech. Knowing that the level of knowledge of the English language of the video listeners varies, the characteristics mentioned above seem to be very important in order for the lecture subject to be effective.

6   Conclusion

In our research we studied the language characteristics that a speaker of an educational video should have in order for the video to be more acceptable by the users of the social media. The whole procedure was based, on the one hand, on the qualitative research of the video transcripts, from which the language characteristics were extracted, and, on the other hand, on the classification of the videos in categories according to their popularity.

The popularity of the videos constituted the first part of an interesting analysis for our research. The formula suggested was based on the likes and views attributes, which besides YouTube, appear in other social media. The classification of the videos in most popular ones and least popular ones, based on the median method, ensured that this classification was objective.

Through our experimental procedure one can find out that in short videos speakers use speech more effectively. Speaking at a faster pace and using sentences with more information do not allow time to be wasted and help keep their listeners' interest. On the other hand, in popular long videos the speech contains more complex sentences than the speakers in popular videos. Finally, we show that both type of speakers in long and short popular videos have similar fast pace of speech. Their common characteristic that concerns the fast pace of speech seems to be interesting and urges us to study further what the pace should be in order for a listener to be satisfied.

The education industry uses videos as a basic tool. Our research shows that new knowledge can be extracted through machine learning techniques from the large quantity of free data in social media. In this way it is possible to identify the factors that can conduce to creating more popular and higher quality educational videos.

References

1. Moel de, E.L.: Expanding the usability of recorded lectures (2010), http://purl.utwente.nl/essays/59431
2. Waldrop, M.M., Nature Magazine: Massive open online courses, aka MOOCs, transform higher education and science (2014)
3. Gibbon, D.C., Liu, Z.: Introduction to Video Search Engines. Springer (2008)
4. Cutting, D.R., Pedersen, J.O., Karger, D.R., Tukey, J.W.: Scatter/gather: a cluster-based approach to browsing large document collections. In: SIGIR, pp. 318–329 (1992)
5. Yang, J., Li, Q., Wenyin, L., Zhuang, Y.: Searching for flash movies on the web: A content and context based framework. World Wide Web 8(4), 495–517 (2005)
6. Cheung, S.C.S., Zakhor, A.: Efficient video similarity measurement with video signature. IEEE Trans. Circuits Syst. Video Technol. 13(1), 59–74 (2003)
7. Hindle, A., Shao, J., Lin, D., Lu, J., Zhang, R.: Clustering web video search results based on integration of multiple features. World Wide Web 14(1), 53–73 (2011)
8. Amine, A., Elberrichi, Z., Simonet, M.: Evaluation of text clustering methods using wordnet. Int. Arab J. Inf. Technol. 7(4), 349–357 (2010)
9. Collobert, R., Weston, J., Bottou, L., Karlen, M., Kavukcuoglu, K., Kuksa, P.: Natural language processing (almost) from scratch. The Journal of Machine Learning Research 12, 2493–2537 (2011)
10. Friederici, A.D.: The brain basis of language processing: from structure to function. Physiological Reviews 91(4), 1357–1392 (2011)
11. Jin, M., Murakami, M.: Authors' characteristic writing styles as seen through their use of commas. Behaviormetrika 20(1), 3–76 (1992)
12. Shehata, S., Karray, F., Kamel, M.S.: An efficient concept-based mining model for enhancing text clustering. IEEE Transactions on Knowledge and Data Engineering 22(10), 1360–1371 (2010)
13. Passini, C., Luiza, M., Estébanez, K.B., Figueredo, G.P., Ebecken, F., Nelson, F.: A Strategy for Training Set Selection in Text Classification Problems. International Journal of Advanced Computer Science & Applications 4(6) (2013)
14. Kiritchenko, S., Matwin, S.: Email classification with co-training. In: Proceedings of the 2011 Conference of the Center for Advanced Studies on Collaborative Research, pp. 301–312 (2011)
15. Filippova, K., Hall, K.B.: Improved video categorization from text metadata and user comments. In: Proceedings of the 34th International ACM SIGIR Conference on Research and Development in Information Retrieval, pp. 835–842 (2011)

16. Gilroy, M.: Higher education migrates to YouTube and social networks. Education Digest 75(7), 18–22 (2010)
17. Selwyn, N.: Social media in higher education. The Europa World of Learning (2012)
18. Chatzopoulou, G., Sheng, C., Faloutsos, M.: A first step towards understanding popularity in youtube. In: INFOCOM IEEE Conference on Computer Communications Workshops, pp. 1–6 (2010)
19. Figueiredo, F., Almeida, J.M., Gonçalves, M.A., Benevenuto, F.: On the Dynamics of Social Media Popularity: A YouTube Case Study. arXiv preprint arXiv:1402.1777 (2014)
20. Liu, B.: Sentiment analysis and opinion mining. Synthesis Lectures on Human Language Technologies 5(1), 1–167 (2012)
21. Chen, H., Zimbra, D.: AI and opinion mining. IEEE Intelligent Systems 25(3), 74–80 (2010)
22. Wattenhofer, M., Wattenhofer, R., Zhu, Z.: The YouTube Social Network. In: ICWSM (2012)
23. Moran, M., Seaman, J., Tinti-Kane, H.: Teaching, Learning, and Sharing: How Today's Higher Education Faculty Use Social Media. Babson Survey Research Group (2011)
24. Padilla, A., DeFields, A.: Beginning Zend Framework. Apress (2009)
25. Weka.: Data Mining Software in Java. University of Waikato (2014), http://www.cs.waikato.ac.nz/ml/weka/
26. Kanungo, T., Mount, D.M., Netanyahu, N.S., Piatko, C., Silverman, R., Wu, A.Y.: The analysis of a simple k-means clustering algorithm. In: Proceedings of the Sixteenth Annual Symposium on Computational Geometry, pp. 100–109 (2000)
27. Kanungo, T., Mount, D.M., Netanyahu, N.S., Piatko, C.D., Silverman, R., Wu, A.Y.: An efficient k-means clustering algorithm: Analysis and implementation. IEEE Transactions on Pattern Analysis and Machine Intelligence 24(7), 881–892 (2002)
28. Vora, P., Oza, B.: A Survey on K-mean Clustering and Particle Swarm Optimization. International Journal of Science and Modern Engineering (IJISME), 24-26 (2013)
29. Färber, I., Günnemann, S., Kriegel, H.P., Kröger, P., Müller, E., Schubert, E., et al.: On using class-labels in evaluation of clusterings. In: MultiClust: 1st International Workshop on Discovering, Summarizing and Using Multiple Clusterings Held in Conjunction with KDD (2010)
30. Wiegand, N.: Creating complex sentence structure. In: Proceedings of the Annual Meeting of the Berkeley Linguistics Society vol. 10 (2011)
31. Beliakov, G., Bustince, H., Fernandez, J.: The median and its extensions. Fuzzy Sets and Systems 175(1), 36–47 (2011)
32. Grzybek, P., Stadlober, E., Kelih, E.: The relationship of word length and sentence length: the inter-textual perspective. In: Advances in Data Analysis, pp. 611–618. Springer (2007)
33. Mahowald, K., Fedorenko, E., Piantadosi, S.T., Gibson, E.: Info/information theory: Speakers choose shorter words in predictive contexts. Cognition 126(2), 313–318 (2013)
34. Hill, R.L., Murray, W.S.: Commas and spaces: The point of punctuation. In: 11th Annual CUNY Conference on Human Sentence Processing (1998)
35. Palmer, D.D.: Tokenisation and sentence segmentation, chapter 2. In: Dale, R., Moisi, H., Somers, H. (eds.) Handbook of Natural Language Processing. Marcel Dekker (2000)

Fall Detection Using Commodity Smart Watch and Smart Phone

Ilias Maglogiannis[1,2], Charalampos Ioannou[3],
George Spyroglou[2,3], and Panayiotis Tsanakas[3]

[1] University of Piraeus, Dept. of Digital Systems, Piraeus, Greece
[2] University of Thessaly, Dept. of Computer Science and Biomedical Informatics,
Lamia Greece
[3] School of Electrical and Computer Engineering, National Technical University of Athens
imaglo@unipi.gr

Abstract. Human motion data captured from wearable devices such as smart watches can be utilized for activity recognition or emergency event detection, especially in the case of elderly or disabled people living independently in their homes. The output of such sensors is data streams that require real-time recognition, especially in emergency situations. This paper presents a novel application that utilizes the low-cost Pebble Smart Watch together with an Android device (i.e a smart phone) and allows the efficient transmission, storage and processing of motion data. The paper includes the details of the stream data capture and processing methodology, along with an initial evaluation of the achieved accuracy in detecting falls.

Keywords: Fall Detection, Smart watch, Activity Recognition, Ambient Assisted Living, Streaming data.

1   Introduction

The introduction of ubiquitous and mobile systems in human centered computing pushes towards achieving availability of invisible software applications and information anywhere and anytime [1]. Applications and interfaces able to automatically process data provided by sensors, exchange knowledge and make intelligent decisions in a given context, are strongly desirable. Natural user interactions with such applications are based on autonomy, avoiding the need for the user to control every action, and adaptivity, so that they are contextualized and personalized, delivering the right information and decision at the right moment. Thus, the development of assistive applications particularly for elderly or disabled people living on their own, has gained significant attention during the last years. The aging population is a significant issue. According to a recent report published by the EU the percentage of European population over the age of 65 is expected to be over 20% in 2020 and over 30% in 2060 [2].

The detection of falls is considered an essential function in the context of such intelligent assistive applications and environments, since approximately 33% of persons over the age of 65 and 50% of persons over the age of 85 experience a fall

L. Iliadis et al. (Eds.): AIAI 2014, IFIP AICT 436, pp. 70–78, 2014.

each year [3]–[4]. The injuries associated with falls can have serious consequences. For example, following hip fracture, 50% of older people are unable to live independently, 25% will die within six months, and 33% die within one year [3]–[4].

The goal of this work is to propose an unobtrusive wearable system capable of capturing and analyzing motion data and reaching a prompt decision for activity recognition and fall detection. For this purpose, the Pebble smart watch [5] has been utilized, as it combines several positive characteristics (low price, water-resistance, user friendliness, long battery life, open SDK). The specific smart watch has a reliable built-in accelerometer to measure human movements, and a personal area network interface in order to transmit data and communicate with other devices. More details about the Pebble device are provided in Section 2.2. The processing of the motion data occurs in a nearby Android device, since the smart watch lacks the necessary computing power to perform data analysis and the wide networking interface for transmitting the fall detection information. The overall architecture of the proposed fall detection application is illustrated in Figure 1.

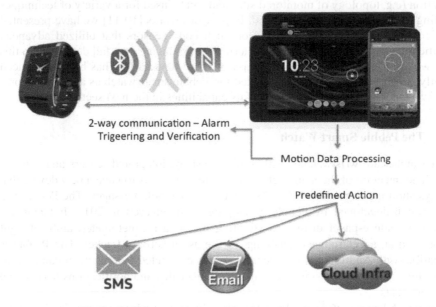

Fig. 1. Architecture of the proposed fall detection application

The rest of the paper is organized as follows: In Section 2, background material and related work is provided. Next, in Section 3 the SW of the proposed application that resides in both the smart watch and the smart phone are presented, while Section 4 is devoted to an initial evaluation. Finally, Section 5 covers concluding remarks and some future research directions.

2   Background Information and Related Work

2.1   Fall Detection Applications

The concept of activity recognition with focus on fall detection is relatively new; however, it is quite popular due to the vast amount of sensors that have flooded the market recently. Several research groups and R&D departments of prominent companies have developed similar applications and there already exists significant related research work in the field. A comprehensive review exists in [6]. Information regarding the human movement and activity in assisted environments is frequently acquired through visual tracking of the subject's or patient's position. Overhead tracking through cameras provides the movement trajectory of the patient and gives information about user activity on predetermined monitored areas [7]. A different approach for collecting patient activity information is the use of sensors that integrate devices like accelerometers, gyroscopes and contact sensors. The latter approach depends less on issues like patient physiology (e.g. body type and height) and environmental information (e.g. topology of monitored site) and can be used for a variety of techniques enabling user activity recognition [8-9]. In previous works [10-11] we have presented a patient fall detection system, based on such body sensors that utilized advanced classification techniques and Kalman filtering for producing post fall detection. In this work we are trying to exploit a new class of smart devices that has been introduced recently: smart watches. We have selected the Pebble smart watch as a popular representative and we present some details of its capabilities in the next section.

2.2   The Pebble Smart Watch

Smart-watches are wristwatches with a micro display, integrated sensors and connectivity. The intention of the manufactures of smart watches is to have a new device that displays short messages like SMS, RSS feeds or Facebook messages. The Pebble is a smart watch developed by Pebble Technology and released in 2013. It features a black and white e-paper display, a vibrating motor, a magnetometer, ambient light sensors and an accelerometer, enabling its use as an activity tracker. The Pebble is compatible with Android and iOS devices. When connected to a phone, it can receive a vibrating alert to text messages, emails, incoming calls, and notifications from social media accounts.

The watch has a 1.26-inch 144 × 168 pixel black and white e-paper display using an ultra low-power "transflective LCD" with a backlight. The communication with an Android or iOS device is implemented using Bluetooth 4.0 (Bluetooth Low Energy) protocol. An important feature of the Pebble smart watch is that it is a solid and strong construction and it has a waterproof rating of 5 atm, which means it can be submerged down to 40m. An open software development kit (SDK) is available to programmers for developing custom applications. The SDK (known also as PebbleKit) allows the two-way communication between Pebble and smartphones running iOS or Android via the AppMessage framework and includes access to the accelerometer, as well as a Javascript API.

3   The Proposed System

As shown in Figure 1 (system architecture) and illustrated also in Figure 2, the proposed activity and fall detection system consists of two software modules, which work in conjunction and communicate with each other using the Bluetooth 4.0 networking interface. The first module runs on the Pebble smart watch, and it is responsible for the user interaction and the data input. The second module is executed on the android device and handles data processing and alarm management. In order the system to be fully functional, the user has to wear the Pebble watch with the fall detection application running and remains within the Bluetooth range of the android device. In case the Bluetooth device is a smart phone the user can carry the device, so it can be always within range; however, at a home setup, the android device could be a tablet, a board PC, or a set-top box.

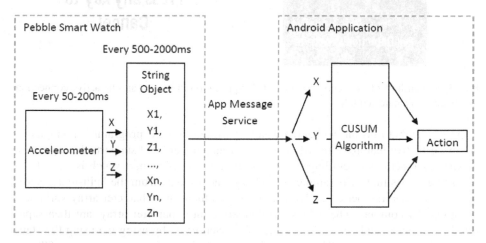

Fig. 2. The proposed fall detection application SW modules

3.1   The Pebble Application

The Pebble SDK 2.0 (until release 6), does not support multitasking, thus the user wearing the Pebble device should have the fall detection app installed in the foreground. Therefore, it is not feasible to run any other application on the background or to interact with other applications. The main screen of the developed app consists of a minimal clock, so the user can be informed about the time. Also at the bottom of the main screen, there is a small text field, which displays the status of the connection with the android device. If the android device is connected with the Pebble, the text field displays "Connected - Fall Detection is Active", whereas if the Pebble is not connected, the text field displays "Disconnected - Fall Detection is not Active" (see Figure 3).

As the Pebble display shows the time, the developed application checks whether an active Bluetooth (BT) connection with the android device is established. If a connection is found, the smart watch transmits the current accelerometer data and the corresponding time stamps. In the case for some reason the Pebble smart watch is disconnected from the android device (for instance, it gets out the BT range), the Pebble application changes the visible status connection, closes the accelerometer, stops transmitting data and notifies the user with a small vibration about the change in the status of the connection. The user has to be notified immediately about the disconnection. Once reconnected, the application starts again automatically.

Fig. 3. Screenshots of the Pebble Smart watch Application in case of an idle active connection and in case of a detected fall

The code that runs on the Pebble watch is written in the C programming language. A very effective feature of the SDK is the "event handler" that supports the triggering of various functions, according to the captured event. The smart watch is capable of receiving 3-axis integer (example: x1, y1, z1) motion data from the built-in accelerometer in a sampling period of 50 ms, and stores them into character array variables, separated by commas. The values are converted into character array, are then separated by commas and copied to a single character array, by using the using the standard C function, strcat(). The final string is as follows, for ten samples or 500ms: [x1,y1,z1, x2,y2,z2, x3,y3,z3, x4,y4,z4, x5,y5,z5, x6,y6,z6, x7,y7,z7, x8,y8,z8, x9,y9,z9, x10,y10,z10].

Every 500ms the Pebble converts the character array into a tuplet string dictionary, by using a characterization key. After the serialization procedure, Pebble sends the data as a single string via the BT interface, while at the same time it continues to receive and store the incoming accelerometer data. The data is then sent to the android device for further processing, according to fall detection algorithms. In case a fall is detected, the android device sends a string value (fall-alarm) to the Pebble, and the proper event handler changes the watch status from watch view and status view, into alarm view. Alarm view consists of a text field, which displays: "A fall was detected - Press any key to Cancel", the Pebble screen's backlight is active and a vibration is triggered every 500ms.

If the user presses any of the Pebble's buttons in 5 seconds after the fall is detected, the fall alarm will be set off and a cancel-alarm string value will be sent to the android device. If the user does not respond within the 5-second interval the Pebble will send a trigger-alert sting value with the proper event handler, to notify about the triggered alarm.

3.2   The Android Application

The Android application has been developed in Java using the Eclipse Android Developer Tools, intended for android devices. This app is launched (in background mode), every time the user launches the app on the Pebble smart watch and vise versa.

When the app is active, whether it runs on background or on foreground, it receives data from the Pebble device. The incoming string using the same tuplet dictionary and characterization key, is converted into comma-separated integer numbers.

According to the corresponding accelerometer axis, the data is parsed into the fall detection algorithm as follows: $X=[x1,x2,x3,x4,x5,x6,x7,x8,x9,x10]$, $Y=[y1,y2,y3,y4, y5,y6,y7,y8,y9,y10]$, $Z=[z1,z2,z3,z4,z5,z6,z7,z8,z9,z10]$. The implemented fall detection algorithm has been proposed in a previous work and is based on the cumulative sum (CUSUM) change detection algorithm [12]. The CUSUM decision function, as calculated for a fall in the X- and Y-axis, is depicted in Figure 4. In case a fall is detected a fall-alarm string will be sent to the Pebble and as already mentioned if the user does not cancel the alarm within the appropriate time interval, the application will initiate an actual alarm, transmitting automatically the fall detection information as defined (i.e. via email or via a call or SMS in the case of a smart phone).

4   The System in Practice - Initial Experimental Results

The main goal of this work is to present a user-friendly and unobtrusive application for fall detection. The proposed system combines the advantages of two simple, low-cost, and widely available devices. The obvious advantages of the smart watch are the ease and comfort of use, the high quality of acquired sensor data, the minimum weight and the low power consumption characteristics. The limited computing resources are replaced by the processing power of a typical android smartphone. The proposed solution is based on the bare minimum sampling rate (i.e. 50-200 ms), in order to minimize power consumption and increase battery life. During our experiments, the Pebble smart watch was proven capable of continuous sampling for at least 30 hours. An improvement in autonomy was achieved taking into consideration the actual time and the frequency of the user's movement, in order to decrease even more the sampling rate during the sleep. That resulted in a combined 35+ hours of continuous user fall tracking with a single battery charging.

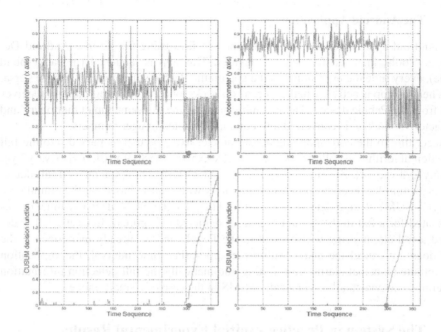

Fig. 4. Accelerometer readings and the corresponding CUSUM calculation in case of a typical fall

A significant parameter of the CUSUM algorithm presented in [12] is the value of the threshold for the CUSUM decision function, which determines a fall. This is symbolized as the h parameter in the specific algorithm. According to the performed experiments, a lower value for h corresponds to smaller delays in fall detection, but produces more false positives. By experimenting with the h-parameter, we tried to minimize false positive alarms, ensuring the minimum possible delay time. The results of the conducted experiments are presented in Table 1.

Table 1. Number of false alarms and the delay for several values of h, for each axis

h	2.5	3.0	3.4	3.6	3.8	4.0	5.0
Accelerometer (X-axis)							
False Alarms	5	3	2	1	0	0	0
Delay (ms)	0	16	46	43	10	10	10
Accelerometer (Y-axis)							
False Alarms	7	5	4	3	3	2	0
Delay (ms)	1	0	10	16	45	18	10
Accelerometer (Z-axis)							
False Alarms	8	6	5	4	4	3	0
Delay (ms)	14	18	31	10	40	45	10

5   Discussion and Conclusions

This paper presents a fall detection application based on the Pebble smart watch, which is capable of transmitting the built-in accelerometer readings via Bluetooth to an android device for further processing. To the best of our knowledge, this is the first integrated application developed for fall detection based on a commodity smart watch. The main advantages of the proposed application are that it is non-obtrusive, low-cost, easy to deploy and use, and suitable for processing streaming data and reaching a prompt decision. In a future work, we intend to combine contextual data of the android device (e.g. location, audio, etc.), in order to decrease false alarms and improve the accuracy of the system and add additional functionalities, such as fall severity estimation. Furthermore the same device will be used for the simultaneously assessing sleep and other vital activities.

Acknowledgment. The authors would like to thank the European Union (European Social Fund ESF) and Greek national funds for financially supporting this work through the Operational Program "Education and Lifelong Learning" of the National Strategic Reference Framework (NSRF) - Research Funding Program: \Thalis \ Interdisciplinary Research in Affective Computing for Biological Activity Recognition in Assistive Environments. Part of the work reported here has been carried out in the framework of national project *Providing Integrated eHealth Services for Personalized Medicine utilizing Cloud Infrastructure (PinCloud)*, led by the University of Piraeus, conducted in the context of the National Strategic Reference Framework NSRF 2007-2013, Cooperation 2011 and co-funded by the European Commission.

References

1. Varshney, U.: Pervasive Healthcare. IEEE Computer Magazine 36(12), 138–140 (2003)
2. 2012 Ageing Report: Economic and budgetary projections for the 27 EU Member States (2010-2060)
3. West, J., Hippisley-Cox, J., Coupland, C., Price, G., Groom, L., Kendrick, D., Webber, E.: Do rates of hospital admission for falls and hip fracture in elderly people vary by socio-economic status? Public Health 118(8), 576–581 (2004)
4. Robinson, B., Gordon, J., Wallentine, S., Visio, M.: Relationship between lower-extremity joint torque and the risk for falls in a group of community dwelling older adults. Physiotherapy Theory and Practice 20, 155–173 (2004)
5. Official Pebble Web Site, https://getpebble.com
6. Mubashir, M., Shao, L., Seed, L.: A survey on fall detection: Principles and approaches. Neurocomputing 100, 144–152 (2013)
7. Doukas, C.N., Maglogiannis, I.: Emergency fall incidents detection in assisted living environments utilizing motion, sound, and visual perceptual components. IEEE Transactions on Information Technology in Biomedicine 15(2), art. no. 5623343, 277–289 (2011)
8. Zhang, Z., Kapoor, U., Narayanan, M., Lovell, N.H., Redmond, S.J.: Design of an unobtrusive wireless sensor network for nighttime falls detection. Proceedings of the Annual International Conference of the IEEE Engineering in Medicine and Biology Society, EMBS, art. no. 6091305, 5275–5278 (2011)

9. Noury, N., Herve, T., Rialle, V., Virone, G., Mercier, E., Morey, G., Moro, A., Porcheron, T.: Monitoring behavior in home using a smart fall sensor and position sensors. In: Proceedings of 1st Annual International Conference on Microtechnologies in Medicine and Biology, pp. 607–610 (2000)
10. Doukas, C., Maglogiannis, I.: Advanced classification and rules based evaluation of motion, visual and biosignal data for patient fall incident detection. International Journal on Artificial Intelligence Tools, 175–191 (2010)
11. Doukas, C., Maglogiannis An, I.: assistive environment for improving human safety utilizing advanced sound and motion data classification. Universal Access in the Information Society 10(2), 217–228 (2011)
12. Tasoulis, S.K., Doukas, C.N., Maglogiannis, I., Plagianakos, V.P.: Statistical data mining of streaming motion data for fall detection in assistive environments. In: Annual International Conference of the IEEE Engineering in Medicine and Biology Society (2011)

Predicting Information Diffusion Patterns in Twitter

Eleanna Kafeza[1], Andreas Kanavos[2], Christos Makris[2], and Pantelis Vikatos[2]

[1] Athens University of Economics and Business, Greece
kafeza@aueb.gr
[2] Computer Engineering and Informatics Department, University of Patras, Greece
{kanavos,makri,vikatos}@ceid.upatras.gr

Abstract. The prediction of social media information propagation is a problem that has attracted a lot of interest over the recent years, especially because of the application of such predictions for effective marketing campaigns. Existing approaches have shown that the information cascades in social media are small and have a large width. We validate these results for Tree-Shaped Tweet Cascades created by the ReTweet action. The main contribution of our work is a methodology for predicting the information diffusion that will occur given a user's tweet. We base our prediction on the linguistic features of the tweet as well as the user profile that created the initial tweet. Our results show that we can predict the Tweet-Pattern with good accuracy. Moreover, we show that influential networks within the Twitter graph tend to use different Tweet-Patterns.

Keywords: Information Diffusion, Machine Learning, Social Media Analytics.

1   Introduction

Social contagion, or as it is often called word-of-mouth, is mainly based on viral marketing strategies. There is a plethora of tools in the area of social media analytics that are available to help marketers to develop a real time viral marketing strategy. A central question when developing such strategies is how information is diffused within the social network. Recently an augmenting body of research has focused on examining the nature and structure of information diffusion from the point of view of an information cascade. For example, in the case of Twitter, an information cascade is the followers of a user that retweeted the initial user message. We look into the problem of defining typical cascades and predict when they will appear in the social graph. We argue that the content of the message and the type of network diffusion are the two main dimensions that lead to robust cascade prediction.

The problem of how information spreads within social media has been examined from different perspectives in the literature. In several cases, information flow has been associated to virus contamination and a diverse set of models

L. Iliadis et al. (Eds.): AIAI 2014, IFIP AICT 436, pp. 79–89, 2014.

have been developed trying to count the spread of information [1], [6], [19]. Yet, there are fundamental differences in the case of information spread as opposed to virus spread mainly because the spread is based on a network structure and is influenced by the message. For instance, it has been observed in studies [19] that social media users who are several steps afar tend not to propagate information of each other. Although probability decay models have been proposed to capture that element, it is not certain that such decay occurs in a homogenous manner. Moreover, existing studies show that large cascades are rare and usually, larger cascades are utilized in terms of width and not depth. These results are in accordance to our findings. Hence, since most cascades appear, we need a more pragmatic approach in order to predict one.

In the problem of Twitter message diffusion, we take a different approach and similarly to [2] we investigate the actual users diffusion. Our objective is to predict the information diffusion of users for messages related to a specific topic. Our contribution is that we present a solid method for predicting tweet cascades within the Twitter graph. We argue that not all users and their connections behave in the exact same way. We assume that there are two aspects that drive the information spread, the message itself and the network structure. So, for different message subjects different diffusion occurs. For a given subject we examine the frequencies of all possible diffusions and thus we extract some simple and basic patterns of diffusion. We then, associate the content of tweets with the propagation patterns, using linguistic analysis and machine learning techniques. Given a tweet of a user, we are then able to predict its pattern path. We do so, with very good accuracy as our results indicate. Hence given a user's tweet on a subject we predict the approximate path that the tweet will take into the Twitter graph. Moreover, we show that influential communities tend to have a diverse set of propagation patterns.

The rest of the paper is organized as follows: Section 2 is the related work. In Section 3 we present the methodology of our approach. The implementation and our results are presented in Section 4. We conclude in Section 5 where we also present our future work.

2   Related Work

In [17], a nonstandard form of Bayesian shrinkage implemented in a Poisson regression is proposed. There, the approach identifies the specific users who most influence others' activity and does so considerably better than simpler alternatives. The authors find that for the social networking site data, approximately one-fifth of a user's friends actually influence their activity level on the site.

Also, in [15], the authors present a study towards finding influential authors in Twitter brand-page communities (many enterprises have set-up their official webpages) where an implicit network based on user interactions is created and analyzed. More specifically, author profile and user interaction features are combined in a decision tree classification framework (DT framework) and thus, a novel objective evaluation criterion is used for evaluating these features.

Another work was [10], where a novel method to find influentials by considering both the link structure and the temporal order of information adoption in Twitter is adopted. Recently, in [9], the authors consider the issue of choosing influential sets of individuals as a problem in discrete optimization. The optimal solution is NP-hard for most models that have been studied, including the model of [5]. The framework proposed in [16], on the other hand, is based on a simple linear model where the solution to the optimization problem can be obtained by solving a system of linear equations. The generality of the mentioned NP-hard models lies between that of the polynomial-time solvable model of [16] and the very general model of [5], where the optimization problem cannot even be approximated to within a non-trivial factor.

In [18], the authors compare a wide assortment of node-level network measures (degree centrality, clustering coefficient, network constraint, and eigenvector centrality), testing their robustness to different forms of measurement error. They also investigated network-structural properties (average clustering, degree distributions) as explanations for the varying effects of measurement error. In addition, Leskovec et al. [11] deal with information cascades; they are phenomena in which an action or idea becomes widely adopted due to influence by others. They develop a scalable algorithm and set of techniques to illustrate the existence of cascades as well as to measure their frequencies. From their experiments, they found that most cascades are small, cascade sizes approximately follow a heavy-tailed distribution, the frequency of different cascade subgraphs depends on the product type and last that these frequencies reflect more subtle features of the domain in which the recommendations are operating.

In article [4], a business process classification framework to put the research topics in a business context is employed while providing a critical perspective on business applications of social network analysis and mining. Furthermore, in [12], authors provide some assumptions concerning epidemic models. They find that the probability of purchasing a product increases with the number of recommendations received, but quickly saturates to a constant and relatively low probability. Also, they present a simple stochastic model that allows for the presence of relatively large cascades for a few products, but reflects well the general tendency of recommendation chains to terminate after just a short number of steps; they observe that the most popular categories of items recommended within communities in the largest component reflect differing interests between these communities.

Another empirical study of user-to-user content transfer occurring in the context of a time-evolving social network in Second Life (e.g. a massively multiplayer virtual world) is presented in [2]. There, adoption rates quicken as the number of friends adopting increases and this effect varies with the connectivity of a particular user. Also, in [3], authors find out that the largest cascades tend to be generated by users who have been influential in the past and who have a large number of followers. The authors in [13] focused on the problem of link prediction in Microblogs and proposed the notion of social distance based on the interaction patterns. The main idea is that social networks exhibit homophily

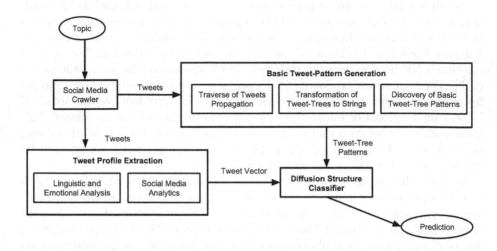

Fig. 1. System Architecture of the Tweet Prediction Pattern System

and that the agents prefer to create ties with other agents who are close to them. In [6], methods for identifying influential spreaders in online social networks are presented and in following, a taxonomy of various approaches employed to address diffusion modeling techniques is proposed.

Finally, Peng et al. [14] introduced the basic characteristics of the diffusion process of multi-source information via real datasets collected from Digg (e.g. a social news aggregation site); in following, a mathematical model to predict the information diffusion process of such multisource news is used.

3   Methodology

In our model we examine information diffusion in Twitter as Tree-Patterns. Our methodology for predicting the tweet cascade is based on the linguistic and emotional content of the tweets as well as the user communication behavior. According to Twitter: "A Retweet is a re-posting of someone else's Tweet. Twitter's Retweet feature helps you and others quickly share that Tweet with all of your followers". Although ReTweets might have other forms as well (for example the use of RT in the beginning of the message denotes that it is a ReTweet), still most of the ReTweets are done according to the formal Twitter definition using the @username convention. This is the conversion we use to retrieve ReTweets. Figure 1 depicts the architecture of our system.

We represent the information cascade as a Tree-Pattern i.e. the set of nodes that retweeted the initial user tweet. Note that all of these nodes are related with the "follow" relationship since only a user's followers can retweet their message.

The Social Media Crawler takes a topic as input and retrieves the information from Twitter. It collects all the Tweets and their corresponding ReTweets on the specific topic. The Tweet Profile Extraction is the module where the Tweets are inserted and they are abstracted as vectors that contain linguistic and emotional information about the Tweet as well information regarding the user behavior as depicted through the Twitter analytics i.e. number of followers etc. The Basic Tweet-Pattern Generation is a module that creates the frequent basic Tweet Propagation Patterns. The Tweet Vector and the Patterns are the input to the Diffusion Structure Classifier that predicts the Tweet-Pattern that will be followed by a given user tweet.

3.1   The Social Media Crawler

In order to retrieve information from the Twitter graph we use a crawler topic-based sampling approach where Tweets are collected via a keyword search query. Our data source is Tweets retrieved from the Twitter through the Twitter API (e.g. Twitter4J[1]) for a specific period of time (e.g. a couple of hours). We retrieved all the Tweets that were relevant to the subject *#MH370* concerning Malaysia Airlines Flight 370 disappearance. Our dataset contains 13000 Tweets that have been done by 11130 users.

The crawler is responsible for sampling and traversing the Twitter media; it also collects information regarding the users' activity. More specifically, we extract 6 basic user features which better describe user communication behavior in Twitter. The set of 6 features contain the number of Followers, the number of Direct Tweets, the number of ReTweets, the number of Conversational Tweets (e.g. if a user replies to a post), the Frequency of user's Tweets (e.g. how "often" an author posts Tweets) and last the number of Hashtag Keywords (e.g. words starting with the symbol # and can specify the thematic category of a specific Tweet) as in [7], [8].

3.2   Generating the Tweet-Pattern

To measure how tweets propagate through the Twitter network we represent tweet propagation as a tree. The root of the tree is the user that posts the initial tweet. Every child node is a follower of the initial user that has retweeted the initial tweet. This continues recursively until there are no nodes that retweeted the initial tweet. Hence each Tweet-Tree represents all the followers that retweeted the same message as the root user.

In Figure 2 we present some Tweet-Trees returned by our crawler. As previous studies have shown and as our results have shown as well, there are no large cascades of information, and Tweet-Trees tend to be small. So, our first objective is to find a set of representative Tweet-Patterns.

We observe that isomorphic Tweet-Trees exhibit the same diffusion. Isomorphic paths have the level number of a node as invariants, the number of paths

---

[1] Twitter4J library: http://twitter4j.org/en/index.html

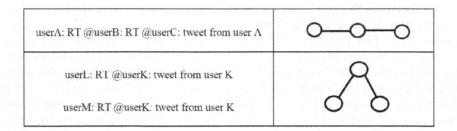

userΛ: RT @userB: RT @userC: tweet from user Λ	
userL: RT @userK: tweet from user K userM: RT @userK: tweet from user K	

Fig. 2. Examples of Tweet-Trees

from the root to the leaves, the number of levels in the tree, the number of leaf descendants of a node and the level number of a node. In information diffusion we are interested in the spread of information. Additionally, we are interested in the number of the leaf descends and not the specific position of the subtree that these descendants come from. Therefore we represent in the same way all isomorphic Tweet-Paths, taking as their representative the left-most variant.

For example in Figure 3, we see three different Tweet-Trees. All of them exhibit the same information diffusion thus we model them choosing one representative; the one depicted in (c).

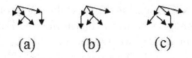

(a)   (b)   (c)

Fig. 3. Isomorphic Tweet-Trees that represent the same information Diffusion

3.2.1   Representing Tweet-Trees as Strings

We encode the diffusion of each tweet as a string and use top-down approach to traverse the tree. The root is the tweet that has been produced by a certain user. Let k be the maximum number of different users that have posted a ReTweet from a user from the whole dataset. Hence given a root tweet, there are at most k different possibilities of diffusion for each node of the tree. The order of these possibilities depends on the depth and width of diffusion. Starting from the root we use symbols to mark the diffusion. The placement of symbols begins from root to the next level. If there is more than one node in the current level we order the created symbols in reverse lexicographic way and we put the symbols respectively.

For each node there are k different possibilities of diffusion to the next level. In total there are k^m different strings where k are different possibilities of diffusion to the next level and m the max depth of the trees. We have to note that based on the above construction, isomorphic trees have the same string representation.

3.2.2 Identifying the Basic Tree-Patterns

We use edit distance as a metric to identify the basic tree patterns. In particular the edit distance between two strings s_1, s_2 of size d_1, d_2 respectively is defined as he number of symbol operations (insertion, deletion and substitution) that must be performed in order to transform one string into the other. The edit distance is equal to the cost of the alignment of the two strings that is defined as the number of character mismatches when putting the one string under the other embedding spaces in both of them in order to produce strings of equal length. The edit distance can be computed with a simple dynamic programming algorithm as it is nicely described in [17] in time $d_1 * d_2$. More specifically a two dimensional table D of size $d_1 * d_2$ is defined where $D[i, j]$ is equal to the edit distance between $s_1[1 \ldots i]$ and $s_2[1 \ldots j]$ and can be expressed as the minimum of $D[i-1, j] + 1, D[i, j-1] + 1, D[i-1, j-1] + (i, j)$, where (i, j) is an indicator variable equal to 1 if $s_1[i]$ and $s_2[j]$ are different, otherwise it is 0. The needed solution is given by $D[d_1, d_2]$ and it can be produced by a simple traversal of the array from top to bottom and from left to right.

Each produced string is compared to others using a table of dynamic programming via edit distance and the frequent strings are gathered in bins. As a result of the above process, a set of basic Tweet Tree-Patterns are extracted. These frequent patterns constitute the pre-assigned labels of each tweet for the training of the classification models.

3.3 Using Linguistics to Represent the Tweet as a Vector

After having specified the basic Tweet-Patterns, we argue that given a tweet on a specific topic the diffusion basic pattern that this tweet will follow is based on the tweet content and on the user profile. Hence we represent each tweet as a vector in which we extract linguistic and emotional information as well as information regarding the user that posted the tweet.

The following are the two main characteristics that we include in the Tweet-Vector:

- Linguistic and Emotional characteristics, which are produced by the Linguistic Inquiry and Word Count (LIWC) software which produce 80 features that include linguistic and psychological use of language as well as personal concerns.
- Social Media Analytics, which can be used to monitor and capture user's behavior. The Followers of a user, the number of Tweets, ReTweets, Conversations and Hashtag Keyword as well the Frequency of Tweets, are some aspects that differentiate user behavior.

Thus every tweet is represented as a vector that contains the 80 features that were extracted when the tweet was processed with LIWC for linguist extractions and the 6 social analytics metrics that represent information about the user that created the initial tweet. We predict the Tweet-Pattern based on that vector using a variety of classification algorithms. The performance is evaluated by the F-Measure metric.

4   Implementation and Results

As already mentioned our dataset contains 13000 Tweets that have been done
by 11130 users. In these Tweets, the maximum depth of the Tweet-Tree was 4
and the maximum width was 92. Thus we verify existing results which claim
that the depth of the Tweet-Tree is small and the width large [10]. We encode
the tweet information diffusion using 92^4 different representations as previously
explained. After applying the methodology presented in Section 3.2, we end-up
with the following basic Tweet-Patterns:

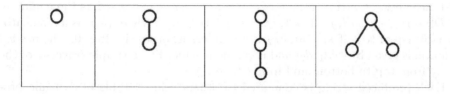

Fig. 4. The Basic Tweet Patterns

As explained in Section 3.2, the next step in our methodology is to represent
the tweet as a vector based on its linguistic characteristics and its user profile
as described in the analytics. We used LIWC as a tool to extract linguistic
characteristics and we used Twitter analytics to extract the user profile. As a
result, we created a vector of 86 characteristics, as presented in Table 1.

Table 1. The Characteristics of the Tweet Vector

Features	#	Description
LIWC	80	4 general descriptor categories (total word count, words per sentence, percentage of words captured by the dictionary, and percent of words longer than six letters), 22 standard linguistic dimensions (e.g., percentage of words in the text that are pronouns, articles, auxiliary verbs, etc.), 32 word categories tapping psychological constructs (e.g., affect, cognition, biological processes), 7 personal concern categories (e.g., work, home, leisure activities), 3 paralinguistic dimensions (assents, fillers, nonfluencies), and 12 punctuation categories (periods, commas, etc)
Twitter Metrics	6	Followers, Tweets, ReTweets, Conversations, Frequency, Hashtag Keywords

We separated dataset to training and test set, using two approaches: a) K-
Fold Cross-Validation (K=10 Fold) and b) Leave-One-Out Cross-Validation. The
concept of using both techniques is that splitting with 10-Fold Cross-Validation,

Table 2. F-Measure for each Classifier

Classifiers	F-Measure
AdaBoost	0.71
IBK	0.71
J48	**0.88**
JRip	**0.89**
Multilayer Perceptron	0.81
Naive Bayes Classifier	0.61
PART	**0.88**
RotationForest	**0.88**
SMO	0.77

important information can be removed from the training set. However, the Leave-One-Out Cross-Validation technique evaluates the classification performance based on one sample.

We used the WEKA library[2] for the classifiers. Table 2 shows the results of F-Measure for each classifier. The classifiers that give us the best results are depicted in bold. We can observe that JRIP achieves the highest F-Measure value. In addition, the next best classifiers are J48, PART as well as RotationForest. Moreover, we see that almost all classifiers achieve an F-Measure above 60%; only one classifier achieves 61%, while the rest are above 70%. Hence the accuracy of our prediction is high which validates our hypothesis that based on the linguistic aspects of the tweet and the user analytics profile, we can predict the Tweet-Tree.

As an application of our approach, we examine the different patterns that occur in highly influential communities. Based on our previous work [8], we extract the existing communities of the graph taking into consideration the personality of the users. We rank these communities based on the number of Tweets. The first in the rank are the most influential communities i.e. the communities where most Tweets occur. We are interested in finding out the type of diffusion that occurs in these communities.

We expect that the top influential communities are those that diffuse more the information. The first Tweet-Pattern which is a node by itself does not exhibit any diffusion since the tweet was not retweeted. Hence we expect that for the influential communities the diffusion patterns should mainly be the other three. Figure 5 verifies that expectation. It shows that when ranking the influential communities in ascending order based on the number of tweets that occurred in them, then the pattern of singe tweet (diffusion class 1) occurs less to the most influential community (com1). While the rest patterns (diffusion classes 2, 3 and 4) occur more in com1 and com2 than in com3. Therefore we conclude that communities where the number of tweets is large (influential communities) have more diffusion where ReTweets occur either as in sequence or in width.

---

[2] Weka toolkit: http://www.cs.waikato.ac.nz/ml/weka/

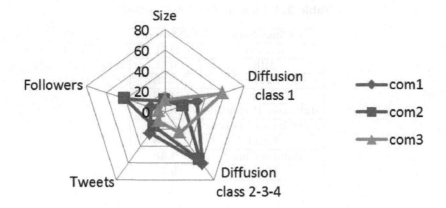

Fig. 5. Information Diffusion in Top Communities

5   Conclusions and Future Work

Recently there has been a growing interest in the literature for examining how information is diffused in social networks especially in the case where information cascades are short and usually with small depth and large width. In our work we use existing evidence supported by our results, to show that these Tweet-Trees have specific patterns. Hence we find the representative Tweet-Tree patterns that information follows when propagated in Twitter. We argue that this Tree-Pattern information diffusion can be predicted given a user's tweet and we use linguistic and user profiling information in order to do so based on machine learning techniques. Our results show that we can predict the basic Tree-Pattern with good accuracy. Our contribution is important especially in cases where marketers are interested in identifying influential users and networks and estimate the propagation of their messages. We show that in influential networks there is a set of different Tree-Patterns that occur.

In our future work, we are interested in examining the overall propagation of information within a whole Twitter network based on our prediction methodology. Also, as a next step we will consider time as a factor that influences the information diffusion.

References

1. Aral, S., Walker, D.: Creating Social Contagion through Viral Product Design: A Randomized Trial of Peer Influence in Networks. ICIS (2010)
2. Bakshy, E., Karrer, B., Adamic, L.A.: Social Influence and the Diffusion of User-Created Content. In: EC, pp. 325–334 (2009)
3. Bakshy, E., Hofman, J.M., Mason, W.A., Watts, D.J.: Everyone's an Influencer: Quantifying Influence on Twitter. In: WSDM, pp. 65–74 (2011)

4. Bonchi, F., Castillo, C., Gionis, A., Jaimes, A.: Social Network Analysis and Mining for Business Applications. ACM Transactions on Intelligent Systems and Technology 2(3) (2011)
5. Domingos, P., Richardson, M.: Mining the Network Value of Customers. In: KDD, pp. 57–66 (2001)
6. Guille, A., Hacid, H., Favre, C., Zighed, D.A.: Information Diffusion in Online Social Networks: A Survey. SIGMOD Record 42(2), 17–28 (2013)
7. Kafeza, E., Kanavos, A., Makris, C., Chiu, D.: Identifying Personality-based Communities in Social Networks. LSAWM, ER (2013)
8. Kafeza, E., Kanavos, A., Makris, C., Vikatos, P.: T-PICE: Twitter Personality based Influential Communities Extraction System. In: IEEE International Congress on Big Data (2014)
9. Kempe, D., Kleinberg, J.M., Tardos, E.: Maximizing the Spread of Influence through a Social Network. In: KDD, pp. 137–146 (2003)
10. Lee, C., Kwak, H., Park, H., Moon, S.B.: Finding Influentials Based on the Temporal Order of Information Adoption in Twitter. In: WWW, pp. 1137–1138 (2010)
11. Leskovec, J., Singh, A., Kleinberg, J.M.: Patterns of influence in a recommendation network. In: Ng, W.-K., Kitsuregawa, M., Li, J., Chang, K. (eds.) PAKDD 2006. LNCS (LNAI), vol. 3918, pp. 380–389. Springer, Heidelberg (2006)
12. Leskovec, J., Adamic, L.A., Huberman, B.A.: The Dynamics of Viral Marketing. ACM Transactions on the Web (TWEB) 1(1) (2007)
13. Liu, D., Wang, Y., Jia, Y., Li, J.: From Strangers to Neighbors: Link Prediction in Microblogs using Social Distance Game. In: Diffusion Networks and Cascade Analytics, WSDM (2014)
14. Peng, C., Xu, K., Wang, F., Wang, H.: Predicting Information Diffusion Initiated from Multiple Sources in Online Social Networks. ISCID 2, 96–99 (2013)
15. Purohit, H., Ajmera, J., Joshi, S., Verma, A., Sheth, A.P.: Finding Influential Authors in Brand-Page Communities. In: ICWSM (2012)
16. Richardson, M., Domingos, P.: Mining Knowledge-Sharing Sites for Viral Marketing. In: KDD, pp. 61–70 (2002)
17. Trusov, M., Bodapati, A.V., Bucklin, R.E.: Determining Influential Users in Internet Social Networks. Journal of Marketing Research (JMR) (2010)
18. Wang, D.J., Shi, X., McFarland, D.A., Leskovec, J.: Measurement error in network data: A re-classification. Social Networks (SOCNET) 34(4), 396–409 (2012)
19. Wu, F., Huberman, B.A., Adamic, L.A., Tyler, J.R.: Information Flow in Social Groups. Physica A, 327–335 (2004)

An Ensemble of HMMs for Cognitive Fault Detection in Distributed Sensor Networks

Manuel Roveri and Francesco Trovò

Dipartimento di Elettronica, Informazione e Bioingegneria
Politecnico di Milano, Milano, Italy
{manuel.roveri,francesco1.trovo}@polimi.it

Abstract. Distributed sensor networks working in harsh environmental conditions can suffer from permanent or transient faults affecting the embedded electronics or the sensors. Fault Diagnosis Systems (FDSs) have been widely studied in the literature to detect, isolate, identify, and possibly accommodate faults. Recently introduced cognitive FDSs, which represent a novel generation of FDSs, are characterized by the capability to exploit temporal and spatial dependency in acquired datastreams to improve the fault diagnosis and by the ability to operate without requiring a priori information about the data-generating process or the possible faults. This paper suggests a novel approach for fault detection in cognitive FDSs based on an *ensemble of Hidden Markov Models*. A wide experimental campaign on both synthetic and real-world data coming from a rock-collapse forecasting system shows the advantages of the proposed solution.

1 Introduction

Distributed sensor networks represent an important and valuable technological solution to monitor and acquire data from an environment, a cyber physical-system or a critical infrastructure system (e.g., water, gas, electric or transport networks). Unfortunately, distributed sensor networks working in (possibly harsh) real-working conditions may suffer from permanent or transient faults, thermal drifts or ageing effects affecting both the embedded electronic boards and the sensors. Fault Diagnosis Systems (FDSs), which are able to detect, identify, isolate and possibly accommodate faults, have been widely studied in the related literature [1, 2]. These systems revealed to be particularly effective in several different application domains provided that a priori information about the system model or the possible faults is (at least partially) available.

Recently, cognitive FDSs have been introduced, representing a novel and promising approach for fault diagnosis [3–7]. The distinguishing features of these FDSs are the capability to work without a priori information about the process or the possible faults (which are learned directly from data) and the ability to exploit temporal and spatial dependencies in the acquired datastreams to improve the diagnosis of faults. Fault detection is obviously a key aspect in cognitive FDSs and, among the available techniques in the cognitive FDSs literature, we focus on

L. Iliadis et al. (Eds.): AIAI 2014, IFIP AICT 436, pp. 90–100, 2014.

the change detection test (CDT) based on a Hidden Markov Model (HMM) suggested in [8]. This HMM-CDT revealed to be particularly effective in detecting faults affecting real-world monitoring applications (e.g., environmental monitoring [8] or water transport network [9]) in the scenario of networked intelligent embedded sensors or distributed sensor networks. Unfortunately, the detection abilities of this CDT are influenced by the initial conditions of the HMM training algorithm, possibly leading to suboptimal HMM parameters estimations [10].

In this paper, we propose an *ensemble-of-models* approach for cognitive fault detection based on HMMs, leading to a novel Ensemble HMM-based CDT (EHMM-CDT). More specifically, the proposed EHMM-CDT relies on a set of HMMs configured on the same training set, but with different randomly generated initial conditions for the HMM training algorithm. The ensemble approach aims at weakening the dependency of the HMM-CDT from the random initial conditions, hence providing better detection abilities. Interestingly, the possibility to combine different models to improve the generalization ability of a single model has been widely studied in the literature (mainly in the regression and classification scenarios) and effectively applied to different application fields [11-13]. Recently, ensembles of models have been successfully considered in time series prediction [14, 15] and on-line missing data reconstruction [16]. Moreover, an ensemble of HMMs within a Bayesian framework for parameter estimation has been derived in [17], while an algorithm for training HMM in the ensemble framework was presented in [18]. Remarkably, for the first time in the literature, in this paper we suggest an ensemble of HMMs for fault detection and we describe and evaluate two different aggregation mechanisms. In the experimental section, the proposed EHMM-CDT has been compared with the HMM-CDT: the advantages provided by the proposed solution have been evaluated both on synthetic and real datasets coming from a rock-collapse forecasting system.

The paper is organized as follows: Section 2 describes the problem formulation, while Section 3 summarizes the HMM-CDT suggested in [8]. Section 4 details the proposed ensemble approach for cognitive fault detection, while Section 5 shows the experimental results.

2   Problem Formulation

Let us consider a distributed sensor network acquiring scalar measurements - or datastreams - from a time invariant dynamic system whose model description is unavailable. Without loss of generality we focus on the relationship between two datastreams; the extension to multiple or multivariate datastreams is straightforward, e.g., see [6].

The unknown relationship $\mathcal{P}$ between the two streams of data is modeled by a Single-Input Single-Output (SISO) linear time invariant (LTI) predictive model belonging to a family $\mathcal{M}$ parametrized in $\theta \in \mathcal{D} \subset \mathbb{R}^p$, where $\mathcal{D}$ is a C^1 compact manifold in $\mathbb{R}^p$ and $p \in \mathbb{N}$ is the dimension of θ. For instance, SISO or Single-output/Multiple-input linear predictive models [19] and Reservoir Computing Networks [20] are viable options for the choice of $\mathcal{M}$. In this paper, we consider linear SISO one-step-ahead predictive models:

$$\hat{y}(t|\theta) = f\left(t, \theta, u(t), \ldots, u(t - \tau_u), \ldots y(t - 1), \ldots, y(t - \tau_y)\right), \quad \forall t \in \mathbb{N} \quad (1)$$

where $\hat{y}(t|\theta) \in \mathbb{R}$ is the prediction provided at time t, $f : \mathbb{N} \times \mathbb{R}^p \times \mathbb{R}^{\tau_u} \times \mathbb{R}^{\tau_y} \to \mathbb{R}$ is the approximating function in predictive form, $u(t) \in \mathbb{R}$ and $y(t) \in \mathbb{R}$ are the model input and output at time t, respectively, and τ_u and τ_y are the orders of the input and output, respectively.

Under mild assumption on $\mathcal{P}$, $\mathcal{M}$ and the estimation procedure of θ over a finite dataset $Z_N = \{(u(j), y(j))\}_{j=1}^N$, it holds that [19]:

$$\lim_{N \to \infty} \sqrt{N} \Sigma_N^{-\frac{1}{2}} (\hat{\theta} - \theta^o) \sim \mathcal{N}(0, I_p) \quad (2)$$

where $\hat{\theta}$ is the estimated parameter vector, θ^o is the parameter optimizing the structural risk in $\mathcal{M}$, Σ_N is a properly defined covariance matrix and I_p is the identity matrix of order p. Interestingly, Equation (2) assures that, given a sufficiently large N, the estimated parameter vector $\hat{\theta}$ follows a multivariate Gaussian distribution with mean θ^o and covariance matrix Σ_N. Moreover, this result remains valid even when $\mathcal{P} \notin \mathcal{M}$, i.e., a model bias $||\mathcal{M}(\theta^o) - \mathcal{P}|| \neq 0$ is present. This justifies the use of LTI models, even when the dynamic system under investigation $\mathcal{P}$ is non-linear.

3   The HMM-Based Change Detection Test

The aim of this section is to summarize the key points of the HMM-CDT suggested in [8], representing the core element of the proposed EHMM-CDT. Under the formulation presented in Section 2, the use of a HMM ruled by a mixture of Gaussian (GMM) over a sequence of parameters $\hat{\theta}$s is the natural solution to model the functional relationship presented in Equation (1) in the parameter space. More formally, a HMM [21] can be defined as a tuple $\mathcal{H} = (S, A, \pi)$, where $S = \{S_1, \ldots, S_s\}, s \in \mathbb{N}$ is the indexed set of the states, each of which has an emission probability defined by a GMM, $A \in \mathbb{R}^{s \times s}, A = [a_{ij}], \sum_{i=1}^s a_{ij} = 1 \; \forall j \in \{1, \ldots, s\}$ is the transition matrix, i.e., a_{ij} is the transition probability from state i to state j and $\pi \in [0,1]^s$, with $\sum_{i=1}^s \pi_i = 1$, is the initial distribution probability over S.

The idea underlying the HMM-CDT is to analyse the statistical behaviour of $\hat{\theta}$s over time: when the statistical pattern of the estimated parameters does not follow what learned during an initial training phase, a change in the relationship is detected. Thus, the HMM-CDT relies on an initial parameter vector sequence $\Theta_L = (\hat{\theta}_1, \ldots, \hat{\theta}_L)$ estimated on a faulty-free training set Z_M, where $\hat{\theta}_j$ is estimated on a subsequence $Z_{j,N} = \{(x(h), y(h))\}_{h=j-N+1}^j$ of the training sequence, $j \in \{N, \ldots, M\}, L = M - N + 1$ (i.e., overlapping windows of length N). The parameter sequence Θ_L is used to train a HMM $\mathcal{H}$, aiming at capturing the statistical behaviour of the estimated parameter vectors $\hat{\theta}_t$s in nominal conditions (without any change/fault). Then, during the operational phase, the statistical affinity between the estimated parameter vectors $\hat{\theta}_t$s and $\hat{\mathcal{H}}$ is assessed by looking at the HMM loglikelihood: changes in the relationship expressed by Equation (1) are detected by inspecting the likelihood of $\hat{\theta}_t$ with respect to $\hat{\mathcal{H}}$. More specifically, given a sequence $\Theta_{t,k} = (\hat{\theta}_{t-k+1}, \dots, \hat{\theta}_t)$, i.e., the last k consecutive parameter vectors at time t, a HMM $\hat{\mathcal{H}}$ provides an estimates on how well the sequence follows $\hat{\mathcal{H}}$ by means of the computation of the loglikelihood:

$$l(\hat{\mathcal{H}}, \Theta_{t,k}) = \log \Pr(\hat{\theta}_{t-k+1}|\hat{\mathcal{H}}) + \sum_{h=t-k+1}^{t-1} \log \Pr(\hat{\theta}_{h+1}|\hat{\theta}_h, \hat{\mathcal{H}}). \tag{3}$$

If the relationship does not change over time, the loglikelihood $l(\hat{\mathcal{H}}, \Theta_{t,k})$ is comparable with the one computed during the training phase on a validation set $Z_O = \{(x(j), y(j))\}_{j=M+1}^{O}$. Otherwise, in case of variations in the relationship, $l(\hat{\mathcal{H}}, \Theta_{t,k})$ decreases, since $\Theta_{t,k}$ is no more compatible with the statistical model characterized by $\hat{\mathcal{H}}$. Further details about the HMM-CDT can be found in [6, 8].

4   Ensemble of HMM for Fault Detection

One of the key aspects of the HMM-CDT is the estimation of a HMM on Θ_L, aiming at characterizing the nominal state. The state-of-the art training algorithm for HMMs is the Baum-Welch (BW) algorithm [22], which aims at finding the maximum likelihood estimates of the HMM parameters given Θ_L. Nonetheless, BW algorithm does not provide any guarantee about the convergence to the global maximum of the likelihood function [10]. Since the BW algorithm requires a random initialization of the parameters, different HMMs can be obtained by repeating the training phase with different randomly initialized parameters on the same training sequence. Hence, a viable solution to weaken the effect of the initialization procedure on the HMM-CDT would be to repeat the BW algorithm and select the HMM guaranteeing the largest likelihood on a validation set. Unfortunately, as pointed out in [17], this solution may lead to overfit the training sequence Θ_L (and this is particular evident for reduced training sets), leading to false positive detections (false alarms) during the operational phase.

In this work, we suggest an ensemble approach for fault detection (EHMM-CDT) based on a set $E = \{\mathcal{H}_1, \dots, \mathcal{H}_e\}, e \in \mathbb{N}$ of HMMs, where e is the ensemble cardinality. The main point of the proposed solution is that the HMMs $\mathcal{H}_i \in E$ are trained on the same training set Z_M with different initialization points of the BW algorithm. This ensemble approach allows to weaken the influence of the initial conditions of the HMM training algorithm, providing a better generalization ability and, consequently, better detection performance.

The proposed EHMM-CDT is detailed in Algorithm 1. To characterize the nominal state we rely on a training set Z_M, assumed to be change-free.

Algorithm 1. EHMM-CDT algorithm

1: **Data:** Training set Z_M, Validation set Z_O, C, $\mathcal{A}$;
2: **Results:** Detection time T;
3: Estimate the sequence Θ_L from Z_M;
4: **for** $i \in \{1, \ldots, e\}$ **do**
5:     Estimate HMM $\mathcal{H}_i$ using Θ_L by considering a random initialization point;
6: **end for**
7: Built the ensemble $E = \{\mathcal{H}_1, \ldots, \mathcal{H}_e\}$;
8: Estimate the sequence $\Theta_{O,O-M+1}$ from Z_O;
9: **for** $h \in \{M + k, \ldots, O\}$ **do**
10:     **for** $i \in \{1, \ldots, e\}$ **do**
11:         Compute $l(\mathcal{H}_i, \Theta_{h,k})$ as in Equation (3);
12:     **end for**
13:     Compute $\mathcal{A}(h) = \mathcal{A}(l(\mathcal{H}_1, \Theta_{h,k}), \ldots, l(\mathcal{H}_e, \Theta_{h,k}))$;
14: **end for**
15: Compute $l_{\min}$ as in Equation (4);
16: **while** a new couple $(x(t), y(t))$ is available at time t **do**
17:     Estimate $\hat{\theta}_t$ by using $Z_{t,N}$;
18:     Built the sequence $\Theta_{t,k} = (\hat{\theta}_{t-k+1}, \ldots, \hat{\theta}_t)$;
19:     **for** $i \in \{1, \ldots, e\}$ **do**
20:         Compute $l(\mathcal{H}_i, \Theta_{t,k})$ as in Equation (3);
21:     **end for**
22:     Compute $\mathcal{A}(t) = \mathcal{A}(l(\mathcal{H}_1, \Theta_{t,k}), \ldots, l(\mathcal{H}_e, \Theta_{t,k}))$;
23:     **if** $\mathcal{A}(t) \leq l_{\min}$ **then**
24:         **return** $T \leftarrow t$;
25:     **end if**
26: **end while**
27: **return** $T \leftarrow \emptyset$;

We estimate the parameter vectors $\hat{\theta}_j$s on overlapping windows of N data $Z_{t,N}, t \in \{N, \ldots, M\}$. Without loss of generality, we assume a step of one sample between two overlapping windows; larger steps could be considered as well (e.g., to reduce the computational complexity of the HMM training phase). The estimation procedure is performed by means of a suitable minimization technique, e.g., Least Square estimation on the empirical risk (see [19] for details), and let Θ_L be the sequence of estimated parameter vectors on Z_M, as shown in Section 3 (Line 3).

We build the ensemble $E = \{\mathcal{H}_1, \ldots, \mathcal{H}_e\}$ by repeating e times the training of a HMM on Θ_L, with random initial conditions for the BW algorithm (Line 5). Note that, since Z_M is assumed to be change-free, the ensemble E aims at modeling the statical behaviour of $\mathcal{P}$ in nominal conditions.

One of the key aspects of ensemble methods is the definition of the aggregation mechanism $\mathcal{A}$, that, in this case, specifies how to aggregate the loglikelihoods of the ensemble elements $\mathcal{H}_i \in E$ to provide the ensemble output. In this work, we considered two different aggregation mechanisms, i.e., $\mathcal{A} \in \{\mathcal{A}_{mean}, \mathcal{A}_{min}\}$:

$$\mathcal{A}(t) = \mathcal{A}_{mean}(l(\mathcal{H}_1, \Theta_{t,k}), \ldots, l(\mathcal{H}_e, \Theta_{t,k}))) = \sum_{i=1}^{e} \frac{l(\mathcal{H}_i, \Theta_{t,k})}{e}$$

$$\mathcal{A}(t) = \mathcal{A}_{min}(l(\mathcal{H}_1, \Theta_{t,k}), \ldots, l(\mathcal{H}_e, \Theta_{t,k}))) = \min_{i \in \{1, \ldots, e\}} l(\mathcal{H}_i, \Theta_{t,k})$$

where $\mathcal{A}_{mean}$ computes the average value of the loglikelihoods, while $\mathcal{A}_{min}$ takes into account their minimum. The last step of the EHMM-CDT training phase is the computation of the threshold l_{min} on a validation set Z_O(Line 15), which is computed as follows:

$$l_{min} = \bar{l} - C \left[\bar{l} - \min_{h \in \{M+k, \ldots, O\}} \mathcal{A}(h) \right] \tag{4}$$

where $\bar{l} = \frac{\sum_{h=M+k}^{O} \mathcal{A}(h)}{O}$ and $C > 1$ is a user-defined coefficient factor.

During the operational phase, when the aggregated loglikelihood decreases below l_{min}, a change in the statical behaviour of $\mathcal{P}$ is detected by the EHMM-CDT. More in details, as soon as a new sample $(u(t), y(t)), t > M + O$ is available, the algorithm estimates a new parameter vector $\hat{\theta}_t$ on $Z_{t,N}$ (Line 17). The likelihoods for the ensemble of HMMs, i.e., $l(\mathcal{H}_i, \Theta_{k,t}) \, \forall \mathcal{H}_i \in E$, are computed (Line 20) and then aggregated, according to the aggregation mechanism $\mathcal{A}$ (Line 22). Afterwards, the aggregated likelihood $\mathcal{A}(t)$ is compared with the threshold l_{min} to asses if a change in $\mathcal{P}$ is detected (Line 23). In case of a change, an alarm is raised and the detection time $T = t$ is returned (Line 24), otherwise the algorithm keeps on monitoring data coming from $\mathcal{P}$.

While ensemble approaches are generally able to increase the generalization ability of single models, they are characterized by an increased computational complexity, that in this case scales linearly with the number of HMMs e. Two comments arise: 1) The most time consuming part of the EHMM-CDT refers to the HMMs training (the loglikelihood computation is much more lighter than training). Interestingly, the training phase is performed only once during the initial configuration of the cognitive FDS, while during the operational life only the likelihoods are computed. In addition, in scenarios where networked embedded systems are operating, the training of HMMs could be performed in a centralized high-powerful unit, leaving only the computation of likelihoods directly at the low-power distributed units of the network. 2) In scenarios where the fault detection ability is a relevant activity, the increase in the complexity induced by the ensemble approach is well compensated by the decrease in false and missed alarms and detection delays, as shown in the experimental section.

5 Experimental Results

To evaluate the performance of the proposed EHMM-CDT we considered both synthetic (Application D1) and real data (Application D2), coming from a rock collapse forecasting system deployed in Northern Italy. In both applications, we considered a faulty free dataset divided into training (to create E), validation (to learn $l_{\min}$) and test sets (to evaluate the performance). In the last one we injected five different kinds of abrupt permanent fault at time $\bar{t}$:

- **A1-A3**: $y(t) = y(t) + \Delta_a \cdot a_{\max}, \forall t > \bar{t}, \Delta_a \in \{0.1, 0.2, 0.3\}$ (additive fault);
- **M** : $y(t) = y(t) \cdot (1 + \Delta_m), \forall t > \bar{t}, \Delta_m = 0.3$ (multiplicative fault);
- **S**: $y(t) = y(\bar{t}), \forall t > \bar{t}$ (stuck-at fault);

where $a_{\max} = \max_{j \in \{1,...,M\}} y(j) - \min_{j \in \{1,...,M\}} y(j)$.

As regards the model family $\mathcal{M}$, we considered SISO LTI AutoRegressive with eXogenous input (ARX) models, whose orders are chosen through a model selection procedure (i.e., minimizing the mean square one-step-ahead prediction error on a validation set). The HMMs are configured by considering batches of $N = 100$ samples, loglikelihood window length $k = 10$ and exploring $s \in \{3, \ldots, 6\}$ and the number of models in each GMM $\in \{1, 2, 4, 8, 16, 32\}$. The cardinality of the ensemble was set to $e = 30$. Finally, the parameter vectors estimation is performed with the Least Square method.

To evaluate the detection ability of what proposed the following figures of merit have been considered:

- false positive (FP) rate: fraction of the experiments where the method detected a change before it actually appears;
- false negative (FN) rate: fraction of the experiments where the method did not detected a change;
- detection delay (DD): the number of samples necessary to detect a change;

The proposed method is compared with the HMM-CDT, where a single HMM was considered (the one with largest loglikelihood on a validation set among those considered for the ensemble).

APP D1: Synthetic Data. The data for the synthetic application have been generated from the following non-linear model:

$$y(t) = \sin(a^T \cdot (y(t-1), y(t-2)) + b \cdot x(t-1)) + \eta(t)$$

where $a \in [0, 1]^2, b \in [0, 1], \eta(t) \sim \mathcal{N}(0, \sigma^2), \sigma \in \{0.01a_{\max}, 0.05a_{\max}\}$. Each experiment lasts 6125 samples (i.e., 3268 for training, 817 for validation and 2040 for testing): faults are injected at time $\bar{t} = 5105$. Results are averaged over 1000 runs.

Figure 1 shows the experimental results in the case of APP D1 and $\sigma = 0.05a_{\max}$. Curves are obtained by considering values of C ranging from 1 to 5.1, where 5.1 is the largest value of C for which $FN \leq 0.1$ for all the faulty scenarios. Interestingly, in all the considered scenarios the ensemble approach is able to improve the detection ability (in terms of both FP and DD) of the single HMM-CDT. These curves show that the performance of EHMM-CDT cannot be achieved by the HMM-CDT by simply tuning the parameter C. In addition, in the faulty scenario **M** where FNs are present, the proposed EHMM-CDT

Table 1. Detection results for the single and ensemble HMM-CDT approaches from applications D1 (with different σs) and D2

		HMM-CDT			EHMM-CDT					
					$\mathcal{A}_{mean}$			$\mathcal{A}_{\min}$		
		FN	FP	DD	FN	FP	DD	FN	FP	DD
$\sigma = 0.01 a_{max}$	A1	0.000	0.007	21.614	0.000	0.007	19.419	0.000	0.006	19.717
	A2	0.000	0.007	17.415	0.000	0.007	15.320	0.000	0.006	15.638
	A3	0.000	0.007	15.396	0.000	0.007	13.724	0.000	0.006	13.998
	M	0.000	0.007	37.057	0.000	0.007	35.355	0.000	0.006	35.794
	S	0.000	0.007	16.186	0.000	0.007	15.335	0.000	0.006	15.648
$\sigma = 0.05 a_{max}$	A1	0.000	0.007	67.939	0.000	0.009	63.745	0.000	0.009	64.708
	A2	0.000	0.007	47.465	0.000	0.009	42.625	0.000	0.009	43.281
	A3	0.000	0.007	39.676	0.000	0.009	34.734	0.000	0.009	35.006
	M	0.077	0.007	185.989	0.057	0.009	161.401	0.055	0.009	166.641
	S	0.000	0.007	45.139	0.000	0.009	42.413	0.000	0.009	42.781
Rialba	A1	0.340	0.000	314.879	0.000	0.000	172.100	0.280	0.000	234.500
	A2	0.000	0.000	152.100	0.000	0.000	148.900	0.000	0.000	154.400
	A3	0.000	0.000	119.700	0.000	0.000	30.700	0.000	0.000	93.800
	M	0.040	0.000	163.250	0.000	0.000	153.200	0.000	0.000	162.100
	S	0.000	0.000	81.500	0.000	0.000	82.300	0.000	0.000	89.600

behaves better than HMM-CDT even with respect to this figure of merit (see legend of Fig.1.d).

Experimental results presented in Table 5 refer to FN, FP and DD with fixed values of C. To ease the comparison we set C in the EHMM-CDT and HMM-CDT as the lowest values guaranteeing $FN = 0$ and $FP \leq 0.01$ in APP D1 with $\sigma = 0.01$ (i.e., $C = 4.814$ for HMM-CDT, $C = 4.629$ for EHMM-CDT with $\mathcal{A}_{mean}$ and $C = 4.917$ for EHMM-CDT with $\mathcal{A}_{\min}$). Remarkably, both EHMM-CDT aggregations provide lower DD than HMM-CDT. Interestingly, the *mean* aggregation mechanism generally outperforms the *minimum* one, since the mean is less influenced than minimum by outliers, which generally induce false alarms. Nonetheless, by comparing these results with those provided in Figure 1, we may also comment that the *mean* aggregation does not provide better results for each value of C: hence the *minimum* is a viable solution too.

APP D2: Rialba Data. We consider data coming from a real-world distributed sensor network for rock-collapse forecasting [23, 24]. This dataset is available at [25]. In particular, we analyzed two temperature datastreams composed by 5303 samples (3000 for training, 1000 for validation and 1303 for testing), coming from two network units. The sampling period is 5 minutes. We injected faults (**A1-A3,M,S**) at $\bar{t} = 4695$. Experimental results in Table 5 are obtained by averaging the results of 50 runs. Interestingly, the results on the real-world datasets are in line with those of the synthetic one: the ensemble approach provides lower FN and DD for most of the considered scenarios.

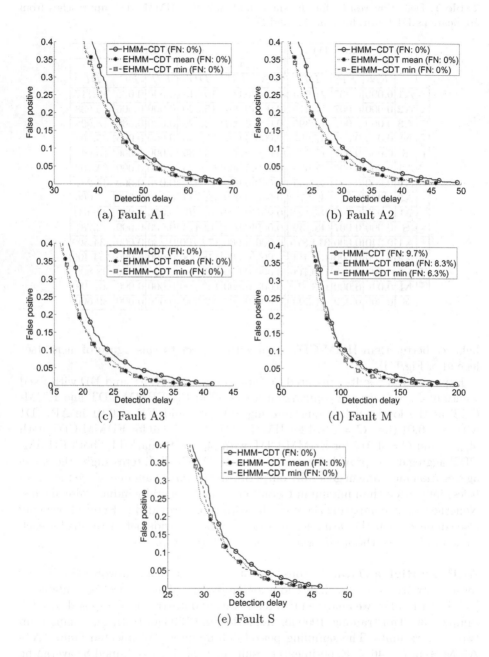

Fig. 1. Experimental results showing the relationship between FP and DD in application D1 with $\sigma = 0.05$, where $C \in [1, 5.1]$

6   Conclusions

We presented an novel approach for cognitive fault detection based on an ensemble of HMMs. The distinguishing feature of the proposed solution is the ability to improve the generalization ability (in terms of detection performance) of a single HMM-CDT by considering an ensemble of HMMs trained on the same training set, but with different initialization points. The effectiveness of the proposed solution has been tested on both synthetic and real datasets coming from a real-world distributed sensor network for rock-collapse forecasting.

References

1. Isermann, R.: Fault-diagnosis systems. Springer (2006)
2. Venkatasubramanian, V., Rengaswamy, R., Yin, K., Kavuri, S.N.: A review of process fault detection and diagnosis: Part I: Quantitative model-based methods. Computers & Chemical Engineering 27(3), 293–311 (2003)
3. Demetriou, M.A., Polycarpou, M.M.: Incipient fault diagnosis of dynamical systems using online approximators. IEEE Transactions on Automatic Control 43(11), 1612–1617 (1998)
4. Farrell, J., Berger, T., Appleby, B.D.: Using learning techniques to accommodate unanticipated faults. IEEE Control Systems 13(3), 40–49 (1993)
5. Trunov, A.B., Polycarpou, M.M.: Automated fault diagnosis in nonlinear multivariable systems using a learning methodology. IEEE Transactions on Neural Networks 11(1), 91–101 (2000)
6. Alippi, C., Ntalampiras, S., Roveri, M.: A cognitive fault diagnosis system for distributed sensor networks. IEEE Transactions on Neural Networks and Learning Systems 24(8), 1213–1226 (2013)
7. Alippi, C., Roveri, M., Trovò, F.: A "Learning from models" cognitive fault diagnosis system. In: Villa, A.E.P., Duch, W., Érdi, P., Masulli, F., Palm, G. (eds.) ICANN 2012, Part II. LNCS, vol. 7553, pp. 305–313. Springer, Heidelberg (2012)
8. Alippi, C., Ntalampiras, S., Roveri, M.: An hmm-based change detection method for intelligent embedded sensors. In: The 2012 International Joint Conference on Neural Networks (IJCNN), pp. 1–7. IEEE (2012)
9. Alippi, C., Boracchi, G., Puig, V., Roveri, M.: An ensemble approach to estimate the fault-time instant. In: 2013 Fourth International Conference on Intelligent Control and Information Processing (ICICIP), pp. 836–841. IEEE (2013)
10. Dempster, A.P.: et al. Maximum likelihood from incomplete data via the em algorithm. Journal of the Royal statistical Society 39(1), 1–38 (1977)
11. Zhou, Z.H.: Ensemble methods: foundations and algorithms. CRC Press (2012)
12. Krogh, A., Sollich, P.: Statistical mechanics of ensemble learning. Physical Review E 55(1), 811 (1997)
13. Perrone, M.P., Cooper, L.N.: When networks disagree: Ensemble methods for hybrid neural networks. Technical report, DTIC Document (1992)
14. Wichard, J., Ogorzalek, M.: Time series prediction with ensemble models. In: The 2004 International Joint Conference on Neural Networks (IJCNN), pp. 1625-1629. IEEE (2004)
15. Zhang, G.P.: Time series forecasting using a hybrid ARIMA and neural network model. Neurocomputing 50, 159–175 (2003)

16. Alippi, C., Ntalampiras, S., Roveri, M.: Model ensemble for an effective on-line reconstruction of missing data in sensor networks. In: The 2013 International Joint Conference on Neural Networks (IJCNN), pp. 1–6. IEEE (2013)
17. MacKay, D.J.C.: Ensemble learning for hidden markov models. Technical report, Cavendish Laboratory, University of Cambridge (1997)
18. Davis, R.I.A., Lovell, B.C.: Comparing and evaluating HMM ensemble training algorithms using train and test and condition number criteria. Formal Pattern Analysis & Applications 6(4), 327–335 (2004)
19. Ljung, L.: System identification. Wiley Online Library (1999)
20. Schrauwen, B., Verstraeten, D., Van Campenhout, J.: An overview of reservoir computing: theory, applications and implementations. In: European Symposium on Artificial Neural Networks. Citeseer (2007)
21. Elliott, R.J., Aggoun, L., Moore, J.B.: Hidden Markov Models. Springer (1995)
22. Baum, L.E., Petrie, T., Soules, G., Weiss, N.: A maximization technique occurring in the statistical analysis of probabilistic functions of markov chains. In: The Annals of Mathematical Statistics, pp. 164–171 (1970)
23. Alippi, C., Camplani, R., Galperti, C., Marullo, A., Roveri, M.: An hybrid wireless-wired monitoring system for real-time rock collapse forecasting. In: 2010 IEEE 7th International Conference on Mobile Adhoc and Sensor Systems (MASS), pp. 224–231. IEEE (2010)
24. Alippi, C., Camplani, R., Galperti, C., Marullo, A., Roveri, M.: A high-frequency sampling monitoring system for environmental and structural applications. ACM Transactions on Sensor Networks (TOSN) 9(4), 41 (2013)
25. isense project website, (March 2014), http://www.i-sense.org/

Modeling ReTweet Diffusion
Using Emotional Content

Andreas Kanavos, Isidoros Perikos, Pantelis Vikatos, Ioannis Hatzilygeroudis,
Christos Makris, and Athanasios Tsakalidis

Computer Engineering and Informatics Department,
University of Patras, Greece
{kanavos,perikos,vikatos,ihatz,makri,tsak}@ceid.upatras.gr

Abstract. In this paper we present a prediction model for forecasting
the depth and the width of ReTweeting using data mining techniques.
The proposed model utilizes the analyzers of tweet emotional content
based on Ekman emotional model, as well as the behavior of users in
Twitter. In following, our model predicts the category of ReTweeting
diffusion. The model was trained and validated with real data crawled
by Twitter. The aim of this model is the estimation of spreading of a
new post which could be retweeted by the users in a particular network.
The classification model is intended as a tool for sponsors and people of
marketing to specify the tweets that spread more in Twitter network.

Keywords: Tweet Emotion Recognition, Microblogging, ReTweet Modeling, Sentiment Analysis.

1 Introduction

During the last years, Twitter has become the most popular microblogging platform all over the world. It is considered to be the most fast spreading and growing social network having more than 500 millions of users. Every day a vast amount of tweet information is published by the users of the platform to the public or to selected circles of their contacts. This information comes in the form of small posts which is allowed to have up to 140 characters each. The rise of Twitter has completely changed the users, transforming them from simple passive information searchers and consumers to active producers. This massive and continuous stream of Twitter data posts reflects the users opinions and reactions to phenomena from political events all over the world to consumer products [17]. It is well pointed that Twitter posts relate to the user's behavior and often convey substantial information about the user's emotional state [2].

The users' posts in Twitter, unlike other networks, have some special characteristics. The short length that the posts are allowed to have, results in more expressive emotional statements. Also, users express their daily thoughts in real time, something that often concludes in posting far more emotional expressions than might normally occur [15]. Analyzing tweets and recognizing their emotional content is a very interesting and challenging topic in the microblogging

L. Iliadis et al. (Eds.): AIAI 2014, IFIP AICT 436, pp. 101–110, 2014.

area. It is necessary for deeper understanding of people behavior but simultaneously for describing public attitude towards different events and topics as well.
It therefore could be helpful in predicting the spread of posts and information
diffusion in the network.

A fundamental way of information spread in the Twitter network is the
ReTweets. A ReTweet happens when a user forwards a message they receive
to circles of their followers. In addition, a ReTweet indicates that the user who
retweets the post, found it very interesting and worth sharing with others [9].
The way that a post is retweeted among users in the network can describe the
amount of interest drawn upon it [10] and also its adoption by the users who
made the ReTweets. Since ReTweets provide the most powerful clue that users
find post information interesting, it would be important to study, analyze and
model the way that information is spread in Twitter through ReTweets. Being
able to model and predict the ReTweet occurrences of a given post is important
for understanding and controlling information diffusion on Twitter [3].

Emotion representation is a key aspect of an emotional recognition system.
The most popular models for representing emotions are the categorical and the
dimensional model ones. The categorical model assumes a finite number of basic,
discrete emotions, each of which serves for a specific purpose. On the other hand,
the dimensional model represents emotions on a dimensional approach, where
an emotional space is created and each emotion lies in this particular space. A
very popular categorical model is the Ekman emotion model [4], which specifies
six basic human emotions: "anger, disgust, fear, happiness, sadness, surprise".
It has been used in several studies/systems that recognize emotional text and
facial expressions related to these emotional states.

In this paper we present a work on modeling and predicting the information
spread on Twitter based on its emotional content. More specifically, we propose
the use of emotional behavior in each ReTweet as an additional parameter to
user's social media analysis. With use of the Ekman emotion model, we can
identify whether one or more out of the six basic human emotions exists or not.
The aim of this system is to model the information spread in form of ReTweets
in the network and in following, to predict the likelihood and the way that a
new post will be retweeted by the users in a particular network. More precisely,
we want to perform the following steps in the below mentioned order: firstly
understanding Twitter topic, then modeling the ReTweet information flow in
Twitter network and finally predicting when a new post will be retweeted and
though the way of how it will be spread in the network (e.g. given a new post with
particular emotional content, we want to predict whether it will be retweeted by
other Twitter users; and if so, we would like to learn its depth and width in the
Twitter network).

The remainder of the paper is structured as follows: Section 2 presents the
related work. Section 3 presents our model while in section 4 we utilize our
experiments. Moreover, section 5 presents the evaluation experiments conducted
and the results gathered. Ultimately, section 6 presents conclusions and draws
directions for future work.

2 Related Work

The spread of Twitter and other social networks has attracted significant interest in sentiment analysis methods for recognizing textual opinions, statement polarity and emotions. Researchers have developed applications to identify whether a text is subjective or objective, and whether any opinion expressed is positive or negative [11]. In [15] authors investigated feature sets to classify emotions in Twitter and presented an analysis of different linguistic styles people use to express their emotions.

There is a lot of research interest in studying different types of information dissemination processes on large graphs and social networks. Naveed et al. [9] analyze tweet posts and forecast for a given post the likelihood of being retweeted on its content. Authors indicate that tweets containing negative emoticons are more likely to be retweeted than tweets with positive emoticons. In [1] authors study the way that information diffuses in Twitter. A dataset of 1.6 million Twitter users was used in order to identify common information patterns of their tweets and many of these patterns involved transmission with third and fourth hop users. Also, the authors indicate that although stronger ties are individually more influential, the more abundant weak ties are responsible for the propagation of novel information.

In [5] the authors study the online social networks of the social news aggregator Digg as well as the microblogging service Twitter, both of which are used by people in order to share news stories and other content with their followers. The work indicates that the spread of information could be modeled as an epidemic process and though could have a non-conservative favor.

In addition, in [2] authors utilized the Profile of Mood States psychometric method so as to analyze Twitter posts and reached the conclusion that "the events in the social, political, cultural and economic sphere do have significant, immediate and highly specific effect on the various dimensions of public mood". Commercial companies and associations could exploit Twitter for marketing purposes, as it provides an effective medium for propagating recommendations through users with similar interests. Moreover, viral marketers could exploit models of user interaction to spread their content or promotions quickly and widely [8].

Peng et al. [12] study the ReTweet occurrences in Twitter and proposed a method utilizing Conditional Random Fields (CRF) so as to predict how likely a tweet will be retweeted by a user. The analyzed tweets and their prediction model take into account network aspects and post content parameters like the topic similarity of posts with the user's interest, the existence of URL and hashtags as well as mentions in the posts. In [3], authors address the ReTweet modeling topic and focus on tweets that contain links to images shared through twitpic.com. They analyze images and extract correlated low-level and high-level image features in their predictive model for ReTweet count.

Finally, Petrovic et al. in [14] predict ReTweet occurrences by using a machine learning approach based on the passive-aggressive algorithm. Authors found that there is a substantial gain in using tweet content specific features and their prediction model takes into account user social features and tweet features. Social features concern user's number of followers, friends, statuses, favorites, the number of the times the user was listed, if the user is verified, and if the user's language is English; while tweet features concern the existence and the number of hashtags, mentions, URLs, trending words in the tweet, the length of the tweet, the novelty, the actual words in the tweet and last if the tweet is a reply. As far as we are aware, our work is primarily to investigate the impact of tweets emotional content, as determined on Ekman's scale, on its diffusion in the network in the form of ReTweets.

3   Model Overview

In our model, we want to predict whether a tweet with specific emotional states based on a user's recent tweets and specific user's Twitter analytics features, can be categorized in a particular class concerning the possibility of being retweeted by other users. We can specify this problem as classification one, because each class will consist of specific thresholds in each category. Furthermore, we want to predict whether the tweets from a specific user can be forwarded deep in the Twitter graph; meaning that many other users will forward this tweet (via ReTweeting).

The overall architecture of the proposed system is depicted in Figure 1 while the proposed modules and sub-modules of our model are modulated in the following steps.

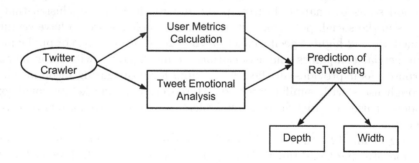

Fig. 1. Architecture of the Model

3.1   User Metrics Calculation

The social media crawler samples the Twitter and identifies a specific number of users who have posted a tweet in a specific topic during a specific time period (e.g. keyword search query). More specifically, the implementation of our method goes as follows: initially, we retrieve users who have posted a tweet within the given time period. Subsequently, we download their k last tweets (e.g. for our experiments $k = 20$), in order to give them as input in our Tweet Emotional Analysis module. Furthermore, from Twitter we can extract 6 basic user features as they can better describe user communication behavior in Twitter. The followers of a user, the number of contributions to the social network and the frequency of contribution are some aspects that differentiate user behavior.

The set of 6 features contain the number of Followers, the number of Direct Tweets, the number of ReTweets, the number of Conversational Tweets (e.g. if a user replies to a post), the Frequency of user's Tweets (e.g. how "often" an author posts Tweets) and finally the number of Hashtag Keywords (e.g. words starting with the symbol # and with use of this symbol, a user can specify the thematic category of their specific Tweet) as in [6], [7]. These metrics describe the user communication behavior in Twitter.

3.2 Tweet Emotional Analysis

In this subsection, the emotional analysis of the tweets based on the tool presented in [13] is described. The tool recognizes the existence of the six basic emotions proposed by Ekman [4] in natural language sentences. Tweets texts are pre-processed (remove images, URLs, etc) and in following, the structure's analysis of the tweet using Stanford parser and Tree Tagger is conducted. Also, lexical resources such as the WordNet Affect [16] are used to spot emotional words. Moreover, the tool analyses each word's emotional dependencies in order to specify its emotional strength and determine the overall emotional status of tweets based on its dependency graph. This module is shown in Figure 2.

We have extracted some tweets during a given period of time for a specific topic (e.g. *#MH370*). In following, we try to identify the emotional state of tweets (or specify their emotional content) through the aforementioned six basic emotions. Tweets are analyzed and for each one the existence of each basic emotion is determined.

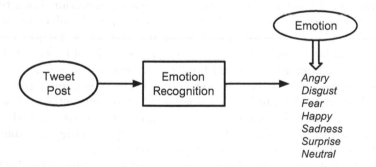

Fig. 2. Tweet Emotional Analysis

Since we ensure whether these tweets have a sentiment or are neutral (a tweet is neutral when no category of Ekman psychometric test appears in this tweet), we calculate the depth as well as the width (if any) of all tweets (e.g. ReTweeting depth and width). More specifically, we count the number that a tweet has been retweeted ($width = 1$); then if someone sees the ReTweet and decides to make a ReTweet of the above ReTweet, the width takes value 2, and so on. Suppose we have a tweet from a user X and many Twitter users want to share this tweet among their followers. So, for example, n users share this post. In the first step, our algorithm has $width = 1$ and $depth = n$. What is more, a user who "follows" one out of these n users wants to share the same ReTweet; they use this ReTweet in their profile and consecutively, the width for the specific post becomes 2.

Table 1. Categories of Features

Features	#	Description
User Twitter Metrics	6	Followers, Direct Tweets, ReTweets, Conversational Tweets, Frequency, Hashtag Keywords
Tweet Emotional Content	6	Anger, Disgust, Fear, Joy, Sadness, Surprise

4   Implementation

We based our experiments on Twitter and used Twitter API to collect tweets and calculate users metrics. We implemented our Twitter network using Twitter4J[1]. We collected tweets published for a specific period of time (e.g. a couple of hours) for the keyword *#MH370* concerning Malaysia Airlines Flight 370 disappearance. Our Twitter data consists of 13000 tweets, authored by 11130 users.

The prediction of ReTweeting is based on the emotional content of the tweets using Ekman model but also the user communication behavior. Each tweet is represented as a vector; this vector includes the existence of each emotion which is produced by Ekman's model as well as Twitter metrics of the user that produce the certain tweet such as their followers, the number of tweets, ReTweets, conversations, hashtag keywords and finally the frequency of posts.

Analysing the retrieved tweets, we extract the length of maximum path of each ReTweet. Moreover, the number of followers that have retweeted each tweet is additionally computed. We use the following approach to classify the ReTweets of a certain tweet. According to the categories of ReTweeting in Table 2, we predict the depth as well as the width of ReTweeting.

Several classifiers are trained using the dataset of vectors. Classification is evaluated based on the separation of the dataset to training and test set. Furthermore, the approach we used was K-Fold Cross-Validation (e.g. $K = 10$ Fold).

---

[1] Twitter4J library: http://twitter4j.org/en/index.html

Table 2. Classes of ReTweeting

Level of ReTweeting	#Instances in Depth ReTweeting	#Instances in Width ReTweeting
○	5192	5192
	5667	4785
	2141	3023

The classifiers that were chosen, are evaluated with the use of F-Measure metric. The classifiers were chosen from "bayes", "functions", "lazy", "trees" and "rules" categories of the Weka library[2], which is a widely used toolkit for machine learning and data mining such as classification, regression, clustering and feature selection. The classifiers from Weka are used with their default settings. The results are introduced in following section.

5   Evaluation

The reported values in the charts for the classification models are recorded as AdaBoost, IBK, J48, JRip, Multilayer Perceptron, Naive Bayes, PART, RotationForest and SMO. The results of F-Measure for each classifier are illustrated in Table 3; where we select the best classifiers either for depth or width, depicted in bold in the table. We can observe that J48 achieves the highest F-Measure value for both the depth as well as the width ReTweeting. In the case of depth, RotationForest is selected as the next best classifier, where in case of width, JRip accomplishes the same performance as J48. Furthermore, Figures 3 and 4 depict the values of F-Measure for each classifier for depth and width accordingly.

Additionally, we evaluate the model using posts with different emotion for the same user's behavior in order to consider the influence of emotion in the diffusion of ReTweets based on our model. Table 4 shows that tweets which are angry or sad are prone to be retweeted rather than tweets that are happy or neutral. In addition, Figure 5 presents the percentage of instances diffusion for each emotion. The bigger a circle is, the bigger diffusion exists for the according Ekman emotional state.

---

[2] Weka toolkit: http://www.cs.waikato.ac.nz/ml/weka/

Table 3. Classification of ReTweeting - Depth and Width

Classifiers	F-Measure (Depth)	F-Measure (Width)
AdaBoost	0.712	0.669
IBK	0.843	0.817
J48	**0.896**	**0.872**
JRip	0.877	**0.873**
Multilayer Perceptron	0.867	0.828
Naive Bayes Classifier	0.683	0.581
PART	0.865	0.862
RotationForest	**0.891**	0.868
SMO	0.687	0.669

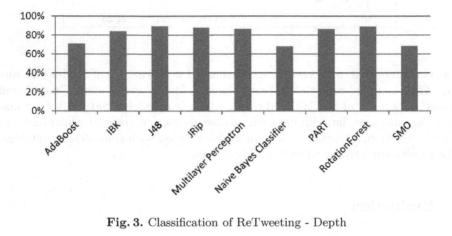

Fig. 3. Classification of ReTweeting - Depth

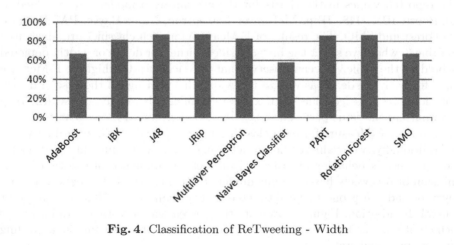

Fig. 4. Classification of ReTweeting - Width

Table 4. Number of Instances per Emotional State

Emotion	Number of Instances	Rate of ReTweeting
Anger	212	78%
Disgust	545	12%
Fear	64	29%
Happiness	344	31%
Sadness	232	65%
Surprise	142	17%
Neutral	631	24%

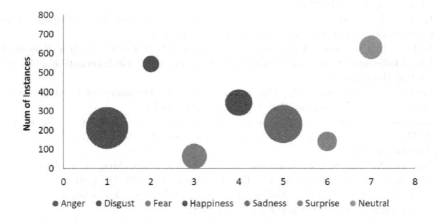

Fig. 5. Rate of ReTweeting

6   Conclusions and Future Work

In our work we present a methodology to model and predict the depth and width of diffusion of a tweet with a machine learning approach. More specifically, based on the result of the Ekman model (e.g. six basic human emotions represented by six emotional states) as well the Twitter analytics features, we can estimate the depth's and width's category of each ReTweet with a classification error; as we have already mentioned, our aim is to investigate the impact that information has in the network in the form of ReTweets. We also examine the influence of emotion in the diffusion, where we find out that posts containing negative emotional states are more likely to be retweeted than tweets with positive or neutral emotional states.

As future work, we plan to incorporate apart from the existence of any basic Ekman's state, also the valence of each emotional state. Next, we intend to make larger scale evaluation and utilize popular Twitter corpuses in order to get a deeper insight of our methodology's performance. Ultimately, we could take into consideration special tweet characteristics such as Hashtag Keywords and URLs in order to improve classification accuracy.

References

1. Bakshy, E., Rosenn, I., Marlow, C., Adamic, L.A.: The Role of Social Networks in Information Diffusion. In: WWW, pp. 519–528 (2012)
2. Bollen, J., Mao, H., Pepe, A.: Modeling Public Mood and Emotion: Twitter Sentiment and Socio-Economic Phenomena. In: ICWSM (2011)
3. Can, E.F., Oktay, H., Manmatha, R.: Predicting ReTweet Count Using Visual Cues. In: CIKM, pp. 1481–1484 (2013)
4. Ekman, P.: Basic Emotions. Handbook of Cognition and Emotion. John Wiley and Sons, Ltd. pp. 45-60 (1999)
5. Goel, S., Watts, D.J., Goldstein, D.G.: The Structure of Online Diffusion Networks. In: EC, pp. 623–638 (2012)
6. Kafeza, E., Kanavos, A., Makris, C., Chiu, D.: Identifying Personality-based Communities in Social Networks. LSAWM, ER (2013)
7. Kafeza, E., Kanavos, A., Makris, C., Vikatos, P.: T-PICE: Twitter Personality based Influential Communities Extraction System. In: IEEE International Congress on Big Data (2014)
8. Leskovec, J., Adamic, L.A., Huberman, B.A.: The Dynamics of Viral Marketing. ACM Transactions on the Web (TWEB) 1(1) (2007)
9. Naveed, N., Gottron, T., Kunegis, J., Alhadi, A.C.: Bad News Travel Fast: A Content-based Analysis of Interestingness on Twitter. Web Science, Article Number 8 (2011)
10. Naveed, N., Gottron, T., Kunegis, J., Alhadi, A.C.: Searching Microblogs: Coping with Sparsity and Document Quality. In: CIKM, pp. 183–188 (2011)
11. Pang, B., Lee, L.: Opinion Mining and Sentiment Analysis. Foundations and Trends in Information Retrieval 2(1-2), 1–135 (2008)
12. Peng, H.-K., Zhu, J., Piao, D., Yan, R., Zhang, Y.: Retweet Modeling Using Conditional Random Fields. In: ICDM Workshops, pp. 336–343 (2011)
13. Perikos, I., Hatzilygeroudis, I.: Recognizing Emotion Presence in Natural Language Sentences. In: Iliadis, L., Papadopoulos, H., Jayne, C. (eds.) EANN 2013, Part II. Communications in Computer and Information Science, vol. 384, pp. 30–39. Springer, Heidelberg (2013)
14. Petrovic, S., Osborne, M., Lavrenko, V.: RT to Win! Predicting Message Propagation in Twitter. In: ICWSM (2011)
15. Roberts, K., Roach, M.A., Johnson, J., Guthrie, J., Harabagiu, S.M.: EmpaTweet: Annotating and Detecting Emotions on Twitter. In: LREC, pp. 3806–3813 (2012)
16. Strapparava, C., Valitutti, A.: WordNet Affect: an Affective Extension of WordNet. In: LREC, pp. 1083–1086 (2004)
17. Suttles, J., Ide, N.: Distant supervision for emotion classification with discrete binary values. In: Gelbukh, A. (ed.) CICLing 2013, Part II. LNCS, vol. 7817, pp. 121–136. Springer, Heidelberg (2013)

A Partially-Observable Markov Decision Process for Dealing with Dynamically Changing Environments

Sotirios P. Chatzis[1] and Dimitrios Kosmopoulos[2]

[1] Department of Electrical Eng., Computer Eng., and Informatics
Cyprus University of Technology, Cyprus
[2] Department of Informatics Engineering, TEI Crete, Greece

Abstract. Partially Observable Markov Decision Processes (POMDPs) have been met with great success in planning domains where agents must balance actions that provide knowledge and actions that provide reward. Recently, nonparametric Bayesian methods have been successfully applied to POMDPs to obviate the need of a priori knowledge of the size of the state space, allowing to assume that the number of visited states may grow as the agent explores its environment. These approaches rely on the assumption that the agent's environment remains stationary; however, in real-world scenarios the environment may change over time. In this work, we aim to address this inadequacy by introducing a dynamic nonparametric Bayesian POMDP model that both allows for automatic inference of the (distributional) representations of POMDP states, and for capturing non-stationarity in the modeled environments. Formulation of our method is based on imposition of a suitable dynamic hierarchical Dirichlet process (dHDP) prior over state transitions. We derive efficient algorithms for model inference and action planning and evaluate it on several benchmark tasks.

1 Introduction

Reinforcement learning in partially observable domains is a challenging and attractive research area in machine learning. One of the most common representations used for partially-observable reinforcement learning is the partially observable Markov decision process (POMDP). POMDPs are statistical models postulating that emission of the observation o_t that an agent receives from the environment at time t follows a distribution $\Omega(o_t|s_t, a_t)$ that depends on the value of some latent (hidden) world-state s_t, and the agent's most recent action a_t. In addition, each action a_t of the agent results in a reward $R(s_t, a_t)$ emitted from the environment, the value of which also depends on the current state s_t, and induces a change in the latent state of the environment, which transitions to a new state s_{t+1}, drawn from a transition distribution $T(s_{t+1}|s_t, a_t)$. Due to the generic nature of their assumptions, POMDPs have been successfully applied to a large number of reinforcement learning scenarios with great success [13].

L. Iliadis et al. (Eds.): AIAI 2014, IFIP AICT 436, pp. 111–120, 2014.

A significant drawback of POMDPs is the large number of parameters entailed from the postulated emission distribution models $\Omega(o|s,a)$, state transition distribution models $T(s'|s,a)$, and reward models $R(s,a)$. These parameters must be learned using data obtained through interaction of the agent with its environment, in an online fashion. However, the combination of the very limited availability of training data with the large number of parameters of the fitted models typically results in highly uncertain trained models, where planning becomes extremely computationally cumbersome. Bayesian reinforcement learning approaches [9,7,10] resolve these issues by accounting for both uncertainty in the agent's model of the environment, and uncertainty within the environment itself. This is effected by maintaining distributions over both the parameters of the POMDP and the latent states of the world s.

A drawback of most Bayesian approaches is their requirement of *a priori* provision of the number of model states: even if the size of the state-space is actually known (which is seldom the case), performing learning for a large number of unknown model parameters from the beginning of the learning process (when no data is actually available) might result in poor model estimates and heavy overfitting proneness. Recently, [3] proposed leveraging the strengths of Bayesian nonparametrics, specifically hierarchical Dirichlet process (HDP) priors [15], to resolve these issues. The so-obtained infinite POMDP (iPOMDP) postulates an infinite number of states, conceived as abstract entities whose sole function is to render the dynamics of the system Markovian, instead of actual physical aspects of the system. Despite the assumption of infinite model states though, at each iteration of the model learning algorithm, only a small number of (actually visited) effective states need to be instantiated with parameters. As such, the model can be initialized with only few parameters, which allows for effectively dealing with overfitting. However, as the agent accumulates experiences (i.e., training data) through interaction with its environment, the number of parameters may increase in an automatic, unsupervised manner, by exploiting the nonparametric nature of the model.

Despite these advances, a significant drawback of existing nonparametric Bayesian formulations of POMDPs consists in their lack of appropriate mechanisms allowing for capturing non-stationarity in the modeled environments, expressed in the form of time-adaptive underlying state transition distributions. Indeed, the problem of capturing time-varying underlying distributions in conventional POMDP model formulations has been considered by various researchers in the recent literature (e.g., [16,5]). In this work, we address this inadequacy by introducing a non-stationary variant of the iPOMDP. Formulation of our model is based on imposition of the dynamic hierarchical Dirichlet process (dHDP) prior [8] over the postulated state transitions in the context of our model. We derive efficient model inference and action planning algorithms, based on a computationally scalable combination of Gibbs sampling interleaved with importance sampling techniques. We evaluate the efficacy of our approach considering a set of well-known benchmark tasks, dealing with partially observable environments.

The remainder of this paper is organized as follows: In Section 2, we introduce our proposed model, and derive its learning and action selection algorithms. In Section 3, we provide the experimental evaluation of our approach, and compare it to state-of-the-art alternatives. Finally, in the last section we summarize our results and conclude this paper.

2 Proposed Approach

2.1 Motivation

The iPOMDP model is based on utilization of an HDP prior to describe the state transition dynamics in the modeled environments. The HDP is a model that allows for linking a set of group-specific Dirichlet processes, learning the model components jointly across multiple groups. Specifically, let us assume C latent model states, and A possible actions; let us consider that each possible state-action pair (s, a) defines a different scenario in the environment. The iPOMDP model, being an HDP-based model, postulates that the new state of the environment (after an action is taken) is drawn from a distribution with different parameters $\boldsymbol{\theta}^{sa}$, which are in turn drawn from a scenario-specific Dirichlet process. In addition, the base distribution of the scenario-specific Dirichlet processes is taken as a common underlying Dirichlet process. Under this construction, the following generative model is obtained

$$s'|s, a \sim T(\boldsymbol{\theta}^{s,a}) \tag{1}$$

$$\boldsymbol{\theta}^{s,a} \sim G_{s,a} \tag{2}$$

$$G_{s,a} \sim \mathrm{DP}(\alpha, G_0) \tag{3}$$

$$G_0 \sim \mathrm{DP}(\gamma, H) \tag{4}$$

As we observe, in the context of the HDP, different state transitions that refer to the same state-action pair (scenario) share the same parameters (atoms) that comprise $G_{s,a}$. In addition, transitions might also share parameters (atoms) across different state-action pairs, probably with different mixing probabilities for each $G_{s,a}$; this is a consequence of the fact that the Dirichlet processes $G_{s,a}$ pertaining to all the modeled state-action pairs share a common base measure G_0, which is also a discrete distribution.

Although the HDP introduces a dependency structure over the modeled scenarios, it does not account for the fact that, when it comes to modeling of sequential data, especially data the distribution of which changes over time, sharing of underlying atoms from the Dirichlet processes is more probable in proximal time points. Recently, [8] developed a dynamic variant of the HDP that allows for such a modeling capacity, namely the dynamic HDP (dHDP). Therefore, utilization of this prior emerges as a promising solution for us to effect our goals.

2.2   Model Formulation

To introduce our model, we have to provide our prior assumptions regarding the state transition distributions, observation emission distributions, and reward emission distributions of our model. Let us begin with the state transition distributions of our model. As we have already discussed, to capture non-stationarity, we model state transitions using the dHDP prior [8].

Let us introduce the notation $\boldsymbol{\pi}_\tau^{s,a} = (\pi_{\tau l}^{s,a})_{l=1}^\infty$. $\pi_{\tau l}^{s,a}$ denotes the (prior) probability of transitioning at some time point t to state l from state s by taking action a, given that the distributions of the various state transitions at that time point are the same as they were at time $\tau = \phi_t$. In other words, the employed dHDP assumes that the dynamics of state transition may change over time, with different time points sharing common transition dynamics patterns. Specifically, following [8], we have

$$s_{t+1} = k|s_t = s, a_t = a \sim \text{Mult}(\boldsymbol{\pi}_{\phi_t}^{s,a}) \tag{5}$$

$$\boldsymbol{\pi}_\tau^{s,a} \sim \text{DP}(\alpha, G_0) \tag{6}$$

and

$$G_0 \sim \text{DP}(\gamma, H) \tag{7}$$

whence

$$\pi_{tl}^{s,a} = \tilde{\pi}_{tl}^{s,a} \prod_{h=1}^{l-1}(1 - \tilde{\pi}_{th}^{s,a}) \tag{8}$$

$$\tilde{\pi}_{tl}^{s,a} \sim \text{Beta}(\alpha_t \beta_l, \alpha_t(1 - \sum_{m=1}^l \beta_m)) \tag{9}$$

$$\beta_k = \varpi_k \prod_{q=1}^{k-1}(1 - \varpi_q) \tag{10}$$

and

$$\varpi_k \sim \text{Beta}(1, \gamma) \tag{11}$$

In the above equations, the latent variables ϕ_t are indicators of state-transition distribution sharing over time. Following [8], their prior distributions take the form

$$\phi_t|\tilde{\boldsymbol{w}} \sim \text{Mult}(\boldsymbol{w}_t) \tag{12}$$

with $\boldsymbol{w}_t = (w_{tl})_{l=1}^t$, and

$$w_{tl} = \tilde{w}_{l-1} \prod_{m=l}^{t-1}(1 - \tilde{w}_m), \; l = 1, \dots, t \tag{13}$$

while $\tilde{w}_0 = 1$, and

$$\tilde{w}_t|a_t, b_t \sim \text{Beta}(\tilde{w}_t|a_t, b_t), \; t \geq 1 \tag{14}$$

As observed from (13), this construction induces a proximity-inclined transition dynamics sharing scheme; that is, $w_{t1} < w_{t2} < \cdots < w_{tt}$. In other words, it favors sharing the same dynamics between proximal time points, thus enforcing our assumptions of transition dynamics evolving over time in a coherent fashion.

Finally, our observation emission distributions are taken in the form $\Omega(o|s,a) \sim H$, and our reward emission distributions yield $R(r|s,a) \sim H_R$. The distributions H and H_R can have any form, with the choice depending on the application at hand. In this paper, we shall be considering discrete reward and action distributions; as such, a suitable conjugate selection for the priors over their parameters is the Dirichlet distribution.

This concludes the formulation of our model. We dub our model the infinite dynamic POMDP (iDPOMDP) model. Our model is a completely non-stationary POMDP model, formulated under the assumption of an infinite space of latent POMDP states, and treated under the Bayesian inference paradigm. Note also that limiting the generative non-stationarity assumptions to the transition functions of our model does not limit non-stationarity *per se* to state transitions. Indeed, the non-identifiability of the postulated model latent states results in the assumed generative non-stationarity of the transition functions being implicitly extended to the observation and reward functions of our model.

2.3   Inference Algorithm

To efficiently perform inference for our model, we combine alternative application of a variant of the block Gibbs sampler of [4], and importance sampling [14], in a fashion similar to the iPOMDP model [3]. Our block Gibbs sampler allows for drawing samples from the true posterior. However, we limit ourselves to using our block Gibbs sampler only on a periodical basis, and not at each time point. In the meanwhile, we use instead an importance sampling algorithm, which merely reweighs the already drawn samples so as to reflect the current posterior as closely as possible. This way, we obtain a significant speedup of our inference algorithm, without compromising model accuracy, since the actual model posterior is not expected to undergo large changes over short time windows.

Block Gibbs Sampler. To make inference tractable, we use a truncated expression of the stick-breaking representation of the underlying shared Dirichlet process of our model, G_0 [12]. In other words, we set a truncation threshold C, and consider $\boldsymbol{\pi}_t^{s,a} = (\pi_{tl}^{s,a})_{l=1}^C$, $\forall t, s, a$ [4]. A large value of C allows for obtaining a good approximation of the infinite underlying process, since in practice the $\pi_{tl}^{s,a}$ are expected to diminish quickly with increasing l, $\forall t$ [4]. Note also that, as discussed in [8], drawing one sample from the dHDP model by means of the block Gibbs sampler takes similar time as drawing one sample from HDP.

Let us consider a time horizon T steps long. We have

$$p(\tilde{w}_t|\ldots) = \text{Beta}(\tilde{w}_t|a + \sum_{j=t+1}^{T} n_{j,t+1}, b + \sum_{j=t+1}^{T} \sum_{h=1}^{t} n_{jh}) \qquad (15)$$

where n_{th} is the number of time points such that $\phi_t = h$. Similar,

$$p(\tilde{\pi}_{tl}^{s,a}|\dots) = \text{Beta}\bigg(\tilde{\pi}_{tl}^{s,a}|\alpha_t \beta_l + \sum_{j=1}^{T} \mathbb{I}(n_{jt} \neq 0)\mathbb{I}(\nu_{jl}^{s,a} \neq 0),$$

$$\alpha_t(1 - \sum_{m=1}^{l} \beta_m) + \sum_{k=l+1}^{C} \sum_{j=1}^{T} \mathbb{I}(n_{jt} \neq 0)\mathbb{I}(\nu_{jk}^{s,a} \neq 0)\bigg) \tag{16}$$

where $\nu_{tk}^{s,a}$ is the number of training episodes where we had a transition from state s to state k, by taking action a at time t.

The updates of the set of indicator variables ϕ_t can be obtained by generating samples from multinomial distributions with entries of the form

$$p(\phi_t = \tau|s_{t-1} = s, a_{t-1} = a; \dots) \propto \tilde{w}_{\tau-1} \prod_{m=\tau}^{t-1} (1 - \tilde{w}_m)\tilde{\pi}_{\tau s_t}^{s,a} \prod_{q=1}^{s_t-1} (1 - \tilde{\pi}_{\tau q}^{s,a})$$

$$\times p(o_{t+1}|s_t, a_t)\, p(r_{t+1}|s_t, a_t), \quad \tau = 1, \dots, t \tag{17}$$

Further, the posterior distribution over the latent model states yields

$$p(s_t = k|s_{t-1} = s, a_{t-1} = a; \dots) \propto \tilde{\pi}_{\phi_t k}^{s,a} \prod_{q=1}^{k-1} (1 - \tilde{\pi}_{\phi_t q}^{s,a})p(o_{t+1}|s_t, a_t)\, p(r_{t+1}|s_t, a_t) \tag{18}$$

As we observe, this expression entails Markovian dynamics. Thus, to sample from it, we have to resort to some method suitable for distributions with temporal interdependencies. In our work, we employ the forward filtering-backward sampling (FFBS) algorithm [1]; this way, we can efficiently obtain samples of the underlying latent state sequences.

Finally, the observation and reward distribution parameters of our model are sampled in a manner similar to the original iPOMDP model [3].

Importance Sampling. At time points when we substitute block Gibbs sampling from the true posterior with importance sampling, we essentially reweigh the samples previously drawn from the true posterior. Initially, all samples have equal weight as they are drawn from the true posterior; this changes when we apply importance sampling, so as to capture small changes in the actual posterior in a computationally efficient manner (possible within short time-windows).

Let us denote as μ a sample of our model with weight $w_t(\mu)$ at time t (all samples have initial weights equal to one). Similar to the iPOMDP model, the weight update at time $t + 1$ yields [3]

$$w_{t+1}(\mu) \propto w_t(\mu) \sum_{\forall s_t} \Omega(o_{t+1}|s_t, a_t)b_\mu(s_t) \tag{19}$$

where $b_\mu(s)$ is the belief (posterior probability) for state s, as determined in the sample μ of the model.

2.4   Action Selection

Once we have obtained a set of samples from the posterior distribution of our model, we can use them to perform action selection. For this purpose, in this work we apply *stochastic forward search in the model-space*, as proposed in [3]. The main concept of forward search is to use a forward-looking tree to compute action-values [11]. Starting from the current posterior (belief) over the model parameters of the agent, the tree branches on each action the agent might take and each observation the agent might see. At each action node, the agent computes the (posterior) expectation of the immediate reward, given the drawn samples, in a standard Monte Carlo-type fashion.

3   Experimental Evaluation

We evaluate our method in several benchmark scenarios and compare its performance to related alternatives, namely Medusa [5] and iPOMDP. Medusa is provided with the true number of states, while iPOMDP determines it automatically, similar to our approach. In all experiments, the iPOMDP and iDPOMDP are initialized in such a way that the observations are given vague hyperparameters, while the rewards are given hyperparameters that encourage peaked distributions. Beliefs are approximated with sample sets of 10 models. In all cases, tests have 200 episodes of learning, which interleave acting and resampling models, and 100 episodes of testing with the models fixed.

The first benchmark scenario considered here, namely Tiger-3, is adopted from [3]; it comprises an environment that changes over time, thus allowing for us to evaluate the capacity of our model to adapt to new situations. Specifically, Tiger-3 is an adaptation of the well-known Tiger benchmark [6]: The agent has to choose one of three doors to open. Two doors have tigers behind them (resulting in the maximum possible penalty, $r = -100$), and one door has a small actual reward ($r = 10$). At each time step, the agent may either open a door or listen for the "quiet" door. It hears the correct door correctly with probability 0.85. The reward is unlikely to be behind the third door ($p = 0.2$). However, during the first 100 episodes, we artificially ensure that the reward is always behind doors 1 or 2. Subsequently, we employ the full probabilistic environmental model, under which the reward may also be behind the third door with some (low) probability. Hence, the dynamics of the environment change after the 100th episode, thus requiring that the trained agent be capable of quickly adapting to environmental changes.

The rest of our considered benchmarks are well-known problems in the POMDP literature. Specifically, we consider:

1. *Tiger*: This benchmark was first proposed in [6]. It is similar to Tiger-3, with the exception of the environment being static. Dynamics do not change after the first 100 episodes of the learning algorithm, but we consider the probability of the reward being behind the third door equal to $p = 0.2$ from the beginning of the simulation.

2. *Shuttle*: This benchmark was introduced in [2]. It considers two space stations separated by a small amount of free space with loading docks located on each station. The task is to transport supplies between the two docks. Each time the agent successfully attaches to the least-recently visited station, it receives a reward equal to $r = 10$. In order to dock, the agent must position itself in front of the station, with its back to the dock, and backup. Whenever the agent collides with the station by propelling forward into the dock, it receives a penalty of $r = -3$. In all other cases, we consider $r = 0$.

3. *Gridworld* [6]: This comprises a simple environment consisting of a 4-by-4 grid, where all cells except for the goal in the lower right-hand corner are indistinguishable. Movement into adjacent cells is permitted in the four compass directions, but an attempt to move off the edge of the world has no effect, returning the agent to its original state with no indication that anything unusual has happened. All states have zero reward, except for the goal (right-hand corner), which has a $r = +1$ reward, and a distinctive appearance. The initial state for this problem is a uniform distribution over all but the bottom right state. Any action taken from the bottom right state results in a transition to any one of the remaining zero reward states with equal probability (i.e., return to the initial distribution).

Our results are provided in Table 1. As we observe, our approach is capable of inferring a smaller number of states than the true count, only retaining states for which adequate information can be derived from the accrued experiences (training episodes); this is attained without any compromises in the yielded accumulated rewards in all scenarios. Given the fact that, as discussed in Section 2.3, drawing one sample from the dHDP by means of the block Gibbs sampler takes similar time as drawing one sample from the HDP, we deduce that our approach allows for obtaining improved total reward compared to the iPOMDP for decreased model complexity and resulting computational costs. Note also that the obtained performance improvement is more prominent in the case of the Tiger-3 problem, where the environment changes over time, thus posing greater learning challenges to the postulated agents. This finding vouches for the capacity of our model to capture non-stationarities in the modeled environments, which is the ultimate goal of this work.

Table 1. Experimental Evaluation: Number of inferred states and total obtained reward

	#States			Total Reward		
Problem	Actual	iPOMDP	iDPOMDP	Medusa	iPOMDP	iDPOMDP
Tiger-3	4	4.1	3.8	-40.26	-42.07	-35.19
Tiger	2	2.1	2.1	0.83	4.06	4.64
Shuttle	8	2.1	2.1	10	10	10
Gridworld	26	7.36	6.82	-49	-13	-12

Table 2. Experimental Evaluation: Execution times (in seconds) of the iPOMDP and iDPOMDP models. For iDPOMDP we evaluate both the proposed inference algorithm as well as applying (exact) Gibbs sampling.

Problem	iPOMDP	iDPOMDP	iDPOMDP-Gibbs
Tiger-3	842.91	913.88	5863.18
Tiger	837.67	912.07	5738.55
Shuttle	839.11	894.32	5707.41
Gridworld	1108.31	1207.15	8094.17

Further, in the first three columns of Table 2, we compare the execution times of our approach and the baseline iPOMDP method[1]. As we observe, our method imposes a modest computational overhead, which averages at less than a 10% increase in computational costs compared to iPOMDP. Based on this result, and taking into consideration the better performance of our algorithm (in terms of total accrued reward), we can claim that our approach yields a favorable trade-off between performance and computational complexity compared to the competition.

Finally, in the last column of Table 2, we show how the proposed inference algorithm, based on an appropriate combination of Gibbs sampling with importance sampling, compares to the more straightforward approach of solely relying on Gibbs sampling to perform inference. As we observe, solely relying on Gibbs sampling to perform inference for our model results in an overwhelming increase in computational costs, which averages at almost one order of magnitude more time. Therefore, we deduce that our strategy of performing inference utilizing a combination of Gibbs sampling interleaved with importance sampling succeeds in considerably accelerating our algorithm, and is crucial for the practical applicability of our method.

4   Conclusions and Future Work

In this paper, we proposed a nonparametric Bayesian formulation of POMDPs that addressed the problem of capturing non-stationarities in the modeled environments. Formulation of our model was based on the imposition of a suitable dynamic prior over the state transitions of our model, namely the dHDP prior. We devised efficient learning and planning algorithms for our model, based on a combination of block Gibbs sampling and importance sampling. We showed that our method outperforms related alternatives, namely Medusa and iPOMDP, in several benchmark tasks, combining increased reward performance with shorter model sizes, and, hence, better computational complexity.

---

[1] Our experiments were executed on a Linux desktop computer, with an Intel Xeon E5 2.10GHz processor. All the considered algorithms were implemented as single-threaded Python applications.

Our future goals in this line of research concern applying our approach to autonomous path planning for aerial robots (drones) operating in unknown, unstructured environments. We are also interested in extending our non-stationary models to address problems dealing with continuous state and actions spaces (continuous POMDPs).

Acknowledgments. This work was implemented under the Operational Program "Education and Lifelong Learning" action Archimedes III, co-financed by the European Union (European Social Fund) and Greek national funds (National Strategic Reference Framework 2007 - 2013).

References

1. Carter, C.K., Kohn, R.: On Gibbs sampling for state space models. Biometrika 81, 541–553 (1994)
2. Chrisman, L.: Reinforcement learning with perceptual aliasing: The perceptual distinctions approach. In: Proc. AAAI. pp. 183–188 (1992)
3. Doshi-Velez, F.: The infinite partially observable Markov decision process. In: Proc. NIPS (2009)
4. Ishwaran, H., James, L.F.: Gibbs sampling methods for stick-breaking priors. Journal of the American Statistical Association 96, 161–173 (2001)
5. Jaulmes, R., Pineau, J., Precup, D.: Learning in non-stationary Partially Observable Markov Decision Processes. In: ECML Workshop on Reinforcement Learning in Non-Stationary Environments (2005)
6. Littman, M.L., Cassandra, A.R., Kaelbling, L.P.: Learning policies for partially observable environments: scaling up. In: Proc. ICML (1995)
7. Poupart, P., Vlassis, N., Hoey, J., Regan, K.: An analytic solution to discrete Bayesian reinforcement learning. In: Proc. ICML. pp. 697–704 (2006)
8. Ren, L., Carin, L., Dunson, D.B.: The dynamic hierarchical Dirichlet process. In: Proc. International Conference on Machine Learning (ICML) (2008)
9. Ross, S., Chaib-draa, B., Pineau, J.: Bayes-adaptive POMDPs. In: Proc. NIPS (2008)
10. Ross, S., Chaib-draa, B., Pineau, J.: Bayesian reinforcement learning in continuous POMDPs with application to robot navigation. In: Proc. ICRA (2008)
11. Ross, S., Pineau, J., Paquet, S., Chaib-Draa, B.: Online planning algorithms for pomdps. Journal of Artificial Intelligence Research 32, 663–704 (2008)
12. Sethuraman, J.: A constructive definition of the Dirichlet prior. Statistica Sinica 2, 639–650 (1994)
13. Shani, G., Pineau, J., Kaplow, R.: A survey of point-based POMDP solvers. Auton Agent Multi-Agent Syst 27(1), 1–51 (2012)
14. Siegmund, D.: Importance sampling in the Monte Carlo study of sequential tests. The Annals of Statistics 4, 673–684 (1976)
15. Teh, Y.W., Jordan, M.I., Beal, M.J., Blei, D.M.: Hierarchical Dirichlet processes. Journal of the American Statistical Association 101, 1566–1581 (2006)
16. Theocharous, G., Kaelbling, L.P.: Approximate planning in POMDPs with macro-actions. In: Proc. NIPS (2003)

Platform for Simulation and Improvement of Swarm Behavior in Changing Environments

Sergi Canyameres and Doina Logofătu

Computer Science Department of Frankfurt University of Applied Sciences 1 Nibelungenplatz 60318, Frankfurt am Main, Germany
logofatu@fb2.fh-frankfurt.de

Abstract. Simulation of particle movement plays an important role when understanding the behavior of natural entities, which can be adopted by many other research fields. The complexity of every environment can cause lots of difficulties to ensure accurate und faithful results, leading to extensive researches but always in very specific conditions. This paper describes a framework which allows wide flexibility on the moving algorithms of the swarm particles as well as an easy modification of the surrounding constrictions. A simple, intuitive interface is implemented and a swarm paradigm is already used. The aim of this tool is to obtain acceptable results in any upcoming test case without the need to run an exhaustive and expensive research.

1   Introduction and Motivation

This paper explains the natural extension of our previous participation in the InformatiCUP challenge [1], a German competition arranged by the "Gesellschaft für Informatik" [2], an organization promoting the transmission of computer science knowledge and research. The goal in this 2014 edition [3] was to implement an ideal simulation of how a set of manganese-collecting machines had to move above a water surface without remote help on how to determine the other's position and direction. Only a small area is reached by radar that provides them information about the relative position of the nearest bots within this range. Although the purpose was to collect as much manganese as possible, the heuristics to determine which factors were to be taken into consideration were not clear. This brought to different paradigm possibilities to develop, and made us think of the need of a general purpose application instead of the strictly constricted projects existing so far. Hence, after focusing on the idea of allowing changing specific requirements, we started to work on the versatile framework that is presented in this paper.

This paper is the basis of a platform which breaks the tendency of thoroughly focusing on carefully fixed environments as done so far [4][5][6]. Here several solutions for different contexts related to the mentioned contest are purposed, even though any other cases are meant to be adopted. The robots can start in several formations, which may be a key fact when determining the movement patterns to be followed. Afterwards they start to recalculate their velocity and direction following one of the most extended swarm paradigms, studied by C. W. Reynolds [7] and consisting of three main rules. During the

L. Iliadis et al. (Eds.): AIAI 2014, IFIP AICT 436, pp. 121–129, 2014.
© IFIP International Federation for Information Processing 2014

simulation any algorithm may be activated or deactivated as well as have their inner parameters modified. Different final events can be handled to conclude the simulation in different ways depending on previously established criteria.

In many occasions it is necessary to know a first approach of the simulation's results, even when the first drafts are not yet designed. For new research fields, gaming industry or any nature-inspired applications it can be interesting to have a tool to find good preview results for simulation and optimization in a dynamic and low cost system without the need to build full applications for that purpose. Many improvements and complements are likely to be developed in the near future, such as distributed computing, more accurate methods for the optimization of the parameters, real-time comparison between two simulations, etc.

Section 2 thoroughly describes the implementation of the main blocks of the application, from the structure of the program until the function to search for the best parameters. Section 3 proposes the possibility of the further parallelization of the optimization function. In section 4 we show and explain some of the first results. Finally, an evaluation of the work done as well as a brief proposal for future research can be found in section 5.

2   Simulation and Optimization Functionalities

Each one of our units is an instance thinking independently and interacting with the simulator class just enough to represent real-life perception of the individual. A coordinate's class is implemented in such a way that all position- and velocity-related data is embedded and can be easily transferred and recalculated. The central processing class organizes the tasks and handles the list of robots, coordinates, parameters and rules to follow, but are the single instances the responsible for taking their own decisions with the limited information provided. The surface where the robots are going to move should be unlimited, but for a comfortable visualization in the GUI this area is constricted to a canvas of 500 by 500 positions or pixels (in this case, representing square meters).

2.1   Deployments

The starting positions of the robots determine the first decisions to take and may condition the overall algorithms to follow. If they start very close to each other, they need to spread in order to cover more area. On the other hand, if they are too distant, many areas may remain uncovered or clearing them would be too inefficient.   In the contest, the initial methodology to distribute the robots was never specified and each simulation uses a different starting set. Therefore, it is important to have the chance to test a same moving paradigm after different deployments. For this, five starting configurations are now implemented, following different regular patterns (square, circular) or mathematic distributions (random, Gaussian and double Gaussian).

The **Random** (Fig. 1.5) deployment drops every robot in a completely random position within the sea limits established in order to focus the simulation in a constricted area. In further simulations much larger limits may be used, taking the risk of dropping robots too far away from the rest and becoming isolated. An interesting line of research could study which algorithms should isolated robots follow. However, a purely random starting set does not bring many possibilities to study constricted behaviors, so more complex irregular deployments are developed.

The **Gaussian** (Fig. 1.1) and *Double Gaussian* (Fig. 1.3) configurations are used to guarantee a certain pattern within the randomness of a non-uniform distribution. With them we can know *a priori* how are the robots going to be distributed, even not their exact coordinates or the distance between them. Obviously, parameters such as the deviation can be manipulated in the every simulating case. In the case of the *Double Gaussian*, this is the name that an evolved deployment received. It consists of two Gaussian deployments, which are independent on number of items and deviation. Both means are calculated in a way that some robots from one kernel can perceive some of the other one, but never all of them. With this, we try to see the strength limits of the algorithms, as both Gaussians will tend to merge and join their own nucleus but at the same time some robots may push towards the other group.

Finally, regular patterns are also implemented to test uniform movements or simulate situations where the initial set can be regular. In this first version, uniformly fulfilled **Square** (Fig. 1.2) and a **Circular** (Fig. 1.4) patterns are available. Due to the variable amount of bots, the shape might slightly vary to try to keep the regularity of the distribution. The square will become a rectangle to ensure equal distance between all the boids, whereas the circle will deploy equally-distant robots on the circumference until this is full and a new ring can be added.

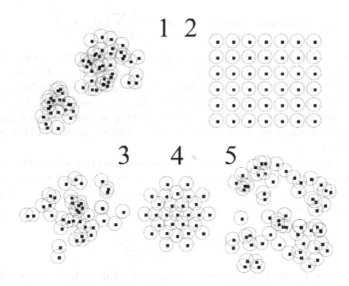

Fig. 1. Captions of the robots with the GUI representation of the different deployment methods: Random (1), Square (2), Gaussian (3), Circular (4) and double Gaussian (5).

2.2   Swarm Behavior: Equations and Weights

The limited communication capacity between the robots makes it too complex to determine the shape or the positions of all the other robots (a strong communicating network should be established) and it would escape from realistic situations. Nature-inspired algorithms for ants [8] or bees [9] are not based on sharing such big amount of data, but on generating a collective behavior out of the minor parts of information shared by each individual [10].

The moving paradigm chosen always follows a same overall concept [Fig.2][7] but at the same time we introduced some regulators which make the decisions experience small variations in order to adapt to the specific situation of each individual within the group. Every single robot analyzes the relative position p_i of all its n surrounding bots and calculates a variation of speed and direction after a weighted average of the values obtained through the main following rules:

$$cohesion(surrounding \quad 1...n) = \frac{1}{n}\sum_{i=1}^{n}(p_{ix}, p_{iy}) \tag{1}$$

Formula (1) calculates the average position of the nearby robots and tends to move there to follow the principle of cohesion and avoid the group spreading around or a robot to travel too far away, which may result in losing contact due to its limited view. After testing, this basic concept was not acting enough as a real-like flow, so a small correction was made to empower the weight of this algorithm when the boid is very far away from the others, and to decrease it when it is too close. To avoid direct contact between the elements, the separation algorithm (2) is also introduced in the final calculation of the new movement.

$$separation(surrounding \quad 1...n) = -\frac{1}{n}\sum_{i=1}^{n}(p_{ix}, p_{iy}) \tag{2}$$

In this case, the algorithm finds the opposite value of the cohesion formula. However, only when the object is very close to the others is when this algorithm should be more important. For this reason, the average point is calculated among the robots positioned in a smaller range than the viewing area used normally. In a similar way as in the first algorithm, this final value is not proportional with distance and gets strongly increased when the robot is too close to others, so it would avoid a final crashing between boids.

Finally, a third formula (3) computes the average direction where the neighboring robots are moving to. This alignment offers a smoother and more realistic movement and provides a non-static understanding of the environment, as it uses the actual velocity instead of the positions.

$$alignment(velocities \quad 1...n) = \frac{1}{n}\sum_{i=1}^{n}(v_{ix}, v_{iy}) \tag{3}$$

The modularity of these implementations allows a full addition or modification of the algorithms to follow. During each iteration of the simulator, which could be understood as a second or as any other time unit, all the robots obtain the list of neighboring boids as well as the algorithms to handle. With this, they decide which algorithms and with weights use, independently from the simulator class. Only once they all have finished the calculation of their decision, is when the actual movement takes place. This avoids that the last robots in the list take decisions with later information about already updated positions, which would not be realistic and may cause synchronization problems as well as the possibility of looped situations.

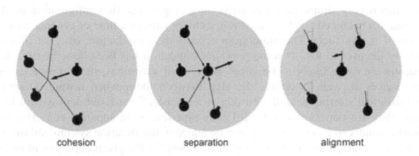

Fig. 2. Visual representation of the principles of cohesion (1), separation (2) and alignment (3) described by Craig W. Reynolds [7]

After the calculation of the final movement, a limiter controls an excessive value of the speed and reduces it to a established maximum, which can be common for all boids or specified for each individual.

Bigger dynamic paradigm modifications are also necessary for the proper simulation of realistic environments. Although the (1) and (2) criteria have small variations for specific situations, sometimes these are not enough and a new value for the algorithm's weight or even a completely new behavior must be done. In this version of the application, a practical modification of the weights allows the creation of the *Recollection* command, which consists on make all the robots meet in a central point. The original instructions from the challenge asked the boids to be collected at a certain point by a mother ship. In any case, the goal was to make them gather, even no optimization method was asked again. A good choice would be to reduce this time to the minimum by meeting them in the central point between the extreme boids. However, our choice was to optimize the fuel consumption so the parameter minimized was the total track covered by all the robots until they reach this point. For this, a signal can command the separation (2) and alignment (3) weights to become 0, so the full decision of the movement relies on the cohesion algorithm (1) and the boids are set to move at full speed. This is represents a realistic situation where the robots are called from the central control of the mother ship, or, with a set of timers, they have the command to gather at a certain execution moment, or after a threshold of inactivity due to the balance of all the robots when they are all close to each other.

2.3   Optimization Function

The aim of this research is to develop a tool which offers different possibilities of simulation, measurement, understanding and comparison, but maybe the most interesting utility is the optimization module. This function can still be improved by using more sophisticated methods such as gradient ascents, neural networks, etc. but the current version already offers a good approximation of the best parameters to use in order to reach a specified goal. Manganese collected, total or unique track covered or time to gather could be some of the goals in our goal to continue with the InformatiCUP context. The goal pursued in all the tests shown in section 4 is the total amount of manganese collected by all the robots combined.

The idea is very simple and it consists on testing a specific configuration until the robots reach a balanced position, or if not, until they reach a time or distance travelled limit (as if it was fuel). The simulation accepts the introduction of a set of values chosen by the user to test, if some previous stipulation has been done because those parameters are more likely to be good. With small variations, the iteration can skip useless values as well as reduce the step between them when testing. Despite of having different starting sets, the current version asks for a fixed deployment to test with, as they are considered fully different experiments in separate contexts. Many different amounts of robots, instead, can be added in the iteration system and different weights can be tested on different number of robots for the given deployment method. In case of choosing non-uniform deployments, the positions of the robots are calculated once and saved for further simulations, in order to guarantee that the results correspond to equal testing conditions.

At the end of each simulation a .txt is generated with a header indicating which range of configurations was used, tested number of robots, and the values which offer the best result obtained. With the purpose of a better understanding of the behavior seen during the variation of the values, a list of all the iterations and the desired results for each combination is now included. A Matlab script processes these outputs and prints a 3D representation of the results, which would be too complex to understand without data processing to summarize them. This graphic can only use two of the parameters simultaneously, as the third axe is always representing the target value. Fortunately, the algorithms used so far are conceptually easy so it is possible to understand the results with the use of two graphs. Further applications will need new solutions for the improvement of these graphs and the output interpretation, but it depends on the algorithms used so no further preparation could be done now.

3   Parallel Computing with MapReduce

The longest simulations iterating over all the parameters can run for some minutes, in the most complex cases with small steps between iterations and a high number of robots, even almost an hour when using a 2GHz i5 laptop. Future applications may need more constrictions, new algorithms and even higher number of individuals, so the computation time can increase exponentially. To ensure an acceptable simulation time it is going to be essential to focus on a big improvement in this aspect.

The number of shared data and parallel applications worldwide is increasing very fast, so we would like to use some of these techniques as a solution to the excessive computation time. Apache offers an open-source implementation of the MapReduce [11] called Hadoop [12]. MapReduce splits the input in the map tasks so the data can be processed in different cores simultaneously. Afterwards, the Reduce tasks collects all the partial results, sorts them and, in our case, search the best configuration among all the simulations run. This model is a good candidate to be implemented in this kind of projects, as has already been done in similar applications [13].

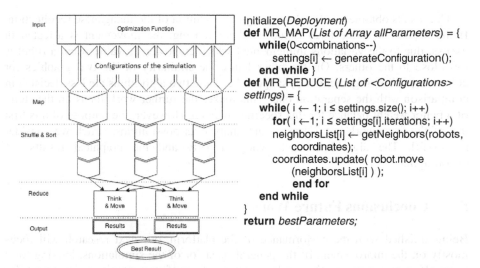

```
Initialize(Deployment)
def MR_MAP(List of Array allParameters) = {
    while(0<combinations--)
        settings[i] ← generateConfiguration();
    end while }
def MR_REDUCE (List of <Configurations>
settings) = {
    while( i ← 1; i ≤ settings.size(); i++)
        for( i ←1; i ≤ settings[i].iterations; i++)
        neighborsList[i] ← getNeighbors(robots,
            coordinates);
        coordinates.update( robot.move
            (neighborsList[i] ) );
        end for
    end while
}
return bestParameters;
```

Fig. 3. Scheme and general pseudo code for one of the next steps in this project, the parallel computation of the simulation. Both diagram and code show how the *Map* splits all the configurations' processes into different chunks which are separately calculated and reunited by the *Reduce* to find the best result.

4   First Experimental Results

The application was first tested by executing simple, understandable parameters with uniform deployments, to see how the system behaved. Balance was easily reached with circular deployment and equal cohesion (1) and separation (2) weights. Once the testing phase proved the correct performance of the simulations through the GUI interface, the first full executions were done.

As explained previously, the need for variations in the algorithms as well as the more detailed output was solved, and finally some applicable results were obtained and shown (Fig. 4) with the Matlab script described in Section 2.3.

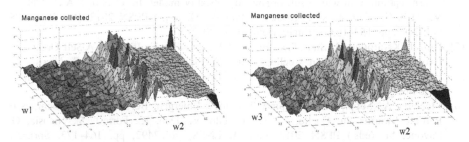

Fig. 4. Two graphics showing the total Manganese obtained (vertical axe) depending on the parameters of cohesion (w1) and separation (w2) on the left, and alignment (w3) and separation (w2) on the right. Both cases have an irregularity at the beginning of the performance.

The results obtained so far show a logical evolution of the manganese recollection. If the separation value (w2) is too small or too big, the movement is affected in excess, due to the variations implemented which empower its value when a robot is too close to the others. On the other hand, the similarity of the two graphics for cohesion (w1) or alignment (w3) suggest that these two algorithms are less decisive in comparison with the separation one. This makes the further work focus on the study of a better weight balance for this specific situation. However, the purpose of this first use of the framework is to observe and analyze its possibilities, which we consider successful. The algorithms used work properly and the graphical results are encouraging.

5   Conclusions Future Work

Being satisfied with the performance of the platform, our next research will focus mostly on the improvement of the general behavior of the simulations, looking for a more efficient execution. The implementation of new functionalities is imminent, for example the possibility of run two different simulations (with different number of robots, deployment method and/or algorithms) at the same time using the visual interface. This can help to visually compare two behaviors at real time, hopefully avoiding the need of result post-processing. Last but not least, as a mid-long term task to do, the parallelization method explained in section 3 to ensure acceptable execution times with larger and demanding simulations.

References

[1] InformatiCUP, http://informaticup.gi.de
[2] Gesellschaft für Informatik, http://www.gi.de/wir-ueber-uns.html
[3] Detailed requirements for the first prototype,
    http://informaticup.gi.de/fileadmin/redaktion/
    Informatiktage/studwett/Aufgabe_Manganernte_.pdf
[4] Fernandes, C.M., Merelo, J.J., Rosa, A.C.: Controlling the parameters of the particle swarm optimization with a self-organized criticality model. In: Coello, C.A.C., Cutello, V., Deb, K., Forrest, S., Nicosia, G., Pavone, M. (eds.) PPSN 2012, Part II. LNCS, vol. 7492, pp. 153–163. Springer, Heidelberg (2012)
[5] Bim, J., Karafotias, G., Smit, S.K., Eiben, A.E., Haasdijk, E.: It's Fate: A Self-Organising Evolutionary Algorithm. In: Coello, C.A.C., Cutello, V., Deb, K., Forrest, S., Nicosia, G., Pavone, M. (eds.) PPSN 2012, Part II. LNCS, vol. 7492, pp. 185–194. Springer, Heidelberg (2012)
[6] McNabb, A., Seppi, K.: The Apiary Topology: Emergent Behavior in Communities of Particle Swarms. In: Coello, C.A.C., Cutello, V., Deb, K., Forrest, S., Nicosia, G., Pavone, M. (eds.) PPSN 2012, Part II. LNCS, vol. 7492, pp. 164–173. Springer, Heidelberg (2012)
[7] Reynolds, C.W.: Flocks, herds, and schools: A distributed behavioral model (SIGGRAPH) (ACM), doi:10.1145/37401.37406 (1987)

[8] Moyson, F., Manderick, B.: The collective behavior of Ants: an Example of Self-Organization in Massive Parallelism. In: Actes de AAAI Spring Symposium on Parallel Models of Intelligence, Stanford, California (1988)

[9] Rodriguez, F.J., García-Martínez, C., Blum, C., Lozano, M.: An artificial bee colony algorithm for the unrelated parallel machines scheduling problem. In: Coello, C.A.C., Cutello, V., Deb, K., Forrest, S., Nicosia, G., Pavone, M. (eds.) PPSN 2012, Part II. LNCS, vol. 7492, pp. 143–152. Springer, Heidelberg (2012)

[10] Tereshko, V., Loengarov, A.: Collective Decision-Making in Honey Bee Foraging Dynamics. Journal of Computing and Information Systems 9(3), 1–7 (2005)

[11] Dean, J., Ghemawat, S.: MapReduce: simplified data processing on large clusters. Communications of the ACM 51(1), 107–113 (2008)

[12] Apache Hadoop open-source implementation, http://hadoop.apache.org/

[13] Logofătu, D., Dumitrescu, D.: Parallel Evolutionary Approach of Compaction Problem Using MapReduce. In: Schaefer, R., Cotta, C., Kołodziej, J., Rudolph, G. (eds.) PPSN XI. LNCS, vol. 6239, pp. 361–370. Springer, Heidelberg (2010)

A Neuro-memetic System for Music Composing

Jacek Mańdziuk[1], Aleksandra Woźniczko[2], and Marcin Goss[2]

[1] Warsaw University of Technology, Faculty of Mathematics and Information Science,
Koszykowa 75, 00-662 Warsaw, Poland
j.mandziuk@mini.pw.edu.pl
[2] Entrepreneurs (private sector)

Abstract. For thousands of years music has accompanied human existence and development. Over the time it turned into a form of art capable of expressing beauty, ideals and emotions. Our previous work has investigated the possibility of automatic generation of music that would be (to some extent) alike to the music created by human composers. Our focus was on romantic era music, in particular a "Chopin-style" compositions. We have proposed a specifically designed memetic algorithm, which operated based on a handful of parameters and rules, which need to be followed when it comes to classical music. In this paper we review the proposed approach and extend it by introducing "a subjective factor" to the system, in the form of a specifically designed neural network. The role of this component is to provide the subjective preferences of the listener, which are taken into account during the music composition process. Preliminary results of this new system are presented.

1 Introduction

The topic of music creation has been of research interest for about 30 years. One of the earliest experiments was the soundtrack to a game called "Ballblazer" generated algorithmically in 1984. The lead melody was assembled from a predefined set of 32 eight-note melody fragments or riffs, which were put together in a random manner. The system had to make several parameter-based choices including the speed and loudness of playing, omitting or eliding notes, or inserting a rhythmic break. The melody was accompanied by bassline, drums and chords, which were also assembled on the fly by a simplified version of the above approach. In effect the music was played forever, without repeating itself, and not straying too far from the original theme [1], [2].

Another interesting example is an interactive music system called $iMUSE$ created in 1990 and used in a number of video games. The system is able to "adjust" or "synchronize" the composed music to the game's action and to make transitions from one musical theme to another [3]. Probably the most similar system to our goal is "Emily Howell" - a computer program with an interactive interface that allows both musical and language communication. The system can be "taught" to compose music more fitting an operator's taste in the process of "encouraging and discouraging". The program uses musical pieces, previously invented by another composing program called Experiments in Musical Intelligence (EMI), as a source database for its musical choices [4], [5].

L. Iliadis et al. (Eds.): AIAI 2014, IFIP AICT 436, pp. 130–139, 2014.

More recent examples of the systems capable of or supporting artificial music composing include ArtSong [6], Symbolic Composer [7], Cybermozart [8] or Lexikon-Sonate [9].

The above-mentioned examples inspired us to develop a system capable of creating pieces of music to some extent similar to those composed by humans and at the same time being able to take into account individual musical preferences of the operator.

The rest of the paper is organized as follows. In the next section we summarize our memetic approach to automatic music creation [10] along with a selective introduction to the theory of music. Section 3 details the role and significance of the neural network component added to the system. Conclusions and directions for possible expansion of this research are placed in the last section.

2   A Memetic Approach To Music Composing

The term *music* has too broad scope to be effectively covered in a single approach. Hence, for the sake of tractability, we decided to restrict our research to the *romantic era*, as music created at that time was still following classical rules, but at the same time, the harmony was not evident. Additionally, we decided to compose pieces for the piano as this reflects the expertise of one of the authors.

Our goal was to create a system that would compose a regular piece of classical music based on user's input. In particular, we aimed at verifying the two following aspects: whether the musical rules and regularities could be flexibly applied and whether it is possible to reflect an individual taste in composed pieces.

Our studies referred to three piano music forms popular in the romantic period, i.e. *nocturne*, *polonaise* and *mazurka*. In particular our focus was on the compositions of Fryderyk Chopin, who wrote many such pieces.

2.1   The Structure of Composed Pieces / Keys / Repetitions

In order to reflect the rules of music composition and structure of musical pieces we distinguished the following elements (listed in a descending order): a part, a period, a sentence, a phrase, and a motive. Each of these elements (except for motives) is divided

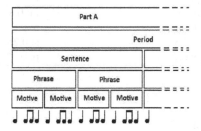

Fig. 1. Exemplary structure of a music piece

into several other smaller elements (see Fig. 1). Each motive consists of two separate lines which are a melody and an accompaniment.

Another important part of the theory of music are keys [11], [12]. Each part of a piece has its base key, but it is possible that the key changes into a temporary key, which may be different from the base key. For each of the keys there are four other related keys: same name key, parallel key, subdominant key and dominant key (see [10] for a detailed explanation). Another crucial aspect of classical music are repetitions. This issue is addressed by creating a repetition matrix defining the probability of repeating each base element within the part as well as between parts. Based on the theory of music different probability schemes were implemented for *nocturne*, *mazurka* and *polonaise*, respectively (see Table 1). In the table m, h, s and p denote motive, phrase, sequence and part, respectively. The row marked with a black square defines the probability of repeating a given element in the same part (e.g. $s = 0.55$ in column A means that on average any sentence in part A is repeated with probability 0.55). The remaining rows define probabilities of choosing a particular part as a source one (for an elements' repetition). For instance, $m = 0.2$ in row B and column C denotes the probability of repeating the a motive form part B in part C. The first column denotes the respective repetition probabilities in parts A^*, B^* and C^*, respectively, i.e. the parts which represent "variations" of the original parts. Based on the theory of music the following schemes were implemented: ABA for *nocturne*, ABA^* for *mazurka* and $ABACA$ for *polonaise*.

Table 1. One of the repetition matrices used in the experiment. See description within the text.

	[]*	A	B	C
■	-	m: 0.8 h: 0.5 s: 0.25 p:0.02	m: 0.8 h: 0.5 s: 0.25 p: 0.02	m: 0.8 h: 0.5 s: 0.25 p: 0.02
A	m: 0.7 h: 0.5 s: 0.3 p: 0.02	m: 0.9 h: 0.7 s: 0.55 p: 0.6	m: 0.2 h: 0.05 s: 0.02 p: 0	m: 0.9 h: 0.7 s: 0.3 p: 0.8
B	m: 0.9 h: 0.8 s: 0.55 p: 0.6	-	m: 0.8 h: 0.5 s: 0.3 p: 0.26	m: 0.2 h: 0.05 s: 0.02 p: 0
C	m: 0.8 h: 0.5 s: 0.3 p: 0.25	-	-	m: 0.8 h: 0.6 s: 0.25 p: 0.02

2.2   Memetic Algorithm

Our music composing system is based on memetic computing which is a combination of genetic algorithm and in-generation, local optimization. The local optimization phase serves as an additional mutation operator which does not mutate the specimen randomly, but changes it according to a set of predefined rules that represent domain knowledge. After a pre-defined number of algorithm's generations the system yields a specimen that represents a musical piece with the highest fitness value.

The algorithm receives a set of parameters as its input, part of which is a set of standard parameters of the evolutionary algorithm and the other part results from the plug-in architecture of the fitness function and local optimization. These parameters are set directly by the user and include: the number of specimens in the population, the number of generations, the percentage of the elitist specimens (transferred to the next generation without mutation), the probability of performing local optimization, the weights of particular components of the fitness function, as well as, parameters responsible for mutation (its range, probability of notes modifications, probability of chords modifications, and permutation range).

Furthermore, the algorithm uses parameters which reflect the rules of composing homophonic music: the key graph, probabilities of occurrence of particular chords in a particular key and probabilities of occurrence of particular notes in particular chords in particular keys. The final set of parameters describes the musical form, like possible structures of the piece, its meter, textures and repetition matrix.

2.3   Generation of a Population

The process of primary specimens generation is analogous to the base methods of (traditional) music composing, i.e. first harmonic structure of the piece is created and then it is filled with particular notes. The initial population consists of some pre-defined number N of specimens generated pairwise independently. In this phase of the algorithm musical rules and form parameters are exploited most intensively.

Generation of a single specimen is quite a complicated process and consists of numerous steps, as shown in [10]. On a general note, first, the biggest elements (parts and periods) of a piece are generated, followed by the smaller ones (sentences, phrases and motives).

2.4   Memetic (Local) Optimization

The goal of local optimization is to improve the quality of a specimen from musical theory point of view. Having in mind that some amount of "randomness" seems to be beneficial in artificial music composing, local optimization may happen in each generation but with some probability defined as one of the input parameters.

The first local optimization consists in adding a pitch to the existing chord which is the sixth or the seventh step of the chord's key. This operation is based on an extended version of a database of chains of chords with three or four elements used in *Harmonia* educational tool [13]. It includes chains of possible chords together with their possible modifications. If a particular chain is present in the database, it is then modified accordingly.

The second optimization operation is focused on technical ability to play the composed piece (i.e. the scale and chords location can be modified, if necessary, so as to actually allow playing the piece with two hands of a pianist).

The third operation allows improving the chords played on either a melody or an ac-companiment so as there are more consonants (concords nice to hear) than dissonances (concords strange to hear). The last local improvement operation provides a way to minimize the number of pauses appearing in the musical piece. As pauses, in general, are in line with musical rules, they are usually "over-represented" in the specimens.

2.5 Fitness Function and Selection

The selection is roulette-elite. The fitness function consists of several components which are independently evaluated and have different weights (depending on the form) configured by the user. Specifically, the following components are considered: *Chains* - evaluates the correctness of chains of chords using pre-defined Markov chains, *Concords* - evaluates concords based on the number of existing consonances and dis-sonances in a strike, *Rhythm* - evaluates the homogeneity of the rhythm in the motives, *Tonality* - evaluates the tonality on the main measures in the meters as high tonality should be maintained, *Pitch* - evaluates the homogeneity of notes' pitches in the motives.

2.6 Mutation

The degree of mutation operator depends on the parameter that sets the percentage of the motives that will be mutated. Mutation may either change the heights of particular notes or change the chords. Moreover, mutation may change the order of motives based on the mixing intensity parameter. If mutation is to change single notes, then a decision is made independently for each of them, with a pre-defined probability threshold. If so, first the origin of the new note is decided (whether it is the current chord, a key, base key or random), then a particular note is chosen. If mutation is to change chords, the procedure is similar to the one used for changing notes, but the origin may be main chord of current key, side chord of current key, main chord of another key, side chord of another key or random.

The degree of the other facet of mutation - changing the order of motives - depends on the mixing intensity parameter, which defines the percentage of motives to be re-ordered. First, a set of motives which will remain in the primary order is chosen. All other motives are inserted in random places between the "stable" motives in the piece.

2.7 Summary of Results

The initial series of tests aimed at optimization of the steering parameters of the memetic algorithm: *% of elite* - percentage of the specimens with the highest fitness automatically promoted to the next generation, *% of optimization* - probability of the local optimiza-tion to take place in the current generation, *mutation range* - the percentage of motives to be mutated, *notes/chords modification factor* - a factor describing the relation be-tween probabilities of mutation of notes and chords, e.g. 3/1 means 75% for notes and 25% for chords, *permutation range* - the percentage of motives that will remain in the primary order. Each combination of parameters was verified in 10 independent experi-ments with population size equal to 150. Each ensemble of 10 tests was performed three times, once per each considered form (*mazurka*, *polonaise* and *nocturne*).

The final selection of the algorithm's parameters was based on the average score of the best specimen evaluations across 10 populations. It turned out that for each of the tested forms different sets of parameters were the ones most effective. The best set of parameters of the *mazurka* form: *% of elite*= 1, *% of optimization* = 15, *mutation range* = 33, *notes/chords modification factor* = 3/1, *permutation range* = 25, has been chosen for further final tests, which involved three human testers.

Each of the testers received the same set of 300 pieces (of *mazurka* form) to listen and evaluate using the scale of -1 (worst) to 1 (best) with a step equal to 0.5. The confusion matrices for each pair of the testers are presented in Fig. 2. Generally speaking, it is

	Mark	Tester 1				
		-1	-0.5	0	0.5	1
	-1	18	10	8	5	1
	-0.5	11	30	46	23	7
Tester 2	0	12	26	31	21	6
	0.5	4	10	8	16	3
	1	0	1	1	2	0

	Mark	Tester 1				
		-1	-0.5	0	0.5	1
	-1	18	12	8	8	2
	-0.5	19	44	38	23	2
Tester 3	0	6	10	34	19	9
	0.5	2	8	11	14	3
	1	0	3	3	3	1

	Mark	Tester 2				
		-1	-0.5	0	0.5	1
	-1	18	11	13	6	0
	-0.5	14	49	47	16	0
Tester 3	0	8	38	21	9	2
	0.5	2	17	11	6	2
	1	0	2	4	4	0

Fig. 2. Comparison of testers' evaluations in the form of confusion matrices. For instance, value 11 in the first table means that there were 11 pieces evaluated as −1 by Testers 1 and as −0.5 by Tester 2.

easy to notice that the testers are often consistent when it comes to the pieces marked lower. This can be seen in the top left corners of the matrices. Furthermore, the testers do not always agree when it comes to the pieces rated with the highest marks which is most probably caused by differences in testers' tastes as well as a small number of the highest rated pieces in the set. A detailed analysis of results is presented in [10].

3   A Neural Network Component

The above described system can be regarded as an "objective" mechanism for composing music according to the rules (parameters) appropriate for a given musical form. In the current version of the system we added a neural network component, which reflects the user's taste and whose score was added to the fitness function. Neural network training is preceded by extracting style parameters from the pieces listened by the testers

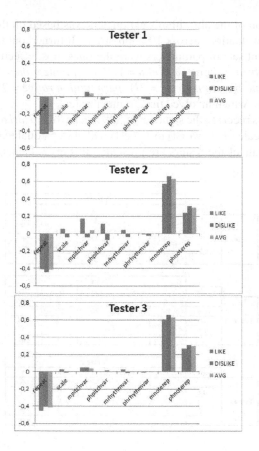

Fig. 3. Average values of style parameters of pieces highly evaluated by the testers (evaluations of 0.5 and 1) and pieces low evaluated (evaluations of -1 and -0.5) in comparison to averages of style parameters in the whole population

(see Fig. 3). The analysis of these graphs allows to observe that the testers have different style parameters in the pieces they prefer. For instance, it is easy to observe that *Tester 2 negatively judges pieces with low repetition rate of motives. The same pieces are evaluated higher by Tester 3* or *Tester 1 positively evaluates pieces with low variance of notes height in motives, unlike Tester 2.*

In the training process the input to the network consists of a set of eight style parameters, depicted in Fig. 3, extracted from the 300 evaluated pieces:

- the ratio of the number of motives versus the number of unique motives
- the difference between the highest and the lowest note in the piece
- average value of the variance of the height of notes in motives
- average value of the variance of the height of notes in phrases
- average value of the variance of the rhythmic values in motives
- average value of the variance of the rhythmic values in phrases

- average ratio of unique heights of the notes versus number of notes in motives
- average ratio of unique heights of the notes versus number of notes in phrases

The choice of parameters reflected our willingness to find such a characteristics of a piece that may take different values for various pieces and may be correlated (positively or negatively) with the individual musical taste of the listener (tester).

For each specimen (musical piece) obtained in the previous experiment, all eight parameters are extracted, normalized to $< -1; 1 >$ and used as an input to a neural network. The hidden layer consists of four neurons and the activation function is bipolar sigmoid. The network's single output belongs to $< -1; 1 >$ and represents the degree to which the piece reflects the listener's taste.

Neural networks (one for each tester) taught in the previous test were included in the algorithm and each of them was used to generate 50 pieces. Additionally, 50 pieces were generated with the help of previously described memetic system, with no use of a neural network.

For each of the testers, after mixing up 50 pieces generated with the use of the respective neural network and 50 ones generated without the network, the 100 pieces set was ready to use in the subsequent tests. Each set was mixed twice (the pieces were shuffled) and given to the respective tester without the labels. As a result each tester had two lists of the same pieces in different order and listened to them with no knowledge which pieces were generated using the neural network and which were not. In order to increase the credibility or the test listening to the first set was followed by a 24 hour break before the tester started to listen to the second set.

The average scores of the testers are presented in Fig. 4. For the sake of clarity it should be noted that among the three testers there were two co-authors of this paper. The third person, Tester 3, was completely unrelated with the paper and implementation of the project and may be found as the most independent source of tests data. On the other hand, due to "blind" construction of tests it was practically impossible to intentionally manipulate the results of Testers 1 and 2, thus we decided to publish them on equal rights as those of Tester 3.

The results of the performed tests may be assessed as moderately positive. Even though the evaluations are not always fully consistent, in all three cases the average evaluation of all listenings is higher for the pieces generated using the neural network. This allows us to say that using the neural network is beneficial for matching the created music to ones taste. On the other hand the observed improvement in the quality of

	AVG Neuro Run 1	AVG Other Run 1	AVG Neuro Run 2	AVG Other Run 2	AVG Neuro	AVG Other
Tester 1	2,78	2,52	3,02	2,72	2,9	2,62
Tester 2	3	2,74	2,96	2,9	2,98	2,82
Tester 3	2,8	2,5	2,58	2,58	2,69	2,54

Fig. 4. Summary of the results of "blind" test

specimens is not substantial. Relatively large differences between two runs of the experiment performed by the same user should, to a large extent, be attributed to changing and uncertain nature of testers' (human) perception.

In summary, the results allow us to say that the approach presented in the paper is a promising idea in the area of computer music generation. The proposed memetic system is able to construct pieces imitating popular musical forms and the additional use of neural network allowed generation of the pieces statistically more attractive to the recipient. This allows to ascertain that generated musical pieces reflect some aspects of human/artificial creativity and extend beyond the formal rules and theory of music.

During the experiments a few hundreds of music pieces of various forms (*polonaises*, *mazurkas* and *nocturnes*) and quality were generated by the system. Some of them clearly sound "artificial" and by no means may pretend to be composed by a (non-novice) human composer. There are, however, also quite many examples of "more human-like" compositions. An example of such a piece (of the form of *mazurka*) in presented in figure 5.

Fig. 5. Example piece of the form of *mazurka*

4   Conclusions and Directions for Future Work

In this work we studied two main aspects of the composed music. The first one refers to the initial stage of generating a musical piece. The hybrid algorithm allows creation of pieces, which are consistent with the user's expectations in the field of musical rules, harmonic and rhythmic restrictions, types of used textures and construction of the whole form. The fitness function evaluates the actual state of the piece and local optimization brings in the "formal order". Repeated execution of these operations allows achieving the goal of a piece containing expected musical substance. The end result is substantially dependent on the user's intentions and ability to translate theoretical knowledge into configuration of individual parameters and the input of the algorithm.

The second relevant aspect that we analyzed in the paper regards reflecting individual taste of particular listeners. Test results clearly show the influence of the use of neural network that learned from a person's preferences on the final quality of the generated piece. Thanks to this mechanism it is possible to not only generate music which fulfills the expectations related to the musical substance, rhythmics, harmonics and texture, but also reflects, to some extent, the sense of aesthetics and beauty of an individual user.

Our current focus is on deeper analysis of the factors defining user's individual preferences, which may potentially enhance the neural network's results. We also plan to perform more test involving other testers possibly with the help of social networks.

Acknowledgements. The authors would like to thank Dr Eddy Chong from NIE (Singapore) for providing them a database of chords chains that was used as part of domain knowledge representation in the local optimization phase.

References

1. Langston, P.S.: Six Techniques for Algorithmic Music Composition. In: 15th International Computer Music Conference, Columbus, Ohio, p. 6 (November 2, 1989)
2. Langston, P.S.: (201) 644-2332 or Eedie & Eddie on the Wire, An Experiment in Music Generation. In: 1986 Usenix Association meeting, pp. 4–5 (Summer 1986)
3. Smith, R.: Rogue Leaders: The Story of Lucas Arts. Chronicle Books, San Francisco (2008) ISBN 978-0-8118-6184-7
4. Cope, D.: Experiments in Musical Intelligence. A-R Editions, Madison (1996)
5. Cope, D.: The Algorithmic Composer. A-R Editions, Madison (2000)
6. Algorithmic Composition Software - ArtSong, http://www.artsong.org
7. Symbolic Composer, http://www.symboliccomposer.com/page_main.shtml
8. CyberMozart, http://www.yav.com/CyberMoz.html
9. Karlheinz Essl, Lexikon-Sonate, http://www.essl.at/works/Lexikon-Sonate.html
10. Mańdziuk, J., Goss, M., Woźniczko, A.: Chopin or not? A memetic approach to music composition. In: IEEE Congress on Evolutionary Computation (CEC 2013). IEEE Press, Cancun (2013)
11. Benade, A.H.: Horns, Strings, and Harmony. Science Study Series S 11. Doubleday & Company, Inc., Garden City (1960)
12. Kostka, S., Payne, D.: Tonal Harmony, 5th edn. McGraw-Hill, New York (2004)
13. Chong Kwong Mei, E.: National Institute of Education in Singapore (2013)

Profound Degree: A Conservative Heuristic to Repair Dynamic CSPs

Yosra Acodad, Amine Benamrane, Imade Benelallam,
and El Houssine Bouyakhf

[1] LIMIARF. Faculty of Sciences Mohammed the Fifth University - Agdal Rabat,
Morocco
{yosra.acodad,benamraneamine}@gmail.com, bouyakhf@fsr.ac.ma
[2] INSEA, National Institut of Statistics and Applied Economic - Irfane Rabat,
Morocco
imade.benelallam@ieee.org

Abstract. For a better treatment of Dynamic Constraint Satisfaction Problems ($DCSPs$), several techniques have been developed to be used in repair algorithms. We cite, for example, the variables/values ordering heuristics and local search techniques.

We distinguish between static heuristics, which calculate their values once at the beginning of the search, and dynamic heuristics that use an expensive intelligence in terms of solving time.

In this paper, we propose a new static variable ordering heuristic, Profound Degree ($pdeg$), based on deg heuristic, which calculates the degree of influence of a given variable, on the whole constraints network, relatively to its position in the network.

We evaluate this heuristic on the Extended Partial-order Dynamic Backtracking ($EPBD$) approach, which is an approach to repair $DCSPs$ solutions, and we compare it to the best-known variables ordering heuristics ($VOHs$) for repairing. The evaluation of performance is on random binary problems and meeting scheduling problems, with the criteria of computation time, number of constraints checks and Hamming distance between the former and the current solution.

1 Introduction

In recent years, several improvements have been made in algorithms for solving and repairing $DCSPs$ [3] and [4]. Among these, we quote $VOHs$ [9], filtering techniques and conflicts analysis.

In fact, the order in which variables are affected by a search algorithm is crucial. Indeed, the use of different $VOHs$ to solve the same CSP can lead to very different results in terms of performance.

The used heuristics can be classified into two broad categories: Static variables ordering heuristics ($SVOHs$) and dynamic variables ordering heuristics ($DVOHs$).

In this paper, we present a new $SVHO$, *profound degree heuristic* ($pdeg$), which is inspired by the *deg* heuristic insofar as it uses the concept of neighborhood.

L. Iliadis et al. (Eds.): AIAI 2014, IFIP AICT 436, pp. 140–149, 2014.

In effect, *pdeg* is not limited to calculate the number of neighbor connections of concerned variable, but it also measures the impact of its eventual disturbances (since we study this heuristic under correction of *DCSPs*) in relation to all constraints network, i.e., it takes into consideration all levels of the network neighborhoods, by giving each level a different weight, then it keeps intact the value of the variable having a strong impact, in order to reduce disturbing the network.

2   Background

2.1   Minimal Perturbation Problem (*MPP*)

Definition 1 (MPP). *A Minimal Perturbation Problem (MPP) is a triple Π = (Θ, α, δ), where:*

- *Θ is a CSP.*
- *α is a partial or complete assignment for Π that is called initial assignment.*
- *δ is a function that defines a distance between any two assignments.*

A solution to a *MPP* is a solution to *Π* with minimal distance from α according to δ.

Definition 2 (Distance Set D). *Let σ and γ be partial assignments for Θ. V is the set of variables and $W(\sigma,\gamma)$ is the set of variables such that the value assignment for v in σ is different from the value assignment for v in γ :*

$$W(\sigma,\gamma) = \{v \in V \mid \langle v, val\rangle \in \sigma, \langle v, val'\rangle \in \gamma, val \neq val'\} \tag{1}$$

$W(\sigma,\gamma)$ *is called a distance set for σ and γ and the elements of the set are called perturbations.*

Definition 3 (Function δ). *In an MPP, the distance function of some assignment σ from α is defined as the cardinal of the set $D(\sigma,\alpha)$:*

$$\delta(\sigma,\alpha) = \mid D(\sigma,\alpha) \mid \tag{2}$$

Otherwise, δ is the hamming distance between σ and α.

Definition 4 (Solution Value Assignment (SVA)). *α is the initial assignment of a CSP. For each variable v_i:*

$$If \langle v_i, val_{i_j}\rangle \in \alpha \Longrightarrow v_i.SVA = val_{i_j} \tag{3}$$

SVA is the acronym of Solution Value Assignment.

2.2   Extended Partial-Order Dynamic Backtracking ($EPDB$)

Unlike several CSP resolution approaches that use a fixed order of variables, the Partial-order Dynamic Backtracking (PDB) [2] uses a partial order, that is built from nogoods and safety conditions, that is dynamically modified during the search process.

Definition 5 (Nogood). *A nogood is a failure justification. It's an expression of the form :*

$$(x_1 = v_1) \wedge ... \wedge (x_k = v_k) \rightarrow x \neq v$$

Definition 6 (Safety Conditions). *A safety condition is defined as an assertion of the form: $x < y$, with x and y variables.*

If S is a set of safety conditions, we denote by $\leq_s$ the transitive closure of $<$, meaning that S is acyclic if $\leq_s$ is antisymmetric.

$$x <_s y \Longrightarrow x \leq_s y, y \nleq_s x \tag{4}$$

In another way: $x <_s y$, if there is a sequence (possibly empty) of safety conditions:

$$x < z_1 < z_2 < ... < z_n < y \tag{5}$$

When having to generate a nogood during the search, due to conflict situation (creating a nogood caused by a violated constraint) or domain wipe out (nogoods resolution), and when no order is imposed in existing nogoods and safety conditions between variables concerned with this conflict, there may be significant choices for the variable that will emerge as the generated nogood conclusion. The performance of this approach depends mainly on the quality of this choice, which is not covered in PDB.

The major goal of $EPDB$ is to extend the classic description of PDB, by exploiting its flexibility to repair assignments using repair-directed heuristics, when having to build an order between some variables, to control changes.

3   Profound Degree Heuristic ($pdeg$)

3.1   Description of Heuristic

The $deg\,VOH$ [5] has a major advantage, it's that it does not compute every time dynamic information about the current state of research (such as the current domains size dom or the current variables degree $ddeg$ [10]), but it operates static information.

However, in the context of repairing solutions, this heuristic appears limited. In fact, no information regarding the position of variables in relation to the entire network is operated, whereby the change of a variable can affect not only the neighbors, but all of the variables in the network with a given probability. For this, we propose to calculate an estimation of this information for each variable

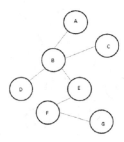

Fig. 1. A simple constraints network

X at the beginning of the research, considering all the network constraints, and by associating to each constraint a weight, higher or lower, in relation to its distance from X. $pdeg$ is considered as the sum of those weights.

Now consider the simple network above (figure 1). Suppose that each variable has a consistent value and due to a perturbation of the $DCSP$, the value of E must be changed. Assume that all constraints have the same tightness p_2. Changing E will generate the following probabilities P:

$$P(violate(C_{EB})) = P(violate(C_{EF})) = p_2$$
$$P(violate(C_{BA})) = P(violate(C_{BC})) = P(violate(C_{BD})) = P(violate(C_{FG})) =$$
$$p_2 * p_2 = p_2{}^2$$

If the probabilities of violating constraints are considered as weights, knowing that $p_2 \in [0, 1]$, the weights of constraints network are:

$$Weight(C_{EB})/E = Weight(C_{EF})/E = p_2 \geqslant Weight(C_{BA})/E =$$
$$Weight(C_{BC})/E = Weight(C_{BD})/E = Weight(C_{FG})/E = p_2{}^2$$

This reflects that, whenever the constraint C is closer to the variable X, C is more influenced by the change of X, since the variables concerned by C are more influenced by changing X.

In fact, given two constraints C_1 and C_2, if the constraint C_1 is closer to X than C_2, then C_1 has a cost in calculation of the heuristic value of X which is more interesting than C_2, since changing X engenders a risk of violating C_1 more important than violating C_2.

In reality, constraints networks are not as simple as that, they are quite as figure 2:

On the other hand, the hardness of constraints network (especially when the constraints used in the modeling are intent) is not always known, reasons why choosing to replace the weight, that corresponds to the tightness of the constraints, by an average hardness of $1/2$.

Inspired by this idea, $pdeg$ heuristic accumulates, for each variable X, the weights of all constraints. These weights are higher each time the constraints are closer to X.

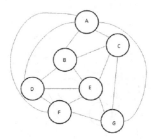

Fig. 2. A constraints network

Function PDEG(X,CSP)
1. $pdeg \leftarrow 0$;
2. $level \leftarrow 0$;
3. $size \leftarrow CSP.Vars.size()$;
4. $actual \leftarrow X$;
5. $total \leftarrow X$;
6.   **while**($totat.size()! = size$)
7. $neighbors \leftarrow getAllNeighbors(actual)$;
8. $filter(neighbors, total)$;
9. $actual \leftarrow getDistinct(neighbors)$;
10. $a \leftarrow neighbors.size()$;
11. $b \leftarrow nbConstraints(actual)$;
12. $total.add(actual)$;
13. $pdeg+ = 1/2^{incLevel()} \times a + 1/2^{incLevel()} \times b$;
14. $Return\ pdeg$;

Fig. 3. Description of *pdeg* heuristic

3.2   Execution of Heuristic

Both of the value of *pdeg* and the *level*, that is incremented each time the constraint is far from the variable in question, are initialized by 0 (lines 1 and 2). The variable *size* contains the number of variables in the CSP (line 3) and *actual* and *total* lists, initiated by the variable for which having to calculate the *pdeg*, serve respectively to store all variables in the current level and all variables already involved in the calculation (lines 4 and 5).

Will not accounting all constraints network variables, the heuristic calculates the neighborhood of the current variables, then this neighborhood is filtered to include only new variables, to not record a same constraint twice (lines 6-8).

Note that *getAllNeighbors()* (line 7) recognizes a neighbor many times as the number of constraints that bind it with the current variables.

The heuristic puts into *actual* current neighbors eliminating redundancies, in *a* the number of neighbors of actual variables with redundancy (which is equivalent to the number of constraints with neighbors) not yet recognized and in *b* the number of constraints between these neighbors (lines 9 and 11). At this point, it adds to *total* the current neighbors and to *pdeg* the weight of neighbors, which is 1/2 to the power the *level* incremented, multiplied by the number of redundant neighbors, and it does the same thing for the number of constraints between neighbors, after incrementing the *level* (lines 12 and 13).

The heuristic leaves the while loop once all variables are accounted, which means it counted all the constraints of the network, and the value of *pdeg* is returned as the profound degree of the variable X (line 14).

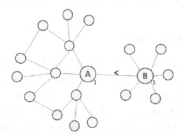

Fig. 4. Graph of constraints

3.3   Example of Simulation

We consider the constraints network above and assume that all variables have a consistent value ($A = 1$, $B = 1$ and all other variables have values consistent with the whole constraints network). We suppose that the $DCSP$ has undergone a disturbance that is the change of the nature of the constraint C_{AB}, that was: $A = B$, and that became: $A < B$. To satisfy this constraint, the $EPDB$ algorithm [1] is restarted in order to repair solution. $EPDB$ chooses to change the value of A, or the one of B, depending on the used heuristic.

Using $EPDB_{deg}$, heuristic values are:
$deg(A) = 4$
$deg(B) = 6$.

So the approach generates the *nogood*: $B = 1 \implies A \neq 1$, since it retains the value of the most relevant variable in the network, i.e., the one whose change can disrupt more the $DCSP$ and delay resolution. For this reason, the value of A is changed.

Now, using $EPDB_{pdeg}$, the values of used heuristic are:
$pdeg(A) = 1/2 \times 4 + 1/2^2 \times 1 + 1/2^3 \times 14 + 1/2^4 \times 2 = 4,125$
$pdeg(B) = 1/2 \times 6 + 1/2^2 \times 0 + 1/2^3 \times 3 + 1/2^4 \times 1 + 1/2^5 \times 9 + 1/2^6 \times 2 = 3,75$

Then the approach generates the *nogood*: $A = 1 \implies B \neq 1$ and changes the value of B.

4   Experimental Results

We carried out a series of experimental tests to compare the integration of *pdeg* heuristic, within the $EPDB$ algorithm, to other $VOHs$.

Heuristics used for repairing are those that improve the best the behavior of $EPDB$, namely the degree (deg), the domain size (dom), the number of conflicts ($conf$) and the ratio between the domain size and the weighted degree ($dom/wdeg$) [6].

We evaluate the performance in terms of constraints checking (CCs), computing time ($Time(sc)$) and Hamming distance (HD) [8] and [7].

All experiments were performed on the Java platform.

4.1 Experiments on Randoms

Random $CSPs$ are characterized by a quadruplet $< n, d, p_1, p_2 >$, where n is the number of variables, d the number of values per variable, p_1 the network connectivity (density) and p_2 the constraints tightness.

The density p_1 of a constraints graph is defined as the the ratio of the number of constraints relative to the maximum number of possible constraints in this network.

The tightness p_2 of a constraint corresponds either to the proportion of unauthorized tuples, or the probability that a tuple is not allowed.

Tests have been performed on sparse and dense problems, respectively $< 20, 10, 0.25, 0.6 >$ and $< 20, 10, 0.75, 0.27 >$. The constraints tightness is selected such as areas are as complex as possible while problems have solution.

For logical reasons, we assume that the maximum rate of added constraints is equivalent to $\sim 25\%$. For each pair, 20 instances were solved using each heuristic, and the results are presented as an average of these 20 instances.

Figure 5 presents the effectiveness of heuristics applied to the $EPBD$, for the most complex regions of the low density selected. In terms of number of constraints tested CCs, Between 10% and 24% of new injected constraints, $pdeg$ is the best, elsewhere, $dom/wdeg$ and dom are the most efficient. Otherwise, the deg is the closest to $pdeg$, which seems the most stable.

In time graph, $dom/wdeg$ recedes (also for dom but not in the same manner), due to the calculation of dom and $wdeg$ heuristics, especially for $wged$. $pdeg$ is almost always the best.

Concerning the Hamming distance HD, there is a competition between heuristics, and $pdeg$ remains a good competitor in most regions.

Figure 6 shows the effectiveness of the heuristics for the most complex areas of the high density $p_1 = 0.75$. In terms of number of CCs and *execution time*, $pdeg$ usually looks the best although not improving too much the behavior of deg. Concerning Hamming distance, no heuristic is absolutely the best.

4.2 Experiments on Dynamic Meeting Scheduling Problems

A Meeting Scheduling Problem MSP is characterized by $< m, p, n, d, h, t, a >$, where m is the number of meetings, p the number of participants, n the number of meetings per participant and d the number of days. Different time slots are available for each meeting, and h is the number of hours per day, t the duration of the meeting and a the percentage of availability for each participant.

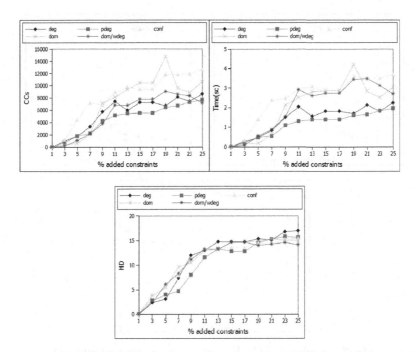

Fig. 5. Number of CCs, Execution time performed and HD ($p_1=0.25$, $p_2=0.60$)

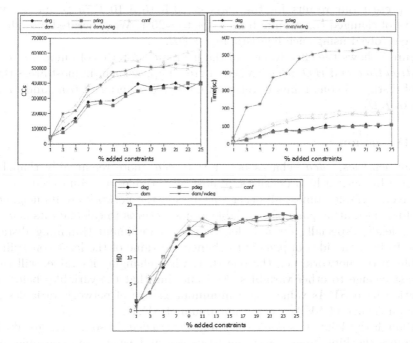

Fig. 6. Number of CCs, Execution time performed and HD ($p_1=0.75$, $p_2=0.27$)

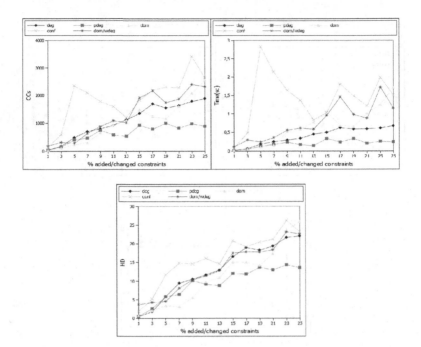

Fig. 7. Number of CCs, Execution time performed and HD

We present our results for the class $< 20, 5, 15, 5, 10, 1, 70 >$ and we vary the rate of changed constraints from $\sim$1% to $\sim$25%. We generated 20 different instances solved using each heuristic in $EPDB$.

Figure 7 shows the effectiveness of the heuristics in terms of number of CCs, *execution time* and HD. *pdeg* leads to a great improvement in most areas (from 10% of disrupted constraints in terms of CCs, 7% in terms of *time* and 11% in terms of HD).

4.3   Discussion

For low densities, *pdeg* behaves better than *deg*, more remarkably than high densities. In fact, in high densities, almost all variables are connected, i.e., for each given variable, almost all other variables in the network are its neighbors. Therefore calculating *pdeg* of a variable almost returns to calculate its *deg*.

pdeg leads, especially for low densities, to a minimum Hamming distance (HD). In fact, the idea of *pdeg* is to change the value of the least constraining variable in the network, i.e., the one that, when changing its value, will cause the least change to other variables. And knowing that the variables before the correction have $SVAs$ values, so a minimum change of network variables is a minimum change of $SVAs$, then a maximum of $SVAs$.

Although the heuristics *dom*, *conf* and *dom/wdeg*, in some regions, do less constraints checking than static heuristics *deg* and *pdeg*, they consume more computational time because of their dynamic aspect.

5   Conclusion

In this paper, we introduce a new VOH, $pdeg$ heuristic, for estimating the influence of variables in the whole constraint network, in the process of repairing solutions in dynamic environments.

Based on the results of experiments on a wide class of problems, we found and proved that this heuristic, used in $DCSPs$ correction, exceeds the heuristics deemed to be efficient when solving $CSPs$.

In fact, the idea of $pdeg$ heuristic is to disturb only the parts of the constraints network, whose probability of influence is minimal. Indeed, for low to medium densities, as seen in the random problems with density $p_1 = 0,25$ and the MSP (and especially for medium densities), using $pdeg$ improves remarkably the repair of solution, not only in relation to research performance ($time$ and CCs), but also with respect to the quality of the solution ($Hamming\ distance$).

As perspectives, we intend to integrate this heuristic in other algorithms for solving and repairing centralized and distributed $CSPs$, to hybridize it with other heuristics according to areas in which they show a remarkable improvement in order to lead to an optimal heuristic, and to experiment it on real problems.

References

1. Acodad, Y., Benelallam, I., Hammoujan, S., Bouyakhf, E.H.: Extended Partial-order Dynamic Backtracking algorithm for dynamically changed environments. In: 2012 IEEE 24th International Conference on (ICTAI), pp. 580–587 (November 7, 2012)
2. Ginsberg, M.L., McAllester, D.A.: Gsat and dynamic backtracking. Journal of Artificial Intelligence Research 1, 25–46 (1994)
3. Dechter, R., Dechter, A.: Belief maintenance in dynamic constraint networks. In: AAAI, pp. 37–42 (1988)
4. Bessiere, C.: Arc-consistency in dynamic constraint satisfaction problems. In: Proc. AAAI 1991, pp. 221–226. AAAI Press (1991)
5. Dechter, R., Meiri, I.: Experimental evaluation of preprocessing techniques in constraint satisfaction problems. In: Proceedings of IJCAI 1989, pp. 271–277 (1989)
6. Hemery, F., Lecoutre, C., Sais, L.: Boosting systematic search by weighting constraints. In: ECAI (August 2004)
7. Zivan, R., Alon, G., Amnon, M.: Hybrid search for minimal perturbation in Dynamic CSPs. Constraints 16(3), 228–249 (2011)
8. Hebrard, E., Barry, O., Walsh, T.: Distance Constraints in Constraint Satisfaction. In: IJCAI (January 6, 2007)
9. Ortiz-Bayliss, J., Terashima-Marin, H., Ender, O., Andrew, J., Santiago, E.: Exploring heuristic interactions in constraint satisfaction problems: A closer look at the hyper-heuristic space. In: 2013 IEEE Congress on Evolutionary Computation (CEC). IEEE (June 20, 2013)
10. Bessiere, C., Jean-Charles, R.: MAC and combined heuristics: Two reasons to forsake FC (and CBJ?) on hard problems. In: Freuder, E.C. (ed.) CP 1996. LNCS, vol. 1118, pp. 61–75. Springer, Heidelberg (1996)

A Greedy Agent-Based Resource Allocation in the Smart Electricity Markets

Armin Ghasem Azar[1], Mansoor Davoodi[1], Mohsen Afsharchi[2],
and Bahram Sadeghi Bigham[1]

[1] Department of Computer Science and Information Technology
Institute for Advanced Studies in Basic Sciences, Zanjan, Iran
{a.ghasemazar,mdmonfared,b_sadeghi_b}@iasbs.ac.ir
[2] Department of Electrical and Computer Engineering
University of Zanjan, Zanjan, Iran
afsharchim@znu.ac.ir

Abstract. The smart grid makes use of two-way streams of electricity and information to constitute an automated and distributed energy delivery network. Coming up with multi-agent systems for resource allocation, chiefly comprises the design of local capabilities of single agents, and therefore, the interaction and decision-making mechanisms that make them create the best or at least an acceptable power allocation. Due to the several issues in providing sustainable and affordable power energy, researchers try to think about creating a decentralized mechanism to be able to manage the entire transactions in retail electricity markets. As a result, this electricity infrastructure is predicted to develop into a market of markets, during which all the trading agents influence on each other and have role in toward an equilibrium one. In these markets, we are interested to minimize the buyers' purchasing cost. Motivating this issue, we model the demand response problem in an evolutionary optimization framework and propose an evolutionary algorithm for handling the *decentralized market-based resource allocation* problem.

1 Introduction

Intelligent power networks, *Smart Power Grids*, are an enhancement of the present century and the result of experts' efforts to improve the power distribution mechanisms. Conventional power grids are mostly utilized to transmit the power from the central generators to an extensive variety of the end users. Conversely, the smart grid uses two-way streams of electricity and information to create an autonomous energy delivery network [1]. Toward this issue, multi-agent systems not only represent the consequent massive steps within the development of the next generation energy delivery network systems, but they also reveal new strategies to design highly flexible ones [2]. Coming up with multi-agent systems for distributed resource allocation, chiefly comprises the design of local capabilities of the trading agents and therefore the interaction and decision-making mechanisms that let them create the best or at least an admissible allocation. One of the challenges in this context is the *Demand Response* problem. Due to the several issues in providing sustainable and affordable power, creating a *decentralized* mechanism that be able to manage the entire transactions dynamically, is

L. Iliadis et al. (Eds.): AIAI 2014, IFIP AICT 436, pp. 150–161, 2014.
© IFIP International Federation for Information Processing 2014

an interesting problem among researchers and practitioners [3]. Proposing a comprehensive approach considering the minimization of buyers purchasing cost, where the decision process of resource allocation is expressed as a demand response optimization under certain constraints, is one of the main contributions of this paper. In this respect, the *"market-based resource allocation"* mechanisms are motivated by economical features [4], in which the resources are allotted utilizing the competition between trading agents.

In order to model such mechanisms, one of the main approaches is to create decentralized markets among trading agents [5]. In this approach, instead of having a dealer, to control the whole system, the environment is divided into diverse non-overlapping regions, in which there is a dealer in each of them who is responsible for controlling all the inside transactions. Motivated this idea, a comprehensive mathematical model to formalize the *decentralized demand response optimization problem in the smart power markets* is proposed. Taking into account the fact that buyer agents are self-interested, they are concerned with minimizing their purchasing costs. In order to satisfy the buyers demands, agents named sellers, are responsible for producing their needed power. Therefore, finding an optimal permutation of buyers to purchase their required power in the electricity markets, necessitates using evolutionary computation techniques.

So far, however, there has been little discussion about using evolutionary computation techniques for the demand response optimization problem in the smart electricity markets. Finding an optimal permutation of buyers requires calculating all the permutations. Accordingly, this problem is similar to *Quadratic Assignment Problem* which is a NP-Hard one [6]. Evolutionary algorithms are typically used to provide good approximate solutions to problems that cannot be solved easily using other techniques. Many optimization problems fall into this category. It may be too computationally-intensive to find an exact solution but sometimes a near-optimal solution is sufficient. In these situations evolutionary techniques can be effective [7].

The rest of this paper is organized as follows. Section 2 contains a review of the related work. In Section 3, we presents the comprehensive description of the mathematical system model with the proposed solution approach. Decentralized greedy resource allocation algorithms along with the evolutionary optimization techniques, definitions and tips are members of the next section. Fifth section demonstrates experimental results. Finally, conclusion and future work are drawn in the last section.

2 Related Work

The initial idea of designing power markets is based on a central system which controls and politicizes all the transactions and exchanges. Many studies have indicated that the presence of only one system or a central manager is a waste of energy and money [5,8]. Furthermore, many studies have been conducted in various aspects of these markets, specially in using multi agent systems and information systems [5,9]. Lamparter et al. have designed an electricity market based on the intelligent agents' technologies. They have presented miscellaneous scenarios of software infrastructure based on the multi agent systems to create this type of markets. Moreover, there is a comprehensive literature review on electricity market designing and interdisciplinary processes required for achieving the decentralized markets [5].

Besides, some of the researches in the field of the energy optimization, including those reported in [10,11], have demonstrated optimal domestic energy management via linear programming, where power required by each device in a house is optimally met by the available resources. Similar works of power matching can be found in [10], in which the proposed market model is based on a centralized approach with only one manager. In this paper, HomChaudhuri et al. have used a centralized market which does not allow buyers or even sellers to have power negotiation with more than one individual. Furthermore, it is assumed that all the sellers' surplus powers are much more than the buyers' demand. Considering these parameters, the proposed market will not confront critical circumstances. In addition, behaving only in the benefit of buyers in line with the sellers' elementary price update mechanism, which depends only on the number of buyers which are desired to purchase power from them, are studied in [11]. The proposed technique optimizes the loss rate of power distribution. Considering their demands and surpluses, matching buyers to sellers is based on the linear programming. Nonetheless, if any of the buyers cannot satisfy its demand in one interval, then, rest of that will be transferred to the next power matching intervals. Finally, in their paper, the fitness is based only on the distance among the buyers and sellers.

This paper follows a case-study design, with in-depth analysis of demand response optimization problem in electricity markets. Proposing an evolutionary greedy power matching algorithm for this concept concerning buyers' profit, is the main contribution of this paper. Besides, proposing a greedy iterative price update rule for seller agents for updating their offered prices for each unit of price in the consecutive time intervals is another contribution.

3 System Model

Visualizing a city with the self-healing power grid necessitates considering the multiple issues that are significant to include. Some belong to the physical infrastructure, for example the generators, transfers and transmission lines. Hereinafter, we focus on the issues in the physical infrastructure which also involve decentralized market-based resource allocation. Consider a city which is composed of a unique main power source, namely macro-station, which is connected to the main power grid of the city, along with a network of buyers and sellers. It is assumed that the macro-station is too far from the main power grid of the city and also purchasing power from this station both costs a lot and incurs high power loss during the power transfer. Furthermore, considering the macro-station is always on-line and has infinite surplus power, will assure us of having a consistent and reliable power source in critical and unpredictable circumstances. Trying to reach the purpose of having a decentralized controlling mechanism of the power exchanges among the distributed buyers and sellers, has motivated us to partition our city into regions. Each of them not only contains a non-empty subset of buyers and sellers, but a power-dealer is also allocated in order to control the power exchanges among the agents. In addition, R_k is denoted as the kth region. Subsequently, the buyers and sellers sets in the kth region are represented by R_k^B and R_k^S respectively. Finally, it is assumed that the system model is studied for consecutive independent intervals, $t = 1, 2, \ldots, T$.

Analyzing essential parameters to obtain an optimal solution in power matching of buyers and sellers, represents the substantial challenges in demand response problem

[12]. Therefore, a mathematical solution approach is applied toward this issue. That the markets behave only in the benefit of buyers is the markets' main perspective. Minimizing the purchasing cost in power negotiations is the buyers' main objective. In this approach, sellers announce power prices in consecutive intervals and buyers choose sellers considering prices along with other parameters.

3.1 Market Mechanism

In all the power markets, each unit of energy has a price for selling. Also, each unit in the electricity markets is in the range of (p_l, p_u). p_l and p_u are constant and belong to minimum and maximum price of each unit of power respectively. Dealers broadcast this range prior to the start of power exchange to the sellers. Considering our market mechanism, sellers announce their own initial prices in the tth interval between (p_l, p_u). Some key factors are included such as financial parameters, general attractiveness of the market and agents' risk factors [1], which influence on both purposes of *picking out an appropriate initial price* and *updating consecutive intervals' prices*. Drawing out a unique parameter for all the sellers which elucidates *malleability rate from the market*, named δ for sellers, is essential for their mentioned purposes. In order to update the power prices in subsequent intervals, they try to alter their prices using *"Iterative Price Updating"* rule. For instance, if some of the sellers announce too high prices, this rule will behave in such a manner as none of them is able to have power negotiations in the next intervals. Hence, they will have to decrease their proposed prices regarding the others' prices.

In time interval t, the $x_{b_i s_j}^t \in \mathbb{Z}^*$ is represented as the number of power units which the ith buyer has purchased from the jth seller. $\mathbb{Z}^*$ is represented as the non-negative integer set. Considering the way which this transfer is happening, there is a main transmission line in each of the regions, which all the buyers and sellers are connected to. Each of the transmission lines ($\tau \iota_k$) has a maximum capacity of power transferring which is denoted by $flow_{\tau \iota_k} \in \mathbb{Z}^*$. Limitation in power transfer using the distribution lines mentions that sellers cannot transfer the power more than the transmission lines' capacity [13].

$$\forall \tau \iota_k \text{ such that } \sum_{\forall b_i \in R_k^B} \sum_{\forall s_j \in R_k^S} x_{b_i s_j}^t \leqslant flow_{\tau \iota_k} \tag{1}$$

$$i=1,2,3,...,N; j=1,2,3,...,M; t=1,2,3,...,T; k=1,2,3,...,K$$

Along with the confined transmission line in each region, any power exchange between a buyer and a seller is accompanied by the power loss over the distribution lines [1]. This exchange is expressed by $loss_{b_i s_j}$. Both of the transmission lines' specification (i.e. *maximum flow capacity* and *distance between the buyer and seller*) have direct impact on the value of the $loss_{b_i s_j}$. For convenience, it is assumed that only the euclidean distance between a buyer and seller has impact on the power loss.

3.2 Agents' Specification

In order to explicate diverse types of the trading agents, which it is assumed to have N buyers and M sellers, following descriptions are presented as follows.

Buyers. Agents who need to satisfy their demands via the sellers are known as the buyers in which the buyer b_i has a predetermined demand $d_{b_i}^t$ in interval t. Thereupon:

$$B = \{b_1, b_2, \ldots, b_N\} \tag{2}$$

$$D_{b_i} = \{d_{b_i}^1, d_{b_i}^2, \ldots, d_{b_i}^T\} \qquad\qquad i = 1, 2, 3, \ldots, N \tag{3}$$

$$d_{b_i}^t \leq \sum_{j=1}^{M} \left(x_{b_i s_j}^t \times \left(1 - loss_{b_i s_j}\right)\right) \qquad i = 1, 2, 3, \ldots, N; t = 1, 2, 3, \ldots, T \tag{4}$$

Equation (2) demonstrates the buyers set. Equation (3) shows the demand set of the ith buyer during all the intervals. Furthermore, $d_{b_i}^t \in \mathbb{Z}^*$ will be the required power unit of the ith buyer in the tth interval. Intuitively, as (4) indicates, due to power loss assumption over the distribution lines, each buyer b_i should buy more than its demand in order to fulfill its requirement completely.

Sellers. Agents who have *Micro-Grids* (e.g. solar panels or wind turbines) [14], and aim on producing power for selling, are known as sellers. Then:

$$S = \{s_1, s_2, \ldots, s_M\} \tag{5}$$

$$Q_{s_j} = \{q_{s_j}^1, q_{s_j}^2, \ldots, q_{s_j}^T\} \qquad\qquad j = 1, 2, 3, \ldots, M \tag{6}$$

$$q_{s_j}^t \geq \sum_{i=1}^{N} x_{b_i s_j}^t \qquad\qquad j = 1, 2, 3, \ldots, M, t = 1, 2, 3, \ldots, T \tag{7}$$

Equation (5) shows the sellers set. Equation (6) demonstrates the set of surplus powers for the jth seller during all the intervals. Furthermore, the $q_{s_j}^t \in \mathbb{Z}^*$ is the number of the power unit that the jth seller has for selling in the tth interval. Equation (7) mentions that the sellers cannot sell more than their surplus powers in the jth interval.

3.3 Iterative Price Updating Rule

Designing a greedy and dominant rule for updating the sellers' prices, has encouraged us to propose an *iterative price updating* rule. $ps_j^t \in \mathbb{Z}^*$ demonstrates the jth seller's surplus power price in the tth interval. Consecutively, the δ_{s_j} explains the malleability rate of the jth seller from the market as mentioned in the previous section. Sellers learn from not only their previous prices, but also the other sellers' amount of power negotiations along with offered prices. Therefore, $\overline{ps}^t$ represents the weighted average of all sellers' offered prices in t^{th} interval, as they is formulated in (8).

$$\overline{ps}^t = \frac{\sum_{j=1}^{M} \left(\left(\sum_{i=1}^{N} x_{b_i s_j}^t \right) \times ps_j^t \times \delta_{s_j} \right)}{\sum_{i=1}^{N} \sum_{j=1}^{M} x_{b_i s_j}^t} \qquad t = 1, 2, 3, \ldots, T \tag{8}$$

Subsequently, after computing the $\overline{ps}^t$, dealers dispense these parameters to their regions' seller agents in order to let them commence with updating their next intervals' price.

$$ps_j^{t+1} = ps_j^t - \left(\delta_{s_j} \times \left(ps_j^t - \overline{ps}^t\right)\right) \qquad\qquad j = 1, 2, 3, ..., M; t = 1, 2, 3, ..., T \quad (9)$$

If any of the agents announce high or low price in an interval, (9) permits them to alter their prices to get close to the weighted average of the power's unit price in the next intervals.

3.4   Solution Approach

In this approach, the markets' purpose is to operate only in the benefit of buyers along with optimizing their objective function (i.e. minimizing the purchasing cost). In this case, sellers announce their prices based on their own malleability rate. The ps_j^1 is denoted as j^{th} seller's initial price, which is within the certain range declared by the dealers. Then, buyers start the purchasing procedure regarding the determined sellers' prices and parameters of the market. As a result, the objective formula with the constraints are as follows:

$$\min f(X) = \sum_{t=1}^{T} \sum_{i=1}^{N} \sum_{j=1}^{M} x_{b_i s_j}^t \times ps_j^t \qquad\qquad (10)$$

subject to:

$$d_{b_i}^t \leq \sum_{j=1}^{M} \left(x_{b_i s_j}^t \times \left(1 - loss_{b_i s_j}\right)\right) \qquad\qquad i = 1, 2, 3, ..., N; t = 1, 2, 3, ..., T \quad (11)$$

$$q_{s_j}^t \geq \sum_{i=1}^{N} x_{b_i s_j}^t \qquad\qquad j = 1, 2, 3, ..., M; t = 1, 2, 3, ..., T \quad (12)$$

$$ps_j^1 = \text{Randomly selected between } (p_l, p_u) \qquad\qquad j = 1, 2, 3, ..., M \quad (13)$$

$$ps_j^{t+1} = ps_j^t - \left(\delta_{s_j} \times \left(ps_j^t - \overline{ps}^t\right)\right) \qquad j = 1, 2, 3, ..., M; t = 1, 2, 3, ..., T \quad (14)$$

$$\overline{ps}^t = \frac{\sum\limits_{j=1}^{M} \left(\left(\sum\limits_{i=1}^{N} x_{b_i s_j}^t\right) \times ps_j^t \times \delta_{s_j}\right)}{\sum\limits_{i=1}^{N} \sum\limits_{j=1}^{M} x_{b_i s_j}^t} \qquad\qquad t = 1, 2, 3, ..., T \quad (15)$$

$$\sum_{\forall b_i \in R_k^B} \sum_{\forall s_j \in R_k^S} x_{b_i s_j}^t \leqslant flow_{T l_k} \qquad {\scriptstyle i=1,2,3,...,N; j=1,2,3,...,M; t=1,2,3,...,T; k=1,2,3,...,K} \quad (16)$$

$$x_{b_i s_j}^t, d_{b_i}^t, q_{s_j}^t, p_l, p_u, ps_j^1, ps_j^t, \overline{ps}^t, flow_{T l_k} \in \mathbb{Z}^*; loss_{b_i s_j}, \delta_{s_j} \in (0, 1)$$

4   Greedy Power Matching Algorithm

Selfish trading agents have caused us to focus on finding a near optimal solution for the demand response problem in retail power markets in a *greedy* approach. Each of the buyers attempts to purchase all of its required power from a seller who is close to it and has announced a low price. In line with this point of view, an *Interest Table* related to all buyers is required prior to start power exchanging in each interval.

Interest Table. Taking into account all the constraints related to the markets and agents, an interest table is created by buyers to demonstrate their desired value in having negotiation with sellers. For instance, in this approach, each buyer's satisfaction is in purchasing the power from the sellers, who have not only been close to the buyer, but also announced low prices. As an example, Fig. 1 illustrates a city, in which there are five buyers and sellers in the third region. The initial prices are distributed haphazardly between the $(1000 \pm 500)\$$ for sellers.

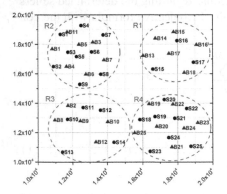

Fig. 1. Buyers' and sellers' location

Table 1. Sample interest table of $R3$

	\multicolumn{5}{c}{**Sellers**}				
b_2	s_{11}	s_{12}	s_{14}	s_{10}	s_{13}
b_8	s_{12}	s_{11}	s_{14}	s_{10}	s_{13}
b_9	s_{11}	s_{12}	s_{14}	s_{10}	s_{13}
b_{10}	s_{14}	s_{12}	s_{11}	s_{13}	s_{10}
b_{12}	s_{14}	s_{12}	s_{11}	s_{13}	s_{10}

(Buyers)

Each element of the interest tables is created by:

$$Interest_{b_i s_j}^t = \frac{d_{b_i}^t \times ps_j^t}{1 + loss_{b_i s_j}} \qquad i=1,2,3,\dots,N; j=1,2,3,\dots,M; t=1,2,3,\dots,T \quad (17)$$

where it shows the ith buyer's desired value of the jth seller, toward purchasing $d_{b_i}^t$ units of power in t^{th} interval. Due to the power loss, it is obvious that a buyer hopes to purchase from a seller who has ability to fulfill its demand completely, with both lower price and power loss attributes. Table 1 demonstrates a sample interest table. Considering all the interest vectors gathered from buyers, dealers combine them to form an interest table, and then commence to run the *Greedy Power Matching* procedure. In each time interval, a random permutation of the buyers is created. Then, each buyer attempts to create its own interest vector. After that, it starts purchasing its required demand from the sellers who are in the first element of its interest vector. After running this procedure for all the buyers, if a buyer fails to satisfy its demand and also is unable to find a seller with available surplus power in its region, then, it will buy the rest from the other sellers in the neighbored regions using the *region to region purchase* sub-algorithm. Similarly in this sub-algorithm, one random permutation of unsatisfied buyers is created at the first and then, each of them creates its own interest vector and search for the appropriate sellers. Finally, if the buyer's demand still is not satisfied yet, it should buy from the macro-station (i.e. *region to macro-station purchase* process).

As a result, finding the optimal sequence of buyers, which want to minimize their objectives, makes it necessary to calculate all the permutations in order to achieve the optimal one. In this regard, the problem is NP-hard and rapidly becomes computationally intractable with growing size of the problem. Thence, evolutionary computation techniques come to help in optimizing the objective and its relevant constraints.

4.1 Evolutionary Computation

Evolutionary Algorithms (EAs) are one of the most well-known meta-heuristic search mechanisms which utilized for the optimization problems [7]. Evolutionary algorithms form a subset of evolutionary computation, in which they generally only involve techniques and implementing mechanisms inspired by biological evolution such as reproduction, mutation, recombination, natural selection and survival of the fittest.

In the evolutionary algorithm, firstly, we generate a random population of feasible solutions for the problem. In this literature, each solution represents the matching tables of the buyers and sellers. Then, these randomly generated solutions try to reach the optimal solution of the problem by using natural parameters as mentioned before in consecutive generations. Considering the randomly generated population in the first step, we used the "decentralized evolutionary greedy power matching algorithm", as mentioned in the previous section. Candidate solutions in the optimization problem play the role of individuals in a population and the cost function. In this process, there are two main forces that form the basis of evolutionary systems [15]:

- **Selection** acts as a force increasing quality.
- **Recombination** and **Mutation** create the necessary diversity and thereby facilitate the solutions' novelty.

In this evolutionary algorithm, we have used a simple tournament selection, with linear crossover and exchange mutation procedures. In addition, an efficient "constraint-handling" sub-algorithm is also presented for enhancing the reliability of feasible solutions [16].

Tournament Selection. Tournament selection is a method of choosing a solution from a set of solutions in the evolutionary computations. It runs a tournament between P randomly chosen solutions (i.e. tournament size=P). Then, the winner (i.e. the one with the best fitness) will be chosen for using in the crossover procedure.

Linear Crossover. Linear crossover procedure gets two solutions and then combines them linearly with the probability of α, in order to produce two other solutions. These new solutions will be the ones in the offspring population. In the revised version of this procedure, we compare the fitness value of these newly generated solutions with their parents and then choose two best ones for using in the mutation procedure.

Exchange Mutation. Mutation is an evolutionary operator used to preserve variety from one solution to the other one, where the solution may change entirely from the previous one. Hence, evolutionary algorithm can come to better solution by using mutation. Mutation parameters depending on the nature of the problem are different. We applied a mechanism, in which the mutation procedure exchanges amount of some negotiated power between two randomly selected buyers and sellers.

Constraint Handling Sub-Algorithm. An efficient constraint-handling technique is a key element in design of competitive evolutionary algorithms to solve optimization problems [16]. We proposed a greedy constraint handling sub-algorithm, in which all solutions of the offspring population will be refined in order to be assure of their feasibility. Regarding the discussing literature, refining is checking the solution in order to find any probable errors (e.g. having "un-satisfied buyers" or "over-sold sellers")in the power matchings.

5 Experimental Evaluation

In this section, the setting of the data used to evaluate the model considering a community of agents and the results obtained from the greedy power matching algorithm are provided.

5.1 Experimental Setup

We begin by clarifying power markets', trading agents' and evolutionary algorithm settings as follows.

Power Markets. A city within a square of 10 km×10 km, in which it encompasses the macro-station which is far from the square, and K regions is simulated. A power market, which has a unique dealer is implemented in each of the regions. All of the dealers cooperate in order to decide on announcing an initial price range such as (1000 ± 500) \$. Due to the problems in actual implementation, the sellers choose an initial price for themselves haphazardly. Having 24 consecutive time intervals, and also analogous to [13], where the surplus and required powers is different for the buyers and sellers, are concerned.

As mentioned previously, there is power loss in power transfer. This loss is calculated as follows:

$$loss_{b_i s_j} = 0.01 \times distance_{b_i s_j} \qquad\qquad i = 1, 2, 3, ..., N; j = 1, 2, 3, ..., M \quad (18)$$
$$loss_{b_i m} = 0.3 \times distance_{b_i m} \qquad\qquad i = 1, 2, 3, ..., N \quad (19)$$

Equation (18) exhibits power loss between i^{th} buyer and j^{th} seller, regarding their distance. Owing to have a *region to macro-station purchase* possibility, (19) is also defined as the power transfer loss between i^{th} buyer and the macro-station.

Trading Agents. N buyers and M sellers are settled randomly in K regions. Each $s \in R_k^S$ and $b \in R_k^B$ has a $q \in Q$ and $d \in D$ respectively. In addition, these parameters are supposed to be uniformly distributed among $((7, 316)$ MW$)$ and $((8, 346)$ MW$)$ for buyers and sellers respectively [13].

Evolutionary Algorithm Setting. We have used *tournament selection* operator (tournament size=2) along with *linear crossover* (crossover percentage=80) and *exchange mutation* (mutation probability percentage=20). Moreover, the generation and population sizes are equal to 400 and 100, respectively.

5.2   Experimental Results

In order to evaluate our proposed approach and algorithms, a community of agents is considered. Hereinafter, all the results' figures are drawn out from the 25×25 community (Fig. 1). Running evolutionary greedy algorithm necessitates randomly generating the first population. Besides, having infinite power production by sellers, informs that this greedy approach results in an efficient and optimal power matching. Unfortunately, this assumption is not practical. Hence, finding the best sequence of buyers in buying procedure is the main challenge in power matching. Finding the optimal solution under these conditions, needs computing all the possible permutation of the buyers.

Figure 2(a) illustrates the fitness diagram obtained from running the evolutionary greedy power matching algorithm on the proposed solution approach. As pictured clearly, the algorithm outputs one solution as the best one in each generation. In the next generations, it tries to improve the initially randomly generated population by the evolutionary parameters to find the better solutions. The descending order of the Fig. 2(a) informs that the algorithm attempts to get as close as possible to the optimal permutation of the buyers. Furthermore, having eagerness to purchase the required powers from the sellers who have announced lower prices is another proof for its descending manner. Although coming to better solution in power matching problem is important, but finding the appropriate and near-optimal solution in a logical and polynomial time is also substantial. Therefore, increasing the population and generation sizes results in increasing the calculation time. As mentioned previously, based on the proposed solution approach and market assumptions, we have used the Integer Linear Programming (ILP) techniques toward solving the matching problem. Hence, due to the NP-Hardness nature of the discussing problem, it is not possible to solve or oven compare with the traditional linear programming techniques.

In order to analyze the robustness of the iterative price updating rule, in which sellers change their announced prices in each interval, we survey the obtained results based on multiple perspective. Figure 2(b) demonstrate all the sellers price changes in successive intervals. This price fluctuations is gathered from the best solution of the last generation. The sellers announced their initial price randomly. Then, in the next intervals, they altered their prices based on their both malleability rate from the market parameter and the weighted-median of other sellers offered prices. For instance, $S9$ and $S24$ have announced their initial prices close to each other. However, having difference in their *malleability rate* has incurred that they could not update their prices in the same order.

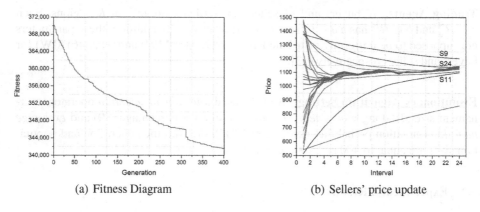

(a) Fitness Diagram   (b) Sellers' price update

Fig. 2. Fitness diagram and sellers' announced price update changes

Along with the malleability rates, the amount of negotiated power is also significant in price updating. Suppose there is a seller which sold all its produced power in a interval. Intuitively, it does not need to modify its next interval's power price rapidly if its malleability rate from the market does not very high. However, there may be a seller which could not sell as much as possible and therefore, it should change its next prices in order to increase its selling amount. Besides, there were some sellers that announced their initial price far from each other (e.g. $S24$ and $S11$). Nonetheless, they attempted to reach the price which the markets preferred. This issue also can be looked at from two perspectives. If a seller has sold all its surplus power at a specific price, probably it has announced a very low price. As a result, the market dealers forced it to change its next prices in order to prevent the market from crisis circumstances. Beside this point of view, the seller might be unable to sell its powers and therefore it has forced to decrease its next intervals' prices to be able to sell its powers. Finally, if an agent, announces an initial price approximately close to the weighted average of the power price, it is not vital to alter its next prices rapidly.

6 Conclusion and Future Work

In this paper, we have proposed a decentralized greedy resource allocation mechanism based on the competition, which manages the demand response problem in power markets. In this regard, a solution approach is proposed, in which electricity markets behave in the benefit of buyers. In this approach, sellers in each interval, based on the markets' available information and their relevant parameters such as malleability rate from the market, offer their own price for each unit of power. Then, buyers create their own interest vectors and start purchasing their required electricity from the sellers who are in the first element of their interest vector. In addition, in order to organize an equilibrium market, considering both of the sellers and buyers who have roles in advancing toward it, a greedy power matching algorithm is also proposed to allocate the energy produced by the sellers and the macro-station to the buyers. Finding a near-optimal power matching solution due to NP-Hardness nature of the matching problems, has motivated us to use evolutionary techniques in demand response optimization regarding the buyers objectives. As some future work, scrutinizing the possibility of having self-learning buyers in power markets along with considering on-line predictive algorithms, considering the markets in the benefit of sellers or in a comprehensive multi-objective manner can be considered.

References

1. Fang, X., Misra, S., Xue, G., Yang, D.: Smart grid the new and improved power grid: a survey. IEEE Communications Surveys & Tutorials 14(4), 944–980 (2012)
2. Rogers, A., Ramchurn, S.D., Jennings, N.R.: Delivering the smart grid: Challenges for autonomous agents and multi-agent systems research. In: AAAI (2012)
3. Albadi, M., El-Saadany, E.: Demand response in electricity markets: An overview. IEEE Power Engineering Society General Meeting, 1–5 (2007)
4. Papavasiliou, A., Hindi, H., Greene, D.: Market-based control mechanisms for electric power demand response. In: 2010 49th IEEE Conference on Decision and Control (CDC), pp. 1891–1898. IEEE (2010)
5. Bichler, M., Gupta, A., Ketter, W.: Research commentary-designing smart markets. Information Systems Research 21(4), 688–699 (2010)
6. Lawler, E.L.: The quadratic assignment problem. Management Science 9(4), 586–599 (1963)
7. Bäck, T., Schwefel, H.P.: An overview of evolutionary algorithms for parameter optimization. Evolutionary Computation 1(1), 1–23 (1993)
8. Dewatripont, M., Maskin, E.: Credit and efficiency in centralized and decentralized economies. The Review of Economic Studies 62(4), 541–555 (1995)
9. Lamparter, S., Becher, S., Fischer, J.G.: An agent-based market platform for smart grids. In: Proceedings of the 9th International Conference on Autonomous Agents and Multiagent Systems: Industry Track, International Foundation for Autonomous Agents and Multiagent Systems, pp. 1689–1696 (2010)
10. HomChaudhuri, B., Kumar, M.: Market based allocation of power in smart grid. In: American Control Conference (ACC), pp. 3251–3256. IEEE (2011)
11. HomChaudhuri, B., Kumar, M., Devabhaktuni, V.: A market based distributed optimization for power allocation in smart grid. In: ASME 2011 Dynamic Systems and Control Conference and Bath/ASME Symposium on Fluid Power and Motion Control, American Society of Mechanical Engineers, pp. 735–742 (2011)
12. Kirschen, D.S.: Demand-side view of electricity markets. IEEE Transactions on Power Systems 18(2), 520–527 (2003)
13. Machowski, J., Bialek, J., Bumby, J.: Power system dynamics: stability and control. John Wiley & Sons (2011)
14. Lasseter, R.H.: Microgrids. In: IEEE Power Engineering Society Winter Meeting, pp. 305–308 (2002)
15. Črepinšek, M., Liu, S.H., Mernik, M.: Exploration and exploitation in evolutionary algorithms: a survey. ACM Computing Surveys (CSUR) 45(3), 35 (2013)
16. Deb, K.: An efficient constraint handling method for genetic algorithms. Computer Methods in Applied Mechanics and Engineering 186(2), 311–338 (2000)

Implicit Predictive Indicators:
Mouse Activity and Dwell Time

Stephen Akuma, Chrisina Jayne, Rahat Iqbal, and Faiyaz Doctor

Department of Computing and the Digital Environment, Coventry University, Coventry, UK
akumas@uni.coventry.ac.uk,
{ab1527,aa0535,aa9536}@coventry.ac.uk

Abstract. Humans vary in their learning behaviour. It is difficult to predict the actual needs of learners through their search activity. It is also difficult to predict accurately the level of satisfaction after the learner finds a perceived relevant document. This research is a preliminary study to examine the predictive strength of some implicit indicators on web documents. An automated study was carried out and 13 participants were given 15 short documents to read and rate according to their perception of relevance to a given topic area. An investigation was carried out to examine if there exists a correlation between user generated implicit indicators and the explicit ratings. The findings show that there is a positive correlation between the dwell time and user explicit ratings. Although there was no significant correlation between mouse movement/distance and user explicit rating, there was a relationship between the homogeneous clusters of the implicit indicators and the user ratings.

Keywords: Implicit indicators, explicit rating, mouse activity, dwell time.

1   Introduction

Information is constantly increasing on the internet leading to information overload. The process of accessing relevant information is not only difficult and time consuming but also cognitively demanding [2]. In order to support users in their search activities, search engine optimisation is required [6]. Implicit feedback and explicit feedback are broadly used to achieve this [6]. Both approaches focus on understanding users' behaviour and interest. In order for users to have recommended information based on their interest, a system must rate every actions of the user. In explicit feedback approach, the most common rating criterion is explicit rating - where users suggest to the system what they think about a given document or information [5]. Predicting users' interest explicitly can be done by users' preference information [19]. Although explicit rating is said to be the most consistent approach in information retrieval, however, its limitation is that it can alter reading and browsing patterns [5]. The implicit approach is therefore more feasible and objective measure. Implicit feedback systems understand the users' interest through the user behaviour. This saves the user the cost of providing feedback [21]. If the implementation of an

L. Iliadis et al. (Eds.): AIAI 2014, IFIP AICT 436, pp. 162–171, 2014.

implicit feedback system is successful, it can be as good as explicit feedback. The advantages of implicit feedback over explicit feedback include:

— Large amount of data can be gathered with no extra cost to the searcher.
— The interaction between a user and the system can be captured at any time.
— With implicit feedback, users need not examine and rate items.

The adaptive engine of most feedback systems uses server-side data like page visited and link selected as its source of implicit data. The challenge in this approach is that the relevancy of a document to be recommended is simply judged by user visit and not by his/her interest on the documents. Client-side or browser approach to data (key, mouse and dwell time) collection provides an efficient alternative for accurately predicting relevant web documents [8]. The aim of interpreting this usage data is to predict user's perceived relevance of a document [4]. Client-side data have become common due to the fact that most browsers use JavaScript as their common technology. Claypool et al. [5] used "The Curious Browser" to examine promising implicit indicators that can predict relevant web documents. They found out that time spent on a page is a good indicator that signifies user's interest on a page. Kim et al [13] also supported this assertion with a focus on user active time on a page and they added that duration on a page is closely related to user's interest. The concept of contextualization was not examined by [5, 13].

In this work, we attempt to examine and correlate some implicit indicators with user explicit ratings based on 15 web documents. We juxtaposed user's interaction on 15 web documents and their explicit ratings of how relevant the documents are to the specific task. We used mouse activity and dwell time to represent the user's interest.

The remaining part of the paper is organized as follows: Section 2 presents a review of related work. Section 3 describes in detail the implicit indicators studied. Section 4 is on the approach used for the study. Section 5 presents and analyse the results. Section 6 is the discussion. Section 7 is the conclusion and future work.

2   Related Work

Morita and Shinoda [15] used implicit approach to transparently capture data. They investigated if the time a reader spends reading Usenet news article can be an important source for measuring relevance. They infer that the longer a user spends reading an article, the more interesting it is to the user.   Nichols [16] discussed the benefits of implicit feedback and he listed some behavioural characteristics that can be used as a source for implicit feedback. He however did not conduct any experiment to examine and validate the efficacy of the implicit indicators listed. Oard et al [18] studied how implicit indicators can be used in place of explicit ratings for a recommender system. They centred their observation on three broad categories of retention, examination and reference as useful criteria for making prediction. They also found that reading time is a good indicator for measuring relevance. This assertion was affirmed by [9, 12, 21].

Claypool et al. [5] studied the relationship that exists between different implicit ratings and explicit ratings on a particular webpage. They developed a web browser ("The Curious Browser") to capture the following implicit indicators: mouse movement, mouse click, scrolling and the time spent on a particular page. The main goal of their study was to collect some implicit indicators that are promising. They found out that although time spent on a particular webpage is a good indicator to predict user interest, a combination of time and scrolling activity gave a better prediction. Joachims [10] study focused on the use of click through as an implicit indicator for measuring relevance. They found out that click through closely follows relevant judgment and can use Ranked SVM algorithm to learn a ranking function. These studies [5], [10] argues that click through by itself is not a significant indicator to measure user interest.

Kim and Chan [13] used experimental measures to examine [16] list of promising indicators. Their focus was to know whether the time spent on a page signifies relevance. 11 students' behavioural characteristics were captured as they search the internet for 2 hours. Their findings suggest that the dwell time and distance of mouse movement are more reliable in predicting user interest than other indicators examined.

Buscher et al [3] examined the correlation between user reading behaviour and user explicit judgment. In the experimentation, participants were given 16 documents to read and rank them in relation to a given task area. An eye tracker was used to measure users' gaze traces on each of the documents. They found out that readers are induced in some way to documents that are of topical relevance. This research is somewhat similar to that of [3]. It however captures user's interest through mouse activity and dwell time.

3   Implicit Feedback Indicators

In this study, a number of implicit indicators were used to capture participant's interest on the given web documents. The implicit behaviour captured by the automated software includes:

- **Active Time Spent on the Document (TS)**
Since users can open multiple browsers and run several applications, it is imperative to note that the active time on a particular window is the period at which the window or web document is in focus. The Active Time Spent (TS) also called Dwell time is the accumulated time spent by a user on an active page during browsing. It starts counting immediately the page is open and stops when focus is moved away from the page or when the page is closed.

Hypothesis
The more the time the page is active, the more it is interesting, informative and of topical relevance [2], [5], [7], [12, 13, 14, 15], [22]. Display time can be used as a substitute for eye tracker for a sophisticated feedback mechanism [4].

Mouse Activities
Most people move the mouse when they read a web document. A user may move the mouse more frequently when viewing a document of interest. The mouse activities are captured only when the current document is in focus. When a user views another application or documents that are not related to the task, the mouse activities are not captured.

- **Distance of Mouse Movement (DMM)**

The distance of mouse movement is calculated by its x and y coordinates on the monitor. The formula is given as:

$$\text{DMM} = \sum_{i=1}^{n} \sqrt{(x - x_i)^2 + (y - y_i)^2} \tag{1}$$

Where x and y are the mouse location along the x and y coordinates of the monitor screen.

Hypothesis
The more the distance the mouse move, the more interesting is the page to the user [5], [11], [13], [20].

- **Total Mouse Movement (TMM)**

This is the total mouse movement calculated by its x and y coordinates on the monitor. The count increment as the mouse hovers on the document.

Hypothesis
The more the mouse hovers on a page, the more interesting the page. Cursor movement complements dwell time information [7].

- **Total Mouse Velocity (TMV)**

This is the total speed covered by the mouse on the monitor. It is given as:

$$\text{TMV} = \sum_{i=1}^{n} \sqrt{(x - x_i)^2 + (y - y_i)^2} \, / (t - t_i) \tag{2}$$

Where x and y are the mouse location along the x and y coordinates. And t is the time of the mouse distance in an interval of 100ms.

Hypothesis
When the mouse speed is low, it shows that the reader is actually reading the document and not skimming it. Therefore, the more the mouse speed the less the user's interest on the page. Cursor movement and cursor speed especially along the vertical axis of a Search Engine Result Page are good predictive indicators of document relevance [7].

- **Explicit Ratings (ER)**

This is the actual user judgement of the visited documents. The buttons for explicit rating was attached on top of each document on a 0 to 5 rating scale. After reading a web document, the user rated the document according to his/her interest and topical relevance.

4   Study Design

The main aim of the study was to capture participant's interest on web documents via some implicit indicators and to correlate the user's interest against their explicit ratings of the given documents. The participants were 4 PhD students, 1 Research assistant, 6 MSc students and 2 undergraduate students. Data for this research was collected by our own automated software developed with JavaScript. The software was injected in 15 web documents to record users' mouse activity, dwell time and explicit rating.   Participants were given a task brief to read and a consent form to complete, after which they were allowed to perform the experiment at their convenient time. They were to login into a website containing links to the 15 web documents and read the documents. The implicit data was captured unobtrusively as the participants read through the documents and then sent to MySQL database when they rated the document by clicking on any of the buttons on the rating scale as shown in fig 2. The task ended after the participants read rated the 15 documents. Fig 1 shows a step to step schema of the task process.

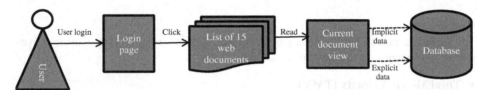

Fig. 1. Step to step schema of the task process

4.1   User Task

The same task was given to all the participants. They were asked to prepare a presentation on the topic - Ethical issues in Big Data. We provided them with 15 documents of equal length containing 350 words with a font size of 20px and a font type of Arial, making the documents one screen view. The documents were created from web articles on ethical issues in Big Data. Two of the documents were however not related to the topic. The participants were asked to read each of the 15 documents and rate them according to how relevant the documents are to the task related topic. The rating was on a scale of 0 - 5. Six buttons were attached on top of the documents and labelled 0 to 5 for explicit rating of the documents: 5 – means very relevant; 4 – means more relevant; 3 – means relevant; 2 – means slightly relevant; 1– means very low relevance; 0 – means not relevant.

The experiment was given a realistic feel by creating a second phase of the task which we called the presentation writing phase. The participants were told that documents will be presented to them according to how they rated them for later use in the presentation writing phase. To avoid Hawthorne effect (the alteration of behaviour by the subjects of a study due to their awareness of being observed), participants were told to do the experiment when and where they are most comfortable. The participants did not actually perform the second phase of the experiment (the presentation writing phase). Fig. 2 shows the user interface of the automated software.

Fig. 2. Login page, Index page with links to the 15 documents and document page with explicit rating buttons

5   Results

This section analyses the data collected from the participants. The analysis is as follows:

5.1   Implicit/Explicit Relationship

Each of the participant's data was analysed separately and as a group. Pearson Correlation Coefficient was used to correlate the parameters of active time, total mouse distance, average mouse velocity, total mouse movement along the x and y axis with user explicit rating. We obtain a correlation coefficient of 0.21 between the user rating and the dwell time. We then ranked the implicit indicators by their predictive strength, using stepwise linear regression to obtain the most promising predictive indicators. The indictors that showed much prominence in relation to the explicit ratings were the dwell time and mouse movement along the x-axis.

The result from the experiment shows a positive correlation between the explicit rating and the dwell time. The dwell time also has a positive correlation with the mouse movement/mouse distance. Fig. 3 shows a Box plot of varying median of the user explicit rating and the dwell time of the participants, and it shows that the values of the ratings for 3 and 4 is the most consistent. The inconsistencies in the other values might be due to noise in the data. The Kruskal-Wallis test on the median for each of the explicit ratings rejected the null hypothesis, meaning that the median values are not the same.

We also found out that although users vary in their reading behaviour, some of them have similar behavioural pattern. We analysed the first two participants' data separately to find out the extent of individual differences. We discovered that they have a relatively similar pattern in dwell time and mouse activity on the documents visited as shown in Fig. 4.

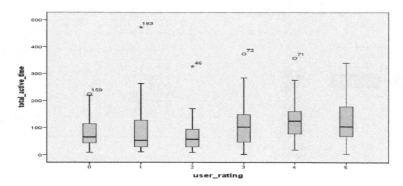

Fig. 3. Graph showing the Boxplot for the combination all the participant's time/explicit rating relationship

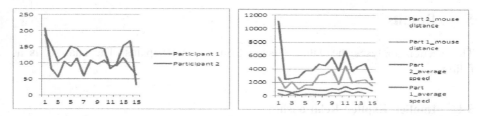

Fig. 4. Graph showing participant 1 and 2 dwell time, mouse distance and average speed on the documents

In order to regroup the user ratings, the five levels of ratings were then reduced into two levels - Relevant and Non-relevant. The user ratings from 0 to 2 were merged together and represented as 0 (Non-relevant) while the ratings from 3 to 5 were represented as 1 (Relevant). A Multilayer Perceptron was used to conduct further analysis on the primary data set (mouse movement along the x and y axis, dwell time and mouse velocity time count) to predict relevant and non-relevant user rating. We obtained a 65% successful mapping with the user ratings after testing the trained data set. Table 1 shows the training and testing of the observed samples.

Table 1. Multilayer Perceptron cluster analysis

Sample	Observed	Predicted		
		0	1	Percent Correct
Training	0	40	27	59.7%
	1	23	45	66.2%
	Overall Percent	46.7%	53.3%	63.0%
Testing	0	13	9	59.1%
	1	10	23	69.7%
	Overall Percent	41.8%	58.2%	65.5%

Dependent Variable: user_rating

6   Discussion

The most promising indicator in the measure of perceived relevance is the active time. Participants spent more time on documents perceived to be of topical relevance. We also found some individual behavioural differences among the participants while reading. The correlation of mouse activity with the explicit rating is relative low probably because of the length of the documents which were of 350 words and could be mostly viewed on screen at once. There is however evidence of a positive relationship between dwell time and explicit user rating. This conclusion is in line with previous research conducted in relation to implicit feedback [1], [12, 13, 14, 15], [17], [22]. The mouse distance/movement are closely related to the dwell time. We can substitute in some way the dwell time by mouse movement or mouse distance in an implicit system.

We can also infer that learners dwell more on documents that are of topical importance and interest to their current activity. The effect of the concept of prior knowledge and cognition on the reader's behaviour was not examined. We assumed that the selected participants barely had knowledge of the task domain. We also observed some individual behavioural differences among the participants. To examine the variety in reading behaviour, we took a closer look at the data for two of the participants (Participants 1&2) and we discovered some level of similarity in their behaviour in terms of dwell time and mouse activity.

Multilayer Perceptron was used to further analyse the primary data set, with user rating as the dependent variable. We obtained a fair result of 65% mapping after training the data set. This suggests a relationship between the user behaviour and their explicit ratings.

Since the experiment was not closely monitored in a controlled environment, we did not get a significant correlation between the user ratings and the mouse activities as hypothesized; we suspect that some of the participants did not follow the instructions carefully as stipulated for the task. This probably added some noise to the data collected.

7   Conclusion and Future Work

Implicit data can be used to predict user's interest on a web documents. When users give explicit feedback, it affects their normal reading or browsing pattern [5]. Implicit methods are a cost effective and objective approach of rating user's interest. This study shows that dwell time on a document is influenced by user perceived relevance and topicality. We also found a relationship between homogenous clusters of user reading behaviour and their explicit ratings.

The next phase of this research is to develop an add-on that will implicitly capture user data on the web as they browse. Additional implicit indicators like copy and paste [6], amount of scroll and keystroke will be examined. The data analysis will be centred on aggregating the most promising indicators and developing a model for effective personalization of relevant documents to learners based on their interest. The effect of document familiarity will also be examined.

References

1. Akuma, S.: Investigating the Effect of Implicit Browsing Behaviour on Students' Performance in a Task Specific Context. International Journal of Information Technology and Computer Science(IJITCS) 6(5), 11–17 (2014)
2. Brusilovsky, P., Tasso, C.: Preface to special issue on user modeling for Web information retrieval. User Modelling and User-Adapted Interaction 14(2-3), 147–157 (2004)
3. Buscher, G., Dengel, A., Biedert, R., Van Elst, L.: Attentive Documents: Eye Tracking as Implicit Feedback for Information Retrieval and Beyond. ACM Transactions on Interactive Intelligent Systems 2(1), 1–30 (2012)
4. Buscher, G., Van Elst, L., Dengel, A.: Segment-level display time as implicit feedback: A comparison to eye tracking. In: Proceedings - 32nd Annual International ACM SIGIR Conference on Research and Development in Information Retrieval, SIGIR 2009, pp. 67–74 (2009)
5. Claypool, M., Le, P., Wased, M., Brown, D.: Implicit interest indicators. In: International Conference on Intelligent User Interfaces, Proceedings IUI, pp. 33–40 (2001)
6. Grzywaczewski, A., Iqbal, R.: Task-Specific Information Retrieval Systems for Software Engineers. Journal of Computer and System Sciences 78(4), 1204–1218 (2012)
7. Guo, Q., Agichtein, E.: Beyond dwell time: Estimating document relevance from cursor movements and other post-click searcher behavior. In: Proceedings of the 21st Annual Conference on World Wide Web, WWW 2012, pp. 569–578 (2012)
8. Hauger, D., Paramythis, A., Weibelzah, S.: Using browser interaction data to determine page reading behavior (2011)
9. Huai, Y.: Study on ontology-based personalized user modeling techniques in intelligent information retrievals. In: 2011 IEEE 3rd International Conference on Communication Software and Networks, ICCSN 2011, pp. 204–207 (2011)
10. Joachims, T.: Optimizing search engines using clickthrough data. In: Proceedings of the Eighth ACM SIGKDD International Conference on Knowledge Discovery and Data Mining (July 2002)
11. Jung, K.: Modeling web user interest with implicit indicators, Master Thesis, Florida Institute of Technology (2001)
12. Kellar, M., Watters, C., Duffy, J., Shepherd, M.: Effect of task on time spent reading as an implicit measure of interest. In: Proceedings of the ASIST Annual Meeting, vol. 41, pp. 168–175 (2004)
13. Kim, H., Chan, P.K.: Implicit indicators for interesting Web pages. In: Proceedings of the 1st International Conference on Web Information Systems and Technologies, WEBIST 2005, pp. 270–277 (2005)
14. Lee, T.Q., Park, Y., Park, Y.: A time-based approach to effective recommender systems using implicit feedback. Expert Systems with Applications 34(4), 3055–3062 (2008)
15. Morita, M., Shinoda, Y.: Information Filtering Based on User Behaviour Analysis and Best MatchText Retrieval, pp. 272–281 (1994)
16. Nichols, D.M.: Implicit Rating and Filtering, pp. 31–36 (1997)
17. Núñez-Valdéz, E.R., Cueva Lovelle, J.M., Sanjuán Martínez, O., García-Díaz, V., Ordoñez De Pablos, P., Montenegro Marín, C.E.: Implicit feedback techniques on recommender systems applied to electronic books. Computers in Human Behavior 28(4), 1186–1193 (2012)
18. Oard, D., Kim, J.: Implicit feedback for recommender systems, pp. 81-83 (1998)

19. Takano, K., Li, K.F.: An adaptive personalized recommender based on web-browsing behavior learning. In: Proceedings of the International Conference on Advanced Information Networking and Applications, AINA, pp. 654–660 (2009)
20. Zemirli, N.: WebCap: Inferring the user's interests based on a real-time implicit feedback. In: 7th International Conference on Digital Information Management, ICDIM 2012, pp. 62–67 (2012)
21. Zhang, B., Guan, Y., Sun, H., Liu, Q., Kong, J.: Survey of user behaviors as implicit feedback. In: 2010 International Conference on Computer, Mechatronics, Control and Electronic Engineering, CMCE 2010, vol. 6, pp. 345–348 (2010)
22. Zhu, Z., Wang, J., Chen, M., Huang, R.: User interest modeling based on access behavior and its application in personalized information retrieval. In: Proceedings of the 3rd International Conference on Information Management, Innovation Management and Industrial Engineering, ICIII 2010, vol. 1, pp. 266–270 (2010)

Normative Monitoring of Agents to Build Trust in an Environment for B2B

Eugénio Oliveira, Henrique Lopes Cardoso, Maria Joana Urbano,
and Ana Paula Rocha

LIACC, DEI / Faculdade de Engenharia, Universidade do Porto
R. Dr. Roberto Frias, 4200-465 Porto, Portugal
eco@fe.up.pt

Abstract. Agents intending to be involved in joint operations need to rely on trust measures pointing to possible future solid partnerships. Using Multi-Agent Systems (MAS) as a paradigm for an electronic institution framework enables both to simulate and facilitate the process of autonomous agents, as either enterprises or individual representatives, reaching joint agreements through automatic negotiation. In the heart of the MAS-based electronic institution framework, a Normative Environment provides monitoring capabilities and enforcement mechanisms influencing agents' behavior during joint activity. Moreover, it makes available relevant data that can be important for building up contextual-dependent agent's trust models which, consequently, also influence future possible negotiations leading to new and safer agreements. To support agents data generation and monitoring, we here present ANTE platform, a software MAS integrating Trust models with negotiation facilities and Normative environments, for the creation and monitoring of agent-based networks.

1 Introduction

In Multi-Agent Systems (MAS), the role of an Environment has been defined by Odel and Weynes et al [1] as a support for agents to exist, and the medium for agents to communicate. However, other than being a kind of a passive component providing services to the agents and facilitating, upon request, the Environment can also be seen as an active component, gathering and exploiting all kinds of information produced during MAS activity. This is mostly the case when the MAS is seen as a support for electronic institutions, a basis for automatic B2B operations [2][5]. In this paper we look at the MAS environment as a component that actively monitors all the relevant information produced by different agents engaging in activities related with both creating joint agreements and coordinating actions that enable their execution. Information, produced along these stages becomes useful for inferring relevant trustworthy capabilities of the agents in the network. In this paper, we highlight the approach taken in ANTE (Agreement Negotiation in Normative and Trust-enabled Environments) platform [4] aiming at creating an active Environment that gathers information related with agents'

L. Iliadis et al. (Eds.): AIAI 2014, IFIP AICT 436, pp. 172–181, 2014.

activity for the sake of decision-making [6]. ANTE encompasses a normative framework guiding agents interaction through norms that apply either sanctions or incentives. Useful information regarding agents' performance will influence future interactions and possibly agents joint work [3].

The structure of the paper is as follows: After this introduction, we first describe how to use MAS environment, based on the power of a normative environment, to produce and gather relevant information about agents activity. Then, in section 3 we introduce a computational Trust Model that benefits from such information in order to assign trust measures to agents, preventing an agent from future involvement with untrustworthy partners. Section 4 briefly presents the ANTE platform that provides all the needed services plus the environment for agents to interact in a trustworthy way. We close the paper with a brief conclusion.

2   The Relevance of the Environment

Many sophisticated problems we usually address can be classified as belonging to the Triple-D class of problems [4]. They reflect a reality that simultaneously is of a Distributed, Decentralized as well as Dynamic nature. This means that, besides input data and output actions being disperse (Distributed) at different nodes of a network, also, and most important, decision-making can be, at least partially, taken at different nodes of the (Decentralized) system. Moreover, the system trying to solve the overall problem at stake, has to deal with a changing, evolving reality (Dynamic). To work jointly in solving this kind of problems, agents need to select, in run-time, their best trustworthy partners for executing the tasks through mutual agreements to coordinate joint work. In the core of software agent's definition we find the "situatedness" property and, thus, the Environment becomes an important component to be specified and designed for multi-agent systems.

As a relevant component of MAS, although not always considered per se, Environment can be seen as an active medium to facilitate the way how to reach certain consensus, agreements or mutual contracts. Instead of looking at the Environment as a regulator through the internal mechanisms of a Normative environment [2][5][3], we here see the Environment as a medium in which notifications about the outcomes of actions agreed upon, after the establishment of joint contracts, appear together with the output of the Normative Environment that assesses them and, then, combines all those pieces of information in meaningful knowledge for future use. More precisely, through the environment embodied in ANTE, the software platform we have developed, relevant data is made available for building up contextual-dependent trust models which, as a consequence, will influence future possible agents negotiation leading to new, and better, agreements.

2.1 Environment as a Facilitator

It is out of the scope of this paper to describe other environment facilities as it is the case of the negotiation protocols and mechanisms we advocate [6] in the ANTE platform, for a dynamic and adaptive convergence towards final agreements then materialized into contracts. We, along this paper, want to mainly highlight the importance of the Environment in recording the relevant events produced by the agents and the normative environment outputs as a result of agents' activities during contract enactment. Appropriately combining the trace of such events, leads to a tightly coupled connection between electronic contract monitoring and a computational trust model for estimating agents' trustworthiness. The Normative dimension of the Environment handles a normative state NS (including all elements that characterize the current state of affairs for each established agreement), a set IR of institutional rules that manipulates NS and a set N of norms, which can be seen as a special kind of rules. NS includes institutional facts that represent institutional recognition of contract related real-world events. The Normative Environment also notifies the agents about relevant contract- related events. The same mechanism conveys information to feed the computation Trust model.

A contract C = $< T; CA; CI >$ is monitored by the normative environment and is defined as follows: A contract C is a relation of type T within which a group of agents in set CA commits to a joint activity, under the normative context using information CI, a set of definitions regarding the role of the participants, the values to be exchanged (e.g. products and money) or any parameters defining their provision. An electronic representation of a contract includes a set of norms that specify how a specific business is to be enacted. A norm is a rule whose conditions analyze the current NS and whose conclusion prescribes obligations agents ought to fulfill. As a consequence, sanctions may be imposed by prescribing obligations upon violation events, thus producing more relevant information in the environment. In the context of a MAS in which agents represent entities looking for partners to establish agreements and commit themselves to joint actions, the role of a normative environment [3] is twofold. Given an agreement, it is necessary to check if the partial contributions of individual agents make their way in enabling a successful execution of tasks leading to the agreed overall goal. In many cases, the execution of the needed tasks is itself distributed, which requires agents to enact by themselves their part of the agreement. Monitoring this phase is therefore an important process and is possible through the analysis of the interactions and facts produced, and made visible, in the environment. Furthermore, it is possible that after successfully negotiating and reaching agreements, self-interested agents are no longer willing to fulfill their commitments, which becomes also visible to the environment, since normative rules generate notifications (prescribing sanctions) in reaction to such events.

As an example, the normative environment may apply norm n enforcing a sanction (e.g. a payment) when a violation regarding delivery has been detected. This puts in evidence the second role of a normative environment, that of enforcing norms by coercing agents to stand for their commitments. At the same time, the whole idea of an active environment is to use the relevant information generated through the normative environment and make it useful for further agents' reasoning strategy.

2.2 Information in the Environment

The way agents enact their contracts provides relevant information for trust building and, as a consequence, may decisively influence partners selection for future electronic contracts. In ANTE, an image of all the relevant real-world transactions between different agents become recognized as "institutional reality" [8]. As stated above, a normative state NS records every element that is relevant in contract enactment and is composed of institutional reality elements. Agents active in the Environment subscribe to the normative environment to be notified about eminent contractual obligations and commitments to be fulfilled. Consider the following representation of a directed obligation with time constraints: $Obl_b, c(l < f < d)$. This is an obligation of agent b (the bearer) towards agent c (the counterparty) to bring about fact f between liveline l and deadline d. An obligation such as this can give rise to a rich set of events, such as temporal violations, due fulfilment or violation. As agents go on interacting according (or not) with pre-established contracts, the normative environment reacts appropriately, and produces more (normative) information about the on-going processes. The challenge is to be able to gather and manage all that produced data, to infer meaningful information about agents' future behavior.

Moreover, other than information generated during two agents' mutual interaction, there is other available information, coming from third parties, that may be relevant to characterize a specific agent. These indirect evidences about an agent's behavior (an "image" the agent builds up in the society of all currently active agents, due to previous interactions), is called reputation. Reputation is seen as the social-based process of transmitting beliefs about a specific agent (the trustee), as a consequence of social evaluation circulating and, according to [9], are represented as reported evaluations. There are computational trust models that use reputation as another piece of evidence to be taken into account about any particular agent. For that purpose, they need to estimate the credibility of both the transmitted information and of those agents reporting that information. Providing that third party agents, not directly involved in a particular negotiation are willing to provide their own perspective on a particular agent, this new piece of evidence will also be present in the environment and can also be identified as relevant to be merged with direct evidences [10]. The perspective an agent is now able to infer, through the combination of the available information, about another agent becomes wider and more independent of local knowledge. However we still did not address the problem that could arise if reputation information and other available trust-based measures happen to be of different nature and formats. How to merge them, in a coherent way, through the computational Trust and Reputation Model is a challenge for our near future research.

3   Trust

We follow a basic and established definition for Trust (although not always consensual). Closely, but not completely following some of the ideas expressed in [7], we propose the following formal definition for trust: Trust(i; j; φ) meaning that the Truster(i) trusts Trustee(j) to do Action(α) leading to the achievement of Goal(φ) if:

$(GOAL_i\varphi)$ $(BEL_iPOWER_j\varphi)$ $(BEL_i\Box(\alpha \models \varphi))$ $(BEL_iINTEND_j\alpha)$,
where $\Box$ is the usual 'necessity' modal operator, $POWER_j\varphi$ states that agent j is capable of achieving φ and $INTEND_j\alpha$ is agent j's intention to execute action α. Intention means choice with commitment according to Cohen and Levesque's well accepted theory.

Agents should thus rely on Trust measures to select their best potential partners, improving their chances to achieve their own future goals. It has been established that, whenever decisions have to be made for selecting potential partners in a future activity, it becomes mandatory to rely on some kind of past evidence pointing to an estimate of the other agents' trustworthiness. We need to remember here that Trustworthiness is an intrinsic characteristic of an entity (here an agent) while Trust measures reflect other ones' (potentially different) assessment of that particular agent under evaluation. Exhaustive discussions are taking place about all the relevant factors that can influence a trust measure characterizing the perspective each agent builds up on other agent internal trustworthiness [7]. We may identify two different categories of such factors: subjective (propensity and disposition) and, most valuable, objective (past experience regarding previous direct interactions, current social image of the agent under scrutiny, the reputation and other indirect evidences). Another perspective over the computational trust model is if, indeed, a single measure is enough to measure an agent's trustworthiness. In fact, although we often look for competence, it sometimes is not the only dimension that matters for successful future joint activities. Other more subtle features like how benevolent and integer an agent can be, may be seen as determinant in deciding on how much shall an agent trust in another one. Another important concern is the contextual nature of trust: Situation, task and time dependence are different factors to take into account when measuring and ranking other agents' trust.

3.1   Computational Trust Models

There has been a number of proposals for computational trust models that rely in one or several of the factors and dimensions referred above. Direct past mutual interaction experience is the most common factor these models take into consideration as it is the case of [11]. A few computational models already include some kind of agent disposition to trust and emphasize the importance of context [12] [13]. Moreover, only recently we have called the attention to the importance of the dynamics of the relationships taking place, along the time, between the different agents, leading to a relative change of the power relationship and the evolution of mutual goodwill [13].

Proposed Trust Model. Our computational Trust model, included in the ANTE software platform, has different components responsible for managing most of the factors recognized as influencing trust-based decision-making. As a model, it encompasses both the different dimensions of trustworthiness (integrity, ability, benevolence) and the truster's disposition to trust. As a software program, implementing most of the referred important aspects, although not yet all of them, it already includes an aggregator of past direct experience, Sinalpha, a Contextual Fitness component, a Social and an Integrity Tuner. Through such model we gain the ability to compute adequate Agent i estimations of Agent j trustworthiness to achieve a specific goal φ, in different environments, including those of high dynamics, where evidence on the agent under evaluation is scarce.

The two main components first aggregate past contractual behavior evidences (Sinalpha) and then consider how agents fit into the context under consideration (Contextual Fitness). The aggregator gets information from the environment about agents' performance regarding their past obligations (either fulfillment, partial fulfillment or violation). It uses a sinus-based curve that is reshaped at both top and bottom extremities [6]. It uses a function of α that presents a sinusoidal shape (see Equation 1). By setting $\delta = +0.5$, the trustworthiness value is restricted to the range [0;1]. $trustworthiness = \delta * (sin\alpha + 1)$ (1) The trustworthiness score of the agent is minimum when $\alpha = 3\pi/2$ and maximum at $\alpha = 5\pi/2$. This score is updated using $\alpha(i+1) = \alpha(i) + \lambda * \omega$, where λ reflects the outcome (either positive or negative) associated with the piece of evidence being aggregated and where parameter ω defines the size of the ascending/descending step in the trustworthiness path. For example $\omega = 6$ means that any trustee that is a newcomer, could be considered by the truster as fully trustworthy after presenting six outcomes of type F (fulfilled) in a row. Asymmetry about gaining and loosing Trust, maturity after reaching a certain status and distinguishing different patterns of past behaviour are properties that are featured by the model. It has been proved that the way the aggregator merges past evidences, gives a realistic idea of, although just for simple situations, agents' expected behaviours in future similar situations.

The second component of the model, responsible for social awareness, counts for the specific appropriateness of the target agent (Trustee) regarding that particular situation in which the other agent (Truster) is willing to get help, to work jointly or to engage in a relationship with [13]. It is based on the concept of a Context that captures and represents the relevant information about the current situation, that Contextual Fitness appreciates how an agent can be, at that specific time, trustworthy or not. A Context is here defined as an N-tuple including agent identity, current time, location and a set of task-related attributes: type, complexity, deadline and expected outcome.

Contextual Fitness component, a situation-aware tuner may downgrade the trustworthiness scores computed by the aggregator (Sinalpha) whenever the agent under assessment has proved to behave poorly in that specific situation. We are using the information gain metric on the previous evidences on the agent different tendencies of failure. This approach differs from other situation-aware computational trust approaches by its flexibility and ability to reason in terms of context even when the evidence on the agent in evaluation is scarce. Briefly, we endow the environment with the capabilities of clustering different potential partners waiting for being selected and extract stereotypes characterizing in what conditions they have either fulfilled or violated their obligations. Therefore, it may be the case that a particular agent has a good trust measure, however it has an handicap regarding a specific feature that is of most relevance for the concrete partnership another agent is interested in (for example, a faster delivery time of a product). The most important to stress here is that the environment is empowered with the means to strongly influence the decision-making process of agents, by providing indicators that, once combined, become relevant for trust evaluation.

4 ANTE Platform

Grounded on the recognition of the Environment as a relevant component to give structure and support situated MAS, we have developed ANTE, a software framework in which agents can both establish agreements (and contracts) through negotiation and, simultaneously, benefit from monitoring capabilities pointing to more transparency of joint activities. Through the development of ANTE we intended to illustrate how an active environment may help in reliable and informed agents decision-making in what joint agreements is concerned [6]. ANTE supports multi-agent collective work leading to agents' intention satisfaction, by providing negotiation as a mechanism for finding mutually acceptable agreements with trusted partners and the enactment of such agreements (contracts). Gathering appropriate data and evaluating the Normative State, enables feeding the Computational Trust model which, in turn, provides information for better selecting partners for future agreements.

In ANTE, the environment actively mediates interaction among agents in several different ways: i) The Environment is, primarily, a Facilitator by providing negotiation protocols, Ontology translation services and Contract building tools. In this perspective we can see it as an Interaction-Mediation level [1] regulating access to resources and mediating interaction between agents leading to possible agreements. ii) The Environment also provides mechanisms for the normative enactment of those agreements (established contracts among agent partners), generating Normative States (NS) and becoming responsible for monitoring those NS in an active way. ANTE includes a normative framework of constitutive and institutional rules, according to the institutional reality model, operating in a rule-based process. iii) Through evaluating the contract enactment phase, the Environment improves the chances for better future negotiations by progressively updating the available Trust measures to help in the partners' selection process. ANTE includes, at present, a computational Trust model made up of two main modules: SINALPHA aggregator and contextual fitness. iv) In the ANTE specification model, another component acts as a "social Environment" and is responsible for adaptive deterrence mechanisms (through dynamically calculated fines imposition) reacting to different agents population behaviors. In Figure 1 we display how the active environment is subject to the simultaneous intervention of the agents activity (negotiation and contract execution) which leads to "institutional facts" monitored in the active environment. Those "institutional facts", in turn, feed and update the computational Trust Model.

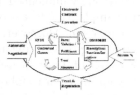

Fig. 1. The Active environment. Facts (clauses violation/fulfillment) activate norms leading to possible sanctions, changing agents trustworthiness and, thus, influencing the way agents negotiate future contracts.

Monitoring rules are responsible for gathering all the needed information that becomes relevant to feed the computational Trust model which dynamically combines that information to infer agents' trust measures. This last component is, however, meaningless in the context of what is the aim of this paper: information merging to dynamically feed an agents' trust model. ANTE is a modular and extensible JADE-based architecture implementation, integrated with JESS rules engine, accessible through several different GUIs for agents to announce their needs as well as for inspecting the whole negotiation process, normative states and contract results. Although ANTE has been targeting electronic contracting in B2B domains, and because it encompasses several agreement technologies like negotiation, normative environments and computational trust models, it was conceived as a more general framework having in mind a wider range of application domains including social networks monitoring.

Experiments Assessment. We have used ANTE platform both as an environment as well as a set of services suitable for achieving agents agreement in a B2B scenario. As an environment it became a useful interface between negotiation services, normative environment services and computational trust services. Many different experiments have been performed, mainly involving agents as enterprise representatives negotiating several different attributes of particular goods (fabric type, quality and time of delivery). Since the description of a full range of different partial and integrated experiments we have made, can be found in [14], we here just stress out a qualitative assessment about the summarized results coming out of the use of Trust during the partners selection negotiation phase together with normative sanctions application. Different experiments enable the agents to use Trust either to pre-select the most promising potential partners for the future joint activity (to supply a certain quantity of a certain type of fabric, until a specific deadline) or to use trust only during the negotiation phase, to better evaluate proposals under scrutiny. Other experiments combine these two possibilities. Also, several different metrics have been used. However, we here only report on the metric reflecting the number of well succeeded contracts that have been established following the use of these referred techniques. After a number of experiments we concluded that, by using the Computational Trust Model, the outcomes were strongly influenced, besides the agents behavior type, by other factors like, for example, the degree of similarity or dissimilarity of the agent proposals under comparison. Moreover, one of the most important conclusions was that using trust for pre-selecting the agents for future negotiation together with a moderate use of Trust measures during the negotiation process itself, was the best policy for an agent to follow. Using Trust for selecting partners increased the number of succeeded contracts in this B2B scenario.

5 Conclusions

Distributed, decentralized systems for dynamic situations, appropriate for Triple D kind of problems, are well represented by means of Multi-Agent System architectures. Distributed Agents, representing different entities (like enterprises), take their own decisions (selecting partners) in a decentralized way. Moreover, they react to dynamic situations (like other agent trustworthiness change) by appropriately using trust measures to help on their own decision-making (which partner to select). Following the same line of those authors who claim that the Environment should be considered as an important, separate component in the specification and design of Multi-agent Systems, we here advocate a set of capabilities that should be made available for more secure, comprehensive and trusted agents mutual activities possibly leading to more fruitful joint agreements, as it is the case of B2B e-contracts. Besides facilitating functionalities like negotiation protocols, ontology services, normative framework application and monitoring, we here, in this paper, strongly emphasize the importance of an Environment active role also as a support for information fusion. Gathering and evaluating information related with normative states and agents behavior regarding norms abidance and contracts fulfillment is of paramount importance for feeding computational Trust models. Combining agents activity-driven information, leading to updated and contextualized agents' Trust measures become crucial to guide future agents' activity and mutual agreements. A "social-based" Environment can also have the privilege of, by reasoning about all agents activity, to detect, prevent (or incentivize), patterns in the society regarding maleficent (or beneficial) agents' behavior. Moreover, and having in mind its active role to ease and promote possible agent agreements, "social Environment" has been empowered with the capability of adaptively impose fines for the sake of better regulate the agent population activities and promoting confidence in the possibility of future better agreements. This aspect of the environment lacks of further investigation. However, in this paper, we mostly emphasize how, through a normative framework responsible for monitoring agents contractual behavior and computational

Trust models, the environment provides relevant evidences that, in the end, enables an agent to better characterize other potential partner agents and better decide about either to engage or not in a future relationship and better agreements, as it this the case for B2B contracts negotiation scenario. ANTE software platform has been developed taking all these features into consideration.

References

1. Weyns, D., Omicini, A., Odell, J.: Environment as a first class abstraction in multiagent systems. J. Autonomous Agents Multi-Agent Systems 14, 5–30 (2007)
2. Dignum, F.P.M., Broersen, J., Dignum, V., Meyer, J.-J.: Meeting the deadline: Why, when and how. In: Hinchey, M.G., Rash, J.L., Truszkowski, W.F., Rouff, C.A. (eds.) FAABS 2004. LNCS (LNAI), vol. 3228, pp. 30–40. Springer, Heidelberg (2004)
3. Cardoso, H., Oliveira, E.: Social Control in a Normative Framework: An Adaptive Deterrence Approach. Web Intelligence and Agent Systems 9(4), 363–375 (2011)
4. Oliveira, E.: Software Agents: Can we Trust Them? In: Proceedings of IEEE 16th Int. Conf. on Intelligent Engineering Systems, INES (2012)
5. Cardoso, H., Oliveira, E.: Electronic Institutions for B2B Dynamic Normative Environments. Artificial Intelligence & Law (Special Issue on Agents, Institutions and Legal Theory) 16(1), 107–128 (2008)
6. Cardoso, H.L., Urbano, J., Brandão, P., Rocha, A.P., Oliveira, E.: ANTE: Agreement Negotiation in Normative and Trust-enabled Environments. In: Demazeau, Y., Müller, J.P., Rodríguez, J.M.C., Pérez, J.B. (eds.) Advances on PAAMS. AISC, vol. 155, pp. 261–264. Springer, Heidelberg (2012)
7. Falcone, R., Castelfranchi, C.: Trust and deception in virtual societies. In: Castelfranchi, C., Tan, Y.-H. (eds.), pp. 55–90. Kluwer A. P., Norwell (2001)
8. Searle, J.R.: The Construction of Social Reality. Free Press, New York (1995)
9. Sabater-Mir, J., Paolucci, M., Conte, R.: Repage:reputation and image among limited autonomous partners. J. of Art. Societies and Social Sim. 9(2), 3 (2006)
10. Urbano, J., Rocha, A.P., Oliveira, E.: A Socio-Cognitive perspective of Trust. In: Ossowski, S. (ed.) Agreement Technologies, ch. 23. Law, Governance and Technology Series, vol. 8, pp. 419–431. Springer (2013)
11. Sabater, J., Sierra, C.: REGRET: Reputation in gregarious societies. In: Proc. of the 5th Int. Conf. on Autonomous Agents, AGENTS 2001, pp. 194–195. ACM (2001)
12. Rehak, M., Gregor, M., Pechoucek, M.: Multidimensional Context Representations for Situational Trust. In: Proc. of the IEEE DIS 2006: Collective Intelligence and Its Applications, pp. 315–320. IEEE Computer Society (2006)
13. Urbano, J., Rocha, A.P., Oliveira, E.: A dynamic agents behavior model for computational trust. In: Antunes, L., Pinto, H.S. (eds.) EPIA 2011. LNCS, vol. 7026, pp. 536–550. Springer, Heidelberg (2011)
14. Urbano, J.: A Situation-aware and Social Computational Trust Model. PhD Thesis Report, Faculty of Engineering, University of Porto (2013)

Extending the Kouretes Statechart Editor for Generic Agent Behavior Development

Georgios L. Papadimitriou, Nikolaos I. Spanoudakis,
and Michail G. Lagoudakis

Technical University of Crete, Chania, Crete, Greece
{gpapadimitriou,nispanoudakis,lagoudakis}@isc.tuc.gr

Abstract. The development of high-level behavior for autonomous robots is a time-consuming task even for experts. The Kouretes Statechart Editor (KSE) is a Computer-Aided Software Engineering (CASE) tool, which allows to easily specify a desired robot behavior as a statechart model utilizing a variety of base robot functionalities (vision, localization, locomotion, motion skills, communication) for the Monas robotic software architecture framework. This paper presents an extension to KSE, which allows defining generic agent behaviors using automatic framework-independent code generation, as long as the underlying framework is written in C++. This way a user can program physical (robots) or software agents that can be executed on any platform using any compatible software framework. This paper demonstrates the transparent use of the extended KSE in the SimSpark 3D soccer simulation and the Wumpus world.

Keywords: agents, robotics, model-based development, CASE.

1 Introduction

The Kouretes Statechart Editor (KSE) [5,7] is a graphical Computer-Aided Software Engineering (CASE) tool, which allows a software developer to define the behavior of an agent (software or physical) using a graphical model and automatic code generation. The graphical model is a statechart, a platform-independent model, modeling the control of the different agent capabilities. A complete statechart is automatically converted by KSE to source code, which can be compiled, linked with the base capabilities, and executed on the target platform. KSE follows ASEME, a model-driven software engineering methodology from the Agent-Oriented Software Engineering (AOSE) domain [6]. An advantage of KSE is the analysis tool, which enables the user to abstractly and compactly define the desired behavior using liveness formulas. A small set of liveness formulas can lead to a large statechart model (design phase), therefore the user can save a lot of time by seeding the design through the analysis tool.

In our recent work, we used KSE extensively for developing the behavior of the Aldebaran Nao humanoid robots of our team Kouretes [www.kouretes.gr] competing in the RoboCup Standard Platform League (SPL). Besides robotic

L. Iliadis et al. (Eds.): AIAI 2014, IFIP AICT 436, pp. 182–192, 2014.

behavior development, KSE has also been used for developing the behavior of software agents for the popular Java-based JADE framework [jade.tilab.com].

The motivation behind the work presented in this paper is the need for using a simulator, when modeling a robotic team behavior. Regularly testing new features on the real robots has several shortcomings, i.e. the robots need maintenance after some hours, a number of people are needed in order to set up an experiment with the robotic team in the lab, and the experiments themselves tend to take quite long to setup and demonstrate. Thus, we decided that we needed to use a simulator for modeling and testing team behavior and, when the simulation proved successful, then move to field tests with the real robots. The SimSpark simulation environment, which is also used for the RoboCup 3D soccer simulation league, was the ideal candidate for our goals. However, the SimSpark platform was not compatible with our Monas robotic software architecture framework [4], which we use for deploying behaviors on the Nao robots. Thus, we decided to develop a platform-independent code generation component for the KSE tool. To this end, we added a few platform-specific parameterization features within code generation. By adopting this parametric approach and providing the correct parameters for the underlying platform, we can now use the KSE tool for deploying platform-independent agents on any platform by exporting the generated code directly in the C++ programming language using the specification provided by the parameters.

This paper focuses on describing the extensions of KSE and demonstrates its practicality on two diverse domains: the SimSpark[1] simulation platform for 3D robot soccer and the Wumpus World[2] simulator, a classical testbed for prototyping Artificial Intelligence programs. In the rest of the paper, after examining the background technologies in Section 2, we present the main features of KSE, focusing on new additions and extensions, in Section 3. Subsequently, we present the results of our application and evaluation on the two aforementioned domains in Section 4. Finally, we outline our findings and future work in Section 5.

2   Background

ASEME [6] is an Agent-Oriented Software Engineering Methodology that supports a modular agent design approach and introduces the concepts of intra- and inter- agent control. The former defines the agents behavior by coordinating the different modules that implement its own capabilities, while the latter defines the protocols that govern the coordination of the society of the agents. ASEME follows the Model-Driven Engineering (MDE) paradigm, where software development is driven by instantiating the appropriate models at each phase.

---

[1] SimSpark is a generic simulator for various multi-agent environments. It supports developing physical simulations for AI and robotics research with an open-source application framework. [simspark.sourceforge.net]

[2] The Wumpus World simulator is a simple C++ framework for simulating the Wumpus World described in Russell and Norvig's classic text book "Artificial Intelligence: A Modern Approach". [www.eecs.wsu.edu/~holder/courses/AI/wumpus]

The transition from one development phase to another is assisted by automatic model transformation leading from requirements to computer programs. The ASEME platform-independent model, which is the output of the design phase, is a statechart [2], and is referred to as the Intra-Agent Control (IAC) model.

KSE adopts the ASEME methodology and assists the developer from the analysis phase to the design and code generation phases. More specifically, KSE supports (a) the automatic generation of the initial abstract statechart model using compact liveness formulas, (b) the graphical editing of the statechart model and the addition of the required transition expressions, and (c) the automatic source code generation for compilation and execution on the robot. KSE has been developed using the Eclipse Modeling Project [www.eclipse.org/modeling] technologies and has been integrated with the Monas software architecture [4], which provides the base functionalities.

Monas modules focus on specific functionalities (vision, motion, etc.) and each one of them is executed independently at any desired frequency completing a series of activities at each execution. Statechart modules are executed using a generic multi-threaded statechart engine, which was built on top of existing open-source projects, provided the required concurrency, and met the real-time requirements of the activities on robots. Code generation in KSE initially targeted specifically the Monas framework.

3   The KSE Generic C++ Generator

The KSE Generic C++ Generator (referred to as *GGenerator* in the rest of the paper) is a tool that extends KSE in order to give the developer the ability to create behaviors for agents, which are automatically transformed to generated C++ code ready to be compiled with the underlying software architecture (assumed to be also in C++) without any additional dependencies. Thus, GGenerator enables the user to create agent behaviors for different platforms and different environments, with the only prerequisite that the underlying software framework accepts C++ code.

We achieve this (a) by enhancing the statechart engine used in KSE with a generic blackboard [3] interface, (b) with a new generic grammar for specifying transition expressions, and, (c) by creating a new source code generator for the extended tool.

The blackboard interface is generated automatically by GGenerator during the source code generation phase and can target any C++ framework, according to its initialization and the parameters given. This interface is used to *bind* our statechart model with any user-specified framework. This way the user is not concerned with model-framework compatibility issues, because the created blackboard interface serves as a middleware between the statechart model and the target environment and handles their communication. This interface, along with our generator, is integrated in KSE.

The initialization of the blackboard interface depends on the *properties files*, which are defined by the user only once during the early stages of the behavior development process. These files are named include_classes.txt and `instances.txt`. Both files are given in plain text format, they are parsed by the KSE GGenerator tool, and provide the required information for generating and setting up correctly a blackboard interface that targets a specific framework. In these files we add functions, variables, and header files needed for the communication between our statechart model and the target environment. The former file lists the C++ header files we need to include, while the latter lists the functions and the variable instances we need to access.

The creation of the properties files requires some knowledge about the framework we target. Prior to generating our blackboard interface, we have to point out the framework's modules and structures that supply the information needed for correct communication between our statechart model and the environment. In most cases, the information needed for creating an agent behavior includes the updated state of the target environment or an updated set of readings from the agent's sensors, without being limited to these examples, since we can always accept any kind of information the framework provides. The properties files are easily created, once the required information has been identified and located in the target framework. In the next section we give two examples of how to create the properties files for two different platforms.

We also allow the user to register user-defined variables in the same blackboard interface using the graphical editing interface of KSE. These variables can be of any type, can be updated in any node of the statechart, and can be used in transition expressions anywhere in the statechart, along with any framework-provided variables. This ability allows the behavior module itself to maintain its own internal memory.

Finally, to execute the generated behavior on the target framework, we need to register our generic blackboard interface along with our statechart model for execution on the target framework. This is the last stage of our behavior development process and needs to be done only once for a target framework; any number of generated behaviors can be tested afterwards on that framework.

As already mentioned, within the KSE tool, the definition of a statechart can be assisted by the use of liveness formulas as a seed. The *liveness formula* [8] is a process model that connects activities using the Gaia operators. Briefly, A.B means that activity B is executed after activity A, A* means that A is executed zero or more times, A+ means that A is executed one or more times, A~ means that A is executed indefinitely (it resumes as soon as it finishes), A|B means that either A or B is executed exclusively, A||B means that A and B are executed concurrently, and [A] means that A is optional. Liveness formulas are automatically converted to a base graphical statechart model.

Every statechart model we create to describe an agent behavior must contain transition expressions, which are responsible for the correct execution of the statechart. These expressions are inserted manually by the user during behavior specification for controlling the statecharts execution. In transition expressions, the user can optionally define *events* and *conditions*, as well as multiple *actions*, if desired, using any framework-specific or user-defined variables. In any case, these variables must be registered in the blackboard's interface for transition

```
transExpression = [ event ] [ "[" condition "]" ] [ "/" actions ]
event           = string
condition       = "(" condition ")" | condition operator condition
                  | operand operator condition | operand operator operand
operator        = ">" | "<" | ">=" | "<=" | "==" | "!=" | "&&" | "||"
operand         = constant | variable
variable        = letter string | letter string "." variable
actions         = timeoutAction | action
timeoutAction   = TimeoutAction "." letter+ "." number
action          = variable actionOp variable | action ";" action
actionOp        = "="
constant        = number | string
number          = digit | digit number
string          = character | character string
character       = letter | digit
letter          = "a" | "A" | "b" | "B" | "c" | "C" | ...
digit           = "0" | "1" | "2" | "3" | ...
```

Fig. 1. Transition expression grammar in EBNF format

expressions to work properly. Note that the user is provided with the ability to create any kind of variables he/she finds necessary for the transition expressions during the graphical editing of the statechart model.

Since we modified the statechart engine used to execute KSE-created statecharts by adding a generic blackboard interface, we also needed a new C++ code generator different from the existing one, which was customized for the Monas framework. At this point, we also changed the transition expressions grammar to be equally generic. The new statechart engine supports a more abstract transition expression grammar compared to the one used for the Monas architecture [5]. This means that it should be relatively easy for a user to edit them, even without having complete knowledge of the underlying framework he/she works with. The new grammar is independent of the underlying platform, and uses common syntactic rules for expressions. It is shown in EBNF format in Figure 1. We created the new C++ generator for our extended statechart engine using the Eclipse Modeling Framework (EMF[3]) along with Xpand[4]. We fully integrated our GGenerator extension in KSE, so that the generated behavior can be easily tested on the user-specified framework.

4   Applications

We use GGenerator to create behaviors for agents in two target different frameworks, mainly to demonstrate the platform-independence property. The two C++-based frameworks we use for our examples are: (a) Simple Soccer Agent

---

[3] The EMF project is a modeling framework and code generation facility for building tools and other applications based on a structured data model.

[4] Xpand is a language specialized on code generation based on EMF models.

Table 1. Comparison of the two different target platforms

SimSpark 3D Soccer Simulator	Wumpus World Simulator
Partially Observable	Fully Observable
Stochastic	Deterministic
Dynamic	Static
Physical representation of the agent	No physical representation of the agent
Uncertainty	No uncertainty
Sequential	Episodic
Continuous	Discrete
Multi-agent	Single-agent
Competitive, Cooperative	non-Competitive, non-Cooperative

framework, which interacts with the Simspark 3D Robot Soccer Simulator, and (b) WumpSim framework, which interacts with the Wumpus World Simulator. The two simulation environments have many differences, shown in Table 1.

4.1 Creating a Behavior for the SimSpark 3D Soccer Simulator

In this example, we use GGenerator to create a behavior for an agent that plays soccer. Our simple behavior can be described as follows: *The agent searches for the ball in the soccer field and, if the ball is found, the robot walks towards the ball and takes it in possession. If the game is over, the robot pauses.* Our purpose is to work only at the high level of agent programming (game strategy) and not at the low level (Walk, Kick, and other actions) or with agent-simulator technical aspects (e.g. communication with the simulator). The underlying framework we chose is based on the software release of the RoboCup team Berlin United - Nao Team Humboldt [1]; it is written in C++ and provides basic activities, such as *Walk, Turn Left, Scan For Ball* or *Stand Up*. The first step is to create the properties files; these files are shown below. In fact, we just need to include the appropriate header files to gain access to those framework-provided variables that allow the user (and the statechart) to gain information about the game and the perception of the ball, but also request the execution of a motion.

include_classes.txt	instances.txt
BallPercept.h	BallPercept theBallPercept;
MotionRequest.h	MotionRequest theMotionRequest;
SimsparkGameInfo.h	SimsparkGameInfo gameState;

The next step is to use the KSE graphical interface to create our behavior. We start by providing the liveness formulas that describe our agent behavior abstractly:

```
LogicalAgent  =  Init . ( Play | NoPlay )+
Play          =  [ StandUp ] . ( PlayBall | ScanBall )
PlayBall      =  Turn | Walk
```

The first formula indicates that our behavior (LogicalAgent) will execute *Init* (for initialization of the player) and then will choose one or more times between *Play* or *NoPlay* exclusively (depending on the current game state). The second formula suggests that our behavior may execute *StandUp* (if needed) and then will choose between *PlayBall* or *ScanBall* exclusively (depending on whether the ball is visible or not). Finally, the third formula indicates that our agent will choose between *Turn* or *Walk* exclusively (depending on where the ball is seen). As soon as we provide the KSE GGenerator with the liveness formulas, the initial statechart model is generated and the user has to associate it to a source code repository that provides the code for the basic agent activities (*Walk*, *Scan*, etc.). If the framework does not provide some activity, our tool generates the corresponding skeleton code and the user is asked to provide the corresponding C++ code using the built-in editor in KSE. Note that the abstract behavior specification of the liveness formula specifies what activities are included in the desired behavior, but gives no information on when execution switches from one activity to another. It is the execution of transition expressions that makes this switching between activities possible.

After the statechart is created, we use the KSE tool to add the necessary variables we are about to use in the transition expressions. In our case these are: `myInertialSens`, `ballFound`, and `horizontalAngle`. The first variable is of type `class` and holds information about the posture of the robot. The second one is of type `bool` and is true, if the ball is seen by the robot and false otherwise. The last one is of type `double` and holds information about the horizontal angle between the robot and the ball. These are user-defined variables, which are updated from the framework-provided variables inside the statechart, and are used in our transition expressions for controlling the statechart execution. They are added in the blackboard interface using the editing tool provided by KSE.

The next step is to fill in the transition expressions, according to the grammar in Figure 1, using the KSE editor. For example, the transition expression:

```
[ (ballfound == true) && (horizontalAngle < 0.1) ]
```

states the conditions (ball is visible straight-ahead) under which the robot will execute the *Walk* activity within the *PlayBall* state. The final statechart model with all the transition expressions is shown in Figure 2.

The last step of our procedure is to use GGenerator to produce the C++ code for this statechart model and transfer the generated files to the source folder of the framework, so that it is compiled and used in the target framework.

To execute the generated agent behavior we have to instantiate it within the target framework. This is a framework-specific step and depends on the requirements set by the framework itself. In the present example, it amounts to simply adding a couple of C++ lines in the configuration files of the framework to register the generated behavior as a new module within the framework. This is the only handwritten C++ code during our behavior generation procedure and is done only once, since any updates to the statechart simply modify the behavior, which remains registered in the framework. In Figure 3 we show three screenshots from the execution of the LogicalAgent behavior. Initially, the agent

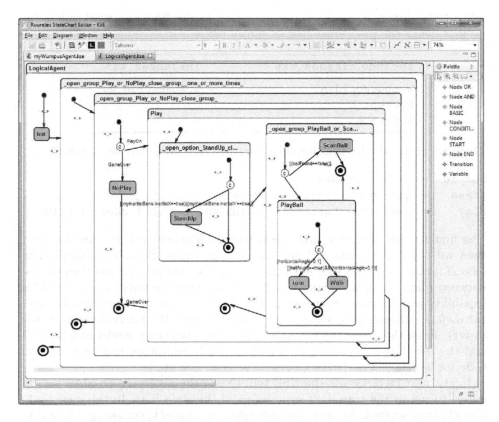

Fig. 2. The complete LogicalAgent statechart within the KSE graphical environment

Fig. 3. Three successive screenshots from the execution of the LogicalAgent

searches for the ball (left), when it finds the ball it walks towards it (middle), and, finally it takes possession of the ball (right).

4.2   Creating a Behavior for the Wumpus World Simulator

In this example, we create a behavior for the Wumpus world: *the agent tries to find the gold in the maze and escape alive.* Our goal is not to solve the maze, but to simply demonstrate how easy it is to create a behavior for this environment

using GGenerator. To connect our statechart to the Wumpus World platform
we first define the properties files, as shown below.

`include_classes.txt`	`instances.txt`
WumpusWorld.h	WumpusWorld ww;
Percept.h	
Action.h	

Then, we create the liveness formulas that describe the behavior abstractly:

```
Wumpie  =  init . Sense+
Sense   =  percept . ( Play | NoPlay )
Play    =  forward | turnLeft | turnRight | grab | shoot | climb
```

The first formula indicates that our behavior (*Wumpie*) will execute *Init* and
then will sense the world and act one or more times (*Sense* capability). The
second formula suggests that our behavior will execute *percept* and then choose
between *Play* or *NoPlay* exclusively. Finally, the third formula extends the *Play*
capability and indicates that our agent will choose one of the moves *forward,
turnLeft, turnRight, grab, shoot, climb* (depending on the information gathered
so far). From these liveness formulas, the initial statechart model is generated
and the user has to associate it to a source code repository that provides the
code for basic agent activities (*forward, percept, init,* etc.).

In the next step, we use the KSE tool to add the necessary variables we are
about to use in the transition expressions of the statechart. In this case these
are: `glitter`, `stench`, `breeze`, `turnedRight`, `turnedLeft`, `scream`, `gold`, `arrow`,
`posX`, `posY`. These are user-defined variables that are registered in the blackboard
during the code generation. Once we add the transition expressions according
to the grammar rules, we are ready to generate code for our behavior. Once we
register the statechart in the Wumpus framework again by adding a couple of
C++ lines in the `Wumpsim` class, we can execute and test our behavior. Figure 4
shows the final statechart for Wumpie.

5   Discussion and Conclusion

In this paper, we presented an extension to KSE that allows to specify statechart-
based agent behaviors independently of the underlying software framework, relying on the C++ programming language, as the common code base. The extended
KSE tool still supports the C++-based Monas framework for which it was originally customized, but now is open to any other C++-based framework, such
as the popular ROS [www.ros.org] framework. Thus, GGenerator increases the
added value of the KSE CASE tool. Its main advantages compared to other similar products, noted also in the past [7], include (a) the use of liveness formulas
that aid and accelerate the initial statechart generation, and, (b) the option to
define the code generation component as an add-on, aiming any target language.

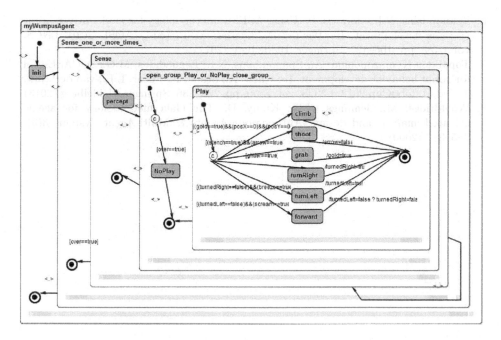

Fig. 4. The complete Wumpie statechart for the Wumpus World

Our future plan, which also motivated this work, is to automatically and massively generate team behaviors for RoboCup using our statechart tools and test them in simulation within an evolutionary framework for discovering suitable soccer team behaviors. We also plan to develop additional code generators to support other programming languages, thus expand the range of frameworks compatible with KSE, such as the Python-based Pyro [pyrorobotics.com].

References

1. Hafner, V., Burkhard, H.D., Mellmann, H., Krause, T., Scheunemann, M., Ritter, C.N., Schütte, P.: Berlin United – Nao Team Humboldt 2013. In: RoboCup 2013 Team Description Papers (2013)
2. Harel, D., Naamad, A.: The Statemate semantics of statecharts. ACM Transactions on Software Engineering and Methodology 5(4), 293–333 (1996)
3. Hayes-Roth, B.: A blackboard architecture for control. Artificial Intelligence 26(3), 251–321 (1985)
4. Paraschos, A.: Monas: A Flexible Software Architecture for Robotic Agents. Diploma thesis, Department of ECE, Technical University of Crete, Greece (2010)
5. Paraschos, A., Spanoudakis, N.I., Lagoudakis, M.G.: Model-driven behavior specification for robotic teams. In: Proceedings of the 8th International Conference on Autonomous Agents and Multiagent Systems (AAMAS), pp. 171–178 (2012)

6. Spanoudakis, N.I., Moraitis, P.: Using ASEME methodology for model-driven agent systems development. In: Weyns, D., Gleizes, M.-P. (eds.) AOSE 2010. LNCS, vol. 6788, pp. 106–127. Springer, Heidelberg (2011)
7. Topalidou-Kyniazopoulou, A., Spanoudakis, N.I., Lagoudakis, M.G.: A CASE tool for robot behavior development. In: Chen, X., Stone, P., Sucar, L.E., van der Zant, T. (eds.) RoboCup 2012. LNCS, vol. 7500, pp. 225–236. Springer, Heidelberg (2013)
8. Wooldridge, M., Jennings, N.R., Kinny, D.: The Gaia methodology for agent-oriented analysis and design. Autonomous Agents & Multi-Agent Systems 3(3), 285–312 (2000)

Hierarchic Fuzzy Approach Applied in the Development of an Autonomous Architecture for Mobile Agents

Márcio Mendonça[1], Esdras Salgado da Silva[1], Karina Assolari Takano[1],
Mauricio Iwama Takano[1], and Lúcia Valéria Ramos de Arruda[2]

[1] Postgraduate Mechanical Engineering Program (PPGEM)
Federal Technological University of Paraná (UTFPR)
Av. Alberto Carazzai, 1640, Centro, 86300-000, Cornélio Procópio, Paraná, Brazil
[2] Postgraduate Eletrical Engineering And Industrial Computers Program (CPGEI)
Federal Technological University of Paraná (UTFPR)
Av. Sete de Setembro, 3165, Rebouças, 80230-901, Curitiba, Paraná, Brazil
{mendonca,takano,lvrarruda}@utfpr.edu.br,
esdras.utfpr@bol.com.br, ka.assolari@gmail.com

Abstract. The development of a controller architecture based on hierarchic fuzzy logic to create the trajectory of a mobile agent in a virtual environment is initially used. The objective of the low cost controller is to regulate an open-source autonomous explorer robot, which moves between two pre-established points. The system establishes a viable route in a scenario with obstacles. Simulations in virtual environment are used to ensure the autonomy of the developed controller. Initial results demonstrate the success of this proposal.

Keywords: Autonomous agent, Fuzzy logic, Virtual kinematic model, Intelligent Systems, Hierarchic Fuzzy.

1 Introduction

Researches about mobile autonomous robots are becoming more important due to technological evolution and high processing computers. Some technological evolutions that can bear witness to are the development of intelligent computational techniques, such as fuzzy logic, artificial neural network, intelligent agents, and others.

A number of challenging problems must be solved for mobile autonomous navigation. The problems can depend on different domains like perception, localization, environment modeling, and decision making, control etc. [1].

Autonomous robots can be used in different fields for instance in high risk areas (to avoid risky situations for humans), in exploration of unattainable places of difficult access or hazardous environment, and in service robots which makes this an emergent field of research and application, as presented by [2, 3, and 4].

Explorer robots or agents can be virtually controlled by humans; this however demands communication systems, cameras and operators. A solution for this is the use of intelligent computational systems which can be applied in the development of robots with the ability of adaptation in different environments by reading signals from low level sensors such as ultrasound sensors [5]. This can be helpful in cases such as

L. Iliadis et al. (Eds.): AIAI 2014, IFIP AICT 436, pp. 193–202, 2014.

the exploration of natural environment (inhospitable) when autonomous decisions are required [6, 7, 8].

The environment exploration has some difficulties to be solved for example uncertainties, inaccuracies and sequential errors. The complexity involved in creating an efficient path is known to be high due to sequential dynamic decisions that have to be made (a wrong decision can compromise the next steps). The navigation tasks are initiated by informing an objective. During the exploration the agent needs to have the ability to deal with unpredictable events like mobile and/or unexpected obstacles [9].

According to Khodayari [10] there are two main strategies to develop autonomous agents. The first one is related to heuristic knowledge and is represented in this paper by fuzzy logic as presented in [11, 12, 13], and adapted to the model used. The second one is based in the control theory, subdivided in linear control theory and nonlinear control theory.

In past decades, researches have developed many obstacle avoidance navigation algorithms. These methods can be divided into the following categories: model-based method [14-19], fuzzy logic method [20-23] and reactive method based on neural network [24, 25]. Of all the methods, the obstacle avoidance method based on fuzzy logic reacts quicker and is less susceptible to sensor measuring errors for the robot's reaction has come from qualitative reasoning of different categories in the fuzzy logic method, so the algorithm proposed in this paper is based on fuzzy logic principles. The difference is the fact that the hierarchical fuzzy quantitatively simplify solving the problem by organizing rules with one fuzzy used for each objective.

The objective of this paper is to develop an autonomous explorer agent (robot) based in a low cost and open source platform with the ability to learn navigation strategies by environment interaction. The development proposal adopted is inspired by Braitenberg [26], who suggests the application of intelligent computational techniques starting up with a simple model with one or only a few functionalities, and gradually adding new objectives to improve the exploration capability of the agent, such as manage the battery charge and navigate back to the recharging spot. In this paper, the agent will have initially the ability to navigate through two pre-established points in a surface. The second objective is to add the ability to deviate from any obstacles that are found in the proposed initial path.

This paper is organized as follows: section 2 briefly describes the fuzzy logic concept; section 3 presents the development of this work (kinematic model for the robot position and controller); sections 4 and 5 show the results and conclusions.

2   Fuzzy Logic Fundamentals

Fuzzy systems according to Zadeh [27] is a set of methods based in the linguistic terms, multivalued logic, systems based in rules that aim to work with failure tolerance, inaccuracy and uncertainty in real systems. These methods and linguist terms proposes the concept of the fuzzy logic which consists of the combination of these theories, allowing the modeling of complex processes (based on inaccurate or approximate information) expressing control rules in linguistic terms. Different areas of knowledge use this technique as a solution for control applications, pattern recognition and function approximation, especially when there is little or no knowledge of the mathematical models.

A fuzzy system consists of three main components: input and output, set of rules and inference mechanisms [28, 29]. A difficulty to implement simple Fuzzy logic is that as you increase the number of variables involved, the number of rules grow exponentially. One possible solution is the development of a fuzzy hierarchical structure in which different rule bases are used to address each goal [30].

3   Development

3.1   Kinematic Model for the Robot Position

There are two ways of representing the robots: kinematic or dynamic. The kinematic models are used to represent the robots according to the velocity and orientation of the wheels, and the dynamic models according to the generalized forces applied by the actuators. There are two kinematic models used to represent the agents: position model and configuration model. The first of them considers only the position and the orientation of the robot, while the second one also considers other internal variables (for example angular movement of the wheels). The position model fulfills the needs of the position control and spatial orientation of the robot. Most of the researches reported in the literature describe the robot according to Cartesian coordinates using just a position kinematic model [31, 32, and 33].

The robot used as a model in this paper consists of three wheels, one of which is mobile (in the front part of the robot) and the other two stationary on the sides (each of them connected to a motor); three ultrasonic sensors placed to collect signals that identify possible obstacles in the path. The physical model of this agent is shown in Fig. 1 (a), and the schematic model in Fig. 1 (b).

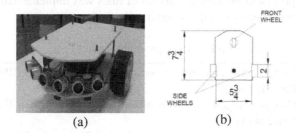

(a)   (b)

Fig. 1. Physical (a) and schematic (b) models of the robot

3.2   Hierarchical Fuzzy Approach

In some situations fuzzy logic using just a set of rules can control processes in different situations, during start up and course of operations with the proper activation of the rules. However when there are different objectives in the same controller, it is necessary to apply a set of rules for each objective.

Summarized when the fuzzy controller requires more complexity to achieve the desired goal, a common strategy is to decompose the controller problem hierarchically, using low-dimensional fuzzy systems. A priority is defined between the fuzzy systems adopted [34]. It is called Fuzzy hierarchical system using more than one fuzzy and establishes a hierarchy between them.

A hierarchical fuzzy architecture based in cognitive architecture [34] especially the subsumption [35], was developed to work with hierarchical decision making and this is due to the need of two opposite main targets for example reach a target and deviate from obstacles in the trajectory.

The subsumption fuzzy architecture proposed has two sets of rules which work independently and with different and hierarchical competence levels. When certain competence level is enabled all other levels are disabled. Thereby the fuzzy controller coordinates the competence levels prioritizing the higher levels (deviate from obstacles is a higher level than reaching the target) [36]. In other words, when the autonomous agent is deviating from an obstacle it stops going after the target, this demands an implemented switching mechanism. In Fig. 2 it is shown the hierarchical fuzzy architecture developed.

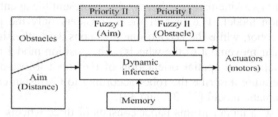

Fig. 2. Hierarchical model of the fuzzy system

Fuzzy I, the initial objective of this paper, uses a decision-making system based in heuristic methods. This step has the objective of determining the shortest route between the initial point and the target. This set of rules was implemented from observations of the dynamic behavior of the agent in virtual scenarios.

Table 1. Example of fuzzy rules I

1. If (LatDist is farLeft) and (FrontDist is farF) then (LeftPulse is medium)(RightPulse is high)
2. If (LatDist is farLeft) and (FrontDist is mediumF) then (LeftPulse is slow)(RightPulse is high)
3. If (LatDist is farLeft) and (FrontDist is closeF) then (LeftPulse is high)

This strategy consists in abstracting the input data (of mathematical simulations of the lateral and frontal distances) and converting them in linguistic variables (sets of rules) and then generating a response (pulses to the motors linked to the left and right wheels).

In Fig. 3 are presented the control surfaces of the fuzzy I corresponding to the left (Fig. 3a) and right (Fig. 3b) wheels.

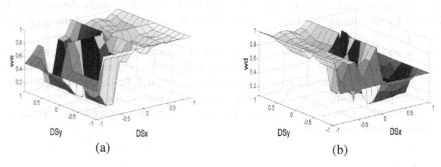

Fig. 3. Fuzzy I control surfaces for left (a) and right (b) wheels

The x-axis represents the lateral distance between the robot and the target (DSx), the y-axis, the front distance (DSy) and the z-axis, the pulses wheels: right to left in Fig. 3a and Fig. 3b.Whenever the sensors detect an obstacle, the hierarchical fuzzy controller prioritizes the deviation of the obstacles using the fuzzy II set of rules. When no obstacle is detected, the agent will keep going after the target using the fuzzy I set of rules. Some of the rules used in fuzzy II are shown in table 2:

Table 2. Example of fuzzy rules II

1. If (sensor_L is far) and (sensor_F is far) then (LeftPulse is medium)
2. If (sensor_L is far) and (sensor_F is medium) then (LeftPulse is high)
3. If (sensor_L is far) and (sensor_F is close) then (LeftPulse is high)

The control surfaces related to the second fuzzy are presented in Fig. 4. Fig. 4a refers to the influence of both left and frontal sensors on the left wheel. Fig. 4b refers to the influence of both right and frontal sensors on the left wheel. In Fig. 4b, "we" represents the pulses on the left wheel with range [0.14 0.19] because when we have obstacles on the right hand front of the robot, it has to turn to left. Therefore there is a low pulse in left wheel. The rules are symmetric, which means the results are the same for the right wheel.

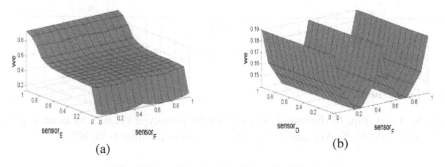

Fig. 4. Fuzzy II control surfaces for left wheel

A two dimensional simulator was implemented in Matlab to study the dynamic behavior of the mobile agent. The scale used for both the prototype and the scenario is 1:100.

Russel and Norvig [06] suggest that in order to consider an autonomous agent, it is necessary to succeed in at least three different simulations. Therefore the arrangement of the obstacles, the initial position of the mobile agent and the target are adjustable in the Matlab program, allowing the construction of different scenarios.

In the simulator developed, the sensors were calibrated to send the signals at a maximum distance of 15 cm.

4   Results

For each scenario simulated, three situations where considered (except 4): scenario with obstacles; rearrangement of these obstacles; and scenario with noise of five percent on the sensors signals. The noise added in the sensors signals represents the uncertainty of the data acquisition, characteristic of real scenarios due to different behavior of the sound wave reflection in different materials.

In the following subsections are presented the simulated scenarios. In the following images, items (a) and (b) indicate tests in obstacles arranged in two different ways, and item (c) indicates the response to noises.

4.1   Scenario 01

In this scenario (Fig. 5) the initial position for the prototype is (0,0), at an angle of zero degrees with the X-axis, and the target is located in position (50,70).

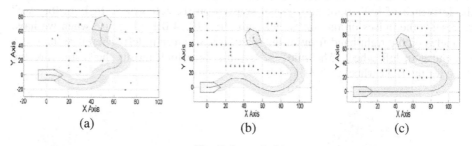

(a)   (b)   (c)

Fig. 5. Scenario 01

4.2   Scenario 02

In this scenario (Fig. 6) the initial position for the prototype is (30, 25), at an angle of 45 degrees with the X-axis, and the target is located in position (-50, 80).

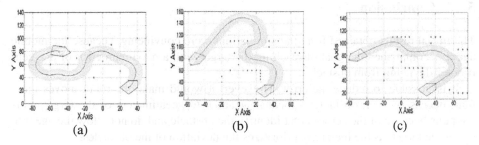

Fig. 6. Scenario 02

4.3   Scenario 03

In this scenario (Fig. 7) the initial position for the prototype is (-80,75), at an angle of zero degrees with the X-axis, and the target is located in position (35,75).

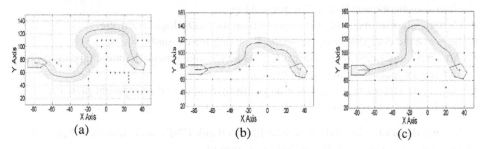

Fig. 7. Scenario 03

4.4   Scenario 04

In this scenario (Fig. 8) the initial position for the prototype is (-50,-90), at an angle of zero degrees with the X axis, and the target is located in position (0,0). In this simulation was considered the complexity of a spiral scenario. In this scenario situation (b) was not considered.

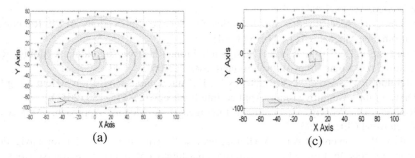

Fig. 8. Scenario 04

5   Conclusion

The initial results obtained from the simulations were convincing, because the mobile agent accomplished the goal of reaching the target with a maximum error of one centimeter deviating from obstacles.

It is possible to reduce the error. However, it would make the robot moves more than necessary if an obstacle is located at about ten centimeters of the target. This happens because of the sensors that identify the obstacle and do not allow the agent to reach the target, as the hierarchy prioritizes the deviation of the obstacles.

Seven different scenarios were successfully simulated. This implies the autonomy of the proposed fuzzy control. The simulations with uncertainties also imply a robust system, because they approach simulations to real environment where data input from the sensors may have noise and errors. Future works are presented:

- Improve the minimum error by changing the hierarchical process;
- Compare the proposed controller with other intelligent techniques like Artificial Neural Network and Fuzzy Cognitive Maps;
- Genetic algorithm can also be used to tune the fuzzy logic controller;
- Improve the complexity of the scenarios using for example, walls;
- Add new functions in the controller: battery management (locate charging points) and reverse gear for extreme situations;
- Apply the system in a real robot using an open source development platform, such as a low cost microcontroller (example, Arduino).

Acknowledgment. The authors would like to thank CNPq and Araucaria Foundation of the State of Paraná, Brazil, for financial support.

References

1. Shaikh, M.H., Kosuri, K., Ansari, N.A., Khan, M.J.: The state-of-the-art intelligent navigational system for monitoring in mobile autonomous robot. In: 2013 International Conference on Information and Communication Technology (ICoICT), pp. 405–409, 20-22 (2013)
2. Asami, S.: Robots in Japan: Present and Future. IEEE Robotics and Automation Magazine 1(2), 22–26 (1994)
3. Schraff, R.: Mechatronics and Robotics for Service Apliccations. IEEE Robotics and Automation Magazine 1(4), 31–35 (1994)
4. Fracasso, P.T., Costa, A.H.R.: Navigation From Reactive Autonomous Mobile Robots Using Fuzzy Logic Rules With Weighted. SBAI (2005)
5. Mendonça, M., Arruda, L.V.R., Neves Jr., F.: Autonomous navigation system using Event Driven-Fuzzy Cognitive Maps. Applied Intelligence 37(2), 175–188 (2011)
6. Russell, S.J., Norvig, P.: Artificial Intelligence: A Modern Approach. Prentice Hall, Englewood Cliffs (1995)
7. Calvo, R., Romero, R.A.F.: A Hierarchical self-organizing controller for navigation of mobile robots. In: Proceedings of International Joint Conference on Neural Netwoks. IEEE World Congress Computational Intelligence, Vancouver (2006)

8. Broggi, A., Zelinsky, A., Parent, M.E., Thorpe, C.E.: Intelligent vehicles. Springer Handbook of Robotics, pp. 1175–1198 (2008)
9. Bakambu, J.N.: Integrated autonomous system for exploration and navigation in underground mines. In: Proc. 2006 IEEE/RSJ Int. Conf. Intelligent Robots and Systems, pp. 2308–2313 (2006)
10. Khodayari, A.G.: A historical review on lateral and longitudinal control of autonomous vehicle motions. In: 2nd International Conference on Mechanical and Electrical Technology (ICMET), Singapore (2010)
11. Chiang, H.-H., Ma, L.-S., Perng, J.-W., Wu, B.-F.: Longitudinal and lateral fuzzy control systems design for intelligent vehicles. In: Proceedings of the 2006 IEEE International Conference on Networking, Sensing and Control (ICNSC), pp. 544–549 (2006)
12. Yang, J.: e Zheng, N.: An expert fuzzy controller for vehicle lateral control. In: 33rd Annual Conference of the IEEE on Industrial Electronics Society (IECON), pp. 880–885 (2007)
13. Cai, L., Rad, A.E., Chan, W.-L.: An intelligent longitudinal controller for application semiautonomous vehicles. IEEE Transactions on Industrial Electronics 57(4), 1487–1497 (2010)
14. Lim, J.H., Cho, D.W.: Sonar based systematic exploration method for an autonomous mobile robot operating in an unknown environment. Robotic (16), 659–667 (1998)
15. Krishna, K.M., Kalra, P.K.: Perception and remembrance of the environment during real-time navigation of a mobile robot. Robotics and Autonomous Systems (37), 25–51 (2001)
16. Maaref, H., Barret, C.: Sensor-based navigation of a mobile robot in an indoor environment. Robotics and Autonomous Systems (38), 1–18 (2002)
17. Chatterjee, R., Matsuno, F.: Use of single side reex for autonomous navigation of mobile robots in unknown environments. Robotics and Autonomous Systems 35, 77–96 (2001)
18. Ge, S.S., Xuecheng, L., Al Mamun, A.: Boundary following and globally convergent path planning using instant goals. IEEE Trans. on Systems, Man, and Cybernetics-Part B: Cybernetics 35(2), 240–254 (2005)
19. Toibero, J.M., Carelli, R., Kuchen, B.: Switching Control of Mobile Robots for Autonomous Navigation in Unknown Environments. In: IEEE International Conference on Robotics and Automation, Roma, Italy, pp. 1974–1979 (2007)
20. Xu, W.L., Tso, S.K.: Sensor-based fuzzy reactive navigation of a mobile robot through local target switching. IEEE Trans. Syst., Man Cybern., C 29(3), 451–459 (1999)
21. John, H.L.: Evolution of a Negative-Rule Fuzzy Obstacle Avoidance Controller for an Autonomous Vehicle. IEEE Trans. on Fuzzy Systems 15(4), 718–727 (2007)
22. Chao, C.-H., Hsueh, B.-Y., Hsiao, M.-Y., et al.: Fuzzy Target Tracking and Obstacle Avoidance of Mobile Robots with a Stereo Vision System. International Journal of Fuzzy Systems 11(3), 183–191 (2009)
23. Abiyev, R., Ibrahim, D., Erin, B.: Navigation of mobile robots in the presence of obstacles. Advances in Engineering Softwares 41, 1179–1186 (2010)
24. Jackson Phinni, M., Sudheer, A.P., RamaKrishna, M., et al.: Obstacle Avoidance of a wheeled mobile robot: A Genetic-neuro-fuzzy approach. In: IISc Centenary – International Conference on Advances in Mechanical Engineering, Bangalore, India, pp. 1–3 (2008)
25. Zhu, A., Simon, X.Y.: Neurofuzzy-Based Approach to Mobile Robot Navigation in Unknown Environments. IEEE Trans. on Systems, Man, and Cybernetics-Part C: Cybernetics 37(4), 610–621 (2007)
26. Braitenberg, V.: Vehicles: Experiments in Synthetic Psychology. The MIT Press (1986)
27. Zadeh, L.A.: Fuzzy algorithms. Info. & Ctl. 12, 94–102 (1968)
28. Passino, M.K., Yourkovich, S.: Fuzzy control. Addison-Wesley, Menlo Park (1997)

29. Kasabov, N.K.: Foundations of Neural Networks, Fuzzy Systems, and Knowledge Engineering. MIT Press (1998)
30. Wang, L.: Analysis and design of hierarchical fuzzy systems. IEEE Trans. Fuzzy Syst. 7, 617–624 (1999)
31. Bloch, A.M., Reyhanoglu, M., McClamroch, N.H.: Control and stabilization of nonholonomic dynamic sytems. IEEE Transactions on Automatic Control 37(11), 1746–1756 (1992)
32. Kolmanovsky, I., McClamroch, N.H.: Developments in nonholonomic control problems. IEEE Control Systems Magazine 15(6), 20–36 (1995)
33. Murata, S., Hirose, T.: On board locating system using real–time image processing for a self–navigating vehicle. IEEE Transactions on Industrial Electronics 40(1), 145–153 (1993)
34. Brooks, R.: A robust layered control system for a mobile robot. IEEE Journal of Robotics and Automation 2(1), 14–23 (1986)
35. Coppin, B.: Artificial Intelligence Illuminated. Jones And Bartlett Publishers, Burlington (2004)
36. Goerick, C.: Towards an understanding of hierarchical architectures. IEEE Trans. Auton. Ment. Dev. 3, 54–63 (2011)

Classification of Mammograms Using Cartesian Genetic Programming Evolved Artificial Neural Networks

Arbab Masood Ahmad, Gul Muhammad Khan, and Sahibzada Ali Mahmud

University of Engineering and Technology, Peshawar, Pakistan
{arbabmasood, gk502, sahibzada.mahmud}@nwfpuet.edu.pk
http://www.nwfpuet.edu.pk

Abstract. We developed a system that classifies masses or microcalcifications observed in a mammogram as either benign or malignant. The system assumes prior manual segmentation of the image. The image segment is then processed for its statistical parameters and applied to a computational intelligence system for classification. We used Cartesian Genetic Programming Evolved Artificial Neural Network (CGPANN) for classification. To train and test our system we selected 2000 mammogram images with equal number of benign and malignant cases from the well-known Digital Database for Screening Mammography (DDSM). To find the input parameters for our network we exploited the overlay files associated with the mammograms. These files mark the boundaries of masses or microcalcifications. A Gray Level Co-occurrence matrix (GLCM) was developed for a rectangular region enclosing each boundary and its statistical parameters computed. Five experiments were conducted in each fold of a 10-fold cross validation strategy. Testing accuracy of 100 % was achieved in some experiments.

Keywords: Mammogram, Image classification, GLCM, CGPANN, Haralick's parameters.

1 Introduction

Breast cancer is a leading cause of death in women worldwide. The disease has often times no symptoms till it advances to a dangerous level. It is therefore highly recommended to carry out screening tests after a certain age. The best screening method to date is the mammography, in which both the breasts are imaged with low dose x-ray radiation. Nowadays digital mammography has become the standard screening practice.

As mammography is highly subject to the expertise of radiologist and prone to errors, a decision support system based on machine learning is highly desirable.

There are two main indicators for breast cancer using mammography. These are masses and microcalcifications. Often times before a suspicious mass appears in the mammogram, very fine specs appear in the image, which is caused by micro-calcifications (MCs). The MCs that are 1 mm or smaller in diameter

L. Iliadis et al. (Eds.): AIAI 2014, IFIP AICT 436, pp. 203–213, 2014.
© IFIP International Federation for Information Processing 2014

and appear in clusters are more likely to be cancerous while those that are larger in size and scattered randomly are often benign. A mass on the other hand is considered benign if its shape is round or oval and the margins are circumscribed, while it is malignant if it is irregular, stellate or micro-lobulated in shape. The density of breast is such that the mammogram shows a very low contrast between parenchymal tissue and a mass and hence very difficult to isolate the two, visually. A number of methods have been developed for computer based classification of the masses and microcalcifications. Before classification the mammogram image is preprocessed using digital filters to enhance the different regions of the image. The masses and microcalcifications are then outlined using segmentation techniques. Before these segmented images could be classified using machine, a machine learning system is trained with a large set of mammograms that are classified by expert radiologists. There are a number of mammogram databases available on the internet. The two most popular of them are Digital Database for Screening Mammography (DDSM) at the University of South Florida[1] and the MIAS [2]

In this paper we present our work in classifying the masses and MCs seen in mammograms, as either benign or malignant. The system that we developed assumes that the physician segments the masses or MCs manually using the **CROP** function, found in most windows based graphic software packages. In order to train our system for classification we had to train it with sample mammogram images from the Digital Database for Screening Mammography (DDSM). These images are available in compressed lossless JPEG format. The Windows versions of the uncompressed images were downloaded under a transfer agreement with Dr. Thomas Deserno (nee Lehmann) Department of Medical Informatics Aachen University of Technology D-52057 Aachen GERMANY, under the IRMA (Image Retrieval in Medical Applications) project. The database contains 2620 cases comprising of Normal, Benign and Malignant, high resolution mammograms. Each case further consists of two views of each breast, the cranio-caudal view and the oblique medio-lateral view. Besides the images, each case has a .ics file that contains information about the patient and the image file. Section 4 gives the details of the work done in preparing the training and testing data for the CGPANN. After training the CGPANN with the former the system is then evaluated with the later.

The paper is organized as follows: The next section titled "Literature Survey" describes the latest research done in the field of mammogram classification. The sections "Cartesian Genetic Programming Evolved Artificial Neural Network (CGPANN)" and "Evolution Strategy" describe the evolutionary computation system that we used for the classification task. The section "Experimental setup" describes the preprocessing and statistical parameters extraction steps needed before mammogram classification. It also describes the network parameters that we chose. The section "Results and Analyses" compares the results of

---

[1] http://marathon.csee.usf.edu/Mammography/Database.html
[2] http://peipa.essex.ac.uk/info/mias.html

our experiments with those of the competitors. Finally, the section "Conclusion and Future work" summarizes the paper and states our intentions for future work in this field.

2   Literature Survey

A brief review of the work done in the field of mammogram classification is presented below. As the abnormality in mammograms can be in the form of microcalcifications or masses, separate subsections review each of them.

2.1   Microcalcifications

In [1] Xuejun et al. presented a number of ANN architectures for detecting microcalcifications (MC) in mammogram images. A back propagation neural network responds to an MC lying in the middle of a region of interest. To reduce the image size, FFT of the image was also determined. Another network with multiple outputs had one of its outputs as 1 when its corresponding input had a microcalcification at its position in the image. A Shift Invariant ANN (SIANN) was also experimented with. The performance of the networks was compared with Triple Ring Filter (TRF) which is a rule based filter [2]. A sensitivity of 95% was achieved by using SIANN and TRF together.

In [3] Issam et al. presented an SVM based microcalcification (MC) classification technique. In [4] Rolando et al. presented a technique in which a mammogram is first preprocessed with a median filter to remove noise. This is followed by binarization. The resulting image is then applied to two Gaussian filters. The optimized difference of Gaussian (DoG) filters is used to enhance those regions in the image that contain bright points. A number of techniques are then applied to this image to detect potential MCs.

In [5] Walker et al. discussed a multi-chromosome Cartesian Genetic Programming in general and its application for classification of microcalcifications (MCs) in mammograms, in particular. The process is termed multi-chromosome Evolution (MCE). About 80% test images were classified correctly. In [6] Zhang et al. presented a technique in which areas marked by radiologists in the DDSM database were resized and fourteen features extracted from each area. These features are represented by genes in a chromosome of a neural-Genetic Algorithm. A subset of features is applied as input and the NN trained. A random population of subsets is generated and each individual is evaluated. Genetic operators of crossover and mutation are applied. The NN with the best classification rate together with the selected feature subset is chosen. It was observed that the highest classification result using the proposed algorithm with NN classifier was 85%.

2.2   Masses

In [7] the authors preprocessed and segmented images from popular databases on the basis of grey level distribution and texture using Otsu's method. They extracted the intensity histogram and Gray Level Co-occurrence Matrix (GLCM) features. To improve prediction accuracy the author used a hybrid of Genetic Algorithm (GA) and Linear Forward Selection (LFS) algorithms. For classification they used the machine learning package WEKA and trained the system using J48 decision tree method. The classification accuracy based on the selected features was 86.7%. In [8] the author presents mammogram classification technique using Linear Vector Quantization (LVQ). Best classification accuracy of 99% was achieved with 100 DDSM cases.

In [9] the authors present an algorithm that is based on the principle that a mass segment has uniform texture, its margins should coincide with the maximum gradient points that corresponds to the edges and that the mass profile shape has minimum changes. In [10], [11] the authors present mammogram image segmentation based on Iso-level contour map representation of the image.

In [12] a detailed review of the work done in segmentation and classification of the different anatomical regions of the breast is presented. In [13] the authors used Gray Level Co-occurrence Matrix (GLCM) to extract second order statistical descriptors of the texture from the masses. These features were applied to error back propagation ANN for classification. The best performance result that they obtained had 91.67% sensitivity, and 84.17% specificity.

In [14] the authors made use of the chain code in the overlay files of mammograms, in the DDSM database. A rectangular image is formed from the ROI defined by the chain code. The image is preprocessed and a GLCM formed from it. The Haralick energy descriptors are then determined from the GLCM. Adaptive thresholding is applied to the image so formed to detect edges. A binary image is thus formed. The morphological operation of image filling is applied to all closed path edges. The filled area that has the largest overlapping with the central part of ROI is the segmented mass.

In [15] the authors present a technique for segmentation and classification of masses. For segmentation the highest intensity point inside the ROI is taken as center and radial lines drawn to its margins. Critical point, the point with highest local variance on each line is detected and new ROI drawn by linearly interpolating in between these points. Sixteen different morphological and textural features of the segmented masses are computed and applied to an ANN for classifying the mass as benign or malignant.

In [16] the authors present a technique focusing on utilization of mass margin descriptors called Fourier Transform of Radial Distance (FTRD). These features are applied to an MLP ANN for classification of the mass as either benign or malignant. The best accuracy that they achieved was 92.8%.

In [17] the authors present a mammogram classification system based on computational and human features. These features include shape, size and margins of the mass, patient age and tissue density. The density values are represented in Breast Imaging Reporting and Data System (BI-RADS). Textural features are found using Spatial Gray Level Dependence Method (SGLDM) and Run Difference Method (RDM). Sequential Forward Selection technique and genetic algorithm technique were used for classification.

In [18] the authors present a mass detection technique in which the pectoral muscles are segmented and removed. The mass inside the breast is then segmented using intensity thresholding. The segmented mass is analyzed for its textural features, proposed by Haralick, using the gray level co-occurence matrix (GLCM). The textural indices are applied to a support vector machine (SVM) for training and classification. An average classification rate of 95% was achieved.

3   Cartesian Genetic Programming Evolved Artificial Neural Network (CGPANN)

In CGPANN a neural network is developed by evolving its inter-node connections, weights, output connection(s) and the activation functions. It's an enhancement to the popular Cartesian genetic programming [19]. The nodes and connections in CGPANN are arranged as a graph with rows and columns. A CGPANN node, shown in fig.1b, consists of a summer and an activation function, similar to an ANN neuron. The summer takes its inputs from other nodes through weighted connections. Each connection additionally has a switch that can be turned On or Off. The number of inputs to each node, number of rows, number of columns and levels-back are predefined.

A CGPANN genotype is a string of integers, each representing one of the network quantities. A typical phenotype and its related genotype can be seen in fig.1a. For each node the letters **F I W C ..** have the following meanings and possible values. **F**: Activation function-sigmoid or Tangent Hyperbolic; **I**: The node connected to an input of this node, only a node in a column on the left or a network input are allowed; **W**: connection weight (-1 to +1); **C**: on-off switch (0 or 1); **O**: Network output, all nodes and network inputs are allowed to connect to it. Nodes of the corresponding phenotype are numbered sequentially, starting with the top node in the first column followed by the nodes down the column, followed by all the columns to the right in the same manner. When the genotype is decoded to phenotype, outputs of the nodes that don't connect to any other node are called inactive nodes. During evolution, a certain percentage of genes are randomly picked and mutated to allowed values only. Unlike other ANN configurations that use both crossover and mutation, CGP and its derivative CGPANN give excellent results with mutation only. During the process many inactive nodes become active and vice versa. CGP architectures and its derivative CGPANN have been investigated on a number of problems in the past[20,21,22].

3.1   Evolution Strategy

We use a 1+ λ (Number of Offspring) evolution strategy in which $\lambda = 4$. The pseudo-code for the algorithm that we use for training and testing is presented below.

- *Prepare the set of parameters to be classified and a set of corresponding target outputs.*

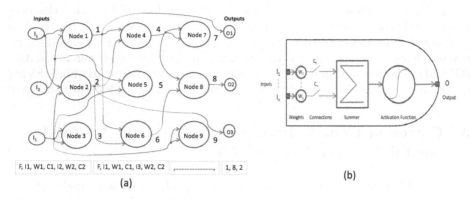

Fig. 1. (a) A typical CGPANN phenotype. Nodes 3, 5, and 9 are inactive and the remaining nodes are active. (b) Internal view of a single CGPANN neuron.

– *Specify the network parameters.*
– *Form the initial population of fifty genotypes by assigning random, yet legitimate values to the genes of each genotype as discussed above.*
– *Generate new population with the fittest genotype and its four replicas. Accuracy of the result has been used as the fitness measure for this application.*
– *Randomly mutate 10% (Mutation Rate) of each replica, forming four offspring.*
– *Determine fitness of the parent and the offspring and select the fittest to make the next parent. If the parent and an offspring have equal fitness, priority is given to the offspring over the parent, [23]. The fitness improves from generation to generation until the iterations are stopped.*
– *Evaluate the performance of the trained network by applying the testing data.*
– *The system is now ready to classify a new pattern.*
– *END*

4   Experimental Setup

We trained our classifier system with 1000 mammogram images of benign and malignant types each. Using the Matlab command **GET_DDSM_GROUNDTRUTH** we extracted the image information from the groundtruth file. The overlay file inside the groundtruth file contains the following information: lesion_type, assessment, subtlety, pathology, annotations and a chain code representing the region of interest (ROI) marked by expert radiologists. The function is normally used to get a binary image of the ROI but we modified it to return a rectangular image that contains this ROI (see fig.2c). The network trained with parameters from these rectangular images, containing masses and MCs, makes it capable to accept parameters from manually segmented images (using the **CROP** function) easier. A Gray Level Co-occurrence Matrix (GLCM) of an image is then formed. The matrix contains information about the frequency of gray level repetition between pixels at a certain offset distance and angle and can be used to extract statistical parameters of the image. Out of the many possible second order statistical texture descriptors, proposed by Haralick [24], we extracted only contrast, energy, homogeneity and correlation. Each of these four parameters were determined for four different angles and a certain distance, using the **graycoprops** function in matlab. The sample parameters were divided into a training set and a testing set for training and testing our CGPANN. See equations (1) to (4) for the four Haralick's parameters.

$$Contrast = \sum_{i,j=0}^{N-1} P_{ij}(i-j)^2 \tag{1}$$

$$Energy = \sum_{i=1}^{N}\sum_{j=1}^{N} p(i,j)^2 \tag{2}$$

$$Homogeneity = \sum_{i,j=0}^{N-1} \frac{P_{ij}}{1+(i-j)^2} \tag{3}$$

$$Correlation = \frac{[\sum\sum(ij)p(i,j)] - \mu_x\mu_y}{\sigma_x\sigma_y} \tag{4}$$

Where $p(i,j)$ is the normalized entry in row i and column j of the GLCM, i is the intensity of one pixel while j is that of the next pixel making the pair for GLCM [13]. We get a total of 16 parameters. A 10-fold cross validation strategy was adapted, where ten distinct data sets are formed. Each set has nine parts for training and one for testing. We experimented with the following six distance offsets (D): 4, 8, 12, 16, 20 and 24, and found that for D=20 we got the best average 10-fold accuracy result i.e. Accuracy=90.85%, Sensitivity=86.2% and Specificity=95.5%. We performed five experiments with this offset value, each time using a different seed for random number generator in the initial population in CGPANN. The tenfold results for these experiments are shown in table 1.

The CGPANN network that we used has the following features: Number of nodes= 100 ($10rows \times 10columns$), Inputs per node= 3 and Number of outputs= 1(Average of ten outputs). An input-output set contains the 16 statistical parameters as inputs and a target output that is 0 for benign and 1 for malignant. In the process of evolution, an initial population of 50 networks was formed randomly. Each network is applied all the 2000 sample parameters and its fitness determined using the following metrics:

True Positive (TP):A=1, T=1
False Positive (FP):A=1, T=0
True Negative (TN):A=0, T=0
False Negative (FN):A=0, T=1
Accuracy=(TP+TN)/N
Sensitivity =TP/(TP + FN)
Specificity=TN/(FP+TN)

Table 1. Average values for training and testing results for five independent evolution runs with offset:20 and 10-fold cross validation; Acc: accuracy, Sen: Sensitivity, Spec: Specificity

Train	Fold-1			Fold-2			Fold-3			Fold-4			Fold-5		
	Acc	Sen	Spec	Acc	Sen	Spec	Acc	Sen	Spec	Acc	Sen	Spec	Acc	Sen	Spec
	91.0	86.34	95.56	90.7	84.52	96.88	90.2	85.06	95.38	89.82	84.34	95.38	93.66	91.42	95.86
	Fold-6			Fold-7			Fold-8			Fold-9			Fold-10		
	Acc	Sen	Spec	Acc	Sen	Spec	Acc	Sen	Spec	Acc	Sen	Spec	Acc	Sen	Spec
	90.28	85.1	95.44	90.08	85.02	95.16	89.75	84.4	95.14	89.78	84.58	94.96	92.82	89.54	96.36
Test	Fold-1			Fold-2			Fold-3			Fold-4			Fold-5		
	Acc	Sen	Spec	Acc	Sen	Spec	Acc	Sen	Spec	Acc	Sen	Spec	Acc	Sen	Spec
	89.1	82.2	96.0	91.5	99.0	84.0	96.1	95.0	97.2	99.0	100.0	98.0	65	30	100.0
	Fold-6			Fold-7			Fold-8			Fold-9			Fold-10		
	Acc	Sen	Spec	Acc	Sen	Spec	Acc	Sen	Spec	Acc	Sen	Spec	Acc	Sen	Spec
	95.4	90.8	100.0	97.5	95.0	100.0	100.0	100.0	100.0	100.0	100.0	100.0	72.2	61.2	83.2

Table 2. Comparative results from other authors and the overall 10-fold average from the proposed method; MC: Micro-calcifications, Acc: accuracy, Sen: Sensitivity, Spec: Specificity

Results from other researchers for comparison		Acc	Sen	Spec
Al Mutaz et al.[13]	(mass)	87.9	91.67	84.17
M Vasantha et al.[7]	(mass)	86.7		
Zhang et al.[6]	(MC)	85		
Amir Tahmasabi et al.[16]	(mass)	92.8		
Fatima Eddaoudi et al.[18]	(mass)	95		
Proposed method (overall 10-fold average)	(Both MC and mass)	90.58	85.32	95.84

Where **N**=Number of samples, **A**=Actual, **T**=Target. We chose accuracy as the measure of fitness for our network. Thus in the initial population the network that has the highest accuracy is chosen as the parent for the next generation. The parent, together with four of its mutated replicas, form the next generation. The fittest of these five networks form the new parent. This process repeats till we get the required fitness. In our case the fitness became almost stable after 100,000 generations so we stopped the experiments at 200,000 generations.

5   Results and Analyses

In this project we tried to provide a user friendly mammogram classification system using image statistical parameters and CGPANN. Table 1 shows that although many of the accuracy, sensitivity and specificity values in the 10-fold cross validation strategy test results are above 90%, some are infact 100%. Table 2 shows the comparative results for a few other authors in comparison to our

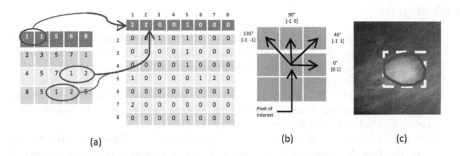

Fig. 2. Conceptual Illustration of (a) Gray Level Co-occurence Matrix (b) GLCM offsets (c) A mammogram with a malignant mass, outlined by expert (Red) and generated by software (White Rectangle). Courtesy of TM Deserno, Dept. of Medical Informatics, RWTH Aachen, Germany for the mammogram image

proposed work. All these authors have tried to classify either masses or microcalcifications alone. In comparison, our method classifies a sample set containing both masses and microcalcifications. Amongst the authors who worked in this area, Al Mutaz et al. [13] used the same Haralick's statistical texture descriptors as we did and the same database for their system training and testing. The main difference however is that they used an MLP ANN as the computational intelligence system for classification while we used CGPANN.

6   Conclusions and Future Work

In this paper we have attempted to classify mammograms for both masses and microcalcifications. The reason for this versatility lies in the fact that irrespective of the type of abnormality the technique relies on the textural characteristics of the abnormality on which our network is trained. Unlike this approach, most of the other techniques that we reviewed in the literature classify only masses or microcalcifications alone. We also developed the method to convert the chain code for ROIs, associated with the images, into rectangular image sections and training our network with the statistical parameters obtained from them. This makes the system simple to use in a windows based graphical environment for real mammogram classification. We evaluated our system with a 10-fold cross validation strategy and got an accuracy of 100% in some experiments (see table 1). In the current work the mammogram is segmented manually. Our system only classifies the region for benignity or malignancy. In future, we intend to work on automatic mammogram segmentation as well. We would then be able to segment and classify the image on a single platform.

Acknowledgments. Many thanks to TM Deserno, Dept. of Medical Informatics, RWTH Aachen, Germany, for providing the windows version of DDSM mammogram images.

References

1. Zhang, X., Fujita, H., Chen, J., Zhang, Z.: Effect of training artificial neural networks on 2d image: An example study on mammography. In: International Conference on Artificial Intelligence and Computational Intelligence, AICI 2009, vol. 4, pp. 214–218. IEEE (2009)
2. Ibrahim, N., Fujita, H., Hara, T., Endo, T.: Automated detection of clustered microcalcifications on mammograms: Cad system application to mias database. Physics in Medicine and Biology 42(12), 2577 (1997)
3. El-Naqa, I., Yang, Y., Wernick, M.N., Galatsanos, N.P., Nishikawa, R.: Support vector machine learning for detection of microcalcifications in mammograms. In: Proceedings of the 2002 IEEE International Symposium on Biomedical Imaging, pp. 201–204. IEEE (2002)
4. Hernández-Cisneros, R.R., Terashima-Marın, H.: Feature selection for the classification of both individual and clustered microcalcifications in digital mammograms using genetic algorithms. In: A Recombination of the 15th International Conference Genetic Algorithms (ICGA) and the 11th Genetic Programming Conference (GP), Seattle, WA, USA (2006)
5. Walker, J.A., Völk, K., Smith, S.L., Miller, J.F.: Parallel evolution using multichromosome cartesian genetic programming. Genetic Programming and Evolvable Machines 10(4), 417–445 (2009)
6. Zhang, P., Verma, B., Kumar, K.: Neural vs. statistical classifier in conjunction with genetic algorithm based feature selection. Pattern Recognition Letters 26(7), 909–919 (2005)
7. Vasantha, M., Subbiah Bharathi, V.: Classifications of mammogram images using hybrid features. European Journal of Scientific Research, 87–96 ISSN
8. Khuwaja, G.A.: Breast cancer detection using mammography. WSEAS Transactions on Mathematics 3, 317–321 (2004)
9. Martí, J., Freixenet, J., Munoz, X., Oliver, A.: Active region segmentation of mammographic masses based on texture, contour and shape features. In: Perales, F.J., Campilho, A.C., Pérez, N., Sanfeliu, A. (eds.) IbPRIA 2003. LNCS, vol. 2652, pp. 478–485. Springer, Heidelberg (2003)
10. Hong, B.-W., Brady, J.M.: A topographic representation for mammogram segmentation. In: Ellis, R.E., Peters, T.M. (eds.) MICCAI 2003. LNCS, vol. 2879, pp. 730–737. Springer, Heidelberg (2003)
11. Hong, B.-W., Sohn, B.-S.: Segmentation of regions of interest in mammograms in a topographic approach. IEEE Transactions on Information Technology in Biomedicine 14(1), 129–139 (2010)
12. Saidin, N., Sakim, H.A.M., Ngah, U.K., Shuaib, I.L.: Segmentation of breast regions in mammogram based on density: A review. arXiv preprint arXiv:1209.5494 (2012)
13. Mutaz, A.A., Dress, S., Zaki, N.: Detection of masses in digital mammogram using second order statistics and artificial neural network. International Journal of Computer Science & Information Technology (IJCSIT) 3(3), 176–186 (2011)
14. Zhang, Y., Tomuro, N., Furst, J., Raicu, D.S.: Image enhancement and edge-based mass segmentation in mammogram. In: Proc. SPIE, vol. 7623, pp. 76234P–8 (2010)
15. Delogu, P., Fantacci, M.E., Kasae, P., Retico, A.: Characterization of mammographic masses using a gradient-based segmentation algorithm and a neural classifier. Computers in Biology and Medicine 37(10), 1479–1491 (2007)
16. Tahmasbi, A., Saki, F., Shokouhi, S.B.: Mass diagnosis in mammography images using novel ftrd features. In: 2010 17th Iranian Conference on Biomedical Engineering (ICBME), pp. 1–5. IEEE (2010)

17. Elfarra, B.K., Abuhaiba, I.S.: Mammogram computer aided diagnosis. International Journal of Signal Processing, Image Processing and Pattern 5(4), 1–30 (2012)
18. Eddaoudi, F., Regragui, F., Mahmoudi, A., Lamouri, N.: Masses detection using svm classifier based on textures analysis. Applied Mathematical Sciences 5(8), 367–379 (2011)
19. Miller, J.F., Thomson, P.: Cartesian genetic programming. In: Poli, R., Banzhaf, W., Langdon, W.B., Miller, J., Nordin, P., Fogarty, T.C. (eds.) EuroGP 2000. LNCS, vol. 1802, pp. 121–132. Springer, Heidelberg (2000)
20. Ahmad, A.M., Khan, G.M., Mahmud, S.A., Miller, J.F.: Breast cancer detection using cartesian genetic programming evolved artificial neural networks. In: Proceedings of the Fourteenth International Conference on Genetic and Evolutionary Computation Conference, pp. 1031–1038. ACM (2012)
21. Ahmad, A.M., Khan, G.M.: Bio-signal processing using cartesian genetic programming evolved artificial neural network (cgpann). In: 2012 10th International Conference on Frontiers of Information Technology (FIT), pp. 261–268. IEEE (2012)
22. Ahmad, A.M., Muhammad Khan, G., Mahmud, S.A.: Classification of arrhythmia types using cartesian genetic programming evolved artificial neural networks. In: Iliadis, L., Papadopoulos, H., Jayne, C. (eds.) EANN 2013, Part I. CCIS, vol. 383, pp. 282–291. Springer, Heidelberg (2013)
23. Miller, J.F., Thomson, P.: Cartesian genetic programming. In: Poli, R., Banzhaf, W., Langdon, W.B., Miller, J., Nordin, P., Fogarty, T.C. (eds.) EuroGP 2000. LNCS, vol. 1802, pp. 121–132. Springer, Heidelberg (2000)
24. Haralick, R., Shanmugam, K., Dinstein, I.: Texture features for image classification. IEEE Transactions on Systems, Man, and Cybernetics 3(6) (1973)

A Face Recognition Based
Multiplayer Mobile Game Application

Ugur Demir, Esam Ghaleb, and Hazım Kemal Ekenel

Faculty of Computer and Informatics, Istanbul Technical University, Istanbul, Turkey
{ugurdemir,ghalebe,ekenel}@itu.edu.tr

Abstract. In this paper, we present a multiplayer mobile game application that aims at enabling individuals play paintball or laser tag style games using their smartphones. In the application, face detection and recognition technologies are utilised to detect and identify the individuals, respectively. In the game, first, one of the players starts the game and invites the others to join. Once everyone joins the game, they receive a notification for the training stage, at which they need to record another player's face for a short time. After the completion of the training stage, the players can start shooting each other, that is, direct the smartphone to another user and when the face is visible, press the shoot button on the screen. Both the shooter and the one who is shot are notified by the system after a successful hit. To realise this game in real-time, fast and robust face detection and face recognition algorithms have been employed. The face recognition performance of the system is benchmarked on the face data collected from the game, when it is played with up to ten players. It is found that the system is able to identify the players with a success rate of around or over 90% depending on the number of players in the game.

Keywords: face recognition, multiplayer mobile game, DCT, LBP, SVM.

1 Introduction

Mobile applications have become more common and popular with the advances in smartphone technologies. They have also become one of the main segments of the digital entertainment. Many successful mobile gaming companies have emerged recently with very high market values. Most of these games require users' focus and attention, isolating them from the real world. However, mobile games can also be designed to provide entertaining human-human interaction experience. In this way, instead of isolating the users from the real world, mobile game applications can convert a smartphone to a tool that allows users to interact with the others in a fun way. With this motivation, in this study, we designed a mobile game application that enables individuals play paintball or laser tag style games using their smartphones[1]. The application benefits from computer vision and

---

[1] A demo video of the system can be found at
http://web.itu.edu.tr/ekenel/facetag.avi

L. Iliadis et al. (Eds.): AIAI 2014, IFIP AICT 436, pp. 214–223, 2014.

pattern recognition technologies, specifically, face detection and recognition to detect and identify the players. In order to have real-time processing capability, which is required for the game, we have employed fast and robust face detection and recognition algorithms.

The developed game consists of two main components, server and mobile application. The server application regulates the game rules and players' network stream. In addition, server application contains the face recognition module. The game application sends camera frames that contain players' faces to the server. Thus, mobile application contains only network and face detection module. At the beginning of the game, players join a game session, which is created by one of the players. After all the players are attended, training stage is started by the server. Mobile applications gather player images as training data and sends them to the server. Each player collects different player's face data. When server application receives all of the training data, it extracts the features and trains the classifiers. Afterwards, the game starts and players attempt to shoot each other. If camera detects a face, while a player is shooting, it sends the detected face image to the server application. The server application assigns an identity to the received face image by using trained classifiers, and then broadcasts the necessary information to the mobile devices. The flow diagram of the game is shown in Fig. 1.

The rest of the paper is organised as follows. In Section 2, the server application architecture is introduced. In Section 3, design of the mobile application is explained. In Section 4, proposed face recognition methods and classifiers are stated. Tests and experimental results are presented in Section 5. Finally, conclusions and future work are given in Section 6.

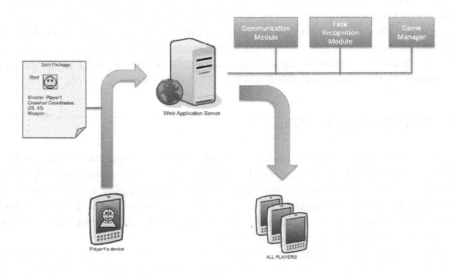

Fig. 1. Multiplayer Mobile Game Application Flow Diagram

2   Server Application

Server application that controls the game flow is the main part of the system and consists of three core modules, which are network module, game flow control module, and face recognition module. All of these core modules work collectively during a game session.

Network module's main purpose is receiving the commands and images from mobile devices, and then deciding to what information should be broadcasted to players. All of the mobile devices connect to the server over a TCP socket before starting to play. Server always listens to the TCP port for incoming client data, which is represented with JSON format. It creates and runs a new thread called network thread for each connected mobile device. These threads manage the network operations. Network module does not interfere the game flow logic. It is only responsible to transfer the received data to game flow module and transmit resulting data to relevant clients. This abstraction enables us to design a more robust multiplayer game server.

Game flow control module executes the client's orders delivered by network module, which are based on the game rules. This module decides to execute necessary function calls depending on the client's request. If an image is received, it is transferred to face recognition module and results are evaluated. Player scores are calculated by this module. If a player shoot another player, game flow control module tells network module to notify these players.

Face recognition module handles training and classification tasks in the server. It has two stages in its life cycle; training and classification. In the training part, face recognition module collects player images and generates a multi-class classifier for each mobile device. We assume that players do not shoot themselves. Thus, each classifier recognise all the players in a game session except the player, who owns the mobile device. After the training stage, flow control module starts the game. Then classification stage takes control in face recognition module. In this stage, player images are queried to determine, which player's face image it is. All modules and server application are implemented in Java programming language. In the face recognition module, OpenCV library Java port is used.

3   Mobile Application

Mobile game application contains user interface and communication module, which delivers user commands and listens for incoming server data. At first, devices send a connection request to the server, when a player starts the game application. If the connection is successfully established, a user menu, which shows the game options, appears on the screen. Then, player can join or create a game lobby via this menu to start the game.

After starting the mobile game application, a thread called communication thread is started to run by the application. Communication thread listens the incoming server data during the application life-cycle to broadcast the necessary data to the application screens, which player interacts with. We apply observer pattern to achieve this scheme. In the observer pattern, objects that display the data is isolated from the object that stores the data and storage object is observed to detect changes on data [1]. Observer pattern is suitable for communication module of the game due to its architecture. Communication thread listens and stores the data, so it represents the observable object in the system. Application screens, which is called Activity in Android SDK [2], stand for the observer object in observer pattern. Server sends state changes via TCP sockets to communication thread and activities observe the communication thread for any data changes. Fig. 2 shows the network architecture.

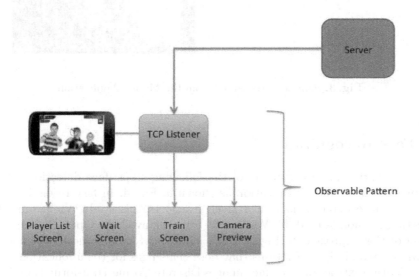

Fig. 2. Mobile Application Flow Diagram

Activities in the application register themselves to observe communication thread before starting and unregister before closing. Thus, communication thread transfer the data to only relevant activities and modules. Observer activities filter and parse the received data and display to user. Mobile application flow consists of three main step; joining the game, collecting the training data, and playing. One player, called creator, creates a game lobby and waits for the other players join the existing lobby. Creator player can see the entered players on the application screen and start, when there are at least two players in the lobby including creator.

After the creator starts the game, the game lobby is configured in the server and server sends a command, which tells training session must be started, to all players. In the training stage, players' face images are collected via smartphones' cameras. Collected images are sent to the server for training. When all players send the images and the training is completed, server starts the game and sends start game command to players' devices. Main game screen basically consists of camera preview and an aim marker. If a player presses the shoot button and the aim marker overlays with a player's face at the same time, detected face image is sent to the server. Server analyses the face image and sends the information about who shoots which player to the corresponding players. Fig. 3 shows same screenshots from the mobile game screen.

Fig. 3. Sample Screenshots from the Mobile Application

4   Face Recognition

Face recognition process consists of the following steps: face detection, feature vector extraction and classification as shown in Fig. 4. In face recognition, the first step is to detect a face in a given frame. In this study, Viola-Jones face detection method is used [3]. We benefit from the available OpenCV implementation of this approach [4]. Fig. 5 shows sample detected faces. Please note the illumination and view variations that pose challenges for the application.

For feature extraction from face images Discrete Cosine Transformation (DCT) and Local Binary Pattern (LBP) are implemented. These two representation methods are proven to be fast and robust in previous studies [5,6]. DCT is popular for being a fast and efficient transformation, mainly used for data compression, due to its capability to represent data in a compact form. As a result, this makes DCT a suitable approach for mobile applications. In this paper, to extract a feature vector from a given image, DCT was applied as explained in reference [7]. The images are first resized to have 64×64 pixel resolution and then divided to nonoverlapping blocks. Each block has 8×8 pixel resolution. Then DCT is applied on these local blocks. Following this process, zig-zag scanning is used to order the DCT coefficients. For each block, M features are picked and normalised. These local features are then concatenated to form the overall feature vector of the image. This obtained feature vector represents the given image. In the implementation M is set to 5, making the resulting DCT feature vector of the length equal to $5 \times 8 \times 8 = 320$.

LBP is widely used for face feature extraction and it has been proven to be very effective and successful approach for many tasks such as face recognition and face verification. In LBP, for every pixel in gray level image, it assigns a label

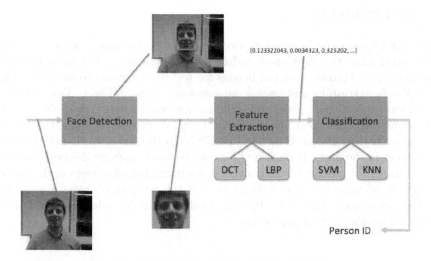

Fig. 4. Face Recognition Process

Fig. 5. Sample Detected Faces

to it, based on its neighborhood [5]. In this study, first the face image is resized to 66×66 pixel resolution -two additional pixels due to the operations at the image boundaries-, then divided to nonoverlapping blocks of 8×8 pixel resolution. For each pixel in a block, the pixel is compared to each of its 8 neighbours and its corresponding LBP code is assigned. Since the 8 neighborhood is considered, a 256-bin histogram of LBP codes is computed. To reduce the dimensionality, the subset of LBP codes, called the uniform patterns, are used [5, 8, 9]. A uniform pattern contains a two or less number of transitions from zero to one or vice versa. As a result, the patterns that has more than two transitions are aggregated into a single bin in the LBP histogram besides the other uniform patterns, reducing the histogram size to 59 bins. The obtained histograms from local blocks are concatenated into a single feature vector that represents the entire image. The resulting LBP feature vector's length equals to $8 \times 8 \times 59 = 3776$.

Following feature vector extraction from the face images, the application employs multi-class classifiers, either Support Vector Machine (SVM) or K-Nearest Neighbor (KNN) classifiers. In SVM the used kernel function is radial basis function and constraint parameter is set to 0.5 [10]. In KNN, K is set to 1 and the L1 and L2 distance metrics are used.

5   Experiments

In order to asses the face recognition system's performance, a face database is generated using the developed mobile game application. For these evaluations ten images for training and ten images for test are collected from twelve players. Samples from training and testing data sets are shown in Fig. 6. The variations in facial expression, head pose, and illumination results in a very challenging task. DCT and LBP are used for feature extraction, respectively, and the extracted features are then classified by SVM or KNN. In the tests, we change the number of players in the game from two to ten players and examine the performance for each number of players in the game. For each number of players in the game, we form ten different player combinations and the recognition rates are calculated as the average of these ten different combinations in order to have a reliable measure for the system performance.

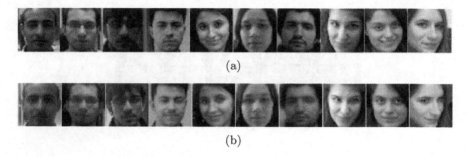

(a)

(b)

Fig. 6. Samples from (a) the Training Data Set, (b) the Testing Data Set

5.1   Experiments on the Standard Data Set

In the first experiment, the evaluation is conducted using the collected data set without doing any data manipulation for additional sample generation. The results of this experiment are plotted in Fig. 7(a). As shown in the figure, the horizontal axis represents the number of the players in the mobile game application, who must be classified, while the vertical axis shows the correct classification rate, which is the average over ten different player combinations. The performance curves give the results for different combination of classification and feature extraction methods. As can be observed from the figure, using LBP features, the best performance is achieved. There is no significant performance difference, when using SVM or KNN. As the number of players in the game increases, the obtained performance decreases, however this decline is slight and the system performance stays around 90% when using LBP features.

5.2   Experiments on the Extended Data Set

In the experiments on the extended data sets, two different scenarios are applied. In these scenarios, either training data set or test data set is extended. In the first scenario, the purpose is to enrich the training data set. As a result, the collected 10 image for training are manipulated to generate 900 images using mirroring, translation and rotation. This extended data set is used for training, while test data set remains as it is. The obtained results are shown in Fig. 7(b). As shown in the performance curve, enriching training data set has improved the recognition rate. Especially, the LBP+SVM combination consistently performs at 100%. However, since SVM's training complexity depends on the square of the amount of samples, that is, it has a complexity of $O(mn^2)$, where m is the dimension of the feature vector and n is the amount of training samples, it is not feasible to use it in this scenario.

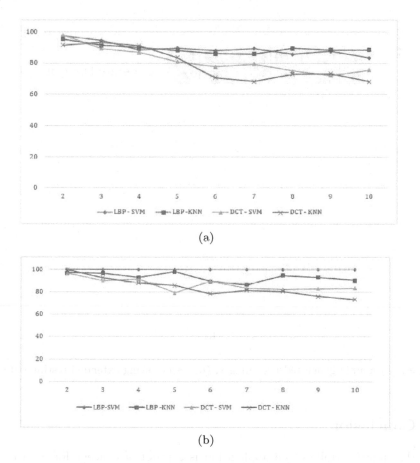

(a)

(b)

Fig. 7. (a) Results when ten images are used for training and test stages, (b) results using extended training data set

In the second scenario, the training data set is used as it is, however, this time the test data set is extended and enriched by operations, such as translation, rotation and mirroring. As a result, the collected ten images for test are extended to 60. Since the testing has to be done in real time, this number is far less than the number of additional samples generated for training in the first scenario. Sample generated test images are shown in Fig. 8(a). The performance curve for this experiment is shown in Fig.8(b). Only the KNN classifier is used for this experiment due to its ease of use in decision fusion among different test samples. Similar to the case in the first scenario, the performance improves when using the enriched test data set.

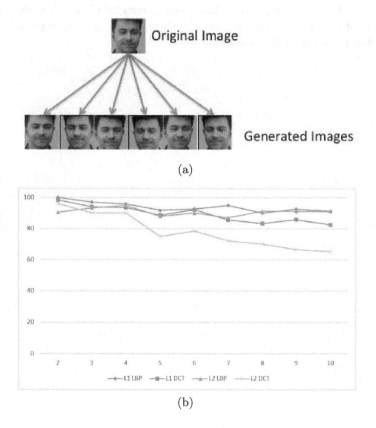

Fig. 8. (a) sample generated test images, (b) results using extended testing data set

6   Conclusion

The developed mobile game application is a proof of concept for a successful application of computer vision and pattern recognition technologies to convert a smartphone to a tool that enables human-human interaction in an entertaining way. The game allows users to play paintball or laser tag style games with their smartphones. Face detection and recognition technologies have been employed to detect and identify the players in the game. The implemented practical and real-time face recognition system is benchmarked within the scope of the application. When using LBP features with ten players in the game, around 90% performance is obtained. SVM and KNN provided similar results. The application is found to be entertaining by the users. It also inspires users about how these technologies can be used in the future. However, we experienced a problem about the usability of the system. It was observed that players keep their smartphones at the face level in order to see the screen, which causes occlusion on the face. To solve this problem, we plan to work on occlusion robust face detection and recognition. In addition, we also plan to design a physical mechanism that can be used in the game to plug the smartphone to, so that the players can have a sufficient distance from their smartphones, while playing.

Acknowledgement. This work was supported by the Avea Labs Research Grant, TUBITAK project no. 113E121; and a Marie Curie FP7 Integration Grant within the 7th EU Framework Programme.

References

1. Cooper, J.W.: Java Design Patterns: A Tutorial. Addison-Wesley (2000)
2. Android SDK Activity,
   http://developer.android.com/reference/android/app/Activity.html
3. Viola, P., Jones, M.: Rapid object detection using a boosted cascade of simple features. In: Proc. of CVPR, pp. 511–518 (2001)
4. OpenCV Android SDK (May 2014), http://opencv.org
5. Ahonen, T., Hadid, A., Pietikäinen, M.: Face Description with Local Binary Patterns: Application to Face Recognition. IEEE Transactions on Pattern Analysis and Machine Intelligence 28(12), 2037–2041 (2006)
6. Ekenel, H.K.: A robust face recognition algorithm for real-world applications, PhD dissertation, University of Karlsruhe, TH (2009)
7. Ekenel, H.K., Stiefelhagen, R.: Analysis of local appearance based face recognition: Effects of feature selection and feature normalization. In: CVPR Biometrics Workshop, New York, USA (2006)
8. Shan, C., Shaogang, G., McOwan, P.W.: Facial Expression Recognition Based on Local Binary Patterns: A Comprehensive Study. Image and Vision Computing 27(6), 803–808 (2009)
9. Choi, S.E., Lee, Y.J., Lee, S.J., Park, K.R., Kim, J.: Age estimation using a hierarchical classifier based on global and local facial features. Pattern Recognition 44(6), 1262–1281 (2011)
10. Burges, C.: A Tutorial on Support Vector Machines for Pattern Recognition. Data Mining and Knowledge Discovery, 121–167 (1998)

A Dynamic Questionnaire to Further Reduce Questions in Learning Style Assessment

Espérance Mwamikazi[1], Philippe Fournier-Viger[1], Chadia Moghrabi[1], and Robert Baudouin[2]

[1] Department of Computer Science, Université de Moncton, Canada
[2] Department of Secondary Education and Human Resources, Université de Moncton, Canada
{esperance.mwamikazi,philippe.fournier-viger,
chadia.moghrabi,robert.baudouin}@umoncton.ca

Abstract. The detection of learning styles in adaptive systems provides a way to better assist learners during their training. A popular approach is to fill out a long questionnaire then ask a specialist to analyze the answers and identify learning styles or types accordingly. Since this process is very time-consuming, a number of automatic approaches have been proposed to reduce the number of questions asked. However the length of questionnaire remains an important concern. In this paper, we address this issue by proposing *T-PREDICT*, a novel dynamic electronic questionnaire for psychological type prediction that further reduces the number of questions. Experimental results show that it can eliminate 81% more questions of the Myers-Briggs Type indicators questionnaire than three state-of-the-art approaches, while predicting learning styles without increasing the error rate.

Keywords: dynamic adaptive questionnaire, classification, learning styles, Myers-Briggs Type Indicator, psychological types.

1 Introduction

Knowing the learning style of learners is important to ensure enhanced and personalized interactions with teachers. Various studies established that presenting information in a manner that is adapted to the student's learning style facilitates learning (e.g. [1, 2]).

Two main approaches are used to carry out learning style or type assessment: the first approach consists of examining the learner's interaction with the system [1, 3]. The second approach entails using a standard questionnaire that is filled out by an individual before a specialist examines the responses and establishes the corresponding learning style.

Even though these approaches can correctly identify the learning type of a learner, they still suffer from various shortcomings. Using the first approach entails that an initial random learning style is assigned to the learner, implying that if the initial guess is wrong, the system will offer wrong interactions with the learner and that could have negative impact on learning. In addition, the interaction will persist until enough data is stored so that ideal learning style can be established [3]. The second

L. Iliadis et al. (Eds.): AIAI 2014, IFIP AICT 436, pp. 224–235, 2014.

approach has various limitations as well. First, the questionnaires tend to be long and time consuming. For instance, Myers-Briggs Type indicator questionnaire has more than 90 questions. Second, the fact that questionnaires are long implies that the person filling the questionnaire might be demotivated to reply to the questions with enough attention [3] which might lead to abandoning the test, skipping questions, answering falsely, etc. Consequently, an incorrect learning style might be adopted [4]. The third limitation of this approach is that in order to assign a learning style, there is a need for a specialist to analyze the learner's answers to questions.

To address these limitations, several approaches [5, 6, 7] have been proposed to automatically reduce the number of questions asked to learners in an effort to identify their learning styles or psychological types. For example, *Q-SELECT* [5] is an adaptive electronic questionnaire that uses association rules to predict part of the answers and hence shorten the questionnaire by up to 30%. However, even when applying these approaches, the length of questionnaires remains a concern. Therefore, an important research question is "could a new method be designed to further reduce the number of questions asked?"

In this paper, we answer this question positively by proposing a dynamic electronic questionnaire *T-PREDICT* that further reduces the number of questions and automatically recognizes the psychological types with high accuracy.

The rest of the paper is organized as follows. The Myers-Briggs Type Indicator model is presented in section 2. Section 3 discusses related work on adaptive questionnaires. Section 4 and 5 respectively present the proposed electronic questionnaire and the experimental results. Finally, section 6 draws the conclusions.

2   Myers-Briggs Type Indicator (MBTI)

The Myers-Briggs Type Indicators (MBTI) is a well-known personality assessment model that has been used for over three decades. It uses Carl Jung's personality type theory to categorize individuals into four dimensions. Each dimension consists of two opposite preferences that depict inclinations and reveal dispositions in personal mindsets and techniques of making decisions [8].

The E-I dimension (*extraverted-introverted*) centers its attention on establishing whether a person's approach is influenced by the outward environment such as other objects and individuals, or it is internally-oriented. The S-N dimension seeks to measure *sensing-intuitiveness*, illustrating the perception approach of an individual. As per the T-F dimension, *thinking* implies coherent reasoning and decision making processes while *feelings* influence personal, objective, and value oriented approach. The J-P dimension involves either a *judging* attitude and quick decision-making or *perception* that demonstrates more patience and information gathering prior to decision-making. Given these four dimensions, an individual's personality type is therefore designated by a four-letter code: {ISTJ, ISFJ, INFJ, INTJ, ISTP, ISFP, INFP, INTP, ESTP, ESFP, ENFP, ENTP, ESTJ, ESFJ, ENFJ, ENTJ}.

While a number of the preferences may be dominant, others are likely to be secondary and can be easily subjected by other dimensions. For instance, the J-P dimension influences the two function preferences namely T or F versus S or N [9].

Though the Myers-Briggs Type Indicator questionnaire has been used widely for a long time, it has some limitations. Its hypothetical and numerical imports are restricted to some extent by the use of dichotomous preference items [10]. In addition, the MBTI questionnaire contains numerous questions, a number which may put off some users. As a result, they may choose to answer the questionnaire without paying much thought or attention to the choices thus raising doubts on the reliability of the assessment [11]. Reducing the number of questions in a questionnaire has been a way to increase its efficiency [3, 12].

3   Related Work

Some major challenges in building an e-learning system that can adapt itself to a learner is giving it the capability of changing the type, the order, or the number of questions presented to the learner [3, 5, 13, 14]. For instance, McSherry [12] reports that reducing the number of questions asked by an informal case-based reasoning system minimized frustration, made learning easier and increased efficiency.

Various methods have been suggested in order to minimize the number of questions that are required to establish the learning style or preference of an individual. The *AH questionnaire* [3] uses decision trees to reduce the questions and categorize the students as per the Felder-Silverman theory of learning styles. From an experiment that had 330 students, it was likely to anticipate the learning styles with precision of up to 95.71% while only asking four or five questions among the eleven questions that are applied in each dimension.

EDUFORM is a software used for adaptation and dynamic optimization of questionnaire propositions for online profiling of learners, which was proposed by Nokelainen et al. [6]. The tool, that uses Bayesian modeling as well as abductive reasoning, minimized the questionnaire items by 30% to 50% while retaining an error rate of 10% to 15%. These results showed that shortening questionnaires does not detrimentally influence the correct categorization of individuals.

In previous work [7], two methods based on back-propagation neural networks and decision trees were proposed to predict learning types and reduce the number of questions asked in the MBTI questionnaire. We refer to these methods as *Q-NN* and *Q-DT*, respectively. The general experimental method tries to identify questions that are less influential in determining the learning types. These questions are then eliminated from the questionnaire and the learning types are predicted. This process is repeated until a maximum error rate of 12% is achieved. In an experimental study, that had 1,931 filled questionnaires, the *Q-NN* method clearly identified and eliminated 35% of the questions while establishing the learning preferences with an error rate of 9.4%. On the other hand, *Q-DT* eliminated 30% of the questions with an error rate of 14%.

Recently, we presented an alternative approach to reduce the number of questions in MBTI questionnaire [5]. This approach (*Q-SELECT*) comprises three modules: (1) an answer prediction algorithm, (2) a dynamic question selection algorithm and (3) an algorithm to accurately predict a person's learning style based on both user supplied and predicted answers. The two first modules rely on association rules between answers from previous users. The third module predicts learning types using a neural network. An experimental study with the same 1,931 MBTI filled questionnaires has shown that *Q-SELECT* reduces the number of questions asked by a median of 30% with an average error rate of 12.1%. The main advantage of *Q-SELECT* compared to *Q-NN* and *Q-DT* is its adaptability, i.e. the ability to ask a variable number of questions and to reorder them depending on the user answers, thus providing a personalized questionnaire to each user.

Reducing the MBTI questionnaire by up to 35% still leaves around 60 questions. This paper presents *T-PREDICT,* a new dynamic approach to further minimize the number of questions in MBTI questionnaires while predicting learning types with a comparable error rate.

Reducing the number of questions might bring to mind algorithms for dimensionality reduction such as PCA, ICA [15]. However, these methods would have reduced the same number of questions for all users as in *Q-NN*. Our choice was to continue our experiments with a dynamic user-adaptive approach. This choice was later confirmed by our results from *T-PREDICT*. We were sometimes able to predict the learning type with as little as six questions out of 92 but with a median of 11 questions.

Another popular theory that deals with questions and answers is *Item Response Theory (IRT)* [16]. It has been applied in education and psychology to assess an underlying ability or trait using a questionnaire. To apply *IRT*, users answers need to be collected and analyzed. Using a technique such as factor analysis, a model (e.g. logistic function) is created for each question to represent the amount of information provided by each answer about the latent trait. The quality of the generated models varies depending on the data available. When a model does not fit well a question, this latter is typically removed, replaced or rewritten [17]. Applying *IRT* can be very time-consuming, since for each modification, more data may need to be collected to update models and human intervention is required to analyze questions and tweak models. Furthermore, *IRT* does not provide means for user adaptability. In contrast, our proposal is automatic and user-adaptive in all steps: selecting important questions, reordering them, and later predicting learning types.

4   The Electronic Questionnaire

The proposed electronic questionnaire *T-PREDICT* comprises two major components: a question sorting algorithm and a learning type prediction algorithm. The philosophy behind this division is that some questions might be more important than others i.e. their answers could easily classify the learners in one or the other personality type. Hence, the sorting algorithm sorts the questions by their ability to discriminate between classes, while the prediction algorithm uses them in this preferential order. An additional module is developed to dynamically select parameters needed for the prediction algorithm.

4.1   The Question Sorting Algorithm

The MBTI questionnaire evaluates each dimension (EI, SN, TF, and JP) by a distinct subset of questions. Thus, we split the questionnaire into the four sets of questions representing these dimensions. Each dimension consists of two opposite preferences, hence the need to classify individuals into one of these two preferences (classes).

Since our goal is to reduce the number of questions asked, it is crucial to recognize which questions are more important for identifying the preferences. We define the importance of a question as its ability to discriminate between the two classes. Let q be a question and $A(q) = \{a_1 \dots a_m\}$ be the set of possible answers to this question. Let T be a training set of filled questionnaires such that each questionnaire belongs to one of the two classes. For any answer a_i ($1 \leq i \leq m$), let $N_1(a_i)$ and $N_2(a_i)$ respectively be the number of questionnaires in T containing the answer a_i that belong to the first class and the second class. The *discriminative power* of the answer a_i is denoted as $DA(a_i)$ and defined as $N_1(a_i) / N_2(a_i)$ if $N_1(a_i) > N_2(a_i)$ or $N_2(a_i) / N_1(a_i)$, otherwise. Intuitively, it represents how many times one class is larger than the other for individuals having answered a_i, and thus how this answer helps to discriminate between the two classes. The *discriminative power* of a question q is denoted as $DQ(q)$ and defined as $DQ(q) = \sum_{k=1}^{|A(q)|} DA(a_i)/ |T|$. The proposed adaptive questionnaire initially calculates the discriminative power of each question in the training set T and sorts them accordingly (see Fig. 1 for the pseudocode). The next subsection describes how the adaptive questionnaire asks the questions by decreasing order of discriminative power and predicts the preference (class) of each individual using the provided answers.

QUESTION_SORT (a training set T, a list of questions Q, a set of possible answers A)
1. **FOR** each question $q \in Q$
2.   **FOR** each possible answer $a_i \in A(q)$
3.   **SCAN** T to calculate $N_1(a_i)$ and $N_2(a_i)$.
4.   **IF** $N_1(a_i) > N_2(a_i)$ **THEN** $DA(a_i) := N_1(a_i) / N_2(a_i)$.
5.   **ELSE** $DA(a_i) := N_2(a_i) / N_1(a_i)$.
6.   **END FOR**
7. $DQ(q) = \sum_{k=1}^{|A(q)|} DA(a_i)/ |T|$.
8. **END FOR**
9. **SORT** Q such that q_a appears before q_b if $DQ(q_a) > DQ(q_b)$, for all questions $q_a, q_b \in Q$.
10. **RETURN** Q.

Fig. 1. The question sorting algorithm

4.2   The Learning Style Prediction Algorithm

The learning style prediction algorithm (see Fig. 2) automatically identifies the preference of a user in a given dimension based on supplied answers and their similarity to answers from previous users. The algorithm takes as input a training set of filled questionnaires T, the list of questions Q for the dimension, sorted by their discriminative power (see section 4.1), a maximum number of questions to be asked *maxQuestions* (by default set to $|Q|$), and two additional parameters *minMargin* and *minQ* that will be defined later.

The algorithm first initializes a set U as empty to store the user answers. Then, a loop is performed to ask the questions to the user in decreasing order of their discriminative power. Each provided answer is immediately added to U. Then, the algorithm attempts to make a prediction by scanning through the set T to count how many users have the same answers as the current user and belong to class 1 or belong to class 2, i.e. $S_1 = |\{X \mid X \in T \wedge U \subseteq X \wedge X$ is tagged as class 1$\}|$ and $S_2 = |\{X \mid X \in T \wedge U \subseteq X \wedge X$ is tagged as class 2$\}|$. Two criteria must be met to make this prediction. First, the number of filled questionnaires, from the training set T, matching the answers of the current user should be higher or equal to a pre-set threshold ($minQ$), i.e. $S_1 + S_2 \geq minQ$. Second, there should be a large enough difference between S_1 and S_2 in order to make an accurate prediction. This difference is defined as $|S_1 - S_2| \geq minMargin$, where $minMargin$ is also a pre-set threshold. If both conditions are met, a prediction is made. The prediction is class 1 if $S_1 > S_2$, and class 2 otherwise. If no prediction can be made, the algorithm continues with other questions, one at a time, until it is able to make a prediction.

PREDICT (a training set T, a list of questions Q sorted by their discriminative power, a maximum number of questions to be asked $maxQuestions$, a minimum margin between classes $minMargin$, a minimum number of questionnaires to be matched $minQ$)
1. $U := \emptyset$.
2.   **FOR** each question $q_i \in Q$ until $maxQuestions$ have been asked
3.   **ASK** q_i to the user and store the provided answer in U.
4. $S_1 := S_2 := 0$.
5.   **FOR** each questionnaire $X \in T$ such that $U \subseteq X$
6.   **IF** X is tagged as class 1 **THEN** $S_1 := S_1 + 1$. **ELSE** $S_2 := S_2 + 1$.
7.   **IF** $S_1 + S_2 \geq minQ$ **AND** $|S_1 - S_2| \geq minMargin$ **THEN**
8.   **IF** $S_1 > S_2$ **THEN RETURN** *"class 1"*. **ELSE RETURN** *"class 2"*.
9.   **END FOR**
10. $Y_1 := Y_2 := 0$.
11. **FOR** each questionnaire $X \in T$ such that $C(X, U) \geq |U|/2$
12.   **IF** X is tagged as class 1 **THEN** $Y_1 := Y_1 + C(X, U)$. **ELSE** $Y_2 := Y_2 + C(X, U)$.
13. **IF** $Y_1 > Y_2$ **THEN RETURN** *"class 1"*. **ELSE RETURN** *"class 2"*.

Fig. 2. The learning type prediction algorithm

After *maxQuestions* questions have been exhausted and no prediction was possible with the exact matching, a prediction is made by considering an *approximate* match between the present user answers and questionnaires from the training set T. A questionnaire $X \in T$ approximately matches U if it shares at least $|U|/2$ answers with it. The number of answers common to X and U is denoted as $C(X, U)$ and defined as $C(X, U) = |X \cap U|$. Let Y_1 and Y_2 be the sets of all questionnaires from T that approximately match U and are tagged as class 1 or class 2. The prediction is class 1 if $Y_1 > Y_2$. Otherwise, it is class 2.

4.3   The Parameters Selection Algorithm

As mentioned earlier, the prediction algorithm needs two preselected parameters *minMargin* and *minQ*. These parameters are not global. They are dynamically selected as per number of questions asked. The *PARAMETERS_SELECT* algorithm (see Fig. 3) attempts to select the best values for these parameters exhaustively by simulating predictions and calculating the corresponding precisions on a test set of filled questionnaires *TS* by using a training set T.

PARAMETERS_SELECT (a training set T, a test set TS, a list of questions $Q= \{q_1, q_2,...q_n\}$ sorted by their discriminative power, a target precision $TargetPrecision$, and the two upper bounds $maxMargin$ and $maxQ$)

1. $GlobalPrecision := 0$.
2. $RequiredPrecision := TargetPrecision$
3. $P := \emptyset$.
4. $RemainingQuestionnaires := TS$.
5.   **FOR** each $q_k \in Q$ and $k := 1, 2...n$
6. $bestScore := 0$.
7. $bestParameters := (0, 0, 0)$.
8. $bestResult := (0, 0, 0)$.
9.   **FOR** each combination of values i, j, α such that $i := 1, 2... maxQ$
        and $j := 1, 5,... maxMargin$
        and $\alpha := 0.5, 0.6 ... 0.9$
10.   Predict a class for each questionnaire $X \in RemainingQuestionnaires$ using the first k questions and the training set T.
        Set parameters $minQ := i, minMargin := j$ for the kth question, while keeping the best previously found parameters for questions 1 to k-1.
        Let $predicted_k$ be the set of questionnaires where a prediction was performed and $precision_k$ be the precision.
11. $score := precision_k \times \alpha + |predicted_k| \times (1- \alpha)$.
12.   **IF** $score > bestScore$ **AND** $precision_k \geq RequiredPrecision$ **THEN**
13. $bestParameters := (k, i, j)$.
14. $bestScore := score$.
15. $bestResult := (k, predicted_k, precision_k)$.
16.   **END IF**
17.   **END FOR**
18. $P := P \cup \{bestParameters\}$.
19. $GlobalPrecision := GlobalPrecision + |bestResult.predicted| \times bestResult.precision$
20. $RemainingQuestionnaires := RemainingQuestionnaires \setminus bestResult.predicted$.
21. $RequiredPrecision := (|TS| \times TargetPrecision - GlobalPrecision)$
        $/ RemainingQuestionnaires$
22. **END FOR**
23. **RETURN** P.

Fig. 3. The parameter selection algorithm

The algorithm proceeds by considering each question from Q in the order that they will be asked by the PREDICT algorithm (in descending order of discriminative power). For every question q_k considered, all combinations of values i and j for parameters $minQ$ and $minMargin$ in their respective intervals [1, $maxQ$] and [1, $maxMargin$] are tested. For each combination, a certain number of predictions $NbPredictions_k$ are obtained with a corresponding precision $precision_k$. The combination of values i and j that is considered the best is the one maximising the function $score(i,j) = \alpha \times precision_k + NbPredictions_k \times (1- \alpha)$, where α is a constant varied in the [0.5, 0.9] interval. The weight α is used to calibrate the relative importance of the precision and the number of predictions on the selection of parameter values.

Furthermore, an additional constraint is that the best combination needs to have a precision higher than a moving threshold *RequiredPrecision*. For the first question, this threshold is equal to a pre-set minimum precision *TargetPrecision*. For any subsequent question q_k, the required precision is recalculated to take into account the precisions obtained with previous questions (called *GlobalPrecision$_k$*). The reason is that if a high precision is obtained for the previous questions, it is possible to accept a lower precision for the next questions, while maintaining a global precision above *TargetPrecision*. The global precision is calculated by weighting the previous precisions with their corresponding number of predictions. Formally, the global precision is defined as: $GlobalPrecision_k = \sum_{f=1}^{k-1} precision_f \times NbPredictions_f)$. The required precision for the kth question is calculated as: $RequiredPrecision_k = (|TS| \times Target Precision - GlobalPrecision_k) / |RemainingQ|$, where *RemainingQ* is the number of unpredicted questionnaires.

When the algorithm terminates, it returns a list *P* of triples of the form (k, y, z) indicating that parameters *minMargin* and *minQ* should be respectively set to values y and z for the kth question to establish a prediction with a global precision not below *TargetPrecision*.

5   Experimental Results

Two experiments were performed. The first one compares the performance of the learning type prediction algorithm *T-PREDICT* with other methods. The second one assesses the influence of question sorting and limiting the number of questions in *T-PREDICT*. A database of 1,931 MBTI filled out questionnaires were supplied for experimentation by Prof. Robert Baudouin, an expert of the MBTI technique at Université de Moncton. The number of samples used for training was 1,000 while 931 were used for testing. The proposed dynamic questionnaire *T-PREDICT* is implemented using Java. The goal of the experiments was to ask as few questions as possible while automatically identifying learning types with a minimum precision of 88 %, i.e. a maximum error rate of 12%. This is the same error rate that was used in *Q-SELECT* [5], thus allowing a more precise comparison base.

Parameters used were automatically selected by the *parameters selection algorithm* (see section 4.4). It varied *minQ* and *minMargin,* for each question, in the [0.01, 0.15] and [0.1, 0.9] intervals respectively and returned their optimal values to be used by the *type prediction algorithm*. These predictions were done for each of the four MBTI dimensions.

5.1   Comparison with Other Methods

The first experiment compared the median number of questions asked in each dimension with results obtained by three specialized methods that have been specifically developed to reduce the number of questions for MBTI. These methods are based on back-propagation neural networks (*Q-NN*) [7], decision trees (*Q-DT*) [7], and association rules (*Q-SELECT*) [5]. Table 1 shows the number of questions asked per dimension, as compared to the original higher number, and the corresponding error rates. Note that the *T-PREDICT* line shows that the number of questions asked for EI, SN, TF and JP dimensions were 2 out of 21 questions, 4 out of 25 questions, 2 out of 23 questions, and 3 out of 23 questions respectively. The corresponding error rates obtained were 10.7% for EI, 13% for SN, 10.3% for TF and 13.1% for JP. Thus, maintaining a weighted average error rate of 12.1% for the four dimensions, which is practically the pre-set error rate of 12% mentioned earlier. All dimensions for the four compared methods maintained error rates below 13.1% except the TF and JP dimensions for the *Q-DT*, which respectively had error rates of 16.3% and 13.7%. Overall, *T-PREDICT* asked a median of only 11 questions, out of 92 questions, to predict learning types with an average error rate of 12.1%, while *Q-SELECT*, *Q-NN*, and *Q-DT* asked 62, 60, and 64 questions to achieve error rates of 11.4%, 9.4%, and 14% respectively. In sum, the number of questions was greatly reduced by *T-PREDICT* while maintaining a very comparable error rate.

Table 1. Median number of questions asked by *T-PREDICT* per dimension, compared to other methods

	EI (21 quest.)	SN (25 quest.)	TF (23 quest.)	JP (23 quest.)	Total (92 quest.)
T-PREDICT	**2**	**4**	**2**	**3**	**11**
Avg. error rate	10.7%	13%	10.3%	13.1%	12.1%
Q-SELECT	14	16	14	18	62
Avg. error rate	9.9%	11.8%	12%	11.7%	11.4%
Q-NN	14	16	16	14	60
Avg. error rate	8.3%	10.9%	9.3%	8.9%	9.4%
Q-DT	14	17	17	16	64
Avg. error rate	12.8%	13.1%	16.3%	13.7%	14%

Since the above table presents median numbers of questions asked, one might wonder how the actual numbers of questions asked are distributed. For example, Table 2 shows the distribution of questions asked for the SN dimension. Note that 34% of individual learning types were predicted using just one question, a cumulative 76% were predicted using four questions, and the remaining 24% were predicted using more questions. All 100% of the predictions were possible with at most thirteen questions. However, the error rate increases to 25% when thirteen questions are used. This is due to the fact that after the maximum number of questions (*maxQuestions*) has been exhausted, predictions are made by approximate matching.

Table 2. Distribution of questions asked by *T-PREDICT* for the SN dimension

Number of questions asked m	Number of learning types predicted using m questions (in %)	Error rate for predictions using m questions	Cumulative percentage of learning types predicted using up to m questions
1	313 (34%)	4.8%	34%
2	0 (0%)	-	34%
3	96 (10%)	5.2%	44%
4	298 (32%)	17.1%	76%
5	0 (0%)	-	76%
6	0 (0%)	-	76%
7	0 (0%)	-	76%
8	47 (5%)	14.9%	81%
9	0 (0%)	-	81%
10	0 (0%)	-	81%
11	0 (0%)	-	81%
12	0 (0%)	-	81%
13	0 (0%)	-	81%
13 (approx. match)	177 (19%)	25%	100%

Note that some questions, such as the second question in Table 2, did not generate any additional predictions. These are the questions where the two established criteria necessary to make a prediction, as set by the *prediction algorithm*, were not met.

5.2   Influence of Sorting and Limiting Number of Questions

The following experiment assesses the influence of two optimizing strategies on error rate and median number of questions asked, these are (1) sorting questions by discriminative power and (2) limiting the maximum number of questions asked. *T-PREDICT* results were compared with a modified version that does not sort questions (*UT-PREDICT*). Moreover, the maximum number of questions asked (*MaxQuestions*) was varied from one to all questions. Table 3 shows the obtained results for the SN dimension. Results for other dimensions were similar, although not shown due to space limitation. It can be observed that the median number of questions is higher and that the error rate is much higher when the questions are unsorted. For example, Table 3 shows that *T-PREDICT* had an error rate of 13% using a median of 4 questions, while *UT-PREDICT* used 5 questions with an error rate of 17.84%, when *MaxQuestions* = 13. It can also be noticed that limiting the maximum number of questions to 13 gave the lowest error rate (13 %).

Table 3. Influence of *MaxQuestions* and sorting questions by discriminative power on error rate and median number of questions asked

Parameter MaxQuestions	Median number of questions (T-PREDICT)	Average error rate (T-PREDICT)	Median number of questions (UT-PREDICT)	Average error rate (UT-PREDICT)
1	1	32,8%	1	32,8%
2	2	32,8%	2	32,8%
3	3	18,1%	3	28,0%
4	4	18,1%	4	32,3%
5	4	15,4%	5	20,8%
6	4	18,4%	5	22,6 %
7	4	16,3%	5	20,0%
8	4	14,7%	5	21,4%
9	4	14,5%	5	17,9%
10	4	15,0%	5	19,2%
11	4	14,0 %	5	17,9%
12	4	13,7%	5	19,2%
13	**4**	**13,0%**	**5**	**17,8%**
14	4	13.9%	5	19.6%
25	4	15.1%	5	18.3%

6   Conclusion

Filling questionnaires for learning style assessment is a very time-consuming task, which might lead to abandoning the test, skipping questions, answering falsely, etc. To address this issue, various approaches have been proposed to reduce the size of questionnaires. In this paper, we have presented a novel dynamic electronic question-naire *T-PREDICT* to further reduce the number of questions needed for learning type identification. It comprises three modules: a question sorting algorithm, a prediction algorithm and an automatic parameter selection algorithm.

Experimental results with 1,931 filled questionnaires for the Myers Briggs Type Indicators show that our novel approach asked a median of only 11 out of 92 questions to predict learning types, with an average error rate of 12.1%, while previous approaches *Q-SELECT* [5], *Q-NN* and *Q-DT* [7] asked between 60 and 64 questions to achieve error rates between 9.4% and 14%. Another defining characteristic of *T-PREDICT* is its ability to ask a variable number of questions, thus providing an automatic personalized questionnaire to each user, like *Q-SELECT* but unlike *Q-NN*, *Q-DT*, *PCA* and *IRT* that apply the same reduction to all users.

References

1. Felder, R.M., Brent, R.: Understanding student differences. J. of Engineering Education. 94(1), 57–72 (2005)
2. Graf, S.: Adaptivity in Learning Management Systems Focusing on Learning Styles. Ph.D. Thesis, Vienna University of Technology, Vienna (2007)
3. Ortigosa, A., Paredes, P., Rodríguez, P.: AH-questionnaire: An adaptive hierarchical questionnaire for learning styles. Computers & Education 54(4), 999–1005 (2010)
4. García, P., Amandi, A., Schiaffino, S., Campo, M.: Evaluating Bayesian Networks' Precision for Detecting Students' Learning Styles. Computers & Edu. 49(3), 794–808 (2007)
5. Mwamikazi, E., Fournier-Viger, P., Moghrabi, C., Barhoumi, A., Baudouin, R.: An Adaptive Questionnaire for Automatic Identification of Learning Styles. In: Ali, M., Pan, J.-S., Chen, S.-M., Horng, M.-F. (eds.) IEA/AIE 2014, Part I. LNCS, vol. 8481, pp. 399–409. Springer, Heidelberg (2014)
6. Nokelainen, P., et al.: Bayesian Modeling Approach to Implement an Adaptive Questionnaire. In: Proc. ED-MEDIA, pp. 1412–1413. AACE, Chesapeake (2001)
7. Barhoumi, A.: Simplification du questionnaire MBTI par apprentissage automatique en vue de faciliter l'adaptabilité des logiciels de formation en ligne. M.Sc. Thesis. Université de Moncton, 117 pages (2012)
8. Boyle, G.J.: Myers-Briggs Type Indicator (MBTI): Some psychometric limitations. Australian Psychologist 30(1), 71–74 (2009)
9. El Bachari, E., Abdelwahed, E.H., El Adnani, M.: Design of an Adaptive E-Learning Model Based on Learner's Personality. Ubiquitous Comp. Comm. J. 5(3), 27–36 (2010)
10. Francis, L.J., Jones, S.H.: The Relationship between MBTI and the Eysenck Personality Questionnaire among Adult Churchgoers. Pastoral Psychology 48(5), 377–386 (2008)
11. Pittenger, D., Measuring, J.: the MBTI and coming up short. Journal of Career Planning and Employment 54(1), 48–52 (1993)
12. McSherry, D.: Increasing dialogue efficiency in case-based reasoning without loss of solution quality. In: Proc. IJCAI 2003, pp. 121–126. Morgan Kaufmann (2003)
13. Abernethy, J., Evgeniou, T., Vert, J.-P.: An optimization framework for adaptive questionnaire design. Technical Report, INSEAD, Fontainebleau, France (2004)
14. Chen, C.-M., Lee, H.M., Chen, Y.-H.: Personalized e-learning system using item response theory. Computers & Education 44(3), 237–255 (2005)
15. Jolliffe, I.T.: Principal Component Analysis, 2nd edn. Springer, New York (2002)
16. DeMars, C.: Item response theory. Oxford University Press, Oxford (2010)
17. Weinhardt, J.M., Morse, B.J., Chimeli, J., Fisher, J.: An item response theory and factor analytic examination of two prominent maximizing tendency scales. Judgment & Decision Making 7(5), 644–658 (2012)

Recognizing Emotions from Facial Expressions Using Neural Network

Isidoros Perikos, Epaminondas Ziakopoulos, and Ioannis Hatzilygeroudis

School of Engineering
Department of Computer Engineering & Informatics
University of Patras
26500 Patras, Hellas, Greece
{perikos,ziakopoul,ihatz}@ceid.upatras.gr

Abstract. Recognizing the emotional state of a human from his/her facial gestures is a very challenging task with wide ranging applications in everyday life. In this paper, we present an emotion detection system developed to automatically recognize basic emotional states from human facial expressions. The system initially analyzes the facial image, locates and measures distinctive human facial deformations such as eyes, eyebrows and mouth and extracts the proper features. Then, a multilayer neural network is used for the classification of the facial expression to the proper emotional states. The system was evaluated on images of human faces from the JAFFE database and the results gathered indicate quite satisfactory performance.

Keywords: Facial Expression Recognition, Feature Extraction, Image Processing, Multilayer Perceptron Neural Network, Human-Computer Interaction.

1   Introduction

The aim of facial expression recognition is to enable machines to automatically estimate the emotional content of a human face. The facial expressions assist in various cognitive tasks and is well pointed that the expressions provide the most natural and powerful means for communicating human emotions, opinions and intentions. In an early work of Mehrabian [8] it has been indicated that during the face to face human communication only 7% of the information of a message is communicated by the linguistic (verbal) part of the message, i.e. spoken words, 38% by paralanguage (vocal part) and the 55% is communicated by the facial expressions. So, facial gestures constitute the most important communication part in face to face communication.

Analyzing the expressions of a human face and understanding their emotional state can find numerous applications to wide-ranging domains. The interaction between human and computer systems (HCI) would become much more natural and vivid if the computer applications could recognize and adapt to the emotional state of the human. Embodied conversational agents can greatly benefit from spotting and understanding the emotional states of the participants, achieving more realistic interactions at an emotional level. In intelligent tutoring systems, emotions and learning are inextricably

L. Iliadis et al. (Eds.): AIAI 2014, IFIP AICT 436, pp. 236–245, 2014.
© IFIP International Federation for Information Processing 2014

bound together, and so recognizing the learner's emotional states could significantly improve the learning procedure delivered to him/her [1, 14]. Moreover, surveillance applications such as driver monitoring and elderly monitoring systems could benefit from a facial emotion recognition system, gaining the ability to deeper understand and adapt to the person's cognitive and emotional condition Also, facial emotion recognition could be applied to medical treatment to monitor patients and detect their status. However, the analysis of the human face characteristics and the recognition of its emotional state are considered to be very challenging and difficult tasks. The main difficulty comes from the non-uniform nature of the human face and various limitations such as lightening, shadows, facial pose and orientation conditions [6].

In order to classify facial human expression into the proper emotional categories, it is necessary to locate and extract the important facial features which contribute in identifying the expression's emotions. In this work, a facial emotion recognition system developed to determine the emotional state of human facial expressions is presented. Initially, the system analyzes the facial image, locates and measures distinctive facial deformations and characteristics such as the eyes, the eyebrows and the mouth. Each part of the face then is deeper analyzed and its features are extracted. The features are represented as information vectors. The classification of the feature vectors into the appropriate expression emotion is conducted by a multilayer neural network which is trained and used for the classification of the facial expression to the proper emotional category.

The rest of the paper is organized as follows. In section 2, basic topics on face image analysis and emotion recognition are described and also related work is presented. In Section 3, the methodology followed and the system developed is illustrated. In Section 4, the evaluation conducted and the experimental results gathered are presented. Finally, Section 5 concludes our paper and presents direction for future work.

2   Background and Related Work

2.1   Background Topics

In the field of facial emotion recognition two types of methods dominate: the holistic methods and the analytical or local-based methods [11]. The holistic methods try to model the human facial deformations globally, which encode the entire face as a whole. On the other hand, the analytical methods observe and measure local or distinctive human facial deformations such as eyes, eyebrows, nose, mouth etc. and their geometrical relationships in order to create descriptive and expressive models [2]. In the feature extraction process for expression analysis there are mainly two types of approaches which are the geometric feature based methods and the appearance based methods. The geometric facial features try to represent the geometrical characteristics of a facial part deformation such as the part's locations and model its shape. The appearance based methods utilize image filters such as Gabon wavelets to the whole face or on specific parts to extract feature vectors.

The way that emotions are represented is a basic aspect of an emotion recognition system. A very popular categorical model is the Ekman emotion model [4], which specifies six basic human emotions: anger, disgust, fear, happiness, sadness, surprise.

It has been used in several studies and systems that recognize emotional text and facial expressions related to these emotional states. Another popular model is the OCC (Ortony/Clore/Collins) model [10] which specifies 22 emotion categories based on emotional reactions to situations and is mainly designed to model human emotions in general. In this paper, an analytical approach is implemented for recognizing the basic emotions as defined by Ekman. More specifically, special areas of interest of the human face are analyzed and their geometrical characteristics such as locations, length, width and shape are extracted.

2.2   Related Work

Over the last decade there are a lot of efforts and works on facial image analysis and emotion recognition. A recent and detailed overview of approaches and methodologies can be found in [17, 18]. The work in [3], a facial expression classification method based on histogram sequence of feature vector is presented. It consists of four main tasks which are image pre-processing, mouth segmentation, feature extraction and classification which is based on histogram-based methods. The system is able to recognize five human expressions: happy, anger, sad, surprise and neutral based on the geometrical characteristics of the human mouth with an average recognition accuracy of 81.6%. In [9], authors recognize Ekman basic emotions sad, anger, disgust, fear, happy and surprise in facial expressions by utilizing Eigen spaces and using dimensionality reduction technique. The system developed achieved a recognition accuracy of 83%. The work presented in [16] recognizes facial emotions based on a novel approach using Canny, principal component analysis technique for local facial feature extraction and artificial neural network for the classification process. The average facial expression classification accuracy of the method is reported to be 85.7%. The authors in the work presented in [12], recognize four basic emotions of happiness, anger, surprise and sadness focusing in preprocessing techniques for feature extraction such as Gabor filters, linear discrimination analysis and Principal component analysis. They achieve in their experiments 93.8% average accuracy in images of the Jaffe face database with little noise and with particularly exaggerated expressions and an average accuracy of 79% in recognition on just smiling/non smiling expressions in the ORL database. In [15], a work that recognizes the seven emotions on Jaffe database using Fisher weight map is presented. Authors utilize image preprocessing techniques such as illumination correction and histogram equalization and the recognition rate of their approach is reported to be 69.7%.

3   Emotion Recognition System

In this section, we present the emotion recognition system developed and illustrate its functionality. The aim of the system is to endow software applications with the ability to recognize users' emotions in a similar way that the human brain does. The system takes as input images of human face and classifies them regarding the emotional state according to the basic emotions determined by Ekman. The system's steps for the automatic analysis of human facial expressions are presented in Figure 1.

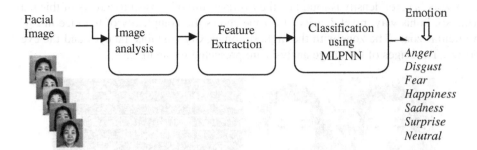

Fig. 1. The work flow chart of the system

Initially, the image analysis processes try to normalize the image and detect special regions of the face which contribute in identifying its emotional content. These regions are named Areas of Interests (AOIs) and are the region of the eyes, the area of the mouth and the eyebrows. The feature extraction process tries to select the proper facial features which describe the characteristic of the facial part and contribute in the classification stage. Since the system follows an analytical local-based approach, the feature extraction is implemented in the specific AOIs of the human face. Finally, the classification process, based on the feature extracted, specifies the appropriate emotional category among the set of different the emotional states defined by Ekman by utilizing a Multi-Layer Perceptron Neural Network (MLPNN) approach.

3.1   Image Analysis and AOIs Determination

Initially, the images of the human face are preprocessed using contrast limited histogram equalization. The histogram equalization is used to enhance the contrast in order to obtain a uniform histogram and also it re-organizes the image's intensity distributions and eliminates lighting effects [13].   A Sobel filter is then applied to emphasize edges and transition of the face and after its application, the image of the human face is represented as a matrix whose elements have only two possible values: zero and one. An element that has a value set to one represents an edge of a facial part in the image, such as an edge of an eye or the mouth. More specifically, the system analyzes a facial image as follows:

1.   Apply histogram equalization to enhance the contrast of the image.
2.   Apply a Sobel filter to emphasize edges and transition of the face.
3.   Cut the face into eight horizontal zones.
4.   Measure each zone's pixel density.
5.   Select the zone with the highest pixel density. This zone contains the eyes and the eyebrows.
6.   Locate the eyes and eyebrows.
7.   Locate the mouth in the center of the zone.

A zone's pixel density is found by the average sum of consecutive rows of this matrix. So, in his way, pixel density of a zone shows the complexity of the face in each particular zone. The zone with the highest complexity contains the eyes and the eyebrows. The stages of the image analysis are presented in Figure 2.

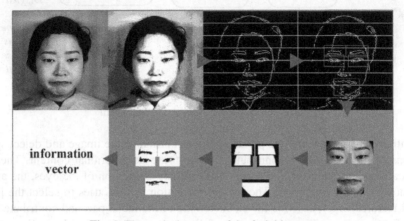

Fig. 2. The analysis stages of the facial image

3.2   AOI Analysis and Feature Extraction

The extraction of the proper facial feature is considered to be the most important step in facial expression recognition and is based on finding sets of features that convey meaningful information about the facial expression. Features extracted represent the geometrical characteristics of the facial part's deformation such as the part's height, width and model the element's shape. After the successful AOI isolation, the analysis of each AOI is performed in an effort to simplify the gathered information and extract the proper feature values. The analysis procedure implemented has seven steps:

1. Measure the average brightness of the face.
2. Set a brightness threshold.
3. Relate the brightness threshold to the average brightness.
4. Find the product of the brightness threshold and the average brightness.
5. Set a brightness filter.
6. Apply the brightness filter to the AOI.
7. Cut of the excess AOI noise using proper masks.

Application of an effective brightness filter has the effect to simplify an AOI in a way so all the data containing the emotional state of the face are left unaltered while all the irrelevant to the task data are discarded. The AOIs, after the filter application, are represented more smoothly and the information vectors can be easily extracted. To measure the average brightness, it is necessary to apply a facial mask to isolate the face area from the background image and the model's hair since these areas do not affect the brightness of the face. A series of 5 more sub masks must also be applied as a final step to reduce excess noise in the marginal areas of the AOI. The features extracted for each AOI are the following (see Figure 3):

Left eyebrow

- H1: The height of the far left part of the eyebrow.
- H2: The height of the part found on the 1/3 of the distance between the far left part and far right part of the eyebrow.
- H3: The height of the part found on the 2/3 of the distance between the far left and far right part of the eyebrow.
- H4: The height of the far left part of the eyebrow.
- L1: The length of the eyebrow.

Right eyebrow

- H5: The height of the far left last part of the eyebrow.
- H6: The height of the part found on the 1/3 of the distance between the far left and far right part of the eyebrow.
- H7: The height of the part found on the 2/3 of the distance between the far left and far right part of the eyebrow.
- H8: The height of the far left part of the eyebrow.
- L2: The length of the eyebrow.

Left eye

- H9: The height of the far left part of the eye.
- H10: The height of the part found on the 1/2 of the distance between the far left and far right part of the eye.
- H11: The height of the far right part of the eye.
- W1: The width of the eye.
- L3: The length of the eye.

Right eye

- H12: The height of the far right part of the eye.
- H13: The height of the part found on the 1/2 of the distance between the far left and far right part of the eye.
- H14: The height of the far left part of the eye.
- W2: The width of the eye.
- L4: The length of the eye.

Mouth

- H15: The height of the far left part of the mouth.
- H16: The height of the part found on the 1/2 of the distance between the far left and far right part of the mouth.
- H17: The height of the far left part of the mouth.
- W3: The width of the mouth.
- L5: The length of the mouth.

As mentioned above the recognition of the expressions' emotional content is determined based on the five AOIs. Each AOI is represented geometrically by five (5) values therefore the length of the information vector representing the facial expression is 25. The information vector contains all the information needed to classify a face into each of the seven basic facial emotional states.

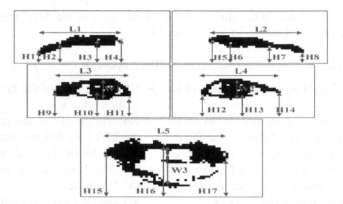

Fig. 3. Features extracted from each AOI

3.3 Neural Network Classification

The classification of the facial expression to the proper emotional category is made based on a Multi-Layer Perceptron Neural Network (MLPNN) which enables higher flexibility in training and also great adaptation to the problem of emotional classification [5]. The MLPNNs are feed-forward neural networks and can be used for the classification of multi-class, nonlinear problems. In our methodology, the MLPNN used for the classification of the facial expressions has three hidden layers, containing 13, 9 and 5 neurons respectively. Since there is no standard method to select the number of hidden layers and neurons due to the nature of the Neural Networks, the selection is often, as in this case, the architecture with the best performance. The input layer has 25 sensory neurons to match the length of the information vectors. The output layer has 3 neurons, that is, to classify the seven different classes. The output neurons combined produce a binary 3-digit sequence and each sequence is mapped to a class. The MLPNN network is trained using the back propagation supervised learning technique which provides a systematic way to update the synaptic weights of multi-layer perceptron networks.

4   Experimental Study

4.1 Data Collection

The methodology and the system developed were evaluated on facial images from the Japanese Female Facial Expression Database (JAFFE) [7]. Jaffe database contains 213 images of seven facial expressions angry, disgust, fear, happy, sadness, surprise and the neutral expression, posed by 10 different Japanese female models of young age. Each image has resolution of 256 x 256 pixels. The seven facial expressions (6 basic facial expressions and 1 neural expression) as posed of a JAFFE model are illustrated in Figure 4.

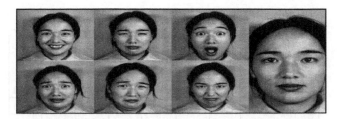

Fig. 4. The seven facial expressions as posed by a JAFEE model

A Face Matrix is represented as 256 x 256 pixel square matrix. The elements of the matrix represent the pixels of the Jaffe model's face. Each element of the face matrix ranges from 0 to 255. An element with a value between 0 and 255 represent a gray pixel of the image. An element with the value 0 represents a completely black pixel of the original image, while an element with the value 255 represents a completely white pixel, thus making the Face Matrix a Brightness Matrix.

4.2 System Evaluation

For the evaluation of the system's performance two datasets were created from the Jaffe database, the training dataset and the test dataset. The training dataset contains 140 facial images, 20 for each one of the seven facial emotions. The test dataset contains a total of 73 images, at least 9 images for each emotional state. The training dataset is inputted into the MLPNN, and the weights of the neurons are calculated in an optimal way. After the training phase is completed, the system's performance is evaluated on the test dataset. The results gathered are presented in Table 1.

Table 1. Performance results of the classification

Average accuracy	76,7 %
Precision	78,9%
F-measure	77,8%

The results show a good performance of the system. More specifically, from the corpus of 73 tested, it correctly identified the emotional state in 56 images. So, the general accuracy of the system is 76,7 % which indicates a good performance.

A noticeable point is that the system had a very good performance in identifying happiness emotion. This is due to the fact that in general and also in the database, this is a very strong emotion and is almost always expressed with vivid facial expression deformations. In contrast, the system lacks in recognizing the anger emotion. We believe that the main reason that the anger emotion is poorly classified is that the Jaffe models express this emotion in an inconsistent way regarding its valence. Specifically, in some cases, the models express anger emotion extremely, in an almost theatrical way while other models express the same emotion more natural in a smooth way.

Table 2. The confusion matrix

	Joy	Sadness	Surprise	Fear	Disgust	Anger	Neutral
Joy	12	0	0	0	1	0	0
Sadness	0	6	0	2	1	1	0
Surprise	0	0	8	0	0	0	1
Fear	0	0	0	9	0	0	2
Disgust	0	1	0	0	8	0	0
Anger	0	1	0	0	3	6	1
Neutral	0	0	0	1	2	0	7

5   Conclusion and Future Work

In this paper, an automatic facial expression recognition system is presented. The aim of the system developed is to analyse and recognize the emotional content of human facial expressions. The system initially analyzes the facial image, locates and measures distinctive human facial deformations such as the eyes, the eyebrows and area of the mouth and extracts the proper features. The features extracted try το model the geometrical characteristics of each area part of the human face. Finally, for the categorization, a multilayer perceptron neural network is trained and used. The experimental study conducted indicates quite promising performance.

However, there are some points that the system could be improved. Initially, evaluating the system on different facial expression databases such as the ORL and the Cohn-Kanade facial expression databases will give us a better insight of or methodology's performance. Moreover, currently the face analysis and the determination of the AOIs work on still frontal images of human face and so, a challenging extension will be to detect and analyse facial expressions from different poses. Furthermore, the system's classification approach is based on a multilayer perceptron neural network. Extending the classification method by using a neuro-fuzzy approach could improve the system's accuracy in recognising the emotional state. Exploring this direction is a key aspect of our future work.

References

1. Akputu, K.O., Seng, K.P., Lee, Y.L.: Facial Emotion Recognition for Intelligent Tutoring Environment. In: 2nd International Conference on Machine Learning and Computer Science (IMLCS 2013), pp. 9–13 (2013)
2. Arca, S., Campadelli, P., Lanzarotti, R.: An automatic feature-based face recognition system. In: Proceedings of the 5th International Workshop on Image Analysis for Multimedia Interactive Services (WIAMIS 2004) (2004)
3. Aung, D.M., Aye, N.: A Facial Expression Classification using Histogram Based Method. In: International conference on Signal Processing Systems (2012)
4. Ekman, P.: Basic emotions. Handbook of Cognition and Emotion, pp. 45–60 (1999)

5. Filko, D., Goran, M.: Emotion recognition system by a neural network based facial expression analysis. Autom. –J. Control Meas. Electron. Comput. Commun. 54, 263–272 (2013)
6. Koutlas, A., Fotiadis, D.I.: An automatic region based methodology for facial expression recognition. In: IEEE International Conference on Systems, Man and Cybernetics, SMC 2008, pp. 662–666 (2008)
7. Lyons, M.J., Akamatsu, S., Kamachi, M., Gyoba, J.: Coding facial expressions with Gabor wavelets. In: Proceedings of the Third IEEE International Conference on Automatic Face and Gesture Recognition, pp. 200–205. IEEE Computer Society, Los Alamitos (1998)
8. Mehrabian, A.: Communication without words. Psychology Today 2(4), 53–56 (1968)
9. Murthy, G.R.S., Jadon, R.S.: Recognizing facial expressions using eigenspaces. In: International Conference on Conference on Computational Intelligence and Multimedia Applications, vol. 3, pp. 201–207. IEEE (December 2007)
10. Ortony, A., Clore, G., Collins, A.: The Cognitive Structure of Emotions. Cambridge University Press, Cambridge (1988)
11. Pantic, M., Rothkrantz, L.J.M.: Automatic analysis of facial expressions: The state of the art. IEEE Transactions on Pattern Analysis and Machine Intelligence 22(12), 1424–1445 (2000)
12. Přinosil, J., Smékal, Z., Esposito, A.: Combining features for recognizing emotional facial expressions in static images. In: Esposito, A., Bourbakis, N.G., Avouris, N., Hatzilygeroudis, I. (eds.) HH and HM Interaction. LNCS (LNAI), vol. 5042, pp. 56–69. Springer, Heidelberg (2008)
13. Gonzalez, R.C., Woods, R.E., Eddins, S.L.: Digital Image Processing using MATLAB. Prentice Hall (2003)
14. Shen, L., Wang, M., Shen, R.: Affective e - learning: Using "emotional" data to improve learning in pervasive learning environment. Educational Technology & Society 12(2), 176–189 (2009)
15. Shinohara, Y., Otsu, N.: Facial expression recognition using fisher weight maps. In: Proceedings. Sixth IEEE International Conference on Automatic Face and Gesture Recognition, pp. 499–504. IEEE (May 2004)
16. Thai, L.H., Nguyen, N.D.T., Hai, T.S.: A facial expression classification system integrating canny, principal component analysis and artificial neural network. arXiv preprint arXiv:1111.4052 (2011)
17. Tian, Y., Kanade, T., Cohn, J.F.: Facial expression recognition. In: Handbook of Face Recognition, pp. 487–519. Springer, London (2011)
18. Verma, A., Sharma, L.K.: A Comprehensive Survey on Human Facial Expression Detection. International Journal of Image Processing (IJIP) 7(2), 171 (2013)

Automatically Detected Feature Positions for LBP Based Face Recognition

Ladislav Lenc[1,2] and Pavel Král[1,2]

[1] Dept. of Computer Science and Engineering, University of West Bohemia, Plzeň, Czech Republic
[2] NTIS, University of West Bohemia, Plzeň, Czech Republic
{llenc,pkral}@kiv.zcu.cz

Abstract. This paper presents a novel approach for automatic face recognition based on the Local Binary Patterns (LBP). One drawback of the current LBP based methods is that the feature positions are fixed and thus do not reflect the properties of the particular images. We propose to solve this issue by a method that automatically detects feature positions in the image. These key-points are determined using the Gabor wavelet transform and k-means clustering algorithm. The proposed method is evaluated on two corpora: AT&T Database of Faces and our Czech News Agency (ČTK) dataset containing uncontrolled face images. The recognition rate on the first dataset is 99.5% which represents 2.5% improvement compared to the original LBP method. The best recognition rate obtained on the ČTK corpus is 59.1% whereas the original LBP method reaches only 38.1%.

Keywords: Automatic Face Recognition, Czech News Agency, Gabor Filter, Local Binary Patterns.

1 Introduction

Recognizing people from digitized images is already an old concept. A great progress was made during the last twenty-five years and the amount of proposed approaches is overwhelming. There are numerous applications of the face recognition such as an access control, tracking people or other systems where an authentication is necessary. There is also a great potential of face recognition in social networks and photo sharing applications.

The recognition of controlled images is considered as a well-resolved problem. However, the face recognition from the ordinary photographs is an open issue yet, because it requires very sophisticated algorithms.

Nowadays, the Local Binary Patterns (LBP) is a very popular algorithm for face representation and recognition. The strength of this algorithm lays in its high ability to capture important information in the images and in its low computational complexity. However, one weakness of the current LBP based methods is that the feature positions are fixed and thus do not respect the properties of the particular images. We would like to solve this issue by proposing an approach to

L. Iliadis et al. (Eds.): AIAI 2014, IFIP AICT 436, pp. 246–255, 2014.

detect feature positions in images automatically. The key-points candidates are determined using the Gabor Wavelet Transform (GWT). K-means clustering algorithm is used subsequently to identify the final points. Another improvement is to compare the features separately instead of concatenating them into one feature as in the case of other existing LBP based methods. To the best of our knowledge, no existing methods use the algorithms in this way. The results of this work will be used by the Czech News Agency (ČTK[1]) to annotate people in photographs during insertion into the photo-database[2].

The rest of the paper is organized as follows. Section 2 summarizes important face recognition methods with a particular focus on the local binary patterns. The next section describes the LBP algorithm and the proposed face recognition method. Section 4 evaluates the approach on the AT&T and ČTK corpora. In the last section, we discuss the results and we propose some future research directions.

2   Related Work

It is possible to divide the existing approaches into two main groups: holistic and feature based methods. The approaches from the first group represent the face image as a whole while methods belonging to the second group use for face representation a set of features. The successful holistic approaches are for example Eigenfaces [14], Fisherfaces [3] or Independent Component Analysis [2]. These methods achieve good results on controlled data. However, their performance decreases significantly when low quality real data are used.

This issue is partly solved by the feature based approaches. The most important representatives are detailed next. As a pioneer method we can consider Elastic Bunch Graph Matching (EBGM) [15]. It is based on features created using Gabor wavelet transform. It is later used with many modifications as shown for example in [6]. This method extracts the feature points automatically based on the Gabor filter responses. Another type of features used for face representation is based on Scale Invariant Feature Transform (SIFT) [9]. The SIFT features are invariant to rotation, scale and lighting conditions. Later a Speeded-Up Robust Features (SURF) is also used for face recognition as shown for instance in [4]. It reaches comparable results as the SIFT, however it has significantly lower computational demands.

In the last couple of years, also Local Binary Patterns (LBP) are successfully used as features for face representation and recognition. The LBP operator was first used for texture representation as presented in [10]. It is computed from the neighbourhood of a pixel and uses the intensity of the central pixel as a threshold. The pixels are marked either 0 or 1 if the value is lower or greater than this threshold. The binary values are concatenated into one binary string and its decimal value is then used as a descriptor of the pixel.

---

[1] http://www.ctk.eu
[2] http://multimedia.ctk.cz/en/foto/

The first application of LBP for face recognition is proposed by Ahonen et al. in [1]. The face is divided into rectangular regions. In each region a histogram of the LBP values is computed. All histograms are then concatenated into one vector which is used for the face representation. A histogram intersection method or Chi square distance are used for vector comparison. A weighted LBP modification is also proposed in this work. It gives more importance to the regions around the eyes and the central part of the face. The reported recognition rate on the FERET dataset [11] reaches 93% for the original method and 97% for the weighted LBP method.

A modification of the original LBP approach called Dynamic Threshold Local Binary Pattern (DTLBP) is proposed in [8]. It takes into consideration the mean value of the neighbouring pixels and also the maximum contrast between the neighbouring points. It is stated there that this variation is less sensitive to the noise than the original LBP method.

Another extension of the original method is Local Ternary Patterns (LTP) proposed in [13]. It uses three states to capture the differences between the center pixel and the neighbouring ones. Similarly to the DTLBP the LTP is less sensitive to the noise.

The so called Local Derivative Patterns (LDP) are proposed in [18]. The difference from the original LBP is that it uses features of higher order. It thus can represent more information than the original LBP.

An important idea which is proposed already by Ojala in [10] are so called uniform Local Binary Patterns. The pattern is called uniform if it contains at most two transitions from 0 to 1 or from 1 to 0. It was proved that approximately 90% of the patterns in facial images are uniform. The histogram can then be shortened from 256 intervals (bins) to 59, where the 59th bin is reserved for the non-uniform patterns.

An interesting method which uses uniform patterns is proposed in [17]. The authors state that the histogram bin containing non-uniform patterns dominates among other bins and gives thus too much importance to this bin. Therefore they propose to assign such patterns to the closest uniform pattern. Hamming distance is used for the face comparison.

Some methods also combine other preprocessing tools with the LBP. In [12] Gabor features and LBP are combined. The Gabor features as well as the LBP features are extracted and transformed using PCA. The features are then combined and used as face representation.

Another method called Local Gabor Binary Pattern Histogram (LGBPH) [16] also combines the Gabor wavelet transform and LBP. It first filter the image with a set of Gabor filters and obtains a set of magnitude images. Then the LBP operator is applied to each of the magnitude images.

Note that the common property of all above described LBP methods is that the images are divided into rectangular regions and histograms are computed in each region. All histograms from one image are concatenated and create the face representation.

3   Method Description

The proposed method is motivated by the assumption that the most representative features must be created in the "important" face points. This set of points where the features are created is calculated dynamically for each image by the Gabor wavelets. Therefore, the feature positions differ for each face image. We also propose to compare the features separately. In order to facilitate the reading of the paper, we shortly introduce next the Gabor wavelets and the LBP algorithm.

3.1   Gabor Wavelets

Gabor filter is a sinusoid modulated with a Gaussian. A basic form of a two dimensional Gabor filter is shown in Equation 1.

$$g(x, y; \lambda, \theta, \psi, \sigma, \gamma) = \exp\left(-\frac{\acute{x} + \gamma^2 \acute{y}^2}{2\sigma^2}\right) \cos\left(2\pi \frac{\acute{x}}{\lambda} + \psi\right) \tag{1}$$

where $\acute{x} = x \cos\theta + y \sin\theta$, $\acute{y} = -x \sin\theta + y \cos\theta$, λ is the wavelength of the cosine factor, θ represents the orientation of the filter and ψ is a phase offset, σ and γ are parameters of the Gaussian envelope, σ is the standard deviation of the Gaussian and γ defines the ellipticity (aspect ratio) of the function.

The Gabor wavelets are often used in image analysis because of their great ability to capture important information in images.

3.2   Local Binary Patterns

The original LBP operator uses the 3×3 neighbourhood of the central pixel. The algorithm assigns either 0 or 1 value to the 8 neighbouring pixels by Equation 2.

$$N = \begin{cases} 0 \text{ if } g_N < g_C \\ 1 \text{ if } g_N \geq g_C \end{cases} \tag{2}$$

where N is the binary value assigned to the neighbouring pixel, g_N denotes the gray-level value of the neighbouring pixel and g_C is the gray-level value of the central pixel. The resulting values are then concatenated into an 8 bit binary number. Its decimal representation is used for further computation. This approach is illustrated in Figure 1.

The original LBP operator was further extended to use circular neighbourhoods of various sizes and also with different numbers of points. A bilinear interpolation is used to compute the values in the points that are not placed in the pixel centres. The LBP is then denoted as $LBP_{P,R}$ where P is the number of points and R is the radius of the neighbourhood.

Figure 2 depicts the original image and the LBP image after applying the original LBP, $LBP_{8,1}$ and $LBP_{8,2}$ operators.

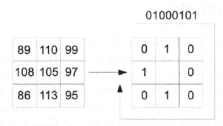

Fig. 1. An example of the feature computing by the original LBP operator

Fig. 2. An example of the original image and the LBP image after applying the original LBP, $LBP_{8,1}$ and $LBP_{8,2}$ operators

3.3 Key-Point Position Determination

The determination of the feature positions is probably the most important part of the whole algorithm, particularly in the case of real world photographs where the faces are extracted automatically. We decide to adopt and extend the idea proposed by Kepenekci in [6]. In that work the fiducial points (positions where the features are created) are extracted automatically by the Gabor wavelets. A set of Gabor filters of different orientations and wavelengths is applied to the original image and then the fiducial points are determined from the filter responses.

The filtered images are scanned using a square sliding window W of the size $w \times w$. The window centre (x_0, y_0) is considered to be a fiducial point iff:

$$R_j(x_0, y_0) = \max_{(x,y) \in W} R_j(x, y) \tag{3}$$

$$R_j(x_0, y_0) > \frac{1}{wi * hi} \sum_{x=1}^{wi} \sum_{y=1}^{hi} R_j(x, y) \tag{4}$$

where $j = 1, ..., N$ (N is the number of Gabor filters) and wi and hi are image width and height respectively.

3.4 K-Means Clustering

The number of points determined in the previous section is usually too high (hundreds) and the points are often concentrated near important facial parts.

Moreover, a high number of the points increases significantly the computation complexity. Therefore, we propose to use clustering to identify only the most important points. This idea is supported by the fact that usual LBP based methods use less than 100 points and achieve very good results. We chose the K-means algorithm to cluster the key-points.

Figure 3 shows points detected with different window sizes (15, 25, 35) and clustered points for the same window sizes and point counts (100, 75, 50).

Fig. 3. An example of the fiducial points determined with different window sizes (15, 25, 35) *(left three ones)* and the clustered points for the same window sizes and point numbers (100, 75, 50) *(right three ones)*

3.5 Feature Construction

After determining the fiducial points, the features are constructed in these points. Each feature is described by its coordinates and the LBP histogram. The histogram is computed from a square window around the feature point. The resulting histogram is then normalized and can then be interpreted as a vector of the length 256.

In our application, we usually have several training images for each person. It is thus beneficial to use all available images to create the face model. We propose a *face specific* schema for model creation. The fiducial points determination algorithm and clustering are applied to each training image separately. Then, all obtained features are put together to create the face model of a given person.

3.6 Face Comparison

The comparison of two faces is based on the image features extracted according to the above described procedure. Each face is thus represented by a set of features that contain also feature point coordinates. Note that the number of the features in a face model may vary because the model is created from multiple images.

Contrary to the most of LBP based approaches we do not concatenate the feature histograms but compare the features separately. The Chi square statistic is used for computation of the distance of the features f and r:

$$\chi^2(f, r) = \sum_i \frac{(f_i - r_i)^2}{f_i + r_i} \tag{5}$$

The distance of two faces is defined as:

$$sim(F, R) = \sum_{f_i} \min_{r_j \in N(f_i)} (\chi^2(f_i, r_j)) \qquad (6)$$

where $N(f_i)$ is the neighbourhood of the feature f_i defined by the *distanceThreshold* that specifies the maximum distance within that the features are compared. It means that for each feature of the face F we find the closest feature within the neighbourhood $N(f_i)$ from the face R. The distance of the two faces is computed as a sum of these minimum distances.

The recognized face $\hat{F}$ is given by the following equation:

$$\hat{F} = \arg \min_{R}(sim(F, R)) \qquad (7)$$

4   Experimental Setup

4.1   Corpora

AT&T Database of Faces. This database was formerly known as the ORL database. It was created at the AT&T Laboratories[3]. The pictures were captured between years 1992 and 1994 and contain the faces of 40 people. 10 pictures for each person are available. Each image contains one face with a black homogeneous background. They may vary due to three following factors: 1) time of acquisition; 2) head size and pose; 3) lighting conditions. The size of pictures is 92×112 pixels. A more detailed description of this database can be found in [5].

Czech News Agency (ČTK) Database. This database is created from real-world photographs owned by the Czech News Agency. The dataset contains the images of 638 people. The database was created automatically from common photographs. The images have significant variations in pose, lighting conditions and also ageing of the objects.

The experiments described in this paper will be realized on the cleaned version of this corpus (see [7]), which contains up to 10 images for each of the 638 individuals. The testing part contains one image for each person whereas the rest of images are used for training. Note that only the testing part was checked manually.

4.2   Experiments on the AT&T Database of Faces

The first series of the experiments was carried out on the AT&T Database of Faces. Its purpose was to show the performance of our method on a controlled database where the face recognition is simple. Ten-fold cross-validation was used for evaluation.

---

[3] http://www.cl.cam.ac.uk/research/dtg/attarchive/facedatabase.html

Table 1 shows the recognition rates of the original LBP approach proposed by Ahonen [1] in comparison with our methods. We evaluated our method with three different types of the LBP operator: Original LBP operator, $LBP_{8,1}$ and $LBP_{8,2}$. The first two rows show recognition rates when 9 out of 10 images for each person are used for training and 1 for testing. The first row depicts the results with k-means clustering (50 clusters). The second row shows recognition rates without the clustering, it means we used directly all key-points detected by the Gabor wavelet based detection algorithm (see Section 3.3). The last row shows results for only one training example. It should demonstrate the differences between the methods when only one image for training is available. In this case, the clustering is also applied for the proposed approach.

Table 1. Recognition rates of the original LBP approach in comparison with our methods

Method	Baseline	Proposed approaches		
	$LBP_{(Ahonen)}$	$LBP_{(original)}$	$LBP_{8,1}$	$LBP_{8,2}$
Training images number	Recognition rate (%)			
9 (k-means)	-	98.0	97.5	99.5
9 (no clustering)	97.0	97.8	97.8	99.8
1	60.8	66.8	66.4	68.8

The obtained recognition rates show that our method outperforms the original LBP method for all test configurations. Using the $LBP_{8,2}$ operator gives the best results. The difference is apparent especially in the case when only one training image is used. If we compare the results with and without clustering, the recognition rates are comparable. However, omitting the clustering causes significantly increased number of features and longer computation time. Therefore the clustering seems to be a good choice for feature number reduction.

4.3 Experiments on the ČTK Database

The remaining experiments were performed on the the ČTK database. As a baseline we chose (similarly as in the previous section) the LBP based face recognition algorithm proposed by Ahonen et al. [1]. The recognition rate obtained by this baseline algorithm is only 38.1%.

In this experiment (see Table 2), we compare this baseline with the recognition results and computation times of our proposed approaches with and without the clustering. The original LBP operator is used.

This table shows clearly that both proposed configurations outperform significantly the original approach. Moreover, the clustering decreases significantly the computation time with almost no negative impact on the classification accuracy.

The last experiment (see Table 3) shows recognition rates of our algorithm with varying number of clusters for three LBP operator types.

This table shows clearly that the $LBP_{8,2}$ operator outperforms significantly the other two ones. The highest recognition rate is obtained with 100 clusters.

Table 2. Recognition rates and computation times of our proposed approaches with and without the clustering, the original LBP operator is used, tested on the Intel Core i5-2300, 2.80GHz, 16GB RAM

	Clustering	No clustering
Recognition rate (%)	56.4	55.3
Computation time (s)	6,905	17,476

Table 3. Recognition rates for different numbers of clusters. Three types of LBP operators are evaluated

LBP type	$LBP_{(original)}$	$LBP_{8,1}$	$LBP_{8,2}$
Cluster count	**Recognition rate (%)**		
50	45.6	45.1	56.4
75	47.8	46.9	58.6
100	47.5	46.1	59.1

5   Conclusions and Perspectives

We proposed and evaluated a new approach for face recognition based on the LBP features. The main contribution is using automatically detected feature positions instead of fixed positions used in usual LBP based approaches. We use Gabor wavelets for key-point detection. Then, these points are clustered by the k-means algorithm. The next contribution is proposal of a new algorithm for face comparison based on the Chi square statistic. It compares individual features within specified distance. The proposed method is evaluated on the AT&T database of faces and on the ČTK dataset. The recognition rate for the AT&T database is 99.5%. For the more challenging ČTK dataset we obtained recognition rate of 59.1%. We significantly outperformed the baseline approach in all cases.

One possible perspective is applying weighting to the automatically detected key-points. Another possible improvement may be using only uniform patterns or some modification of LBP algorithm such as LTP or LDP.

Acknowledgements. This work has been partly supported by the UWB grant SGS-2013-029 and by the project "NTIS - New Technologies for Information Society", European Centre of Excellence, CZ.1.05/1.1.00/02.0090. We also would like to thank ČTK for support and for providing the photographic data.

References

1. Bronstein, A.M., Bronstein, M.M., Spira, A., Kimmel, R.: Face recognition with local binary patterns. In: Pajdla, T., Matas, J(G.) (eds.) ECCV 2004. LNCS, vol. 3022, pp. 225–237. Springer, Heidelberg (2004)

2. Bartlett, M.S., Movellan, J.R., Sejnowski, T.J.: Face recognition by independent component analysis. IEEE Transactions on Neural Networks 13(6), 1450–1464 (2002)

3. Belhumeur, P.N., Hespanha, J.P., Kriegman, D.: Eigenfaces vs. fisherfaces: Recognition using class specific linear projection. IEEE Transactions on Pattern Analysis and Machine Intelligence 19(7), 711–720 (1997)

4. Dreuw, P., Steingrube, P., Hanselmann, H., Ney, H., Aachen, G.: Surf-face: Face recognition under viewpoint consistency constraints. In: BMVC, pp. 1–11 (2009)

5. Jain, A.K., Li, S.Z.: Handbook of face recognition. Springer (2005)

6. Kepenekci, B.: Face recognition using gabor wavelet transform. Ph.D. thesis, Citeseer (2001)

7. Lenc, L., Král, P.: Face recognition under real-world conditions. In: International Conference on Agents and Artifitial Intelligence, Barcelona, Spain (2013)

8. Li, W., Fu, P., Zhou, L.: Face recognition method based on dynamic threshold local binary pattern. In: Proceedings of the 4th International Conference on Internet Multimedia Computing and Service, pp. 20–24. ACM (2012)

9. Lowe, D.G.: Distinctive image features from scale-invariant keypoints. International Journal of Computer Vision 2 (2004)

10. Ojala, T., Pietikäinen, M., Harwood, D.: A comparative study of texture measures with classification based on featured distributions. Pattern Recognition 29(1), 51–59 (1996)

11. Phillips, P.J., Wechsler, H., Huang, J., Rauss, P.: The FERET database and evaluation procedure for face recognition algorithms. Image and Vision Computing 16(5), 295–306 (1998)

12. Tan, X., Triggs, B.: Fusing gabor and LBP feature sets for kernel-based face recognition. In: Zhou, S.K., Zhao, W., Tang, X., Gong, S. (eds.) AMFG 2007. LNCS, vol. 4778, pp. 235–249. Springer, Heidelberg (2007)

13. Tan, X., Triggs, B.: Enhanced local texture feature sets for face recognition under difficult lighting conditions. IEEE Transactions on Image Processing 19(6), 1635–1650 (2010)

14. Turk, M.A., Pentland, A.P.: Face recognition using eigenfaces. In: Proceedings of the IEEE Computer Society Conference on Computer Vision and Pattern Recognition, CVPR 1991, pp. 586–591. IEEE (1991)

15. Wiskott, L., Fellous, J.M., Kuiger, N., Von Der Malsburg, C.: Face recognition by elastic bunch graph matching. IEEE Transactions on Pattern Analysis and Machine Intelligence 19(7), 775–779 (1997)

16. Xie, Z.: Single sample face recognition based on dct and local gabor binary pattern histogram. In: Huang, D.-S., Bevilacqua, V., Figueroa, J.C., Premaratne, P. (eds.) ICIC 2013. LNCS, vol. 7995, pp. 435–442. Springer, Heidelberg (2013)

17. Yang, H., Wang, Y.: A lbp-based face recognition method with hamming distance constraint. In: Fourth International Conference on Image and Graphics, ICIG 2007, pp. 645–649. IEEE (2007)

18. Zhang, B., Gao, Y., Zhao, S., Liu, J.: Local derivative pattern versus local binary pattern: face recognition with high-order local pattern descriptor. IEEE Transactions on Image Processing 19(2), 533–544 (2010)

Limited Generalization Capabilities of Autoencoders with Logistic Regression on Training Sets of Small Sizes

Alexey Potapov[1,2], Vita Batishcheva[2], and Maxim Peterson[1]

[1] St. Petersburg National Research University of Information Technologies,
Mechanics and Optics, Kronverkskiy pr. 49,
197101 St. Petersburg, Russia
[2] St. Petersburg State University, Universitetskaya nab. 7-9, 199034, St.Petersburg, Russia
{pas.aicv,elokkuu,maxim.peterson}@gmail.com

Abstract. Deep learning is promising approach to extract useful nonlinear representations of data. However, it is usually applied with large training sets, which are not always available in practical tasks. In this paper, we consider stacked autoencoders with logistic regression as the classification layer and study their usefulness for the task of image categorization depending on the size of training sets. Hand-crafted image descriptors are proposed and used for training autoencoders in addition to pixel-level features. New multi-column architecture for autoencoders is also proposed. Conducted experiments showed that useful nonlinear features can be learnt by (stacked) autoencoders only using large training sets, but they can yield positive results due to redundancy reduction also on small training sets. Practically useful results (9.1% error rate for 6 classes) were achieved only using hand-crafted features on the training set containing 4800 images.

Keywords: autoencoders, logistic regression, overlearning, generalization, image categorization.

1 Introduction

Classes of patterns in pattern recognition are rarely linearly separable, and nonlinear methods should be used. Classical nonlinear methods of pattern recognition can be treated as linear methods operating in some extended feature space. However, these methods use fixed sets of nonlinear transformations to map initial features into new space, and these transformations can be inappropriate to separate classes in a specific task. Thus, learning appropriate nonlinear features (or data representations in general) is the central problem in pattern recognition and particularly in its application to computer vision (e.g. [1]). One modern approach to learn such features is deep learning.

Deep learning exploits the fact that adding extra hidden layers in a multi-layer classifier helps to learn more complex features and to improve their representational power [2, 3]. The key problem here is to train such classifiers. Earlier, training of shallow feed-forward networks with the back-propagation algorithm yielded better results than training deep networks with it, since bottom layers are difficult to train because of exponentially quick gradient vanishing, so these layers usually correspond

L. Iliadis et al. (Eds.): AIAI 2014, IFIP AICT 436, pp. 256–264, 2014.

to random (not useful or even harmful) nonlinear feature mappings. Successful training of such networks was achieved [4, 5, 6] with the help of consequent unsupervised pre-training of each hidden layer, and use of supervised training afterwards.

However, successful supervised training of multi-layer perceptrons (resulting in a very low 0.35% error rate on the MNIST benchmark) was also achieved recently [7]. The way of achieving this is remarkable. It consists in intensive training of perceptrons using relevantly (with affine and elastic image deformations) transformed patterns. Even overlearning is avoided in spite of extremely large number of neurons (free parameters), because "the continual deformations of the training set generate a virtually infinite supply of training examples" [7]. One can also see [8] that many good results for MNIST are obtained using similar deformations of training patterns and convolutional nets, which additionally exploit shift invariance. It is obvious, that if one substitutes non-image patterns for training such classifiers, their results will be rather poor. Thus, this is not an appropriate way to learn (problem-specific) invariant features, which are not known a priori.

Necessity of extremely large (with account for deformations of training patterns) training sets indicates that deep networks don't really generalize, but only approximate invariants. Moreover, classifiers with many free parameters appear to be better than with fewer parameters. It can be compared with approximation of exponent with polynomials (or some other basis functions). Very many points should be used to get precise approximation with very complex model, but only few points are enough for exact generalization (if one has means to represent and find it). At the same time, humans have capabilities to learn complex invariant features from few examples [9]. Thus, interesting question consists in learning capabilities of deep networks on small training sets, that is, whether they are able to learn useful representations from such sets.

Here we analyze this question on example of (stacked) autoencoders with logistic regression as the classification layer applied to the task of image categorization. Pixel-level and hand-crafted features are used for experiments.

2   Autoencoders

We selected stacked autoencoders as the unsupervised layers with logistic regression as the classification layer. Single autoencoder has input, hidden, and output (reconstruction) layers. The input layer accepts a vector $\mathbf{x} \in [0,1]^N$ of dimension N, which is transformed to activities of neurons of the hidden layer $\mathbf{y} = s(\mathbf{Wx} + \mathbf{b})$, $\mathbf{y} \in [0,1]^d$, corresponding to new features (hidden representation), where $\mathbf{W}$ is a $d \times N$ matrix of connection weights, $\mathbf{b}$ is a bias vector, and s is the activation (sigmoid) function. Activities of neurons of the last layer are calculated similarly as $\mathbf{z} = s(\mathbf{W'y} + \mathbf{b'})$, $\mathbf{z} \in [0,1]^N$. Autoencoders differ from other feed-forward networks in that they are trained to minimize difference between $\mathbf{x}$ and $\mathbf{z}$ (for patterns from a training set), that is $\mathbf{W'}$ is the matrix of the reverse mapping. $\mathbf{W'}$ is frequently taken as $\mathbf{W}^T$. One can calculate gradient of the reconstruction error relative to connection weights and bias vectors and to train the autoencoder using stochastic gradient descent.

In the case of stacked autoencoders, each autoencoder on the next level takes outputs of the hidden (not reconstruction) layer of the previous autoencoder as its input performing further nonlinear transformation of constructed latent representation of the previous level. Each next autoencoder is trained after training its preceding autoencoder. Outputs from the hidden layer of the last autoencoder are passed to the supervised feed-forward network.

Multinomial logistic regression computes probabilities of a pattern to belong to different classes based on softmax applied to linear combinations of features.

$$p(y = c \mid \mathbf{x}, \mathbf{W}) = \frac{\exp(\mathbf{w}_c^T \mathbf{x} + b_c)}{\sum_{c'=1}^{C} \exp(\mathbf{w}_{c'}^T \mathbf{x} + b_{c'})},$$ (1)

where $\mathbf{x}$ is the input vector, c is the class index, C is the total number of classes, $\mathbf{W}$ is the weight matrix composed by C vectors $\mathbf{w}_c$, b_c are biases. Parameters of the classifier are learnt using stochastic gradient descent minimizing negative log-likelihood.

One common modification is denoising autoencoders [10], in which reconstruction during training is calculated using patterns with introduced noise, but reconstruction errors are calculated relative to initial patterns without this artificial noise. Thus, a denoising autoencoder learns to reconstruct a clean input from a corrupted one. This is a way to deform input patterns efficiently increasing the training set size, but in the least problem-specific fashion.

We introduce some additional modifications into denoising stacked autoencoders. One modification consists in that only the last (regression) layer is trained in the supervised fashion. This helps to check usefulness of unsupervised features.

The second modification (which is to be compared with the basic version) consists in constructing multi-column autoencoders. Multi-column deep neural networks were already used (e.g. [11]), but each column in such networks is usually trained as a separate deep network, and predictions of all columns are then averaged. In our modification, the number of columns corresponds to the number of classes. Each column is trained for its own class to produce some non-linear features. These features are then gathered and passed as the input to the logistic regression layer.

3   Image Features

We took the task of image categorization for investigating behavior of autoencoders. In this task, images can vary considerably, and special hand-crafted features are usually used, that doesn't eliminate necessity to learn non-linear features, especially when different hand-crafted features are combined.

We used three types of image features for image description: color histograms, edge direction histogram (EDH) [12] and Haralick's textural features based on gray-level co-occurrence matrices (GLCM) [13].

Color features are built as concatenation of three 1D histograms calculated for LAB components. Histogram for each component contains N_c=32 bins, so the total size of the color vector is $3N_c$=96.

In order to build an edge direction histogram we apply Deriche filtration [14] at the first step. Then we get an assessment of orientation distribution by computing the direction of gradient vector at local maximums of gradient magnitude. Initial range of edge orientation angles [0°, 360°) is quantized to N_q bins (we use 72 bins). The last bin in histogram represents the amount of image pixels that don't belong to edges.

Finally, because of possible difference in image sizes, all bins are normalized by corresponding values [12]

$$\mathbf{EDH}(i) = \mathbf{EDH}(i)/N_{edge}, \quad i \in [0, N_q - 1];$$

$$\mathbf{EDH}(N_q) = \mathbf{EDH}(N_q)/N_p, \tag{2}$$

where $\mathbf{EDH}(i)$ – value in bin i of the histogram, N_{edge} is total amount of edge points in the image, N_p is total number of pixels.

Unfortunately EDH gives no information about location of contours on the image plane, so we tried to compensate this by splitting image into K parts ($K=3$) and computing the histogram for each of them. Thus, the EDH descriptor size is $K(N_q+1)=219$.

The last part of our image descriptor is computed from normalized GLCM. Co-occurrence matrices represent texture properties, but they are inconvenient for matching two textures, because of big number of elements, thus we used five features of normalized GLCM $\mathbf{C_d}$, proposed in [13]:

- angular second moment: $\sum_i \sum_j \mathbf{C_d^2}(i, j)$;

- entropy: $-\sum_i \sum_j \mathbf{C_d}(i, j)\log \mathbf{C_d}(i, j)$;

- contrast: $\sum_i \sum_j (i - j)^2 \mathbf{C_d}(i, j)$;

- inverse difference moment: $\sum_i \sum_j \dfrac{\mathbf{C_d}(i, j)}{1 + |i - j|}$;

- correlation: $\dfrac{\sum_i \sum_j (i - \mu_i)(j - \mu_j)\mathbf{C_d}(i, j)}{\sigma_i \sigma_j}$,

where μ_i, μ_j, σ_i, σ_j are the means and standard deviations of rows and columns of $\mathbf{C_d}$, $\mathbf{d}$ is a displacements vector, for which co-occurrence matrix is computed (we used D options of $\mathbf{d}$ for each image).

In order to consider color information, initial RGB images were transformed to HSV representation, so GLCM was computed for hue, saturation and value components, which in turn were preliminarily normalized to fit [0, 255] range. The size of modified GLCM descriptor (five GLCM based features for each of H, S and V component for 5 displacement vectors and 3 directions) appeared to be 5×3×5×3=225.

Thus, the image descriptor is constructed by concatenation of LAB histogram and modified EDH and GLCM parts. The total size of the image descriptor is 96+219+225=540. Reduced descriptor containing only two sets of features (96+219=315 features) was also considered for comparison.

4   Experiments

We conducted experiments with the neural networks composed of autoencoders and logistic regression layer described above. For experiments, two training sets were constructed. Variability of images in these training sets was different. The first set (DB1) contained heads of cats of 5 breeds – European Shorthair, British Shorthair, Exotic Shorthair, Burmese, Abyssinian; examples of two of them are presented in fig. 1. The second set (DB2) contained images of 6 categories – city, flowers (garden), indoor, mountains, forest, sea (beach); see fig. 2. DB1 was composed of 330 images, from which small training set was extracted with 50 (10 per class) images. DB2 was composed of 6000 images, and experiments with two training sets containing 60 (10 per class) and 4800 (800 per class) images were conducted.

Fig. 1. Examples of images from two classes (breeds of cats) of DB1

Fig. 2. Examples of images from two classes (categories of scenes) of DB2

Logistic regression with and without autoencoders was tested separately using RGB values of pixels (on images rescaled to sizes of 64x64, 28x28, and 20x20) and hand-crafted descriptors as initial features on these two databases (three training sets). Results are presented in table 1.

It can be seen that hand-crafted features are not always very informative. Reduced image descriptors (315 features) gave worse results and enhanced descriptors (540 features) gave similar results in comparison with pixel-level features on DB1. However, for more complex image classes (DB2) these features yield better results, and results are greatly improved (in contrast to pixel-level features) with increase of the size of training sample. Of course, pixel-level features contain more information, but it is clearly seen that classes under consideration are not linearly separable in this feature space (but better linearly separable in the space of hand-crafted features), so one might expect that autoencoders would help to improve recognition rate for pixel-level features more than for hand-crafted features. However, conducted experiments showed that this assumption is not correct. The results are presented in table 2.

It appeared to be very difficult to train autoencoders in the case of pixel-level features and small training sets. Too small or too large number of training epochs leaded to recognition of all patters as belonging to one same class. The best achieved error rates for DB1 and DB2 (using 10 training images per class) are 0.418 and 0.601 correspondingly with the use of multi-column autoencoder with one layer containing 1000 neurons (with 20x20x3=1200 inputs). That is, autoencoders failed to learn useful features from small amount of raw data, and they only didn't ruined initial features in fortunate cases.

Table 1. Recognition error rates of the logistic regression classifier

	DB1 (10 per class)	DB2 (10 per class)	DB2 (800 per class)
Hand-crafted features (540)	0.356	0.505	0.118
Hand-crafted features (315)	0.436	0.535	0.188
64x64x3	0.361	0.620	0.583
28x28x3	0.382	0.598	0.580
20x20x3	0.401	0.615	0.571

Table 2. Best recognition error rates of the autoencoders with logistic regression classifier

	DB1 (10 per class)	DB2 (10 per class)	DB2 (800 per class)
Hand-crafted features (540)	0.323	0.41	0.091
Hand-crafted features (315)	0.359	0.445	0.122
20x20x3	0.418	0.601	0.353

Training multi-column one layer autoencoders on 800 per class DB2 images resulted in 0.353 error rate (one-column autoencoders gave 0.383 error rate), which is much better than the error rate of the logistic regression on pixel-level features, but is much worse than the error rate of the logistic regression on hand-crafted features. That is, some useful nonlinear features were learnt, but they were worse than both (reduced and enhanced) hand-crafted features. It should be noted that we also tested autoencoders on the MNIST database with 50000 training images, and they considerable reduced error rate relative to the logistic regression (from 8% to 2% only for two layers and can be reduced further) even without using convolution autoencoders or additional image transformations during training. This implies that more complex nonlinear features indeed can be learnt by autoencoders, but this result can be achieved in the case of very low variability of patterns and using large training sets.

Autoencoders trained using patterns represented by hand-crafted features appeared to be more useful (see table 2). The achieved error rate (~9%) for the scene classification task with 6 classes is rather high (e.g., in [12], 6% error rate is achieved while distinguishing only two classes – city and landscape images, and 9% error rate is achieved for forest-mountain-sunset classification), so this result is interesting by itself.

However, it is also interesting to compare results in tables 1 and 2. In the case of hand-crafted features, larger improvement is achieved in the case of small training sets. In the case of pixel-level features, results are opposite implying that nature of this improvement may be different. Possibly, dimension reduction of linear features has stronger influence in the first case (and this is achievable on small training sets), while only construction of nonlinear features helps to increase recognition rate in the second case (and this requires large training sets).

Consider also results in fig. 3-6. In all cases, error rates vary insignificantly for a wide range of hidden layer sizes. Best results are achieved on intermediate sizes, while the error rate rapidly increases for small hidden layers (in which too much information is lost), and gradually increases with the number of hidden neurons exceeding the number of features meaning that reduction of features redundancy is more useful than construction of complex nonlinear approximation on small training sets.

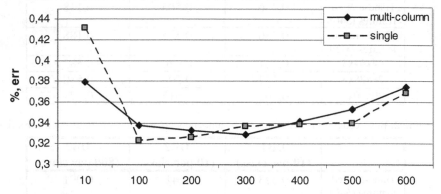

Fig. 3. Error rates of one-layer autoencoders of different hidden layer sizes with logistic regression layer on DB1 (540 features)

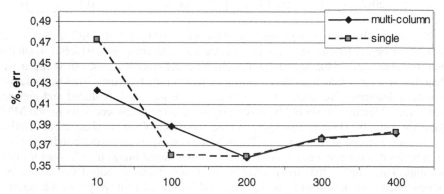

Fig. 4. Error rates of one-layer autoencoders of different hidden layer sizes with logistic regression layer on DB1 (315 features)

It can also be seen that multi-column autoencoders outperform one-column autoencoders on more difficult image classes (DB2) and have similar performance on simpler classes. Although multi-column autoencoders contain much more hidden neurons, they don't suffer from overlearning more than one-column autoencoders (with the same number of neurons per column).

Finally, stacked autoencoders were tested. However, they yielded very small decrease of the error rate. For example, multi-column autoencoders with two layers (400, 300) helped to decrease the error rate from 0.329 to 0.324, and for one-column autoencoders the error rate decreased from 0.323 to 0.319 on DB1. One might expect that multi-level autoencoders will be more useful in the case of pixel-level features and larger training sets, but adding the second layer was unsuccessful resulting in increase of the error rate from 0.353 to 0.373. Possibly, larger training sets or training image deformations are necessary for multi-level autoencoders to learn useful complex features.

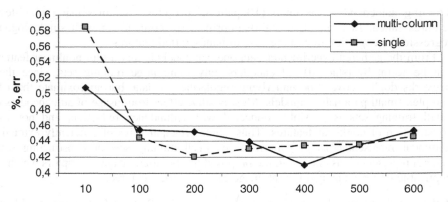

Fig. 5. Error rates of one-layer autoencoders of different hidden layer sizes with logistic regression layer on DB2 (540 features)

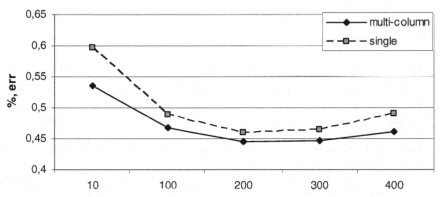

Fig. 6. Error rates of one-layer autoencoders of different hidden layer sizes with logistic regression layer on DB2 (315 features)

5   Conclusion

Stacked autoencoders with logistic regression layer for classifying images were considered. Original modification of multi-column autoencoders was proposed. Their performance was evaluated on pixel-level features and special hand-crafted image descriptors. These descriptors were composed of color, gradient and texture features. Tests were conducted on two data sets – cat breeds and image categories (city, indoor, forest, mountains, features. Thus, some principal ways of efficiently extracting nonlinear representations of patterns from small training sets or constructing some general representations (similar to the proposed hand-crafted image descriptors which are applicable for recognizing different image categories) are required.

Training on hand-crafted features appeared to be much more successful than on pixel-level features (without introducing deformations of training images). Error rates of 9.1% and 35.3% were achieved on the training set containing 4800 images for the task of recognizing 6 image categories using hand-crafted and pixel-level features correspondingly. Thus, only hand-crafted features yielded practically usable results.

Training autoencoders on small (10 patterns per class) samples using pixel-level features failed. In contrast, it stably improved (in comparison with single logistic regression) recognition rate in the case of hand-crafted features.

From the gathered experimental data one can conclude that useful nonlinear features can be learnt by (stacked) autoencoders only using large training sets, since these networks don't recover true underlying regularities in data, but approximate them by complex (multi-parametric) models. Most positive effect from usage of autoencoders on small training sets is possibly connected with redundancy reduction, and not with construction of nonlinear features. Thus, some principal ways of efficiently extracting nonlinear representations of patterns from small training sets or constructing some general representations (similar to the proposed hand-crafted image descriptors which are applicable for recognizing different image categories) are required.

Acknowledgements. This work was supported by the Russian Federation President's grant Council (MD-1072.2013.9) and the Ministry of Education and Science of the Russian Federation.

References

1. He, Y., Kavukcuoglu, K., Wang, Y., Szlam, A., Qi, Y.: Unsupervised Feature Learning by Deep Sparse Coding. arXiv:1312.5783 [cs.LG] (2013)
2. Le Roux, N., Bengio, Y.: Representational Power of Restricted Boltzmann Machines and Deep Belief Networks. Neural Computation 20(6), 1631–1649 (2008)
3. Gregor, K., Mnih, A., Wierstra, D.: Deep AutoRegressive Networks. arXiv:1310.8499 [cs.LG] (2013)
4. Hinton, G.E., Osindero, S., Teh, Y.: A Fast Learning Algorithm for Deep Belief Nets. Neural Computation 18, 1527–1554 (2006)
5. Bengio, Y., Lamblin, P., Popovici, D., Larochelle, H.: Greedy Layer-Wise Training of Deep Networks. In: Advances in Neural Information Processing Systems (NIPS 2006), vol. 19, pp. 153–160 (2007)
6. Ranzato, M.A., Poultney, C., Chopra, S., LeCun, Y.: Efficient Learning of Sparse Representations with an Energy-Based Model. In: Advances in Neural Information Processing Systems (NIPS 2006), vol. 19 (2007)
7. Ciresan, D.C., Meier, U., Gambardella, L.M., Schmidhuber, J.: Deep Big Simple Neural Nets Excel on Handwritten Digit Recognition. arXiv:1003.0358 [cs.NE] (2010)
8. LeCun, Y., Cortes, C.: MNIST handwritten digit database, http://yann.lecun.com/exdb/mnist/
9. Tenenbaum, J.B., Kemp, C., Griffiths, T.L., Goodman, N.D.: How to Grow a Mind: Statistics, Structure, and Abstraction. Science 331(6022), 1279–1285 (2011)
10. Vincent, P., Larochelle, H., Bengio, Y., Manzagol, P.-A.: Extracting and composing Robust Features with Denoising Autoencoders. In: Proc. 25th International Conference on Machine Learning, pp. 1096–1103 (2008)
11. Cireşan, D., Meier, U., Masci, J., Schmidhuber, J.: Multi-Column Deep Neural Network for Traffic Sign Classification. Neural Networks 32, 333–338 (2012)
12. Vailaya, A., Jain, A., Zhang, H.J.: On Image Classification: City Images vs. Landscapes. Pattern Recognition 31(12), 1921–1935 (1998)
13. Haralick, R.M., Shanmugam, K., Dinstein, I.: Textural Features for Image Classification. IEEE Transactions on Systems, Man, and Cybernetics. SMC-3(6), 610–621 (1973)
14. Deriche, R.: Using Canny's Criteria to Derive a Recursively Implemented Optimal Edge Detector. International Journal of Computer Vision 1(2), 167–187 (1987)

A Pattern Recognition Approach for Peak Prediction of Electrical Consumption

Morten Goodwin[1,2] and Anis Yazidi[3]

[1] Department of ICT, University of Agder, Grimstad, Norway
[2] Teknova AS, Grimstad, Norway
[3] Institute for ICT, Oslo and Akershus University College of Applied Sciences, Norway

Abstract. Predicting and mitigating demand peaks in electrical networks has become a prevalent research topic. Demand peaks pose a particular challenge to energy companies because these are difficult to foresee and require the net to support abnormally high consumption levels. In smart energy grids, time-differentiated pricing policies that increase the energy cost for the consumers during peak periods, and load balancing are examples of simple techniques for peak regulation. In this paper, we tackle the task of predicting power peaks prior to their actual occurrence in the context of a pilot Norwegian smart grid network.

While most legacy studies formulate the problem as time-series-based estimation problem, we take a radically different approach and map it to a classical pattern recognition problem using a simple but subtle formulations. Among the key findings of this study is the ability of the algorithms to accurately detect 80% of energy consumption peaks up to one week ahead of time. Further, different classification methods have been rigorously tested and applied on real-life data from a Norwegian smart grid pilot project.

Keywords: Peak Prediction, Energy Consumption, Classification.

1 Introduction

Changes in the consumers electrical power consumption influences demand peaks, which are difficult to predict due to their seemingly random occurrence. This is because consumption peaks are consequences of aggregated and collective behaviour of several customers, and are influenced by many external factors. A simple example of typical consumer's behavioral influence is when people turn on their heater when the weather is cold. When multiple consumers do this at the same time it, which is natural since a drop in temperature affects them all, the aggregated behavior causes a peak in the overall electrical consumption. However, since there are many factors other than temperature that influences a user electrical consumption, it is far from trivial to foresee what the consumption will be and in turn predict high loads.

Both consumption peaks and peak supplies cause challenges for electrical service providers because they need to design their grid for the maximum potential

L. Iliadis et al. (Eds.): AIAI 2014, IFIP AICT 436, pp. 265–275, 2014.

load. This is crucial since energy scarcity has severe consequences such as power outages. An alternative approach to over dimensioning electrical grids for the absolute maximum potential consumption peaks is evening out the peaks with peak controlling methods. The difficulty lies in that to control consumption, smart grid systems need know peaks prior to their occurrence, and there is no existing method that can accurately do so. There is however no doubt that the existence of such a method would help power companies, such as Agder Energy, to design intelligent strategies for reducing the overall consumption. The overall impact of successful peak prediction is enabling for electrical grids with less load and in turn collectively less energy usage.

This paper addresses predicting electrical power consumption peaks in small communities where there are no external data available, such as weather predictions or detailed customer data. This contributes to a crucial part of larger smart grid energy system which performs load balancing and avoids energy over-supply to keep grid voltage at an acceptable level. The peak prediction algorithms are meant to be used in a smart grid installation to: (1) Automatically turn off smart electrical appliances. (2) Automatically ask consumer power cells to be turned on. (3) Carry out local level load balancing. Hence, a potential peak at one customer household will avoid contributing to an overall demand peak in the system. This is a realistic scenario and such algorithms are planned to be part of Smart Village neighbourhood in Arendal, Norway.

This is particularly difficult as peaks are influenced by complex behaviour and typically occur in abnormal situations. Further, since this is planned to be installed in a new Smart Village neighbourhood, there is no external data available. We do not have access to other data such customer behavior, and since this is a small relatively secluded area weather reports is expected to have an equal impact on all customers and is therefore not particularly valuable[1]. The latter is different most approaches in the literature [1].

Our main contributions in this paper are: In contrast to most other approaches, mapping peak detection into a two-labelled classification problem using statistical terms. We further test it out with data from a real Norwegian pilot smart grid project, and are able to get results in line with state-of-the-art with relatively simple classification algorithms.

This paper is organised as following. Section 2 formally defines the problem to be solved as a two-labelled classification problem. Section 3 continues with the most relevant development in the area from two research areas: peak prediction/detection and rare item classification. The data used in these experiments are described in section 4 and the methods used for peak predictions are described in section 5. Section 6 continues with the results and findings from these methods. Finally, conclusion and future work is outlined in section 7.

---

[1] The authors acknowledge that a change in temperature will influence electricity usage and in turn yield peaks. However, if there is an over all energy usage increase for all customers, techniques such as local load balancing will have limited value.

2   Problem Setting

The goal of this research is to predict load peaks prior to their occurrence. Formally, given a time sequence **t** let $p(t_i, n)$ be a statistical function asserting whether t_i is a peak where $t_i \in$ **t** and n tells the extremity of the peak as following:

$$p(t_i, n) = \begin{cases} peak \ if \ t_i > \mu + n * \sigma \\ not \ peak \quad otherwise \end{cases} \tag{1}$$

where μ and σ are the arithmetic average and standard deviation of **t**, and n is a number defined as the extremity of the peak. Thus, a value is defined as a peak if it has a value more than n times the standard deviation above the mean value. Is it possible to predict the output from $p(t_i, n)$ using only part of the sequence occurring prior to t_i ($t_{<i} \in$ **t**)? Hence, the peak detection problem is deducted to a standard two-labelled classification problem. This is less trivial than most two-labelled classification problems because the peaks are rare and "abnormal". Rare abnormalities are particularly hard to predict.

In layman's terms: Is it possible to predict future load peaks using only past data on electrical consumptions?[2]

3   State-of-the-Art

This section presents the most relevant research in the area of peak prediction. Section 3.1 presents the research on prediction electrical loads, and section 3.2 presents research on classifications when there is a skewed division of labelled data — i.e. classification when the positive labels are so few that it has a significant negative impact on the classification results.

3.1   Electrical Consumption Prediction

Many methods for load forecasting is available in the literature [1, 3–5]. Common for many of these is a reliance on some sort of third-party data for predictions and/or try to predict the actual power load. There are obvious similarities between peak prediction and load forecasting, but the objective is notably different.

A common approach is to use linear regression models in combination with other factors such as heat data. This is shown very useful to predict short-term peak loads [6, 7] and annual loads[8]. This research uses various functional linear regression models and support vector regression for the actual prediction, and clusters the data and assign one regression model per cluster.

Another common approach of load forecasting is to find a similar day based on the available data, such as weather reports, and assume that similar days

---

[2] Note that is very different from the seemingly similar peak detection (positive outlier detection). For outlier detection it is possible calculate whether t_i is an outlier using t_i itself.[2]

have similar loads. Most relevant for this work is hierarchical two-step classifiers used for short-term load forecasting [9] which first predicts the "type of day" then an hourly forecast.

Some work exists without reliance on third party data with the assumption that the data has some internal structure than can be learnt.[3] Perhaps most well known is the use of evolutionary algorithms and fuzzy logic approach to predict hourly peaks from one to seven days ahead [10],wavelet transformation techniques to accurately forecast power consumptions about 4 hours ahead of time [11] and short-term predictions using exponential smoothing methods[12]. Further, other machine learning techniques been attempted such as non-linear curve-fitting [13, 14] and Support Vector Machines (SVM) [15, 16]. .

3.2 Rare Item Classification

Rare item classification is a field within pattern recognition where the majority of the data is impertinent, typically the normal situation, and consequently the interesting data appears much rarer. Often times even the labels to be classified are not known in advance. Further, there is an imbalance in the data available for training (a priori). Exhaustive labelling is commonly required to build up a proper training database. Both labelling and corresponding classification is inevitably costly. Much research exists on this, but, to the best of our knowledge, none of these directly address the peak prediction problem.

An effective technique when working with rare item classification is combination methods, such as hierarchical classifiers with more than one classification technique [17, 18]. Generally this means organising local classifiers in an hierarchical structure and defining rules giving a global consensus. Based on techniques such as majority voting, the classifiers yields significantly better results for rare item classification than comparable algorithms, even with a flat structure [19]. Others use clever re-sampling, such as random under- and oversampling [17].

4   Data with Electrical Consumption Peaks

The data used in this project is from a summer cabin area in Hvaler, Norway for six weeks in mid 2012 [20]. In total there are more than 7900 installations with varying degree of activity — making the data very vacillated. From each installation, accumulated hourly energy consumptions (kWh) was collected throughout the six weeks.

An examples of electrical consumption for one installation is presented in Figure 1(b). This shows averaged values of 7 days — displaying some clear trends: A rise and fall of the consumption with its highest around Norwegian dinner time and lowest at night, repeated nightly.

Figure 1(a) shows an example of consumption with peaks for a specific 24 hour period. The threshold level, n, for the peak data in Figure 1(a) is set to 2

---

[3] This is the same assumption we have in this paper.

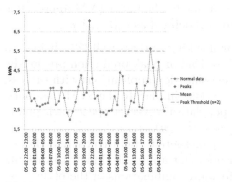 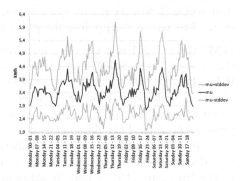

(a) Example of peak in the consumption data with n set to 2

(b) Consumption averaged per 7 days

Fig. 1. Consumption patterns and peaks

yielding two peaks: around 05-03 22:00 and 05-04 16:00. Overall, by varying n in the equation 1, the percentage of peaks varies from 15% for $n = 1$ down to 0.8% for $n = 3$.

5 Approach

This sections presents approaches for to predict $p(t_i, n)$. The first approach is a straightforward solution introduced for comparison purposes. This is continued with standard classification techniques inspired by similar work on load forecasting [6, 15, 16, 21]. Lastly, we present hierarchical classification solutions inspired by techniques in the literature on rate item classification[17–19]. All classifiers aim at predicting $p'(t_{<i}, n)$ as close to $p(t_i, n)$ as possible.

Majority voting(c_m):Firstly, a simple majority voting classifier was implemented that counts the number of peaks in the training data and checks is there are more peaks the previous day than could be expected based on simple statistics.

Consistent Peaks.(c_c):Secondly, another classifier was implemented that always predict the peak equal to the previous peak value as following:

$$c_c(t_{<i}, d, n) = p'(t_{<i}, n) = p(t_{i-d}, n). \tag{2}$$

where t_i is the data to predict and d is how many hours in advance t_i is.

SVM(c_l,c_p):The purpose of the SVM is to create a function, linear(c_l) or polynomial(c_p), that separates data that results in peaks from the data that does not. Upon predicting $p(t_i, n)$ the vector space consists of a defined number of points prior to $t_i \in \mathbf{t}$, namely $t_{<i}$. Each individual point in $t_{<i}$ is then a separate

dimension. Two variants of SVM are implemented; The linear SVM approach is defined as $c_l(t_{<i}, d, n)$, while the polynomial is defined as $c_p(t_{<i}, d, n)$,

GMM(c_g):GMM is a classifier known for handling outliers well [22]. It has, to the best of our knowledge, not been used for load or peak prediction previously. It has however been used for many other classification applications with success, and is specially suited for rare item classification. Similar, to the SVM classifiers, when classification of t_i with GMM we populate the vector space with points prior to t_i ($t_{<i}$). This classifier is defined as $c_g(t_{<i}, n)$.

Two variants of hierarchical classifiers are presented: Hierarchical and Hierarchical with Consistent Peaks. They both use a flat approach similar to variants in the literature [19] — namely straightforward combinations of the other classifiers.

Hierarchical (c_{h1}): The hierarchical classifier is simply a combination of the linear SVM and GMM similar to the common practice in the literature as following:[23, 24].[4]

$$c_{h1}(t_{<i}, d, n) = c_l(t_{<i}d, , n) \mid c_g(t_{<i}d, n) \tag{3}$$

Hierarchical with Consistent Peaks (c_{h2}): An additional classifier was created utilizing the data from the consistent peaks (see equation 2). This extends equation 3 as following.

$$c_{h2}(t_{<i}, d, n) = c_l(t_{<i}, d, n) \mid c_g(t_{<i}, d, n) \mid c_c(t_{<i}, d, n) \tag{4}$$

6   Experiments

This section presents empirical results from running the predictive algorithms introduced in section 5. In line with common practice in the field [10], we present predictions starting from one hour into the future and up to 7 days (168 hours).

Section 6.1 presents a comparative experiments when n set to 1 — predicting moderate peaks. Additionally, section 6.2 shows the behaviour when varying n — making the peaks to detect more extreme — influences the algorithms.

All results presented in this chapter use a vector size of 24.[5]

Further, several metrics exists to evaluate information retrieval approaches. This paper promotes classification methods that favour false positives over false negatives. This is because a false positive, predicting a peak that is not present, has significantly less consequences than not predicting peaks which are present. Thus, high recall is much more sought after than high precision.

---

[4] Note that the decision rule is simpler than most other setups.
[5] Several experiments were carried out with various vector sizes concluding with a vector size of 24 units. These results are not crucial to the understanding of the approach and therefore omitted.

6.1   Predicting Small Consumption Peaks

This section presents peak predictions with n in equation in $p(t_i, n)$ (equation 1) set to 1. This is the most straightforward approach where peaks are defined as simple values larger than the standard deviation.

Figure 2(a) and 2(b) show the recall and precision of the applied algorithms. These figures all show predictions 1 to 168 (7 days) into the future with n set to 1. This means that each time instance in the data, the algorithms try to predict from 1 to 168 hours ahead of time and the average is shown in the graph.

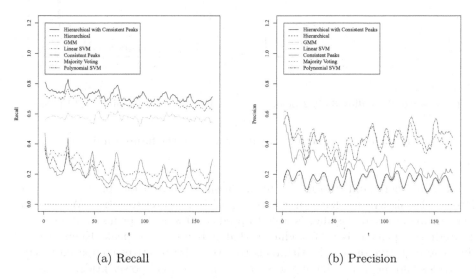

(a) Recall   (b) Precision

Fig. 2. Recall and precision for predicting 1 to 168 hours into the future with n set to 1

All pattern recognition algorithms, SVM, GMM and Hierarchical has a decrease in recall as the t increases. This means that, in line with intuition, the difficulty increases the further into the future the algorithm try to predict peaks. However, when t reaches 24 the quality of the measure increases. Hence, it is easier to predict 24 hours into the future than it is to predict other values. This suggests that people behave similarly in 24 hour cyclic intervals (cook dinner, eat, sleep at roughly the same intervals). This is supported by "Consistent peaks"-classifier is surprisingly accurate.

The figure shows that the hierarchical classifiers have an excellent recall (Figure 2(a)). Thus, these algorithms are able to detect about 80% of the peaks in the data. Further, the recall decreases as t increases for all algorithms. Hence, prediction becomes more difficult the further into the future the algorithms aim at predict. This is however less true for hierarchical classifiers and GMM than for comparative algorithms.

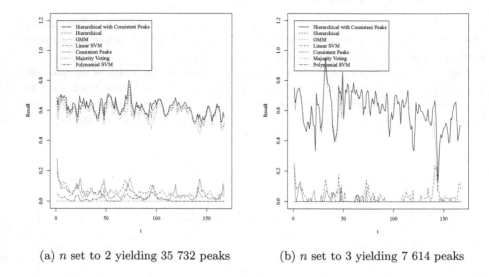

(a) n set to 2 yielding 35 732 peaks   (b) n set to 3 yielding 7 614 peaks

Fig. 3. Recall predicting 1 to 168 hours into the future varying n

On the other hand, the hierarchical classifiers have a low precision (Figure 2(b)). The algorithms have many false positives (close to 80%). We can say that an hierarchical classifier is able to predict 80% of the peaks, and whenever it predicts a peak it is a 20% chance that it is an actual peak. Keep in mind that the objective of the algorithms is to favor false positives compared to false negatives false negatives having less impact in electrical power grids.

6.2 Predicting Larger Consumption Peaks

This section presents results upon varying n in $p(t_i, n)$ (equation 1). By increasing n the peaks become more extreme and (presumably) more difficult to predict.

Figure 3(a) and 3(b) show recall over time when n is set to 2 and 3. This is same setup as shown in figure 2(a) with n set to 1, but the behaviour is notably different. First and foremost, all classifiers have a decreased recall. Most notably, the SVM classifiers drop fast to a recall close to 0. In this scenario, SVM is only able to predict close to 24 hours into the future, and not very well. Consequently, the hierarchical classifiers are only slightly better than GMM — which means than SVM does not contribute much in an increased recall in the hierarchical classifiers.

The trend in Figure 2(a) (n=1) and 3(a) (n=2) is continued in figure 3(b) (n=3), but is more extreme. The SVM classifiers more quickly drops to 0 and the difference between the hierarchical and GMM classifiers are much less. Still, the best algorithms are able to reach a recall more than 70%. [6]

Another trend which is more visible is that the results are more chaotic which may be caused by only some of these extreme peaks are predictable, while others are not. With peaks as n increase this becomes more apparent in the data rendering more challenging results.

A conclusion to be drawn from this is that an increase in the extremities of the peaks makes the classification more difficult. However, GMM and corresponding hierarchical classifiers are still able to detect more than 70% of the peaks. All other classifiers fall short with a recall close to 0.

7   Conclusion and Future Work

This paper deals with predicting electrical consumption peaks as input to load balancing and/or smart pricing strategies. This is done in a completely new way by mapping the prediction activities into a two-labelled classification problem. Further, in contrast to most existing approaches, the features are based solely on previous consumptions, avoiding reliance on third party data for the predictions.

Several classification algorithms are implemented and successfully applied on real-life data from a Norwegian smart-grid pilot project. The most promising results are produced by a simple hierarchical classifier combining linear support vector machines, Gaussian mixture models and deterministic assumption of consistent peaks. This solution is able to detect 80% of consumption peaks up to one week ahead of time. This promising result shows the usefulness of our approach.

This paper is part of an ongoing research project aiming at a developing smart energy housing technology with peak prediction as an essential component. We plan to carry out the following research: Comparison with additional classifiers, such as graph based classification approaches and more advanced hierarchical classifiers. We also plan to include weather information data as commonly done in the literature in order to improve the prediction results. Most importantly, this work is planned to be used in the scheduled Smart Village Skarpnes with 40 passive houses.

Acknowledgements. This project is partially funded by Agder Regional Research Fund with the project "Grid Operation and Distributed Energy Storage: Potential for improved grid-operation efficiency" and the Norwegian Research Council funded project "Electricity Usage in Smart Village Skarpnes". These projects are carried out together with University of Agder, Teknova, Agder Energi Nett and Eltek. Data is collected through the external DeVID project.[7]

---

[6] The trend continues when increasing n even further. This is deliberately left out of the paper since this does not contribute to further insight of the algorithms.

[7] http://www.sintef.no/Projectweb/DeVID/Prosjektpartnere/

References

1. Feinberg, E., Genethliou, D.: Load forecasting. In: Chow, J., Wu, F., Momoh, J. (eds.) Applied Mathematics for Restructured Electric Power Systems. Power Electronics and Power Systems, pp. 269–285. Springer US (2005)
2. Hodge, V., Austin, J.: A survey of outlier detection methodologies. Artificial Intelligence Review 22(2), 85–126 (2004)
3. Campbell, P.R., Adamson, K.: Methodologies for load forecasting. In: 2006 3rd International IEEE Conference on Intelligent Systems, pp. 800–806. IEEE (2006)
4. Weron, R.: Modeling and forecasting electricity loads and prices: A statistical approach, vol. 403. Wiley. com (2007)
5. Mohsenian-Rad, A.H., Leon-Garcia, A.: Optimal residential load control with price prediction in real-time electricity pricing environments. IEEE Transactions on Smart Grid 1(2), 120–133 (2010)
6. Goia, A., May, C., Fusai, G.: Functional clustering and linear regression for peak load forecasting. International Journal of Forecasting 26(4), 700–711 (2010)
7. Pardo, A., Meneu, V., Valor, E.: Temperature and seasonality influences on spanish electricity load. Energy Economics 24(1), 55–70 (2002)
8. Wang, J., Li, L., Niu, D., Tan, Z.: An annual load forecasting model based on support vector regression with differential evolution algorithm. Applied Energy 94, 65–70 (2012)
9. Ilić, S.A., Vukmirović, S.M., Erdeljan, A.M., Kulić, F.J.: Hybrid artificial neural network system for short-term load forecasting. Thermal Science 16(suppl. 1), 215–224 (2012)
10. Yang, H.T., Huang, C.M.: A new short-term load forecasting approach using self-organizing fuzzy armax models. IEEE Transactions on Power Systems 13(1), 217–225 (1998)
11. Annamareddi, S., Gopinathan, S., Dora, B.: A simple hybrid model for short-term load forecasting. Journal of Engineering 2013 (2013)
12. Taylor, J.W.: An evaluation of methods for very short-term load forecasting using minute-by-minute british data. International Journal of Forecasting 24(4), 645–658 (2008)
13. Zhang, G., Eddy Patuwo, B., Hu, M.Y.: Forecasting with artificial neural networks: The state of the art. International Journal of Forecasting 14(1), 35–62 (1998)
14. Suganthi, L., Samuel, A.A.: Energy models for demand forecasting–a review. Renewable and Sustainable Energy Reviews 16(2), 1223–1240 (2012)
15. Mohandes, M.: Support vector machines for short-term electrical load forecasting. International Journal of Energy Research 26(4), 335–345 (2002)
16. Li, Y.C., Fang, T.J., Yu, E.K.: Study of support vector machines for short-term load forecasting. Proceedings of the Csee 6, 10 (2003)
17. Kotsiantis, S., Kanellopoulos, D., Pintelas, P., et al.: Handling imbalanced datasets: A review. GESTS International Transactions on Computer Science and Engineering 30(1), 25–36 (2006)
18. Silla Jr., C.N., Freitas, A.A.: A survey of hierarchical classification across different application domains. Data Mining and Knowledge Discovery 22(1-2), 31–72 (2011)
19. Xiao, Z., Dellandrea, E., Dou, W., Chen, L.: Hierarchical classification of emotional speech. IEEE Transactions on Multimedia (2007)

20. Sand, K., Foosnas, J., Nordgard, D.E., Kristoffersen, V., Solvang, T.B., Wage, D.: Experiences from norwegian smart grid pilot projects. iN: 22nd International Conference and Exhibition on Electricity Distribution (CIRED 2013), pp. 1–4. IET (2013)
21. Chen, B.J., Chang, M.W., et al.: Load forecasting using support vector machines: A study on eunite competition 2001. IEEE Transactions on Power Systems 19(4), 1821–1830 (2004)
22. Jain, A.K., Duin, R.P.W., Mao, J.: Statistical pattern recognition: A review. IEEE Transactions on Pattern Analysis and Machine Intelligence 22(1), 4–37 (2000)
23. Uguroglu, S.: Robust Learning with Highly Skewed Category Distributions. PhD thesis, Carnegie Mellon University (2013)
24. Huang, H., He, Q., Chiew, K., Qian, F., Ma, L.: Clover: a faster prior-free approach to rare-category detection. Knowledge and Information Systems 35(3), 713–736 (2013)

Semi-automatic Measure and Identification of Allergenic Airborne Pollen

Antonio García-Manso[1,*], Carlos J. García-Orellana[1], Rafael Tormo-Molina[2],
Ramón Gallardo-Caballero[1], M. Macías-Macías[1],
and Horacio M. Gonzalez-Velasco[1]

[1] Pattern Classification and Image Analysis Group, University of Extremadura,
Avda. de la Universidad S/N, 10003 Cáceres, Extremadura, Spain
[2] Department of Vegetal Biology, Ecology and Earth Sciences,
University of Extremadura, Avda Elvas S/N, 06006 Badajoz, Spain
antonio@capi.unex.es

Abstract. Current lifestyle in developed countries makes the practice of outdoor activities to be almost mandatory. But, since these practices such as trekking, biking, horseback, or simply running or walking in urban parks, are made in nature (at least outdoors) not everyone can practice them in optimal physical conditions at any time of the year. We are referring to those who suffer from pollinosis or *"hay fever"*.

This work present the first stages in the development of a semi-automatic system for counting and identifying airborne pollen, using artificial intelligence techniques for recognizing four of the most representative allergenic pollen types. The system consists of a first stage for the location of pollen grains in the slides, and a second whose goal is the identification using Independent Component Analysis (ICA) and neural nets or SVM. The overall success results achieved with our system are about 88%, averaging for all classes.

Keywords: Independent Component Analysis (ICA), airborne pollen, pollen allergy, Neural Networks, SVM.

1   Introduction

Modern cities have good urban parks or nearby natural environments where its citizens can practice outdoor activities. But, both in these urban parks and in the nearby natural environments, there may be plants producing allergenic pollen for many people. These plants can be natives or even exotic plants used for ornamental purposes, because landscape artists tend to use them in new urban parks. This can cause that many people may not practice outdoor activities, at least not in full physical conditions at any time of year. We are referring to those who suffer from pollinosis or *"hay fever"*, that is, presenting allergy to certain types of airborne pollen at certain periods of the year in concrete geographical locations. In [1] it can be found a review of the main types of allergenic pollen and

---

* Corresponding author.

L. Iliadis et al. (Eds.): AIAI 2014, IFIP AICT 436, pp. 276–285, 2014.
© IFIP International Federation for Information Processing 2014

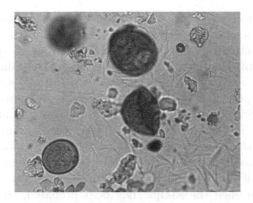

Fig. 1. The slices used to collect the pollen are subjected to a dyeing process. In that way, the pollen grains appear colored over the rest of the surface and others deposed particles in it. Here a frame extracted from a video sequence recorded with a digital camera through the microscope is shown. One can see in the picture the effect of staining the pollen grains. Since, four grains appear fully differentiated from other particles deposited on the adhesive.

the rates of pollen allergy in Europe. There, authors state that the prevalence of pollen allergy is presently estimated to be up to 40%.

Count and classification of airborne pollen is a very laborious task that require a lot of time and has to be made by skilled professionals. It is necessary a high level of training to obtain accurate classification results. The study of a preparation (Fig. 1) normally require the identification of a huge number of pollen grains. These analysis can take 2 hours or more, depending on pollen concentration in the sample. Another problem to consider is that the pollen identification can involve some error, because this task is subject to personal perceptions. Normally, the experts work with 400x optical microscopes and when they locate a pollen grain, they need to change the focal plane many times. The pollen grain is a tri-dimensional particle which have an aerodynamic size of $15 - 40 \ \mu m$. Therefore, its appearance in the microscope image changes dramatically depending on the chosen focal plane. Furthermore, the pollen grains can present some degree of translucence, depending on the type of treatment they have undergone. And, in this case, it is possible to see its apparent inner structure. In Fig. 1 (image extracted from a video sequence) there are four pollen grains of different sizes and, therefore, differently focused. It can be appreciated that the one located at the upper left corner is fully unfocused, whereas the other three can be seen clearly. In subsequent sequences of the video, this pollen grain will appear well focused and completely defined while the other three will be unfocused. This makes pollen identification a very complicated and challenging task, where it is necessary to consider a number of features such as the grain shape, the polarity, the number of openings and their arrangement and shape, texture features,..., and finally, this is translated to a number of perceptions that each expert can have [2].

In recent decades many works have been proposed aimed at locating and identifying pollen using computer vision techniques. In [3], the authors assess the potential of IR spectroscopy of the Fourier transform as an alternative method for fast and realizable identification of some of the pollen grains types that cause a higher rate of allergy. In [4], a prototype of a full self-station for counting and identifying of pollen on standard glass slides is presented. In [5] a system for automatic detection and classification of pollen on standard slides is also proposed. And, as in the previous case, it also uses parameters and texture analysis to obtain the feature vectors. As can be seen, several approaches have already been proposed to get an automatic or semi-automatic counting station and classification of airborne pollen. It is relatively easy to locate the pollen grains on the slide once the sample has been prepared, that is, it has been subjected to a dyeing process so that the pollen grains are colored (Fig. 1).

Currently, we are developing a self-station where a video camera linked to a microscope is controlled by a computer to get the sample images. The microscope digital camera set can be moved in the direction of Z axis to fit the focus, while the sample to be analyzed is placed on a standard microscope slide and it can be moved in the XY plane, so that the system can move along the entire surface of the slide capturing conventional light microscope video images. Then, each one of the captured videos is analyzed to select the frames with better definition, which will provide the main information. The selected frames are subjected to a process of feature extraction using Independent Component Analysis (ICA), resulting in a feature vector that it will be the input to a classifier. ICA has been widely used in the last decade as a feature extraction technique for its ability to get a base of functions adapted to every problem [6], especially for natural images [6,7]. Different approaches for the analysis of frames are given in [4] and [5], where dark field images (photograph of the silhouettes) are captured and then analyzed together with shape and texture features.

2   Dataset

Our aim is to develop a self-station for counting and identifying airborne potentially allergenic pollen. For this reason, we have opted for capturing prototypes from pollen images taken under the same conditions in which palynologists, responsible for this task, carry it out, and not to take the prototypes from a standard database as the European Pollen Database [8]. The samples were collected by an aerobiological sensor (7 days recording volumetric spore trap), Hirst type, Burkard 7-day. Once they were treated, we used those samples to record the videos. And from these videos, the pollen prototype images used in this work (Table 1) were captured by means of an algorithmic method. As will be explained in detail in the next section, the algorithm, looks for the frame with higher contrast (larger number of edges), for each pollen grain present in the sample. That will be equivalent to having the better definition, showing a greater amount of characteristic details of each particular pollen grain.

It can be seen in Table 1 how there are different classes that seem very similar. This could be because, in those cases, the pollen grains came from the same

Table 1. Examples of pollen of each class and of unwanted particles (trash)

class	samples	class	samples
Avena sativa		Avena sterilis	
Calocedrus decurrens		Olea europaea	
Cypress		Phalaris minor	
Dactylis glomerata		Lolium rigidum	
Quercus rotundifolia		trash	

family: *Lolium rigidum, Dactylis glomerata, Phalarais minor* belong to Gramineae family. However, this is not true for *Avena sativa* and *Avena sterilis* which also belong to the Gramineae family. Besides, *Calocedrus decurrens* and *Cypress* belong to Cupressaceae family and in the images appear quite different. For this reason, it is very difficult to discriminate among pollen of the same family, even for skilled persons. Therefore, in some experiments the pollen belonging to the same family is treated as the same class.

3   Methods

Two processes have to be differentiated: first the process followed to segment the pollen grain images in the video sequences, and second the process followed to classify these images.

3.1   Segmentation

Videos from standard glass microscope slides containing the grains of pollen were recorded. These videos were recorded by moving the microscope (that is, changing the focal plane) in the direction perpendicular to the slide (Axis Z) for about 100 μm, depending on the pollen sizes and the slides conditions. Furthermore, these conditions determined the speed at which the microscope was moved, because, at this stage of the project development, the movement was carried out by hand. In this way, each video has about 120 to 180 frames. Afterwards, these videos were analyzed to segment the pollen grains contained in them, with the ultimate goal of finding the best image for each pollen grain that appears in each video. Due to the change in the focal plane, it is possible that, when a video has more than one pollen grain, the best frame for segmenting each grain does not match, as can be seen in Fig. 1.

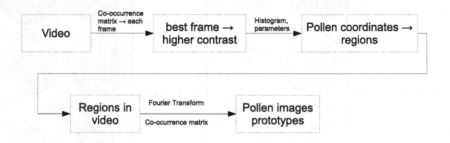

Fig. 2. Outline of segmentation algorithm of pollen grains

The process is shown in Fig. 2. First, the algorithm calculates the co-occurrence matrix for each frame in each video. The contrast of each frame can be measured from this co-occurrence matrix and, thus, it is possible to select the frame with higher contrast. This frame is used to segment the possible grains of pollen contained in the image. And after that, once the position of each grain of pollen is known, the algorithm analyzes again the video, but now looking each time in the position where the grains of pollen are.

To start the process of segmentation of pollen grains, the frame (color image) with higher contrast is converted to grayscale. And, the pollen grains are segmented by using the histogram and taking into account a series of parameters such as shape factor (pollen grains have a more or less rounded shape), size for each pollen class and others based on the gray value features in each region as deviation and entropy (pollen grains have a semi-translucent inner structure (cytoplasm) and it is not homogeneous in its gray levels).

To find the best frame for each pollen grain the algorithm uses two different processes. One of them consists in using the co-occurrence matrix, equal than to look for the frame with higher contrast, but now, focusing at the regions where the pollen grains are. And the other one consists in using the Fourier transform to find the frame with higher contrast focusing also at the same regions that the previous one. In that way, the algorithm obtains two different sets of images for grains pollen prototypes. Obviously, artifacts (unwanted particles as dust and detritus) are taken as well in this process. These artifacts will be used as image prototypes for background and unwanted particles.

3.2 Classification Algorithms

For training our system, a classical scheme to train classifiers was followed, which consists in to use three independent sets of patterns: one for training, one for avoid the overtraining and another for testing [9]. For generating each of these sets was applied ICA to the pollen images prototypes. We have used Neural Networks and Support Vector Machine (SVM) [10] classifiers.

Independent Component Analysis, (ICA). ICA is a statistical generative model whose objective is to explain the original data ($\mathbf{X}$) using statistically independent random vectors ($\mathbf{S}$). $\mathbf{X}$ can be modeled as $\mathbf{X} \approx \mathbf{AS}$, where $\mathbf{S}$ is the matrix of latent independent components and $\mathbf{A}$ is the mixing matrix.

This technique can be used for feature extraction since the components of $\mathbf{X}$ can be regarded as characteristics representing the objects (patterns) [6]. FastICA algorithm [11] was used to build ICA base functions of the pollen image space. Once a suitable ICA model has been created, it can be expressed, as also with other transformations (wavelets or Gabor filters [12]), each sample of an image (I) located in (x, y), that is, the pixel gray-scale values (point luminances) in an image, as a linear superposition of some base functions $A_j(x, y)$ (rows of the mixing matrix $\mathbf{A}$) called features,

$$I(x,y) = \sum_{j=1}^{q} A_j(x,y)s_j \tag{1}$$

where the s_j are image-dependent coefficients. While the features $\mathbf{A}_j$ are the same for all patches, all prototypes used to generate the ICA base functions. That is, by estimating an image basis using an ICA algorithm, it can be obtained a basis adapted to the data that can model the image space.

The optimal ICA model order is estimated based on the performance of a classifier. Where, the input vectors to this classifier are generated in the following way: first, two different input window sizes, 32×32 and 64×64, are considered; hence each sample used to build the ICA model was resized accord to the size of the model. The resizing was done using the bilinear interpolation algorithm provided by the OpenCV library [13]. For each input size were built 11 ICA models providing a given number of features in the range of 10 to 60.

Neural Networks, (NN). The neural classifier was built with a feed-forward multilayer perceptron which has a single hidden layer. The neural network weights were adjusted with a variant of the classical Back-Propagation (BP) algorithm named Resilient Back-Propagation (Rprop) [14]. Rprop is a local adaptive learning scheme performing supervised batch-learning in a multilayer perceptron with faster convergence than the standard BP algorithm.

In this work, the Stuttgart Neural Network Simulator environment SDK [15] was used to generate and train the neural network classifiers. To avoid local minimum during the training process, each setting was repeated four times, changing the initial weights in the net at random. Furthermore, the number of neurons in the hidden layer was allowed to vary between 50 and 650 in steps of 50 selecting the network that provides the highest success rate over the test subset.

Support Vector Machines, (SVM). LibSVM [16] library was used with the radial basis function kernel showed in eq. 2. To find the optimal configuration of the classifier, the best values to use for parameters γ and C were studied, varying them between 0.125 and 4 in a nested loop which doubles the value of the corresponding parameter in each round.

$$K(x_i, x_j) = \exp(-\gamma \|x_i - x_j\|^2, \gamma > 0) \tag{2}$$

4   Results

Due to the nature of the dealt problem was decided to do two experiments. In the first (Experiment 1) all classes shown in Table 1 are considered independently each from others. That is, in Experiment 1, ten classes are considered. While in the second (Experiment 2), *Dactylis glomerata*, *Phalaris minor* and *Lolium rigidum* are grouped in a same class named *Gramineae*; *Avena sativa* and *Avena sterelis* are also grouped in a same class named *Avena*, although, they also belong to the *Gramineae* family. And finally, *Calocedrus decurrens* and *Cypress* are grouped in a same class named *cypress*, although images of those classes appear somewhat different. In that way, only six classes are considered in the Experiment 2.

A double training process was done, on the one hand NN classifiers were trained and, on the other hand, SVM classifiers. After training process were obtained the results shown in Tables 2 and 3, where can be seen the results obtained over the test subsets. To perform the test, the same sets prototypes in a *10-fold cross validation* configuration for the two types of classifiers were used. Therefore, the results are shown using box plots. The total number of prototypes was 1660 with a average distribution in the 10-fold as follow: 167.9 for *trash*, 134.4 for *Cypress*, 174.7 for *Calocedrus decurrens*, 175.1 for *Lolium rigidum*, 76.5 for *Dactylis glomerata*, 141.2 for *Phalaris minor*, 121 for *Avena sativa*, 105.6 for *Avena sterelis*, 298.1 for *Quercus rotundifolia* and 264.9 for *Olea europeae*.

Table 2. Classification results depending on components number obtained for the **Experiment 1**

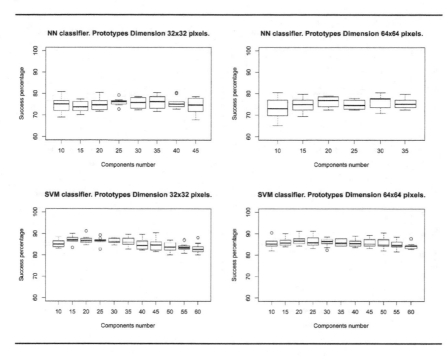

First of all, it has to be said that the training times for the neural nets classifiers were longer than for the SVM classifiers (in average, about 25 minutes for SVM against more than 48 hours for NN). For this reason, after to do a significant number of trainings with NN and to test that the results were worse than with SVM, it was decided not to do more trainings.

In tables 2 and 3 one can see the performance obtained by the neural and SVM classifiers depending on the number of features in the inputs vectors. In that way, the results obtained, choosing the median as a standard to select the best value, are shown in Table 4.

As can be seen in these tables the performance of the SVM classifiers is better than that of the neural classifiers, about 10 percentage points in all cases. This along with the training time do that, for this problem, the election of the SVM classifiers is much more convenient than NN classifiers. Another points to consider are, on the one hand, that the success results do not seem to depend on the prototypes dimension. At least for the two sizes that were considered. Which indicates that there is information enough in the prototypes of less size. And, in the other hand, the grouping of the classes in Experiment 2 does not seem to improve very much the average success results.

Table 3. Classification results depending on components number obtained for the **Experiment 2**

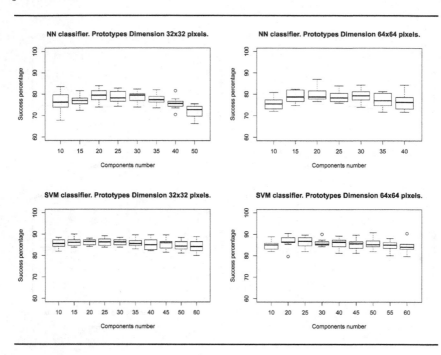

Table 4. Summary of Tables 2 and 3. Maximum value of the median in each case.

	SVM		NN	
Prototypes dim.	Experiment 1	Experiment 2	Experiment 1	Experiment 2
32	87.16→15 comp	86.59→20 comp	76.24→25 comp	79.50→20 comp
64	86.59→20 comp	86.78→20 comp	77.58→30 comp	79.31→30 comp

5 Conclusions

The aim of this study was to test if by mean of an algorithm we were able to count and classify the allergenic airborne pollen in a semi-automatic way. Since, we need the samples already prepared. And from the obtained results, we could say that we have developed an algorithm able to segment well the pollen in the image samples. And the classifier algorithm using SVM carries out about a 90% success. Which, taken into account nature of the problem, is a very promising result. As future work, we have to improve the algorithm to segment the pollen images doing it more adaptable to the samples conditions. And, with the station built, we have also to capture each time more image prototypes of all classes to do more robust the classifier.

Acknowledgments. Presented work in this paper has been funded by *"Research program at the Uni. of Extremadura, 2013. Action VII: P. Int. to Research and Technological Development"* and "Junta de Extremadura" through GR10018, partially by ERDF.

References

1. D'Amato, G., Cecchi, L., Bonini, S., Nunes, C., Annesi-Maesano, I., Behrendt, H., Liccardi, G., Popov, T., Cauwenberge, P.V.: Allergenic pollen and pollen allergy in Europe. Allergy 62, 976–990 (2007)
2. S-otero, M.P.D., Gonzlez, A., Rodrguez-Damin, M., Cernadas, E.: Computer-aided identification of allergenic species of Urticaceae pollen. Grana 43(4), 224–230 (2004)
3. Gottardini, E., Rossi, S., Cristofolini, F., Benedetti, L.: Use of Fourier transform infrared (FT-IR) spectroscopy as a tool for pollen identification. Aerobiologia 23(3), 211–219 (2007)
4. Holt, K., Allen, G., Hodgson, R., Marsland, S., Flenley, J.: Progress towards an automated trainable pollen location and classifier system for use in the palynology laboratory. Review of Palaeobotany and Palynology 167(34), 175–183 (2011)
5. Rodríguez-Damin, M., Cernadas, E., Formella, A., Fernndez-Delgado, M., S-Otero, P.D.: Automatic detection and classification of grains of pollen based on shape and texture. IEEE Transactions on Systems, Man and Cybernetics-Part C: Applications and Reviews 36(4), 531–542 (2006)
6. Hyvärinen, A., Hurri, J., Hoyer, P.O.: Natural Image Statistics. A Probablisctic Approach to Early Computational Vision. Springer-Verlag London Limited (2009)
7. Anjali, P., Ajay, S.: A review on natural image denoising using independent component analysis (ICA) technique. Advances in Computational Research 2(1), 6–14 (2010)
8. Fyfe, R.M., de Beaulieu, J.L., Binney, H., Bradshaw, R.H.W., Brewer, S., Flao, A.L., Finsinger, W., Gaillard, M.J., Giesecke, T., Gil-Romera, G., Grimm, E.C., Huntley, B., Kunes, P., Kuhl, N., Leydet, M., Lotter, A.F., Tarasov, P.E., Tonkov, S.: The European Pollen Database: past efforts and current activities. Vegetation History and Archaeobotany 18(5), 417–424 (2009)
9. Bishop, C.M.: Pattern Recognition and Machine Learning. Springer (2006)
10. Vapnik, V.N.: The Nature of Statistical Learning Theory, 2nd edn. Statistics for Engineering and Information Science. Springer (2000)
11. Ripley, B.: FastICA Algorithms to perform ICA and Projection Pursuit (February 2009), http://cran.r-project.org/web/packages/fastICA/fastICA.pdf
12. Gonzalez, R.C., Woods, R.E.: Digital image processing, 3rd edn. Prentice-Hall, Upper Saddle River (2008)
13. Bradski, G., Kaehler, A.: Learning OpenCV: Computer Vision with the OpenCV Library, 1st edn. O'Reilly Media (2008)
14. Riedmiller, H., Braun, H.: A direct adaptive method for faster backpropagation learning. In: IEEE International Conference on Neural Networks, vol. 1, pp. 586–591 (1993)
15. Zell, A., Mache, N., Huebner, R., Mamier, G., Vogt, M., Schmalzl, M., Herrmann, K.: SNNS (Stuttgart Neural Network Simulator). Neural Network Simulation Environments 254, 165–186 (1994)
16. Chang, C.C., Lin, C.J.: LIBSVM: A library for support vector machines. ACM Transactions on Intelligent Systems and Technology 2(3), 27:1–27:27 (2011)

Enhancing Growth Curve Approach Using CGPANN for Predicting the Sustainability of New Food Products

Jawad Ali, Gul Muhammad Khan, and Sahibzada Ali Mahmud

Department of Electrical Engineering,
University of Engineering and Technology Peshawar, Pakistan 25000
{jawad.ali,gk502,sahibzada.mahmud}@nwfpuet.edu.pk

Abstract. An enhancement to the growth curve approach based on neuro evolution is proposed to develop various forecasting models to investigate the state and worth of the producer, to market a new product. The forecasting model is obtained using a newly introduced neuro evolutionary approach called Cartesian Genetic Programming based ANN (CGPANN). CGPANN helps in obtaining an optimum model for all the necessary parameters of an ANN. An accurate and computationally efficient model is obtained, achieving an accuracy as high as **93.37%** on the time devised terrains, providing a general mechanism for forecasting models in mathematical agreement to its application in econometrics. Comparison with other contemporary model evidences the perfection of the proposed model thus its vital power in developing the growth curve approach for predicting the sustainability of new products.

Keywords: Growth curve approach, Cartesian Genetic Programming (CGP), Artificial Neural Network (ANN), Product sustainability.

1 Introduction

Marketing a product needs knowledge about the anticipation of the end user and forecasts make it possible to respond to these changes. Different types of forecasts are made in order to mend the stability of a firm in the market including: demand forecast, supply forecast, sales forecast, production forecast, profit forecast, advertisement and promotional expense forecast.

Food products are always in the phase of innovation so that the user remains adhered to the products of a specific company. It is the job of a food company to know the exact limits of demand for a new product before its commissioning, machinery planting and management of other capital resources [10]. Trends in the previous products and their variation in different ingredients results a timely development or evolution of new products. Firms having past experience and marketing setup therefore always keep using previous knowledge for establishing a new product, that is a resultant of some alteration in the present or previously developed product(s).

L. Iliadis et al. (Eds.): AIAI 2014, IFIP AICT 436, pp. 286–297, 2014.
© IFIP International Federation for Information Processing 2014

Data from the previous products that correlates the developing products is of great importance while dealing demand forecast. Without proper formulation of the past facts, the theories constituted for product forecast are meaningless [1]. Economists use different traditions strategies and probabilistic models to evaluate the core consumption of a product before its initiation [2,3] . The general concept of economists circle around one point that is the net income from a product which is to be commercialized [9]. So the financial market can be studied completely by its two major aspects: fundamental and technical analysis. The fundamental analysis involves the study of economic factors of the product on a firm, while the technical analysis has to do with the future worth of a product and its sustainability. Joel Dean [4] has explained the achievement of these two goals in a six point formula by exploring evolutionary approach, substitution approach, opinion poll approach, sales experience approach, growth curve approach and vicarious approach.

This paper is using the growth curve approach [4] for the prediction of the sustainability of new products. The uniqueness in the work is that the growth curve of the future sales and stock is used for the prediction of the sustainability of the product by utilizing a neuro evolutionary approach of Cartesian Genetic programming evolved ANN (CGPANN). The paper is organized in the following manner. Section 2 provides an overview of the past work done in the field. It explains factors involved in product design and prior marketing steps. Section 3 gives an insight of the basic strategies followed in the development and commercialization of market products whereas section 4 covers the concept of CGP evolved ANN and their implementation mechanism. The Experiments, Results confirmed by the experiments and concluding remarks are added in later sections of the paper.

2   Literature Review

Expected future market of products and financial time series forecasting has been explored in the past using various methods including econometric, statistical, probabilistic, expert systems and various other ANN based techniques. The most popular econometric model is the logit model [13], based on qualitative analysis of the product. Mixed logit with repeated choices that is an extension of the logit model dealing with the dependencies within each nest is proposed in [16]. The logit model and nested logit [1] model neglects the dependency of the prediction by using independence from irrelevant alternatives property. [3] deals the innovative perspective of new product by measuring the impact of society in choosing the right size of automobile and issues like downsizing of cars etc. [2] deals real world cases, including stock problem and inventory issues, while applying statistical approaches for making estimation about the consequences and likelihood of some future events. Fuzzy logic based production time-lining is developed in [15] for rescheduling of the production in the presence of varying perturbations.

---

[1] http://www.statsoft.com/Textbook/Demand-Forecastin

Neural Networks are always comparable to other prediction models when it comes to time series analysis. Back-propagating Neural Network model is used in [12] for forecasting the demand of sea food material demand. SVM is used for financial time series forecasting in [8]. Financial time series analysis has been covered in [6] using Support Vector Machine (SVM) with modified features. The prediction model in [5] uses neural networks while it takes time series of bankruptcy data, predicting likelihood of bankruptcy. The stock market forecaster design is also obtained using Modular Neural Network in [5]. Neural Networks are employed in stock market for future trading volumes in [9]. This work is different from the past applications of the Neural Networks to the industrial development in the sense that it uses the trading volume and the change in the market value of the production of the firms for predicting the sustainability of the firm. This whole technique is collectively termed as growth curve approach.

3   New Product Development - An Insight

Product design is the core issue for the deployment of new product in the market. New product should be consummate in terms of its function, environment, cost, ergonomics, material, customers need and industrial production feasibility[2]. It should be unique and carry the identity of the producer in case of cash-cow[3] companies. The more relevance of the product to the discussed attributes makes it more user friendly, affordable and attractive. In case, the product that is to be commercialized, is the renovation of some already existing product(s) or product that were developed in the past, the deployment should follow the production and consumption trends of these previous versions of the products. Forecasting the peak consumption era depends on the need of a product, for which the producer must be ready to place the product in the market at the exact time for the exact duration. The key discrepancies can be removed easily if the products are ready in their development.

Investigating the company's ability to introduce a successful product can be done by analyzing the past statistics of the firm [8]. The variation in the production of the company can be studied from the daily market values, variation in the sales with time and the total development to the date of announcing new products. The behavior of the time series is the resultant of the features that constitutes the growth of a product or company. This particular time series is thus used for predicting the near future and ensembles of the product profile.

4   Cartesian Genetic Programming Evolved Artificial Neural Network (CGPANN)

The dynamicity of Artificial Neural Network is enhanced using different evolutionary techniques. These techniques including NEAT [7], TWEANN [17] and

---

[2] http://www.deaconlloyd.com/services/market-feasibility-studies.html
[3] http://www.wikinvest.com/wiki/Cash_cow_business

co-evolutionary method including adaptive and self-adaptive differential evolution systems [18]. The application space of ANN is increased when it is evolved using CGP.

Architectural overview of CGPANN. A typical CGP evolved ANN consists of CGP Nodes connected with each other and to the inputs/output(s). The nodes or neurons thus are used as processing elements. A node can either be in input layer, it can form hidden layer or it can be a part of the output layer. In feed-forward CGPANN, this node takes its input either from the preceding sister nodes or from the system input, depending upon its location in the network. Each node takes a defined number of inputs that is specified in the initialization step for the network evolution. These input are then scaled (defined by Eq. 1) and summed before passing through an activation function.

$$\forall W_\varphi \epsilon PRG([1, -1]) \wedge \varphi = [1, 2, 3, ..., n] \tag{1}$$

φ is the total number of input connections to a node and the weights that are generated by PRG are assigned to these input connections respectively. All the connections are scaled before applying to a random node. Input nodes take their inputs from the system inputs directly without scaling. Intermediate nodes take its input(s) from preceding nodes in case of feed forward ANNs. The activation $f(x)$ is chosen to be log-sigmoid for all nodes in this network, defined by the Eq. 2. Junk nodes have no connectivity hence contribution to the network output.

$$f(x) = \frac{1}{1 + e^{-x}} \tag{2}$$

Here, x is the sum of scaled inputs of the randomly chosen inputs. The x is given by the Eq. 3

$$x = \sum_{i=1}^{n} [I_i] \times [W_i] \tag{3}$$

Where the number of inputs ranges from 1 to n, given by matrix $[I]$ in Eq. 4. The matrix $[W]$ is the collection of respected random weights to the node, given by Eq. 5. Eq. 3 adds the product of all inputs and their respected weights.

$$I = [i_1, i_2, i_3, ..., i_n] \tag{4}$$

$$W = [w_1, w_2, w_3, ..., w_n] \tag{5}$$

Any summing junction in ANN can be represented by

$$y' = \sum_{i=1}^{N} x_i \tag{6}$$

where x_i is the input to the junction. If the same input is scaled with a randomly assigned weight w_i then Eq. 6 takes the form

$$y' = \sum_{i=1}^{N} x_i w_i \tag{7}$$

Let for N inputs to a node, we have y_j output such that

$$y_j = f^j(y'_j) = f^j\left(\sum_{i=1}^{N} x_i w_i\right) \tag{8}$$

Here f is an activation function, particular to node j. If total number of nodes in the network are N_T then j is defined by

$$\{j|j\epsilon N \wedge 1 \leq j \leq N_T\} \tag{9}$$

Let I be the input set to a unique genotype network G_k, consisting unique entries i such that, for all inputs

$$i\epsilon R \wedge 0 \leq i \leq 1 \tag{10}$$

in

$$I = \{i_1, i_2, i_3, ...i_n\}$$

so, network G_k is the a set of random selection from inputs $[I]$, outputs of nodes y_j, for a single output O_p such that

$$O_p = \frac{1}{n}\sum_{i=1}^{N}\left(f\left(\sum(y_j W_j + y_{j-1} W_{j-1} + ... + y_1 W_1 + IW_k)\right)\right) \tag{11}$$

here, $W_j, W_{j-1}, ... W_1$ are Random subsets of W such that W_k is a subset W_j and

$$W_j = \{w_k|w_k\epsilon R \wedge -1 \leq w_k \leq 1\} \tag{12}$$

Now

$$G_k = \{I, y_j, y_{j-1}, ..., y_1, O_p\} \tag{13}$$

$$y_j = \left(f\left(\sum(y_{j-1} W_{j-1} + y_{j-2} W_{j-2} + ... + y_1 W_1 + IW)\right)\right)$$
$$y_{j-1} = \left(f\left(\sum(y_{j-2} W_{j-2} + y_{j-3} W_{j-3} + ... + y_1 W_1 + IW)\right)\right)$$
$$\vdots$$
$$y_2 = \left(f\left(\sum(y_1 W_1 + IW)\right)\right)$$
$$y_1 = \left(f\left(\sum IW\right)\right)$$

$$\tag{14}$$

Let G_k and G_l be the two successive genotypes. G_l is produced from G_k by mutating $\mu\%$ weights of connections in G_k, nodal connectivity, functions defining each node or the combination of these.

Let the total entries in G_k, including weights, connections and functions be N_{cwf} then

$$N'_{cwf} = \mu \times N_{cwf} \tag{15}$$

N'_{cwf} represents genotype entries that are to be randomized to get G_l.

Let r be a unique entry to the set ζ that contains value to be changed to get mutated G_k or G_l defined by r.

$$\{r | r\epsilon\zeta \forall (W, I, y_j)\} \tag{16}$$

each value in r is defined by a Pseudo Random Generator (PRG) that takes N'_{cwf} values from available N_{cwf} entries using the relation

$$\left\{ r_i | r_i\epsilon\{1, 2, 3, ..., N_{cwf}\} \wedge r_i \subset \{1, 2, 3, ..., N_{cwf}\} \right\}, i = 1, 2, 3, ...N'_{cwf} \tag{17}$$

where r_i belongs to $\{1, 2, 3, ...N_{cwf}\}$ so each entry in ζ is

$$\Gamma(i) = \left\{ r_i | r_i\epsilon\{1, 2, 3, ...N_{cwf}\} \wedge \{1 \le r_i \le N_{cwf}\} \right\} \tag{18}$$

The value that is replaced in G_k for N'_{cwf} values uses Pseudo Random Number Generator (PRG). Here, each input I_i is the output of some proceeding node or the network input given by the Eq. 19.

$$I(G, L, N) = PRG([O(G, L_p, N_p)] : [I(G, L_0)]) \tag{19}$$

G represent a specific genotype, L is the Layer in which the node N is situated. L_p and N_p are representing the preceding Layers and Nodes respectively whereas L_0 represent the input matrix to the Genotype G.

The mutation rate is kept 10% for its better results in [11]. In each generation, connections are altered and weight are varied according to the relevance of the network output with the desired values that are also given to the network for its training. Junk nodes that contribute none to the network and having no influence on the network performance are exscinded using CGP in an efficient way. Thus the final phenotype might have node number less than the number of nodes that are initialized at the start of evolutionary phase. The data is taken for three different dairy product producing companies i.e. Nestle (NSRGY)[4], Dean Food Company (DF)[5] and Saputo (SAPIF).

The growth curve estimation model works in two major steps. The first step involves the training of the model. The daily stock market values of NSRGY for the years 1998-99 are used to train the particular CGPANN model. The model develops an intrinsic intuition for predicting the next half day stock value while taking 7 days data as its input sequence. **10** different networks, on the basis

---

[4] http://money.rediff.com/companies/Nestle-India-Ltd/11120007/nse/
   day/chart
[5] https://finance.yahoo.com/q?s=DF

of the number of their nodes, were initialized for this purpose. The number of nodes were varied between 50 and 500 with a step size of 50. Inputs are taken from the provided stock value time series to the network where the CGPANN calculates the output $V_{forecasted}$. This output is compared with the given desired value V_{actual} and the generation proceeds to minimize this difference between the output from the network and the desired output. The evolution stops after reaching to the maximum number of generation, initialized at the start of this training session or if the desired fitness is achieved. The fitness of 10 mutants is measured at the end of each generation using Eq. 20. This training phase took 20 hours when performed with 3.4GHz corei7-2600 processing unit with 16GB RAM and 8MB L3 Cache.

$$Fitness = 100\% - MAPE \tag{20}$$

Where MAPE is the mean absolute percentage error that is calculated as

$$MAPE = \left| \frac{V_{Actual} - V_{forecasted}}{V_{Actual}} \right| \times 100\% \tag{21}$$

After selecting the best mutants among the evolved genotypes, these networks are evaluated with the time series data of normalized stock values, change in stock and total sales to data. Sliding window mechanism is used for the prediction of whole sequence. The model takes **14** inputs and gives forecast value for the nested inputs.

5   Experiment Results

Selecting Network: MAPE values of the networks for the years 1998-1999 given for each network in Table 1. Table 1 is the numerical evidence of the fact that although all networks are giving a minimal MAPE due to their better adaptability features while each network is evolved for its 1M generations, the network that is initialized with 450 nodes outperforms over other networks. Although the error increases when the forecasting horizon is elevated yet the model is useful as this MAPE value is competent to other models quoted in the literature. Table 2 contains MAPE results for the year 2002 for different companies with various data sets. The model takes 14 instant inputs and gives single output. Results

Table 1. MAPE values for training session - 14 instants input

Nodes	Single output	7 outputs	14 outputs	Nodes	Single output	7 outputs	14 outputs
50	1.8016	2.5912	2.681	300	1.8045	2.6114	2.694
100	1.8046	2.6059	2.692	350	1.8109	2.6222	2.705
150	1.8107	2.6000	2.687	400	1.8040	2.6228	2.703
200	1.8028	2.5981	2.703	450	**1.7992**	2.6041	2.694
250	1.8072	2.5929	2.687	500	1.8198	2.6018	2.713

Table 2. MAPE results for Stock value, Change in stock and Cumulative stock with network inputs ranging from 50 to 500 with step size of 50

Nodes	50	100	150	200	250	300	350	400	450	500
SAPIF(S)	6.087	5.372	5.588	5.478	5.523	5.685	6.105	5.788	**5.365**	5.730
SAPIF(CH)	12.806	11.101	11.810	11.441	11.535	11.994	13.094	11.970	**11.042**	12.08
SAPIF(CS)	9.334	8.104	8.519	8.261	8.333	8.666	9.386	8.843	**8.089**	8.750
DF(S)	7.624	6.633	7.000	6.806	6.864	7.117	7.709	7.186	**6.624**	7.178
DF(CH)	11.603	10.068	10.701	10.370	10.456	10.869	11.772	10.877	**10.032**	10.96
DF(CS)	9.310	8.069	8.514	8.254	8.327	8.661	9.38	8.802	**8.064**	8.746
NSRGY(S)	7.861	6.875	7.215	7.023	7.084	7.342	7.898	7.459	**6.867**	7.417
NSRGY(CH)	6.663	6.4366	6.528	6.462	6.449	6.542	6.554	6.564	**6.382**	6.521
NSRGY(CS)	8.949	7.7498	8.182	7.934	8.003	8.323	9.010	8.451	**7.747**	8.402

are tabulated with respect to the initialized nodes for network.
here: S=Stock value, CH=Change in stock and CS=Cumulative stock

The Standard deviation and Peak to Average ratio (P/A) values in Table 3 describe the distribution of the data on the given time line. It is the beauty of the model to predict up to the excellent level of accuracy as given in Table 2. It is evident from the Table 2 that different data sets (of exclusive data trends) gives best results on network with 450 nodes. The best results obtained in all the models for the year 2002 for different data trends are given by in Table 4. The comparison of the proposed CGP evolved ANN with different other forecaster models is performed in Table 5.

Table 3. Standard deviation and Peak to average ration of the data sets

Attribute	DF(S)	DF(CH)	NSRGY(S)	NSRGY(CH)	SAPIF(S)	SAPIF(CH)
P/A	1.766	1.282	1.799	1.825	2.008	1.226
SD	0.234	0.035	0.225	0.0775	0.206	0.0440

Table 4. MAPE results for different data sets - 450 initial nodes - 14 instances input - Single output

Time series	Company		
	SAPIF(%)	DF(%)	NSRGY(%)
Stock Value	5.365	6.624	6.867
Change in Stock	11.042	10.032	6.382
Total Sales to date	8.089	8.064	7.747

294   J. Ali, G.M. Khan, and S.A. Mahmud

Table 5. Comparison of proposed CGPANN model with different models[6,14]

Model(s)	Testing Accuracy (%age)	Model(s)	Testing Accuracy (%age)
MDA	79.13	RBF SVM	83.06
Logit	78.30	Polynomial SVM(1)	82.01
BPNN(1)	81.48	Polynomial SVM(2)	77.24
BPNN(2)	71.16	Sigmoid SVM	71.16
Linear SVM	77.24	**CGPANN**	**93.37**

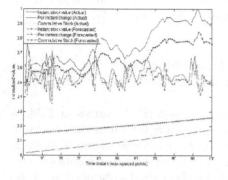

Fig. 1. Nestle - Instantaneous stock value, per instant change and cumulative stock prediction (14 inputs(7 days) 1 output)

Fig. 2. Dean - Instantaneous stock value, per instant change and cumulative stock prediction (14 inputs(7 days) 1 output)

It can be inferred from Table 5 that CGPANN is a better techniques that can be used to evolve Artificial Neural Network for human biased certainty reluctant time series such as stock value and productions due to its self-adaptability and fast learning features.CGPANN, in comparison with other prediction models, performs better for its non-linear characteristics and non-conventional behavior. The non-conventional behavior of CGPANN increases its MAPE to 93% that is 22.21% in addition to the MAPE value of Sigmoid SVM, a non-linear model, conventional model. On the other hand, the Linear SVM that follows linear approach and possesses conventional behavior, gives 6.08% more effective MAPE values for a given timely devised data.

Fig. 1, 2 and 3 graphically explain the accuracy of the model by plotting the estimated stock values, instantaneous changes and the cumulative stocks of each company. The Fig. 1, 2 and 3 explain progress of each producer as well as the change in their market stock time by time. It can be inferred from the estimated and the actual values of the entities that the proposed forecaster has learned the pattern that is experienced by each producer. The time series thus obtain is authenticated one and can be used for econometric analysis.

Cumulative stock or the growth curve defines the exact position of the firm after each and every time instant which has been made possible by using the newly developed CGPANN. Basic fact about SNRGY food producing company can be explained from its cumulative stock value. Based on cumulative stock, SNRGY is leading among its competitors, shown in Fig. 4. The variations in the market stock value (defined by peak to average ratio and standard deviation) is also compared in Fig. 4. More fluctuation in the time series of the per-instant change of the company stock, as given by Table 3, reflects the *dynamicity* in the products of the company.

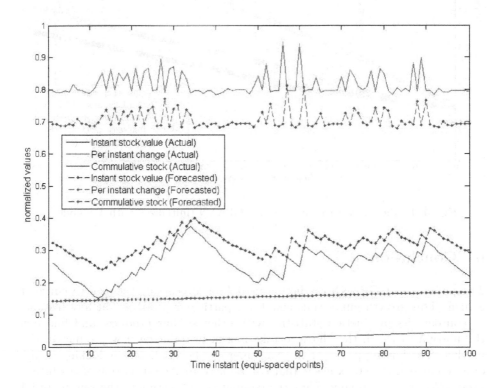

Fig. 3. Saputo - Instantaneous stock value, per instant change and cumulative stock prediction (14 inputs(7 days) 1 output)

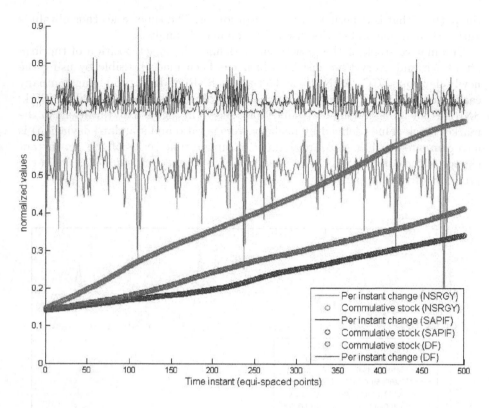

Fig. 4. Predicted growth curve using CGPANN (initialized with 450 nodes)

6   Conclusion

The paper uses CGPANN prediction model for forecasting the growth curve of
a firm. This growth curve thus enables a particular producer, at any instant,
to compare its production statistics with other similar producer and find out
the margin to which they have to take risk in introducing new product in the
market. The buying power of the consumer can be predicted by looking at the
daily change in the stock usage that can be used in the estimation of the future
production by a company. Thus the companies can estimate the raw material
and manage their stock according to the future buying capacity of the present
users. The model further enables newly developed firms to calculate their increment in market reputation while looking at the daily change in stock value. The
model can be much effective if the consumption time series of a single product
is employed for its future use. The producers can then manage their inventory
while knowing the exact number of future consumer of each and every single
product, supplied by them.

References

1. Ben-Akiva, M., Bolduc, D.: Multinomial probit with a logit kernel and a general parametetric specipication of the covariance structure. Working paper, Department of Economics. MIT (1996)
2. Brownstonel, D., Train, K.: Forecasting new product penetration with flexible substitution patterns. Journal of Econometrics, 109–129
3. Cardell, N., Dunbar, F.: Measuring the societal impacts of automobile downsizing. Transportation Research 14(5,6), 423–434
4. Dean, J.: Demand forecasting for a new product, http://www.entranceguruji.in/read_matirial.php
5. Fletcher, D., Goss, E.: Forecasting with neural networks and application using bankruptcy data. Information and Management 24, 159–167 (1993)
6. Francis, E.H., Cao, L.: Modified support vector machines in financial time series forecasting. Neurocomputing 48, 847–861 (2002)
7. Hasanat, A.: Object class recognition using neat-evolved artificial neural network. In: Fifth International Conference on Computer Graphics, Imaging and Visualization, CGIV, pp. 271–275 (2008)
8. Cao, J., Tay, L., Support, H.: vector machine with adaptive parameters in financial time series forecasting. IEEE Transactions on Neural Networks 14(6), 1506–1518 (2003)
9. Kaastra, I., Milton, S.: Forecasting futures trading volume using neural networks. J. Futures Markets 15, 853–970 (1995)
10. Kempf, K.G., Keskinocak, P.: Uzsoy: Planning production and inventories in the extended enterprise. International Series in Operations Research and Management Science 152(2), 588–589 (2011)
11. Khan, G.M., Khan, S., Ullah, F.: Short-term daily peak load forecasting using fast learning neural network. In: Intelligent Systems Design and Applications (ISDA), pp. 843–848 (2011)
12. Lo, C.-Y.: Back propagation neural network on the forecasting system of sea food material demand. In: Zhou, M., Tan, H. (eds.) CSE 2011, Part II. CCIS, vol. 202, pp. 147–154. Springer, Heidelberg (2011)
13. McFadden, D.: Conditional logit analysis of qualitative choice behavior. Frontiers in econometrics. Academic Press, New York (1973)
14. Min, J.H., Lee, Y.C.: Bankruptcy prediction using support vector machinewith optimal choice of kernel function parameters 48, 847–861 (2002)
15. Petrovic, D., Duenas, A.: A fuzzy logic based production scheduling/rescheduling in the presence of uncertain disruptions. Fuzzy Sets and Systems 157(16), 2273–2285 (2006)
16. Revelt, D., Train, K.: Mixed logit with repeated choices: Households choices of appliance effciency level. Review of Economics and Statistics 80(4) (1998)
17. Yao, X.: Evolving artificial neural networks. Proceedings of the IEEE 87(9), 1423–1447 (1999)
18. Yao, X., Islam, M.M.: Evolving artificial neural network ensembles. IEEE Computational Intelligence Magazine 3(1), 31–42 (2008)

DX-IFD: An Intelligent Force Deployment System

Junpeng Bao[1], Yuepeng Zhang[2], Wenqing Wang[1], Jun Zeng[2], De Zhang[2], Ruiyu Yuan[1]

[1] Department of Computer Science & Technology, Xi'an Jiaotong University,
Xi'an 710049, P.R. China
baojp@mail.xjtu.edu.cn
[2] Xi'an Institute of Surveying & Mapping, Xi'an 710000, P.R. China

Abstract. It is easy to gather a huge amount of information from a battlefield by satellites, manned or unmanned aerial vehicles. But it becomes an important issue to effectively deal with the tremendous information and create reasonable solutions based on that. This paper introduces an Intelligent Force Deployment system DX-IFD, which combines topography heuristic information and military principles with the Genetic Algorithm to automatically identify the battlefield features and make the optimized force deployment in terms of the topography, the enemy status and our mission. A Relative Position Code method is presented to encode the deployment, which ensures the optimized formation will be kept after genetic operations. According to basic military principles, we propose a force deployment assessment scheme which evaluates a deployment solution from three aspects: unit fitness, formation fitness and relationship fitness. The DX-IFD system can quickly response to the change of battlefield situation, evaluate and create the optimized deployment in different configuration conditions. It greatly improves the force ability to quickly react and occupy the dominant positions in a battle.

Keywords: Intelligent Force Deployment, Genetic Algorithm, Topography Analysis.

1 Introduction

The modern information technology and computer science have made a deep impact on military. A variety of automatic weapons and systems are changing the way of war. A huge amount of information, which is gathered from battlefields, enemies and ourselves, has to be properly treated on time, otherwise any opportunity is fleeting. In contrast with computer, mankind is good at fuzzy decision making based on knowledge rather than precise computation. It is obviously impossible for a person to manually deal with a large scale of various data in minutes. However, artificial intelligence applications can help us to quickly process big data, analyze miscellaneous situations and find appropriate candidate solutions. Hence, a commander will decide promptly on correct actions with substantial rational power rather than temerity.

This paper introduces an Intelligent Force Deployment system DX-IFD, which aims to automatically identify the battlefield features and make the optimized force deployment in terms of the topography, enemy status and our mission. This system can quickly response to the change of battlefield situation, evaluate and create the optimized deployment in different configuration conditions. As a result, DX-IFD improves the force ability to quickly react and occupy the dominant positions in a battle.

L. Iliadis et al. (Eds.): AIAI 2014, IFIP AICT 436, pp. 298–306, 2014.

2   Related Work

Since 1980s, United States Department of Defense has been studying the application of expert system in the military. Franklin et al.[1] introduced some applications for the military expert system and pointed out the key problems existed in the representation and reasoning about the uncertain knowledge and machine learning. In 2012, Haberlin et al.[2] presented a battle management system based on event simulation and optimization for military decision support. The Turkish army [3,4] researched the simulation system for military deployment and military logistics transport deployment optimization. The system exploited the probability statistics and regression tree to analyze the principal influence factors. Kettani et al.[5] in Canadian Department of National Defense proposed a spatial reasoning method to analyze battlefield situation and support military decision.

White et al.[6] presented a hybrid heuristic/evolutionary algorithm for automatically generating spatial deployment plans that minimize power consumption. Aleti et al.[7] put forward Heuristic for Component Deployment Optimization (BHCDO), which constructs solutions based on a Bayesian learning mechanism to improve quality of new deployment architectures. Lian et al.[8] proposed a method for 3D deployment optimization of sensor network based on an improved Particle Swarm Optimization (PSO) algorithm. Guan et al.[9] also proposed a PSO algorithm to minimize the average outage area rate of the LED deployment scheme in VLC systems.

Chinese researchers studied more on air defense force deployment issues than the land force deployment. The stochastic optimization method is a very popular way. Wang[10] and Xiong et al.[11] separately introduced the PSO based algorithm for the optimized hybrid deployment of air defense troops. Kong et al.[12] presented a grey comprehensive correlation analysis method and a double bee population evolutionary genetic algorithm for air defense force position optimization problem. Chen [13] combined a Genetic Algorithm with a local neighborhood search strategy for the firepower unit deployment issue.

3   The DX-IFD System

3.1   System Architecture

The DX-IFD system is an intelligent force deployment system based on the topography analysis. It exploits empirical military knowledge to automatically and intelligently deal with the topography information and create optimized force deployment solutions according to the military requirements.

The figure 1 shows the DX-IFD system main workflow. The system is composed of five steps. First, it collects battlefield picture photos by satellites or unmanned aerial vehicles. Second, it analyzes the features in these photos to identify objects, such as plain, hill, river, residential area etc. Third, it determines the fitness value for different types of troops in an area, such as armored troops, artillery troops, and infantry troops etc. At this step, the system employs a rule based uncertainty reasoning scheme to take both military knowledge and geographic into account. Fourth, it makes a digital grey map for each troop

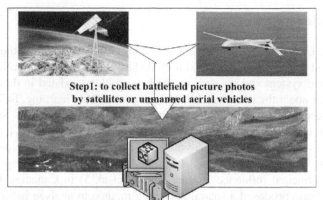

Step1: to collect battlefield picture photos
by satellites or unmanned aerial vehicles

Step2: to analyze the features in the photos to identify objects

Step3: to determine the fitness value for different types of
troops in an area

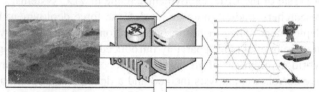

Step4: to make a digital grey map for each troop type to
illustrate its fitness at anywhere of the battlefield

Step5: to create the optimized force deployment solutions based
on the color fitness maps according to military requirements

Fig. 1. The DX-IFD system main workflow

type to illustrate its fitness at anywhere of the battlefield, and then grey maps are combined into color maps according to a special order. Finally, it creates the optimized force deployment solutions based on the color fitness maps according to military requirements.

This paper concentrates on the last step, especially on the GA based intelligent force deployment method and the force deployment assessment scheme.

3.2   The Optimization Goal

The force deployment task is to assign blended troop units into most appropriate areas in the given geographic space and satisfy the given military requirements. The troop units are of various types, including headquarters, armored units, artillery units, and infantry units etc. Apparently, the force deployment task is a Constraint Satisfaction Problem in theory, which is a typical Non-deterministic Polynomial (NP) problem. We abstract the problem as follows.

1. The troop units of various types are represented by different sizes of squares that located somewhere in the battlefield map.
2. The fitness of a troop unit in an area is measured by the Relative Average Grey (RAG) value of the unit in the square.
3. The spatial location relationship among all of the deployed units should conform to a certain formation.
4. The spatial relationship between our deployment and the enemy has to satisfy some constraints.

Thus, the optimization goal of the force deployment is to locate k squares on the given color map so as to maximize the entire assessment value on the whole. Namely, it can be described in the following formulas.

$$Max\{F(U)\}, \quad U = [u_1, u_2, \cdots, u_k],$$
$$s.t. \quad U \subset G$$
$$s.t. \quad T(u_i) \in \{Headquarters, Armored, Infantry, \cdots\}, \quad i = \{1, 2, \cdots, k\}$$
$$s.t. \quad S(U) \rightarrow D$$
$$s.t. \quad R(U, E) \rightarrow C$$

Where $u_i (i=1,\ldots,k)$ denotes a troop unit, U denotes the whole deployment, $F(U)$ means the entire fitness of the whole deployment, G denotes the given battlefield color map. $T(u_i)$ means the troop type of the u_i. $S(U)$ means the spatial location relationship of the U, $R(U,E)$ means the spatial relationship between our deployment U and the enemy E.

3.3   Intelligent Deployment Algorithm

Genetic Algorithm (GA) is a popular method to effectively solve optimization problem. It focuses on the representation and assessment of solutions to the problem, rather than precisely mathematical analysis process. The Intelligent deployment algorithm indeed exploits the GA to make the optimized force deployment solutions. It releases us from plenty of complicated uncertain deduction of deployment transformation. The key points of the Intelligent deployment algorithm are the representation and the assessment of force deployment solution.

3.3.1   The Representation of Force Deployment Solution

A force deployment solution is represented by a k-dimension vector, called deployment vector, in which each dimension is responsible for a specific troop unit, and the type of a unit corresponds to its position in the vector. The construction of the deployment vector is based on prescient military knowledge that is stored in a database. Naturally, it depends on specific military requirements.

In theory, a unit area consists of 2-dimensional space coordinates and the radius. But for a given level of the troop unit, the area size is fixed, which can be obtained from the database. In order to reduce the length of the vector and improve the computational efficiency, 2-dimensional space coordinates is only used to represent a unit.

In Genetic Algorithm, a deployment vector (i.e. individual) is encoded into a sequence of code, i.e. a genome. A directly coding strategy takes a troop unit's absolute position as a gene code, which corresponds to the unit's Cartesian coordinates in the battlefield map. But this strategy would destroy deployment formation and is not conducive to a better formation. For example, there are two deployment solutions A and B. A is on the left of the battlefield while B is on the right, then they exchange half of the vector respectively and generate another two new solutions whose part of the units are on the left and the others on the right. Obviously, it is not a valid formation because a deployment cannot be scattered across the whole battlefield.

We propose a Relative Position Code (RPC) method, which takes the relative coordinates, i.e. polar coordinates, to code a genome. A genome consists of two parts: polar coordinates of the headquarters and the other deployment units' polar coordinates, as shown in the figure 2. The headquarters polar coordinates are relative to the enemy position, but the other units' polar coordinates are relative to the headquarters.

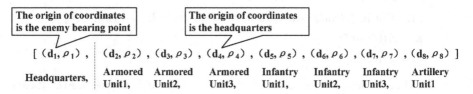

Fig. 2. A deployment vector and Relative Position Code

Formation is actually the spatial location relationship among each unit. It is a very important constraint of force deployment. Because units have to keep reasonable distance and effective connection from each other. A wrong formation causes deadly chaos. The RPC method accurately keeps this relationship so as to avoid the formation damaged by absolute position coding. No matter how to cross and exchange the old genome sequences, the new created genome sequences are valid because each gene is still relative to the origin of polar coordinate, i.e. the headquarters, in the new genome. The RPC method ensures that the optimized formation will be kept after crossing operation.

3.3.2   The Assessment of Force Deployment Solution

We present a 10-point score scheme to assess an individual and conform to generally grading habits. A force deployment solution U is assessed from three aspects: unit fitness $A(U)$, formation fitness $D(U)$ and relationship fitness $R(U)$, i.e.

$$F(U) = w_0 A(U) + w_1 D(U) + w_2 R(U) \tag{1}$$

Where the range of each fitness value is $[0,1]$ and the weights (w_0, w_1, w_2) are adjustable parameters, such as $w_0=4.5$, $w_1=2.5$, $w_2=3.0$.

The unit fitness evaluates how much each deployment unit suits for specific topology. Namely, is a unit deployed to a suitable topography? In practice, it is often not the best solution to put every unit on its best position. When the whole solution is the best, many parts of the solution, i.e. units, may not locate at the best position. A reasonable deployment solution not only makes a tradeoff between all units but also considers the overall formation and the relationship with enemy. It is a complicated multi-parameters optimization problem to find a reasonable deployment solution.

The formation fitness evaluates whether the overall formation of a deployment is reasonable. It refers to the space geometry relation among the deployment units. There is a basic principle, i.e. "Act according to circumstances", the optimized deployment solution must face to the specific topography. Moreover, it must consider the spatial relationship among all units, satisfy certain geometric constraints and follow the basic military principles.

The relationship fitness assesses whether a deployment solution is effective against the enemy according to enemy information. Our deployment should be revised when the enemy position changes. The most basic requirement is that our offensive firepower should be toward the enemy and do not put our back to the enemy. Additionally, we should keep a proper distance with the enemy, neither too close nor too far.

3.3.3   The Unit Fitness

The unit fitness of a deployment solution is defined as follows.

$$A(U) = \frac{1}{|U|} \sum_{u_i \in U} \left(\frac{1}{255 \times |u_i|} \sum_{p \in u_i} Grey(p) \right) \tag{2}$$

Where $|U|$ denotes the number of units in an individual U. $|u_i|$ denotes the area of the unit u_i. $Grey(p)$ is the actual fitness of a unit type at the point p, i.e. how much the point point p is suitable for the unit u_i, which is in the range of $[0,255]$.

The $Grey(p)$ value is read from the fitness grey map of the specific type of unit. For a given battlefield, the grey map depends on the type of deployment unit. For example, an armored unit's grey map is obviously different from an infantry unit on the same field. Because an area where is suitable for infantry, may be adverse to the armored, such as hillside. These fitness grey maps are made at the step 4 of the DX-IFD system main workflow, as shown in the figure 1.

In fact, a pixel grey value needs only 1 byte. But for a colorful RGB image, a pixel has three bytes. Therefore, three grey maps of different unit types can be compounded in a colorful RGB image. It is convenient for data management.

3.3.4   The Formation Fitness

The formation of a deployment solution should try to meet some rules that describe relatively fixed geometric relationship. The position of the headquarters is defined as the formation's origin. Thus, a formation is described by a serial of unit's polar coordinates that is relative to the headquarters.

The formation fitness of a deployment solution is measured by the deviation between the polar coordinate of the unit and the ideal polar coordinate, i.e.

$$D(U) = \frac{1}{|U|-1} \sum_{i=2}^{|U|-1} \left[0.5\exp\left(-\frac{(d_i - ed_i)^2}{\sigma_{di}^2} \right) + 0.5\exp\left(-\frac{(r_i - er_i)^2}{\sigma_{ri}^2} \right) \right] \quad (3)$$

Where d_i and ed_i respectively denote the straight-line distance and ideal distance between the i-th unit and the headquarters. r_i and er_i respectively denote the angle and ideal angle between this unit direction and horizontal direction. σ_{di} and σ_{ri} respectively denote the standard deviation of ideal distance and ideal angle. It is defined that first dimension in a deployment vector corresponds to the headquarters.

3.3.5   The Relationship Fitness

The figure 3 illustrates relations between our deployment and the enemy. The enemy bearing point (the o point) is regarded as the origin of polar coordinate. The distance between o and the deployment center (the c point), denoted as d_c, should be as close as possible to an ideal distance d_e. The c point is better on the way to the front of enemy. Thus, the direction angle from o to c, denoted as r_c, should be opposite to the enemy direction angle r_e. The head unit (the h point) is expected to point to the enemy. So the head unit angle r_h should equal r_c, the distance between o and h, denoted as d_h, should less than d_c. Considering these factors comprehensively, the relationship fitness is defined as follows.

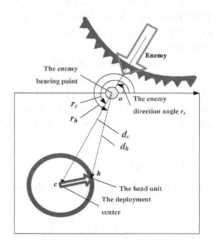

Fig. 3. The spatial relations between the deployment and enemy

$$R(U) = a\exp\left(-\frac{(r_h - r_c)^2}{\sigma_h^2} \right) \times \mu(d_c - d_h) + b\exp\left(-\frac{(r_c - r_e - 180)^2}{\sigma_r^2} \right)$$
$$+ c\exp\left(-\frac{(d_c - d_e)^2}{\sigma_d^2} \right) \quad (4)$$

Where σ_h, σ_r and σ_d are standard deviation of head unit angle, deployment center angle and deployment center distance respectively. Angles are measured in degree. μ is the sign whether the deployment direction is facing to or back to the enemy, it is defined as follows.

$$\mu(d_c - d_h) = \begin{cases} 1, & d_c > d_h \\ -1, & d_c < d_h \\ 0, & otherwise \end{cases} \qquad (5)$$

4   Experimental Results

The figure 4 (a) shows a randomly created deployment solution whose final assessment value is 2.9187. Obviously, its formation is so poor that it cannot effectively attack enemy at all. In the figure, the blue arrow marks the enemy bearing point and the direction. Each deployment unit is marked by a red rectangle with a unit type label, such as "H" for headquarters, "AR" for armored unit, "INF" for infantry unit and "AA" for anti-aircraft artillery unit.

The figure 4 (b) shows a good deployment solution recommended by the DX-IFD system algorithm, of which the assessment value is 7.7757. This deployment is reasonable and the spatial relationships between each unit are in line with military principles. Moreover, it can effectively attack the enemy.

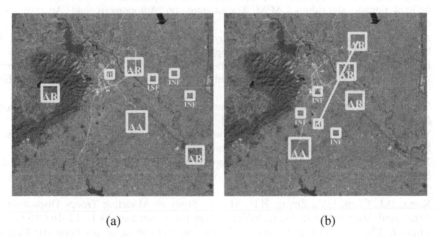

(a)   (b)

Fig. 4. (a) A worse deployment solution ($F(U)$=2.9187). (b) A good deployment solution recommended by the DX-IFD system algorithm ($F(U)$=7.7757).

5   Conclusions

Intelligent force deployment is a complicated multi-parameter optimization process. The DX-IFD system combines topography heuristic information and military principles with the Genetic Algorithm to automatically deal with the topography information and create optimized force deployment solutions according to the military requirements. The RPC method is presented to encode the deployment, which ensures the optimized formation will be kept after genetic operations. According to basic military principles, we propose a force deployment assessment scheme which evaluates a deployment solution from three aspects: unit fitness, formation fitness and relationship fitness. The test result shows that the system can automatically generate the optimized deployment solution for the given topography and enemy information. However, some system parameters need to be further refined.

References

1. Franklin, J.E., et al.: Expert System Technology for the Military: Selected Samples. Proceedings of the IEEE 76(10), 1327–1366 (1988)
2. Haberlin, R., Brodsky, A.: Battle Management System (BMS): An Optimization for Military Decision Makers. In: Proceedings of the 2012 IEEE ICDE, pp. 78–81 (2012)
3. Yıldırım, U.Z., Balcıoğlu, A.: A Design of Experiments Approach to Military Deployment Planning Problem. In: Proceedings of the 2008 Winter Simulation Conference, pp. 1234–1241 (2008)
4. Akgün, İ., Tansel, B.: Optimization of Transportation Requirements in the Deployment of Military Units. Computers and Operations Research 34(4), 1158–1176 (2007)
5. Kettani, D., Roy, J.: A Qualitative Spatial Model For Information Fusion and Situation Analysis. In: Proceedings of the ISIF 2000, pp. 16–23 (2000)
6. White, J., et al.: ScatterD: spatial deployment optimization with hybrid heuristic/evolutionary algorithms. ACM Transactions on Autonomous and Adaptive Systems 6(3), 18–43 (2011)
7. Aleti, A., Meedeniya, I.: Component deployment optimization with Bayesian learning. In: Proceedings of the 14th ACM SIGSOFT, pp. 11–20 (2011)
8. Li, X.Y., et al.: Three-dimensional deployment optimization of sensor network based on an improved Particle Swarm Optimization algorithm. In: Proceedings of the 10th IEEE WCICA, pp. 4395–4400 (2012)
9. Guan, R., et al.: PSO-based LED deployment optimization for visible light communications. In: Proceedings of 2013 IEEE WCSP, pp. 1–6 (2013)
10. Wang, Y.B., Wang, J.J., Wang, Y.L.: Study of Hybrid Force Disposition Based on Particle Swarm Optimization. Tactical Missile Technology (3), 44–47 (2009)
11. Xiong, Z., Xu, R.M., Tian, F.: Improved Particle Swarm Optimization algorithm applied to hybrid force disposition optimization. Computer Engineering and Applications 48(6), 235–237 (2012)
12. Kong, D.J., Chen, L.Y., Zhang, H.F., Ma, Y.: Study on Modeling Troops Disposition of Army Anti-Aircraft Missile Unit. Ordnance Industry Automation 30(1), 12–16 (2011)
13. Chen, J., Chen, C., Zhang, J., Xin, B.: Deployment Optimization for Point Air Defense Based on Memetic Algorithm. ACTA Automatica Sinica 36(2), 242–248 (2010)

A Practical Application of Evolving Fuzzy-Rule-Based Classifiers for the Development of Spoken Dialog Systems*

David Griol, José Antonio Iglesias, Agapito Ledezma, and Araceli Sanchis

Computer Science Department
Carlos III University of Madrid
Leganés, Spain
{dgriol,jiglesia,ledezma,masm}@inf.uc3m.es

Abstract. In this paper we present a statistical approach based on evolving Fuzzy-rule-based (FRB) classifiers for the development of dialog managers for spoken dialog systems. The dialog managers developed by means of our proposal select the next system action by considering a set of dynamic fuzzy rules that are automatically obtained by means of the application of the FRB classification process. Our approach has the main advantage of taking into account the data supplied by the user throughout the complete dialog history without causing scalability problems, also considering confidence measures provided by the recognition and understanding modules. The use of EFS also allows to process streaming data on-line in real time, thus dynamically evolving the structure and operation of the dialog model based on the interaction of the dialog system with its users. We also describe the application of our proposal to develop a dialog system providing railway information.

Keywords: Spoken Dialog Systems, Evolving Classifiers, Spoken Human-Machine Interaction, Statistical Methodologies, Dialog Management.

1 Introduction

Spoken Dialog Systems (SDSs) are computer programs that receive speech as input and generate synthesized speech as output, engaging the user in a dialog that aims to be similar to that between humans [1]. These interfaces make technologies more usable, as they ease interaction, allow integration in different environments, and make technologies more accessible, especially for disabled and elderly people [2].

Usually, SDSs carry out five main tasks [1,3,4]: Automatic Speech Recognition (ASR), Spoken Language Understanding (SLU), Dialog Management (DM), Natural Language Generation (NLG), and Text-To-Speech Synthesis (TTS).

---

* This work has been supported by the Spanish Government under i-Support (Intelligent Agent Based Driver Decision Support) Project (TRA2011-29454-C03-03).

L. Iliadis et al. (Eds.): AIAI 2014, IFIP AICT 436, pp. 307–316, 2014.

These tasks are typically implemented in different modules of the system's architecture.

Although dialog management is only a part of the development cycle of spoken dialog systems, it can be considered one of the most demanding tasks given that it decides the next action of the system and encapsulates the logic of the speech application [5]. The selection of a specific system action depends on multiple factors, such as the output of the speech recognizer (e.g., measures that define the reliability of the recognized information), the dialog interaction and previous dialog history (e.g., the number of repairs carried out so far), the application domain (e.g., guidelines for customer service), knowledge about the users, and the responses and status of external back-ends, devices, and data repositories. Given that the actions of the system directly impact users, the dialog manager is largely responsible for user satisfaction. This way, the design of an appropriate dialog management strategy is at the core of dialog system engineering.

As an attempt to reduce the time and effort required for system implementation and carry out rapid system prototyping, statistical approaches for dialog management are gaining increasing interest. These approaches enable automatic learning of dialog strategies, thus avoiding the time-consuming process that hand-crafted dialog design involves. Statistical models can be trained from real dialogs, modeling the variability in user behaviors. Although the construction and parameterization of these models depend on expert knowledge about the task to be carried out by the dialog system, the final objective is to develop systems that are more robust for real-world conditions, and that are easier to adapt to different users and tasks [3].

In this paper we present a statistical approach for the development of dialog managers, in which the next system prompt is selected by means of a novel approach for online classifying based on Evolving Fuzzy Systems (EFS) [6]. These systems have an evolving structure which is updated according to the input samples. A prototype for this kind of classifiers is a data sample that groups several samples. The classifier is initialized with the first data sample and then, each data sample is classified into one of the existing prototypes. Finally, based on the potential of the new data sample to become a prototype, form a new prototype, or replace an existing one. Unlike other statistical approaches, our approach has the advantage of taking into account the data supplied by the user throughout the complete dialog without causing scalability problems. Confidence measures provided by the recognition and the understanding modules are also taken into account in the selection of the next system response.

The most widespread methodology for machine-learning of dialog strategies consists of modeling human-computer interaction as an optimization problem using Markov Decision Processes (MDP) and reinforcement methods [7]. The main drawback of this approach is that the large state space of practical spoken dialog systems makes its direct representation intractable. As described in [8], Partially Observable MDPs (POMDPs) outperform MDP-based dialog strategies since they provide an explicit representation of uncertainty. This enables the dialog manager to avoid and recover from recognition errors by sharing and shifting probability mass between multiple hypotheses of the current dialog state. Other interesting approaches for statistical dialog management are based on modeling the system by means of stochastic Finite-State Transducers [9], or Bayesian Networks [10].

The use of EFS allows us to cope with huge amounts of data, process streaming data on-line in real time, and evolve the structure of a dialog model that defines an activity based on the human-computer interaction. The dialog model is designed and treated as a changing model which constantly reflects the changes in the way the dialog system interacts with the users. Thus, it can start learning "from scratch" since the number of fuzzy rules do not need to be defined *a priori*. However, EFS can also be used to evolve an initial dialog model which has been previously defined taking into account expert knowledge about the specific interaction domain. In addition, the proposed solution addresses other important related requirements such as processing huge amounts of sequential data in an optimized way.

After this introduction, the remainder of the paper is organized as follows. Section 2 describes the proposed methodology for dialog management and its specific application for the *eclass0* classifier. This section also presents a practical application of our proposal to develop a spoken dialog system providing railway information and the results of a preliminary evaluation. Section 3 presents the conclusions and suggests some future work guidelines.

2   Our Proposed Methodology for Dialog Management

The process followed by a statistical dialog manager for the selection of the next system response is summarized in Figure 1 [11]. As this figure shows, the user has an internal state S_u corresponding to a goal that is trying to accomplish and the dialog state S_d represents the previous history of the dialog. Based on the user's goal prior to each turn, the user decides some communicative action (also called intention) A_u, expressed in terms of dialog acts and corresponding to an audio signal Y_u. Then, the speech recognition and language understanding modules take the audio signal Y_u and generate the pair $(\tilde{A}_u, C)$.

This pair consists of an estimate of the user's action A_u and a confidence score that provides an indication of the reliability of the recognition and semantic interpretation results. This pair is then passed to the dialog model, which is in an internal state S_m an decides what action A_m the dialog system should take. This action is also passed back to the dialog manager so that S_m may track both user and machine actions. The language generator and the text-to-speech synthesizer take A_m and generate an audio response Y_m. The user listens to Y_m and attempts to recover A_m. As a result of this process, users update their goal state S_u and their interpretation of the dialog history S_d. These steps are then repeated until the end of the dialog.

In our statistical approach for dialog management, we propose that, given a new user turn, the statistical dialog model makes the assignation of a system response A_m according to the result of a classification process. The classification

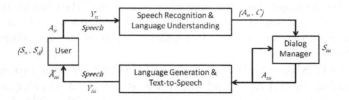

Fig. 1. Set of internal processes in a spoken dialog system

function can be defined in several ways. We propose the use of the *eClass* (evolving Classifier) family. During the training of these classifiers, a set of fuzzy rules that describes the most important observed features for the classification of each class (i.e., system prompt) is formed. These rules can also be constantly adjusted to the available training data. One of the advantages of *eClass* is that it does not require parameter optimization as its only parameter 'scale' can be directly inferred from the training data. This technique [12] is based on partitioning the data space into overlapping local regions through *Recursive Density Estimation* (RDE) and associating clusters (respectively fuzzy sets) to them.

The *eClass* family includes two different architectures and on-line learning methods:

- *eClass0* with the classifier consequents representing class label [12].
- *eClass1* for regression over the features using first order eTS fuzzy classifier.

In this paper, we focus on the *eClass0* classifier. *eClass0* possesses a zero-order Takagi-Sugeno consequent, so a fuzzy rule in the *eClass0* model has the following structure:

$$Rule_i = IF(Feature_1 \ is \ P_1) \ AND \ldots$$
$$\ldots AND \ (Feature_n \ is \ P_n) \tag{1}$$
$$THEN \ Class \ = \ c_i$$

where i represents the number of rule; n is the number of input features (observations corresponding to the different attributes and concepts defined for the semantic representation of the user's utterances); the vector $Feature$ stores the observed features, and the vector P stores the values of the features of one of the prototypes of the corresponding class $c_i \in \{$set of different classes$\}$. Each class is then associated to a specific system action (response).

The dialog manager must consider the concepts and values for the attributes provided by the user throughout the previous history of the dialog to select the next system action. For the dialog manager to take this decision, we have assumed that the exact values of the attributes are not significant. They are important for accessing databases and for constructing the output sentences of the system. However, the only information necessary to predict the next action by the system is the presence or absence of concepts and attributes. Therefore, the codification we use for each concept and attribute provided by the SLU module is in terms of three values, $\{0, 1, 2\}$, according to the following criteria:

- (0): The concept is unknown or the value of the attribute is not given;
- (1): The concept or attribute is known with a confidence score that is higher than a given threshold;
- (2): The concept or attribute has a confidence score that is lower than the given threshold.

The *eClass0* model is composed of several fuzzy rules per class (the number of rules depends on the heterogeneity of the input data of the same class). During the training process, a set of rules is formed from scratch using an evolving clustering approach to decide when to create new rules. The inference in *eClass0* is produced using the "winner takes all" rule and the membership functions that describe the degree of association with a specific prototype are of Gaussian form.

The *potential* (Cauchy function of the sum of distances between a certain data sample and *all* other data samples in the feature space) is used in the partitioning algorithm. However, in these classifiers, the potential (P) is calculated recursively (which makes the algorithm faster and more efficient). The potential of the kth data sample (x_k) is calculated by means of equation 2 [6]. The result of this function represents the *density* of the data that surrounds a certain data sample.

$$P(x_k) = \frac{1}{1 + \frac{\sum_{i=1}^{k-1} distance(x_k, x_i)}{k-1}} \tag{2}$$

where *distance* represents the distance between two samples in the data space.

The potential can be calculated using the euclidean or the cosine distance. In this case, cosine distance (*cosDist*) is used to measure the similarity between two samples; as it is described in equation 3.

$$cosDist(x_k, x_p) = 1 - \frac{\sum_{j=1}^{n} x_{kj} x_{pj}}{\sqrt{\sum_{j=1}^{n} x_{kj}^2 \sum_{j=1}^{n} x_{pj}^2}} \tag{3}$$

where x_k and x_p represent the two samples to measure its distance and n represents the number of different attributes in both samples.

Note that the resolution of equation 2 requires all the accumulated data sample available to be calculated, which contradicts to the requirement for real-time and on-line application needed in the proposed problem. For this reason, in [6] it is developed a recursive expression for the cosine distance. The proposed formula is as follows:

$$P_k(z_k) = \frac{1}{2 - \frac{1}{(k-1)\sqrt{\sum_{j=1}^{n}(z_k^j)^2}} B_k}; k = 2, 3 ...$$

$$where: B_k = \sum_{j=1}^{n} z_k^j b_k^j \; ; \; b_k^j = b_{(k-1)}^j + \sqrt{\frac{(z_k^j)^2}{\sum_{l=1}^{n}(z_k^l)^2}} \tag{4}$$

$$and \; b_1^j = \sqrt{\frac{(z_1^j)^2}{\sum_{l=1}^{n}(z_1^l)^2}} \; ; \; j = [1, n+1]; \; P_1(z_1) = 1$$

where z_k represents the kth data sample (x_k) and its corresponding label $(z = [x, Label])$. Using this expression, it is only necessary to calculate $(n+1)$ values where n is the number of different subsequences obtained; this value is represented by b, where $b_k^j, j = [1, n]$ represents the accumulated value for the kth data sample.

In this case, a specific system action can be represented by several rules, depending on the heterogeneity of the samples that represent the same action. Thus, a class could be represented by one or several prototypes. The different prototypes that represent a system action are obtained from the input data and they are updated constantly. However, an initial rule-based model can be defined (if necessary) by hand as start point of the classifier. In this sense, new prototypes are created or existing prototypes are removed if necessary.

2.1 Practical Application: A Spoken Dialog System Providing Railway Information

We have applied our proposal to develop a mixed-initiative spoken dialog system to provide railway information system using spontaneous speech in Spanish. The system integrates the CMU Sphinx-II system speech recognition module[1]. As in many other conversational agents, the semantic representation chosen for dialog acts of the SLU module is based on the concept of frame [13]. This way, one or more concepts represent the intention of the utterance, and a sequence of attribute-value pairs contains the information about the values given by the user. For the task, we defined eight concepts and ten attributes. The eight concepts are divided into two groups:

1. *Task-dependent concepts*: they represent the concepts the user can ask for (*Timetables, Fares, Train-Type, Trip-Time*, and *Services*).
2. *Task-independent concepts*: they represent typical interactions in a dialog (*Acceptance, Rejection*, and *Not-Understood*).

The attributes are: *Origin, Destination, Departure-Date, Arrival-Date, Class, Departure-Hour, Arrival-Hour, Train-Type, Order-Number*, and *Services*.

A total of 51 system responses were defined for the task (classified into confirmations of concepts and attributes, questions to require data from the user, and answers obtained after a query to the database).

Using the previously described codification for the concepts and attributes, when a dialog starts (in the greeting turn) all the values are initialized to "0". The information provided by the users in each dialog turn is employed to update the previous values and obtain the current ones, as Figure 2 shows.

This figure shows the semantic interpretation and confidence scores (in brackets) for a user's utterance provided by the SLU module. In this case, the confidence score assigned to the attribute *Date* is very low. Thus, a "2" value is added in the corresponding position for this attribute. The concept (*Hour*) and

---

[1] cmusphinx.sourceforge.net

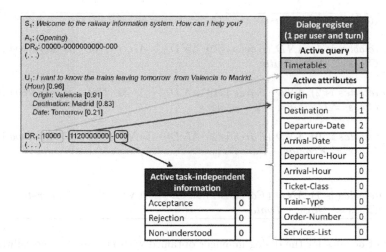

Fig. 2. Excerpt of a dialog with its correspondent representation of the task-dependent and active task-independent information for one of the dialog turns

the attribute *Destination* are recognized with a high confidence score, adding a "1" value in the corresponding positions.

The set of features for the classifier includes the codification of the different concepts and attributes that can be provided by the user and the task-independent information provided in the last user turn (none in this case). A total of 49 rules for the task were obtained with *eClass0*. Figure 3 shows the structure of these rules. Using them, the dialog manager would select the class *'SystemResponse23'*, which corresponds to a system confirmation of the departure date. This process is repeated to predict the next system response after each user turn.

An initial corpus of 900 dialogs (10.8 hours) was acquired for the task by means of the Wizard of Oz technique with 225 real users, for which an initial dialog strategy was defined by experts [4]. A set of 20 scenarios was used to carry out the acquisition. Each scenario defined one or two objectives to be completed by the user and the set of attributes that they must provide. The corpus consists of 6,280 user turns, with an average number of 7.7 words per turn. The corpus was split into a training subset of 4,928 samples (80% of the corpus) and a test subset of 1,232 samples (20% of the corpus).

We defined the following three measures to evaluate the statistical dialog model: i) *Matching*: percentage of responses provided by the system that match exactly the initial dialog strategy defined by the experts; ii) *Coherence*: percentage of responses that are coherent with the current dialog state but do not match the initial strategy; and iii) *Error*: percentage of responses that could cause a dialog failure.

We have test the behavior of our proposal comparing it with different definitions of the classification function used to determine the next system response.

$FRB - RailwayTask(eClass0)$:

IF $(Timetables$ is 1) AND $(Fares$ is 0) AND $\cdots$ AND $(Not - Understood$ is 0)
$THEN$ $Class ='$ $Ask - Date'$

IF $(Timetables$ is 2) AND $(Fares$ is 0) AND $\cdots$ AND $(Not - Understood$ is 1)
$THEN$ $Class ='$ $Confirm - Timetables'$

IF $(Timetables$ is 0) AND $(Fares$ is 1) AND $\cdots$ AND $(Not - Understood$ is 0)
$THEN$ $Class ='$ $Provide - Fares'$

$\ldots$

IF $(Timetables$ is 1) AND $(Fares$ is 2) AND $\cdots$ AND $(Not - Understood$ is 1)
$THEN$ $Class ='$ $Close - Dialog'$

Fig. 3. Set of rules for the dialog manager obtained with *eClass0* for the railway task

In this work, we have used three approaches for the definition of the classification function: a multilayer perceptron (MLP), a multinomial naive Bayes classifier, and finite-state classifiers. We also defined three types of finite-state classifiers: bigram models, trigram models, and Morphic Generator Grammatical Inference (MGGI) models [14].

Table 1 shows the results obtained. As it can be observed, the Fuzzy-rule-based classifier provides satisfactory results in terms of the percentage of correct responses selected (*Matching* and *Coherence* measures) and responses that could cause the failure of the dialog (*Error* measure). With regard the rest of classifiers, the MLP classifier is the one providing the closest results to our proposal. The table also shows that among the finite-state model classifiers, the bigram and trigram classifiers are worse than the MGGI classifier, this is because they cannot capture long-term dependencies. The renaming function defined for the MGGI classifier seems to generate a model with too many states for the size of the training corpus, therefore, this classifier could be underestimated.

Table 1. Results of the evaluation of the different classification functions

Dialog manager	Matching	Coherence	Error
Fuzzy-rule-based (FRB) classifier	76.7%	89.2%	5.6%
MLP classifier	76.8%	88.8%	5.8%
Multinomial classifier	63.4%	76.7%	10.6%
Bigram classifier	28.8%	37.3%	42.2%
Trigram classifier	31.7%	42.1%	44.1%
MGGI classifier	46.6%	67.2%	24.8%

Secondly, we have evaluated our proposal with the acquisition of 100 dialogs by means of a user simulator [4]. We considered the following measures: i) Dialog success rate (*Success*); ii) Average number of turns per dialog (*nT*); iii) Confirmation rate (*Confirmation*); and iv) Error correction rate (*ECR*). The confirmation rate was computed as the ratio between the number of explicit confirmation turns and the total number of turns in the dialog. The *ECR* was computed as the number of errors detected and corrected by the dialog manager divided by the total number of errors.

The results presented in Table 2 show that in most cases the automatically learned dialog model has the capability of correctly interacting with the user. The dialog success depends on whether the system provides the correct data for every objective user's query. All of the objectives defined are achieved in 93.5% of the dialogs. The analysis of the main problem detected in the acquired dialogs shows that, in some cases, the system did not detect the introduction of data with a high confidence value due to errors generated by the ASR that were not detected by the dialog manager. However, the evaluation confirms a good operation of the approach since the information is correctly given to the user in the majority of cases. The confirmation and error correction rates have also a remarkable impact on the described system performance.

Table 2. Results of the objective evaluation with real users

	Success	nT	Confirmation	ECR
Fuzzy-rule-based (FRB) dialog manager	93.5%	13.8	22%	0.87%

3 Conclusions and Future Work

In this paper, we have presented a statistical methodology for the development of dialog managers and the optimization of dialog strategies in spoken dialog systems. The selection of the following system response is based on a classification process that takes into account the history of the dialog. The most important contribution of our work consists of the use of Evolving Fuzzy Systems (EFS) to complete this classification. As a result of the application of our proposal, the dialog model is modeled by a set of automatically obtained dynamic rules that are applied to select the next system response. The use of EFS allows us to cope with huge amounts of data, and process streaming data on-line in real time.

We have described a practical application of our proposal to develop and evaluate a spoken dialog system providing railway information. A codification of the information sources has been proposed to facilitate the correct operation of the *eClass0* classification function. This representation allows the system to automatically generate a specialized answer that takes the current situation of the dialog into account. Task-dependent information is isolated from the model taking into account whether the user has provided a given piece of information related to the task and also the confidence scores assigned by the ASR and NLU modules.This allows not only to cope with the situations observed the training corpus, but also to manage unseen situations by selecting the most convenient system action. The results of the evaluation shows the correct operation of the learned dialog manager with regard other definitions of the classification functions.

Future work will be oriented to deploy and evaluate our proposal in additional domains. As the dialog model is learned from a corpus of training samples, the performance of the dialog manager depends on the quality and size of the corpus used to learn the model. For this reason, we also want to evaluate the influence of the main features of the training corpus in the quality of the dialog model obtained by means of our proposal.

References

1. Pieraccini, R.: The Voice in the Machine: Building Computers that Understand Speech. The MIT Press (2012)
2. Vipperla, R., Wolters, M., Renals, S.: Spoken dialogue interfaces for older people. In: Advances in Home Care Technologies, pp. 118–137. IOS Press (2012)
3. Schatzmann, J., Weilhammer, K., Stuttle, M., Young, S.: A Survey of Statistical User Simulation Techniques for Reinforcement-Learning of Dialogue Management Strategies. Knowledge Engineering Review 21(2), 97–126 (2006)
4. Griol, D., Hurtado, L., Segarra, E., Sanchis, E.: A Statistical Approach to Spoken Dialog Systems Design and Evaluation. Speech Communication 50(8-9), 666–682 (2008)
5. Wilks, Y., Catizone, R., Worgan, S., Turunen, M.: Some background on dialogue management and conversational speech for dialogue systems. Computer Speech and Language 25, 128–139 (2011)
6. Angelov, P., Zhou, X.: Evolving fuzzy-rule-based classifiers from data streams. IEEE T. Fuzzy Systems 16(6), 1462–1475 (2008)
7. Levin, E., Pieraccini, R., Eckert, W.: A stochastic model of human-machine interaction for learning dialog strategies. IEEE T. Speech and Audio Processing 8(1), 11–23 (2000)
8. Young, S., Schatzmann, J., Weilhammer, K., Ye, H.: The Hidden Information State Approach to Dialogue Management. In: Proc. ICASSP 2007, pp. 149–152 (2007)
9. Planells, J., Hurtado, L., Sanchis, E., Segarra, E.: An Online Generated Transducer to Increase Dialog Manager Coverage. In: Proc. Interspeech 2012 (2012)
10. Meng, H.H., Wai, C., Pieraccini, R.: The Use of Belief Networks for Mixed-Initiative Dialog Modeling. IEEE T. Speech and Audio Processing 11(6), 757–773 (2003)
11. Young, S.: The Statistical Approach to the Design of Spoken Dialogue Systems. Technical report, Cambridge University Engineering Department (2002)
12. Angelov, P., Filev, D.: An approach to online identification of Takagi-Sugeno fuzzy models. IEEE T. Systems, Man and Cybernetics 34(1), 484–498 (2004)
13. Minsky, M.: A Framework for Representing Knowledge. In: The Psychology of Computer Vision, pp. 211–277. McGraw-Hill (1975)
14. Segarra, E., Hurtado, L.: Construction of Language Models using Morfic Generator Grammatical Inference MGGI Methodology. In: Proc. Eurospeech 1997, pp. 2695–2698 (1997)

Sequence Matching Genetic Algorithm
for Square Jigsaw Puzzles

Josef Hynek

Faculty of Informatics and Management, University of Hradec Králové,
Rokitanskeho 62, 500 03 Hradec Králové, Czech Republic
Josef.Hynek@uhk.cz

Abstract. Our paper presents a new method for solving the rectangle piece jigsaw puzzle problem. The puzzle image is RGB full color and because of uniform shape of the individual pieces the process of puzzle assembly is based on
information of the pixel values along the border line of the piece only. We have
utilized a genetic algorithm that searches for the optimal piece arrangement using dissimilarity between adjacent pieces as the measure of progress. Unlike the
previous attempts to utilize genetic algorithms to solve the problem, we have
proposed a new heuristic asexual operator that aims at identification of points of
fraction within partially assembled picture, extraction of supposed sequence of
correctly joint pieces, and its insertion into a new position in such a way that, if
possible, the segment is enlarged. Our approach has been successfully tested
and the algorithm is capable of solving puzzles consisting of several hundred
pieces.

Keywords: genetic algorithms, image reconstruction, pattern matching, puzzle
solving.

1   Introduction

Solving puzzles is usually viewed as a special example of pattern matching recognition problem. At the first sight it might be viewed as a pleasant and very popular
widespread game for players of all ages but we should not forget the complexity of
jigsaw puzzles. It is a well-known fact that jigsaw puzzles and their equivalents belong to NP-complete class [1] and thus provide serious obstacles for various attempts
to solve them automatically. Moreover, there are many practical applications ranging
from image mosaicking up to the reconstruction of archaeological artifacts and assembly of shredded documents [2]. The first jigsaw puzzle was produced in 1760 in
London [3] and then the game of puzzle gradually expanded all over the world. There
are various types of jigsaw puzzle and an integrated classification terminology is
missing. However, there are some basic classification terms that are commonly used
and we will mention here at least the most important of them.

The term standard jigsaw puzzle is usually used to describe puzzles created by cutting pictures printed on firm stiff paper into interlocking patterns of pieces. The pieces
have a distinctive geometrical shape that provides an important piece of information

L. Iliadis et al. (Eds.): AIAI 2014, IFIP AICT 436, pp. 317–324, 2014.
© IFIP International Federation for Information Processing 2014

that is used together with image information in the process of puzzle assembly. It is necessary to point out that there are also so cold apictorial types of puzzle that do not have any picture not chromatic information and the shape of the pieces is the only clue to solve the problem.

We are interested in square jigsaw puzzles where the individual pieces lack characteristic curvilinear shape because all the pieces have straight borders and uniform rectangular size. In this case pictorial information provides us only information and guidance that is available there. This type of puzzle is sometimes called as edge-matching puzzle [1] in order to emphasize that the adjacent pieces match along their respective edges. These puzzles are very challenging as there is no guarantee that two pieces that fits together should be together and the only way to proof the correctness of individual matches is the completion of the entire solution.

Of course, the simple categorization of different kinds of puzzle above is very superficial and definitely incomplete, but for the sake of this paper we do not need to go further in this direction. For more details concerning different kinds of puzzle and its classification please refer to [3].

The rest of the paper is organized as follows. Section 2 summarizes various approaches to the automatic solution of jigsaw puzzles with the particular focus on utilization of genetic algorithms. Section 3 is the key part of our paper and the newly proposed solution is described there. Our experimental results are briefly described in Section 4 and it is followed by discussion and conclusions.

2   Previous Work

Many attempts to utilize various problem solving techniques were made since 1964 when the first paper describing computer solution of apictorial jigsaw puzzle was published [4]. Is has been pointed out that the difficulty encountered when attempting to program a computer to solve jigsaw puzzles relates to three different aspects of the problem: 1) the description of the pieces, 2) the manipulation of the pieces (rotating and matching), and 3) the evaluation of the correct matching of the individual pieces. The complexity of the problem implies that a brute-force approach, which may work for small puzzles, becomes impractical as the number of pieces is increased [4].

It is necessary to emphasize that majority of attempts that dates more that fifteen years back focused on apictorial puzzles. It means that the relevant experiments are concerned with geometric shape information of the puzzle pieces and chromatic (pictorial) information is neglected (see, for example, [5] or [6]). These articles brought up gradually improved methods and techniques for representation of boundaries of puzzle pieces and their subsequent efficient pertaining and correct matching.

Perhaps the most efficient algorithm for automatic solution of jigsaw puzzles was presented by Goldber et al. [7]. It is based on the similar ideas as the already mentioned algorithm described in [5], but it utilizes better and more global matching techniques. In particular, use of fiducial points to align adjacent pieces of puzzle together with optimization of partial solutions based on the method of global relaxation worked surprisingly well and allowed authors to solve the puzzles consisting of more than two hundred pieces [7].

An obvious way to improve efficiency and robustness of the algorithm is a suitable utilization of additional pieces of information on the top of a distinctive geometric shape of the puzzle pieces. Colours or textures on the boundaries of these pieces could provide the algorithm further clues and direction if such information is available (see [8] or [9]). However, as we have mentioned above, there are also so called edge-matching puzzles where all the pieces have straight borders and uniform rectangular size and pictorial information provides us the only information and guidance that can be used in order to solve it.

Toyama et al. [10] proposed a method for solving the rectangle piece jigsaw puzzle problem using a genetic algorithm. The picture of their puzzle is painted only in black and white and so the pieces of the puzzle are represented as binary images. The assembly of the puzzle is then performed using information of the pixel value on the border line of the adjacent pieces. They utilised the populations of 200 randomly generated individuals representing candidate solutions and employed two genetic operators (2-point crossover and self-crossover) to produce offspring and run the evolutionary process. They reported that the method described there correctly assembled all pieces in the 8x8-piece puzzle.

This team of researchers collaborated on another interesting paper [11] and developed an improved method for solving the rectangle piece jigsaw puzzle assembly problem, but no evolutionary algorithm is used in this process any more. The assembly of the puzzle is once again performed only using information of the pixel value on the border line of the pieces and this time a puzzle image is RGB full colour. Pieces are connected by a matching function between two pieces and a simple method connects together a single piece to a block that is defined as a group of already connected pieces. According their results the proposed method correctly assembled all pieces in 16x12-piece puzzles.

Alajlan [12] experimented with gray picture puzzles and he used dynamic programming to facilitate non-rigid alignment of border pixels for local matching of the puzzle pieces. Moreover, instead of the classical best-first search, his algorithm simultaneously positioned the neighbours of a puzzle pieces during the search using the so-called Hungarian procedure. This procedure begins with a starting piece of the puzzle and then locates four adjacent pieces of puzzle to the initial piece in the way the sum of border distances is minimal. In order to make the algorithm robust, every puzzle piece was considered as starting piece at various starting locations. Experiments using several images demonstrated that the proposed algorithm correctly assembled puzzles up to 8×8 pieces. However, the author of this article acknowledged that its performance deteriorates as the number of pieces exceeds 64.

Finally, we would like to mention one completely different approach that was described by Gindre et al. [3]. Their algorithm is a part of the development of Intelligent Robotic System that solves an unknown jigsaw puzzle. The system uses pattern recognition techniques as edge and feature detection in conjunction with genetic algorithms. The novelty of this as approach is based on a completely different encoding scheme. The candidate solutions are represented by relevant graphs describing the interconnections between puzzle pieces and that is why the corresponding chromosomes are defined by the relevant adjacency matrix.

The common limitation of the above presented approaches is that just small scale puzzles could be tackled and solved. That is the fact that fostered our interest in this area and based on a detailed analysis of the methods that were already tested we have designed a hybrid genetic algorithm for jigsaw puzzle problem.

3   Problem Definition and Representation

We have shown that there are various types of jigsaw puzzles and therefore it is necessary to start with its definition. We have decided to experiment with rectangular pictorial puzzle as it is described in [10]. Therefore, we assume that it consists of *NxM* puzzle pieces that do not rotate and the assembly of the puzzle is performed by using the pixel values on the border lines of the individual pieces.

Whereas Toyama worked with black and white pictures only where the pixel values are 0 and 1 respectively, we have decided to use color 24-bit RGB images. Despite the fact that the comparison of border lines is very easy and straightforward, there is no guarantee that there is an absolute correspondence between border lines of the neighboring pieces. It depends on the particular picture and irregularities along the border make the problem difficult. It is clear that algorithm could be even mislead in the situation when wrong piece of the puzzle is regarded as more suitable from the point of borders matching than the right one. Hence, it is evident that the problem cannot be solved by the simple local piece matching. The example of the puzzle used in our experiments is shown on Fig. 1.

Having defined the problem we can decide what kind of representation of partial (or candidate) solutions would be suitable for our algorithm. We have realized that graph representation used by Gindre et al. [3] is rather inconvenient for genetic operators and the pertinent manipulations and that is why we have chosen simple *NxM* matrix for each individual and each cell $c_{i,j}$ corresponds to one piece of puzzle. The matrix depicts the relevant arrangement of puzzle pieces in a very straightforward manner that the left/right/up/down adjacency in the matrix means the same interrelationship between the relevant puzzle pieces. This representation is very convenient for fitness evaluation as well as for utilization of genetic operators.

Each piece of the puzzle is represented by matrix *LxLx3*, where *L* is the width and length of the piece (i.e. the number pixels along the border). The fitness function of our algorithm is based on two measures of dissimilarity. Horizontal dissimilarity between two pieces x_i and x_j, where x_j is placed to the right of x_i is defined as

$$D_h(x_i, x_j) = \sqrt{\sum_{l=1}^{L}\sum_{k=1}^{3}(x_i(l,L,c) - x_j(l,1,c))^2}, \tag{1}$$

where c stands for intensity of red, green, and blue color and each value is integer from 0 to 256. Likewise, vertical dissimilarity between two pieces x_i and x_j, where x_i is placed on the top of x_j is defined as

$$D_v(x_i, x_j) = \sqrt{\sum_{l=1}^{L}\sum_{k=1}^{3}(x_i(L,l,c) - x_j(1,l,c))^2}, \tag{2}$$

where square difference between pixels of the last row of x_i and the first row of x_j is computed. These measures of dissimilarity will be calculated many times during the process of evaluation of thousands of individuals and therefore it is advantageous to create a lookup table covering all the relevant values at the beginning of run.

Fig. 1. Example of the puzzle used for the algorithm testing (10x10-piece puzzle)

The fitness function of each individual is calculated as the reverse value of the sums of horizontal and vertical differences of the relevant puzzle pieces. Providing that individual a is represented by NxM matrix C as we have described it above, then its fitness is defined as

$$f(a) = \frac{1}{1 + \sum_{i=1}^{N}\sum_{j=1}^{M-1} D_h(c_{i,j}, c_{i,j+1}) + \sum_{i=1}^{N-1}\sum_{j=1}^{M} D_v(c_{i,j}, c_{i+1,j})}. \tag{3}$$

It is evident that the important feature of fitness function $f(a)$ is that individual with smaller sum of dissimilarities is fitter and thus will have a higher chance of being selected. We have employed standard biased roulette wheel selection method together with an elitist approach when the best individuals (10%) from current population are copied without any change to the newly created population.

The novelty of our approach is based on the utilization of problem specific operator. Besides the common partially-matched crossover (PMX) that assures legitimacy of created offspring we have employed a new operator that helps to preserve the already created segments of the picture being assembled. The main drawback of PMX is that crossover points are selected randomly and therefore it is easy to split a sequence of correctly assembled pieces. Nevertheless, it provides us necessary variability, exchanges pieces of information between different individuals and thus it is important for the whole process of evolution.

We have experimented with asexual heuristic operator that is designed to enlarge the already created segments of the picture. These segments are not only scattered over the whole candidate solution but they are usually located at incorrect locations. We are unable to assign them to their correct place within the picture as the algorithm does not use this piece of information, nevertheless we can try to connect them with other segments believing that larger segments will finally settle at the right positions.

First of all, we identify likely points of fraction where the horizontal measure of dissimilarity is relatively large. For *NxM* pieces of puzzle we identify N points of fraction with the greatest value of D_h function using (1). We select one of these points of fraction randomly using the biased roulette wheel selection mechanism so the point of fraction with the greatest dissimilarity has the highest probability to be selected. When the point of fraction has been determined we follow the horizontal direction till the next point of fraction is identified and the sequence of pieces between these two points of fraction constitutes the segment. An example of distribution of points of fraction as well as the relevant segments is on Fig. 2.

The final phase of this procedure is an attempt to allocate the segment into a new position within the picture using D_h function value (1) for the very first or very last piece of the segment. Moreover, we must not forget that albeit the point of fraction cannot occur at the beginning or at the end of rows (it results from the definition of D_h), the right place of segment could be at the beginning of the row and such a placement will be indicated by D_v according to (2)).

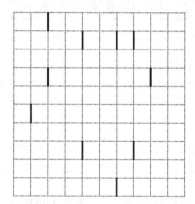

Fig. 2. Points of fraction and prospective segments (10x10-piece puzzle)

Of course, there is no guarantee that the sequence selected is a real and correct segment of the picture. However, as the points of fraction present very likely some faults within current arrangement of pieces, there is a chance to improve the overall composition.

It is clear that the above described asexual operator is efficient especially in latter phases of evolution when suitable building blocks (segments) exist within relevant individuals. That is why we have started the evolution process with 1000 individuals and carried it on for 50 generations. Then we employed decimation operation and discarded most of the population. Decimation is a secondary genetic operation that is normally performed at the beginning of a run and it is used to increase the proportion of fitter individuals in the population. It has been shown [13] that decimation provides improvement in performance that justifies the additional computation at the beginning of the run. Hence, taking advantage of the decimation operation only 100 individuals survived (we made sure that these individuals are unique in order to provide enough diversity) and then we start using the newly devised operator.

4   Experiments

We have experimented with three different pictures that were divided into NxN pieces. The size of the pieces was 80x80 pixels and so the size of the picture changed accordingly the number of pieces taken into account. The algorithm run for 200 generations, the first 50 generations using PMX operator with p_c=0.8 and simple mutation (changing mutually places of two randomly selected pieces) with p_m=0.05. For the next 150 generations the fraction point operator was applied to 50% of population whereas frequency of PMX slowed down to p_c=0.3. Taking into account that the best individuals (10%) are copied automatically to the new population, the size of it remains constant for the rest of the evolution process.

We ran our algorithm 50 times for each image and the results are given in Table 1. We can see that correct solution has been found for all the cases that were tested and the efficiency of the algorithm clearly deteriorates with increasing number of pieces. We have not included puzzles with more than 625 pieces (25x25) into our results table as we were unable to get a correct result within the given size of population and limited number of generations.

Table 1. Experimental Results.

No. of pieces	Test 1	Test 2	Test 3
15x15	100%	94%	100%
20x20	92%	74%	86%
25x25	56%	42%	52%

5   Discussion and Limitations

Our results show that the method using points of fraction is promising. There are still some opportunities to improve the way of points of fraction location as well as the way of matching the individual segments. Nevertheless, using this approach we were able to assemble puzzles of larger sizes than our predecessors.

In several cases our algorithm got stuck when it reached a partial solution in a form of horizontally cut picture that is glued together in a completely wrong way. We have analyzed this malfunction and we have realized that the algorithm was simply misled by a low level of dissimilarity along the cutting line. Because our algorithm does not make any use of the correct order and/or location of the individual pieces within the original picture, there is no way to avoid this kind of fault.

6   Conclusions

We have described the new algorithm for solving the rectangle piece color puzzle problem. Genetic algorithm has been augmented by new heuristic asexual operator that aims at identification of points of fraction within partially assembled picture, extraction of supposed sequence of correctly joint pieces, and its insertion into a new position in such a way that, if possible, the segment is enlarged. This method promotes the gradual enlargement of correctly connected blocks and the search for an optimal position of the block within the frame of the assembled picture is enabled and supported at the same time too. Our approach has been successfully tested and the algorithm is capable of solving puzzles consisting of several hundred pieces.

References

1. Demaine, E.D., Demaine, M.L.: Jigsaw Puzzles, Edge Matching, and Polyomino Packing: Connections and Complexity. Graphs and Combinatorics 23, 195–208 (2007)
2. Richter, F., Ries, C.X., Cebron, N., Lienhart, R.: Learning to Reassemble Shredded Documents. IEEE Transactions on Multimedia 15(3), 582–593 (2013)
3. Gindre, F., Trejo Pizzo, D.A., Barrera, G., Lopez De Luise, M.D.: A Criterion-based Genetic Algorithm Solution to the Jigsaw Puzzle NP-Complete Problem. Lecture Notes in Engineering and Computer Science, vol. 2186 (1) (2010)
4. Freeman, H., Garder, L.: Apictorial Jigsaw Puzzles: The Computer Solution of a Problem in Pattern Recognition. IEEE Transactions on Electronic Computers EC-13(2), 118–127 (1964)
5. Wolfson, H., Schonberg, E., Kalvin, A., Lamdan, Y.: Solving jigsaw puzzles by computer. Annals of Operation Research 12(1-4), 51–64 (1988)
6. Webster, R.W., LaFollette, P.S., Stafford, R.L.: Isthmus critical points for solving jigsaw puzzles in computer vision. IEEE Transactions on Systems, Man and Cybernetics 21(5), 1271–1278 (1991)
7. Goldberg, D., Malon, C., Bern, M.: A global approach to automatic solution of jigsaw puzzles. Computational Geometry 28(2-3), 165–174 (2004)
8. Chung, M.G., Fleck, M., Forsyth, D.: Jigsaw puzzle solver using shape and color. Proc. of 4th International Conference on Signal Processing 2, 877–880 (1998)
9. Makridis, M., Papamarkos, N., Chamzas, C.: An innovative algorithm for solving jigsaw puzzles using geometrical and color features. In: Sanfeliu, A., Cortés, M.L. (eds.) CIARP 2005. LNCS, vol. 3773, pp. 966–976. Springer, Heidelberg (2005)
10. Toyama, F., Fujiki, Y., Shoji, K., Miyamichi, J.: Assembly of puzzles using a genetic algorithm. In: Proc. of 16th International Conference on Pattern Recognition, Quebec, Canada, vol. 4, pp. 389–392 (2002)
11. Murakami, T., Toyama, F., Shoji, K., Miyamichi, J.: Assembly of puzzles by connecting between blocks. In: Proc. of 19th International Conference on Pattern Recognition, Tampa, FL, pp. 1–4 (2008)
12. Alajlan, N.: Solving Square Jigsaw Puzzles Using Dynamic Programming and the Hungarian Procedure. American Journal of Applied Sciences 6(11), 1941–1947 (2009)
13. Nanduri, D.T., Ciesielski, V.: Comparison of the effectiveness of decimation and automatically defined functions. In: Khosla, R., Howlett, R.J., Jain, L.C. (eds.) KES 2005. LNCS (LNAI), vol. 3683, pp. 540–546. Springer, Heidelberg (2005)

Scalable Distributed Genetic Algorithm for Data Ordering Problem with Inversion Using MapReduce

Doina Logofatu[1] and Daniel Stamate[2]

[1] Computer Science Department of Frankfurt University of Applied Sciences 1 Nibelungenplatz
60318, Frankfurt am Main, Germany
[2] Department of Computing, Goldsmiths College, University of London,
London SE146NW, UK
logofatu@fb2.fh-frankfurt.de

Abstract. We present in this work a scalable distributed genetic algorithm of Data Ordering Problem with Inversion using the MapReduce paradigm. This specific topic is appealing for reduction of the power dissipation in VLSI and in bioinformatics. The capacitance and the switching activity influence the power consumption on the software level. The ordering of the data sequences is an unconditional consequence of switching activity. An optimization problem related to this topic is the ordering of sequences such that the total number of transitions will be minimized – Data Ordering Problem (DOP). Adding the bus-invert paradigm, some sequences can be complemented. The resulting problem is the DOP with Inversion (DOPI). These ordering problems are NP-hard. We establish a scalable distributed genetic approach - MapReduce Parallel Genetic Algorithm (MRPGA) for DOPI, MRPGA_DOPI and draw comparisons with greedy algorithms. The proposed methods are estimated and experiments show the efficiency of MRPGA_DOPI.

Keywords: Data Ordering with Inversion, Low Power, Distributed Algorithm, Evolutionary Approaches, Transition Minimization, Greedy, MapReduce, Apache Hadoop.

1 Background and Motivation

Quicker access time and bigger storage capacity are actual demands for electronic devices ([10]). That's why the low-power topics have to be considered at the beginning of the design process ([11], [12], [14], [17]). The software component for the design of embedded systems is of importance for the power consumption of the circuit, since its optimization could lead to significant improvements ([12], [14]).

Often, the design complexity describes directly the power consumption of a circuit. It is a fact that nowadays the power consumption has grown, bringing also more complex design approaches. Reliability, cooling costs, packaging, and computer battery life are some domains with positive effects regarding a design based on low power consumption. For handling the power management issue, new directions and approaches are necessary. The capacitance and the switching activity influence the (software level) power consumption. An important design metric, the switching

L. Iliadis et al. (Eds.): AIAI 2014, IFIP AICT 436, pp. 325–334, 2014.

activity, characterizes the quality of an embedded system-on-chip design. The switching activity proves to be an important design metric, which describes the quality of an embedded system-on-chip design. It is directly related to the ordering of data strings. One first problem related to this topic is the ordering of data words for minimizing the total number of transitions *Data Ordering Problem* (DOP). This problem could be of interest for various practical situations, e.g. reordering and optimize DNA sequences or in the linguistic leveling [4, 7].

2   Problem Domain

Definition 2.1. *Hamming distance.* If a word w_r of length k is followed by the word w_s of the same length k, the total number of transitions is given by the number of bits that change. This is:

$$d(w_r, w_s) = \sum_{j=1}^{k} w_{rj} \oplus w_{sj} \tag{1}$$

also known as the *Hamming distance* between w_r and w_s. Here, the w_{rj} denotes the jth bit of w_r, and $\oplus$ the XOR operation. For instance, $d(1010, 0100) = 3$. The number of transitions is, in fact, the *Hamming distance* between w_r and w_s.

Definition 2.2. *Total number of transitions.* The *total number of transitions* is the sum of number of transitions needed for the transmission of all the words w_1, w_2, ..., w_n. It is denoted with N_T. If σ is a permutation of $\{1, 2, ..., n\}$, then the total number of transitions will be:

$$N_T = \sum_{j=1}^{n-1} d(w_{\sigma(j)}, w_{\sigma(j+1)}). \tag{2}$$

Definition 2.3. *Phase-assignment.* The *phase-assignment* (polarity) δ is a function defined on $\{1,..., n\}$ with values in $\{0, 1\}$, which specifies if a word is sent as it is or complemented (negated, inverted).

For example, the complemented word of 10011 is 01100. That is, if $\delta(i)=0$ then the word w_i is sent as it is, otherwise is sent his complement.

Definition 2.4. The Data Ordering Problem *(DOP)* is defined as the problem of finding a permutation σ of the words w_1, w_2, ..., w_n such that the total number of transitions:

$$N_T = \sum_{j=1}^{n-1} d(w_{\sigma(j)}, w_{\sigma(j+1)}) \tag{3}$$

is minimized.

Definition 2.5. The *Data Ordering Problem with Inversion (DOPI)* is defined as the problem of finding a permutation σ and a phase-assignation δ of the words w_1, w_2, ..., w_n such that the total number of transitions:

$$N_T = \sum_{j=1}^{n-1} d(w_{\sigma(j)}^{\delta(j)}, w_{\sigma(j+1)}^{\delta(j+1)}) \tag{4}$$

is minimized.

The problems DOP and DOPI are NP-hard [11, 12].

Note that the Hamming distance between two words with the same length provides the total number of dissimilarities between these words.

We denote the generalizations of these problems also with DOP and DOPI. Given a set of n words with the same length k over an alphabet $\mathcal{A}$, and in the case of DOPI also a permutation of length k which denotes the inversion (complementation) function, it is requested to provide a permutation of the given words (in the case of DOPI also an assignation of them) which minimizes the total number of dissimilarities (transitions).

3   Previous Work

The optimal method and two evolutionary approaches, MUT_DOPI (with mutation) and GA_DOPI (hybrid genetic algorithm) are proposed in [6], together with three new genetic operators Simple Cycle Mutation (SIM), Cycle OX and Cycle PMX. The two ordering problems are very similar to the *Traveling Salesman Problem* (TSP). For DOP, DOPI and TSP a efficient ordering of elements regarding to a given weight has to be determined. Since the two problems DOP and DOPI are NP-hard [11, 12], the optimal algorithms can only manage small instances. In the past years some approaches were introduced for the problems DOP and DOPI:

> *Spanning Tree/Minimum Matching* (ST-MM) [11]
> *Double Spanning Tree* (DST) [11]
> *Greedy Simple* (GMS) [12]
> *Greedy Min* (GM) [12]
> *Evolutionary Approaches* [6]

The *Evolutionary Algorithms* (EAs) provide the best results quality. Such evolutionary methods return better outcome than the above-presented *Greedy Min*, but with much more time resources. Further, they have no capability to cope with very large data sets. EAs which produce high-quality optimizations are presented in [6], where are presented evolutionary algorithms for both DOP and DOPI.

4   Scalable Distributed Algorithm Using MapReduce

Our scalable distributed evolutionary algorithm uses the genetic algorithm GA_DOPI [6] and is realized with *Hadoop*. Apache [18] proposes Hadoop as an OpenSource

MapReduce [3] framework implementation. Hadoop is a batch data processing system for running applications and process large amounts of data in parallel, in a reliable and fault-tolerant manner on large clusters of compute nodes, usually running on commodity hardware. This system offers monitoring and status tools and provides an abstraction model for programming tasks. It supports automatic distribution and parallelization. Apache's *Hadoop* offers a distributed file system (*HDFS*) which creates multiple duplicates of data sets and distributes them on compute nodes throughout the cluster to enable reliable, rapid computations. The nodes for computation and for storage are usually the same. The application specifies the input/output locations, supply *map* and *reduce* methods and possibly invariant data.

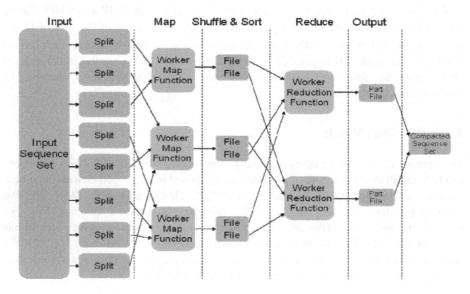

Fig. 1. MapReduce Dataflow

The scalable distributed evolutionary algorithm performs for a given number of generations. During every generation, it creates an arbitrary number of permutations of the current set of sequences, splits the set of permutations in subsets (*map* operation) and performs GA_DOPI [6] on all of them during the *reduce* operation. Lastly, it collects the best permutations (individuals) provided on each reducer node in the cluster and stick them together. Groups of initial individuals are initialized using the greedy approaches GMS and GM [7, 12]. The outcome represents the set of sequences for the next generation. The *map* operation associates for each given permutation a subset index in range 0 to *reductionFactor*−1 randomly, as output key. During the *Shuffle & Sort* action, the sequences get sorted after their indices, and each *reducer* node gets and executes GA_DOPI in one call over its subset of sequences with the same index.

ALGORITHM_ MRPGA_DOPI
    Initialization(*size_of_population, rate_crossover, rate_mutation*)
    Initialization(*factior_reduction*)
    Initialisation_GreedyMinSimplified_individuals()
    Initialization_GreedyMin_individuals()
    Initialization_Random_individuals()
    def MRPGA_MAP(*initialIndex*, pair<*permutation, assignation*>) = {
        reductionSetIndex = random(0.. *reductionFactor*–1)
        context.write(*reductionSetIndex*, pair<*permutation, assignation*>)
    }
    def MRPGA_REDUCE
        (*reductionSetIndex*, Iterable<pair<*permutation, assignation*>>)=
    {
        run_crossover(*number_crossovers*);
        run_mutation(*number_mutatios*);
        get_fitness(*all_new_individuals*);
        removal_WorstInviduals (*number_crossovers+number_mutations*);
        for (pair<*new_perm, new_assign*>: *all_new_individuals*)
            context.write
            (reductionSetIndex,pair<*new_perm, new_assign*>)
    }
    for(j← 1; j ≤ *number_Generations*; **step** 1)
        job← NewJob(MRPGA_MAP, MRPGA_REDUCE)
        job.configuration.set(*populationHDSFPath*)
        job.configuration.set(*crossoverRate*)
        job.configuration.set(*mutationRate*)
        job.configuration.setNumberReduceTasks(*reductionFactor*)
        job.submit_and_wait()
        collect_new_population
    end_for
    return *bestIndividual*
END_ALGORITHM_MRPGA_DOP

Fig. 2. Pseudocode for MRPGA_DOPI

5   Realization Specifics

The Hadoop implementation for DOPI is the *Open Source* project "dopiSolver" on *sourceforge.net*. "dopiSolver.jar" contains the java binary application, as well as all the libraries depends on. The jar's main method (*dopisolver.DopiSolverDriver*) submits the DopiSolver's Hadoop Job (*dopisolver.DopiSolverJob*). I order to launch the application use "hadoop jar dopiSolver.jar" and supply afterwards the solvers name: *dopi*.

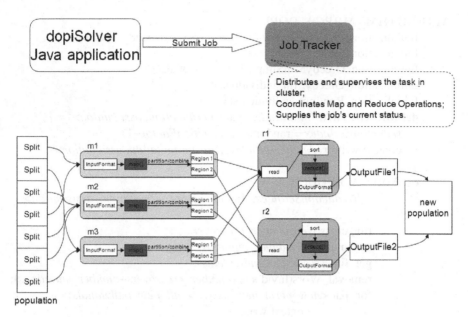

Fig. 3. Implementation of dopiSolver using Java and Apache's Hadoop

With "hadoop jar dcpSolver.jar dopi -h" you get the description of the application supported options. The source code is available on *sourceForge* under

https://dopisolver.svn.sourceforge.net/svnroot/dopisolver

(Subversion repository). The application depends on the libraries JUnit 4 (for test units), Apache Commons CLI2, and Apache Commons Math Version 2.0.

6   Experimental Results

Various sets of data were generated artificially with different parameters, like e.g. sequences set dimension (n, that is the number of words), length of the words (k). We start by performing the two greedy algorithms GM and GMS [6] on 421 different artificial generated data sets with $1000 \leq n \leq 7000$, $50 \leq k \leq 1000$. We do some experiments with the Greedy methods because we use them also in the initialization of the distributed algorithm. In Figure 5 we show the mean values for the difference GMS–GM on intervals and the mean (standard deviations $\pm$) for the execution time for all these experiments.

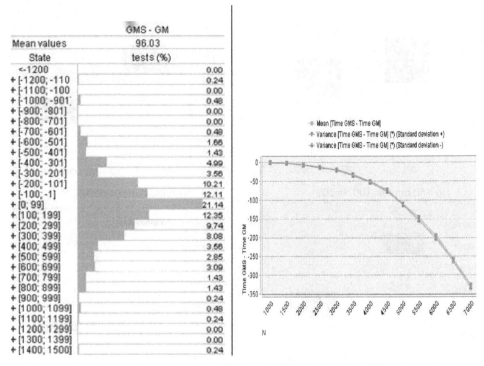

	GMS - GM
Mean values	96.03
State	tests (%)
<-1200	0.00
+ [-1200; -110	0.24
+ [-1100; -100	0.00
+ [-1000; -901]	0.48
+ [-900; -801]	0.00
+ [-800; -701]	0.00
+ [-700; -601]	0.48
+ [-600; -501]	1.66
+ [-500; -401]	1.43
+ [-400; -301]	4.99
+ [-300; -201]	3.56
+ [-200; -101]	10.21
+ [-100; -1]	12.11
+ [0; 99]	21.14
+ [100; 199]	12.35
+ [200; 299]	9.74
+ [300; 399]	8.08
+ [400; 499]	3.56
+ [500; 599]	2.85
+ [600; 699]	3.09
+ [700; 799]	1.43
+ [800; 899]	1.43
+ [900; 999]	0.24
+ [1000; 1099]	0.48
+ [1100; 1199]	0.24
+ [1200; 1299]	0.00
+ [1300; 1399]	0.00
+ [1400; 1500]	0.24

Fig. 4. (a) Allocation on intervals of the difference GMS–GM within 421 experiments with $1000 \leq n \leq 7000$, $50 \leq k \leq 1000$ (b) Mean and variance for the execution time difference (GMS-GM)

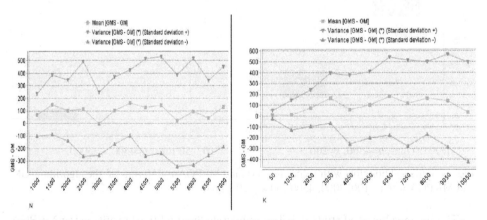

Fig. 5. Mean and variance for the difference GMS-GM for (a) values of n and b) values of k

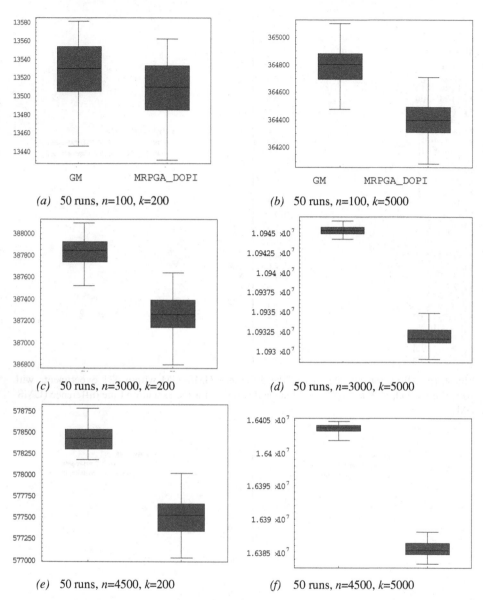

(a) 50 runs, *n*=100, *k*=200

(b) 50 runs, *n*=100, *k*=5000

(c) 50 runs, *n*=3000, *k*=200

(d) 50 runs, *n*=3000, *k*=5000

(e) 50 runs, *n*=4500, *k*=200

(f) 50 runs, *n*=4500, *k*=5000

Fig. 6. Statistical boxplots with values for GM and MRPGA for 50 runs of the algorithms GM and MRPGA with n=100, 3000, 4500 and k = 200, 5000.

The mean values and student T-Tests illustrate the quality of results using MRPGA algorithm versus GM. Comparison in pairs (GM vs. MRPGA) concludes the null hypothesis rejection with significance level = 0.001.

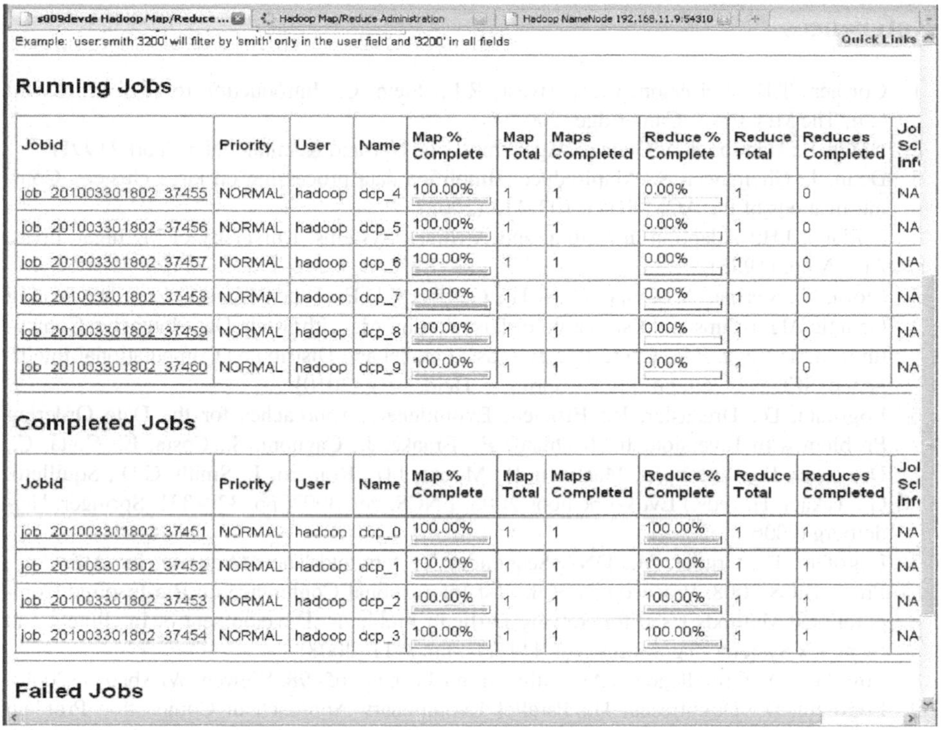

s009devde Hadoop Map/Reduce ... Hadoop Map/Reduce Administration Hadoop NameNode 192.168.11.9:54310

Example: 'user:smith 3200' will filter by 'smith' only in the user field and '3200' in all fields Quick Links

Running Jobs

Jobid	Priority	User	Name	Map % Complete	Map Total	Maps Completed	Reduce % Complete	Reduce Total	Reduces Completed	Jol Scl Inf
job_201003301802_37455	NORMAL	hadoop	dcp_4	100.00%	1	1	0.00%	1	0	NA
job_201003301802_37456	NORMAL	hadoop	dcp_5	100.00%	1	1	0.00%	1	0	NA
job_201003301802_37457	NORMAL	hadoop	dcp_6	100.00%	1	1	0.00%	1	0	NA
job_201003301802_37458	NORMAL	hadoop	dcp_7	100.00%	1	1	0.00%	1	0	NA
job_201003301802_37459	NORMAL	hadoop	dcp_8	100.00%	1	1	0.00%	1	0	NA
job_201003301802_37460	NORMAL	hadoop	dcp_9	100.00%	1	1	0.00%	1	0	NA

Completed Jobs

Jobid	Priority	User	Name	Map % Complete	Map Total	Maps Completed	Reduce % Complete	Reduce Total	Reduces Completed	Jol Scl Inf
job_201003301802_37451	NORMAL	hadoop	dcp_0	100.00%	1	1	100.00%	1	1	NA
job_201003301802_37452	NORMAL	hadoop	dcp_1	100.00%	1	1	100.00%	1	1	NA
job_201003301802_37453	NORMAL	hadoop	dcp_2	100.00%	1	1	100.00%	1	1	NA
job_201003301802_37454	NORMAL	hadoop	dcp_3	100.00%	1	1	100.00%	1	1	NA

Failed Jobs

Fig. 7. Sample Hadoop Web Admin. The system offers monitoring and status tools and provides an abstraction model for programming tasks. It supports automatic distribution and parallelization.

7 Conclusions

Experiments executed on randomly generated data input indicated the efficiency of the proposed approach - MRPGA_DOPI - especially for specific attributes of the benchmarks. An application was written for the scalable distributed algorithm using the Hadoop implementation for MapReduce [11, 12]. Scalability, status reporting and monitoring, the fault toleration, the facility to work on commodity hardware, performing the computation there the data is, robustness, are all benefits of the Hadoop implementation. Next steps would be to perform experiments on larger data inputs, with various distributions, but also variants for the map or reduce functions. By performing the algorithms on real data inputs, e.g. from bioinformatics or system-on-chip design, will provide further knowledge. E.g. in bioinformatics, DOPI problem could be formulated using the Watson-Crick complements, for every nucleic acid:

$$\overline{A} = T, \overline{G} = C, \overline{C} = G, \overline{T} = A . \quad (5)$$

References

1. Cormen, T.H., Leiserson, C.E., Rivest, R.L., Stein, C.: Introduction to Algorithms, 3rd edn. The MIT Press, Cambridge (2009)
2. Davis, L.: Handbook of Genetic Algorithms. van Nortand Reinhold, New York (1991)
3. Dean, J., Ghemawat, S.: MapReduce: simplified data processing on large clusters. Communications of the ACM 51(1), 107–113 (2008)
4. Holland, J.H.: Adaption in Natural and Artificial Systems. University of Michigan Press, Ann Arbor (1975)
5. Llorà, X., Verma, A., Campbell, R.H., Goldberg, D.E.: When Huge is Routine: Scaling Genetic Algorithms and Estimation of Distribution Algorithms via Data-Intensive Computing. In: de Vega, F.F., Cantú-Paz, E. (eds.) Parallel and Distributed Computational Intelligence. SCI, vol. 269, pp. 11–41. Springer, Heidelberg (2010)
6. Logofatu, D., Drechsler, R.: Efficient Evolutionary Approaches for the Data Ordering Problem with Inversion. In: Rothlauf, F., Branke, J., Cagnoni, S., Costa, E., Cotta, C., Drechsler, R., Lutton, E., Machado, P., Moore, J.H., Romero, J., Smith, G.D., Squillero, G., Takagi, H. (eds.) EvoWorkshops 2006. LNCS, vol. 3907, pp. 320–331. Springer, Heidelberg (2006)
7. Logofătu, D., Gruber, M.: DNA Sequences Vectors and Their Ordering. In: AIP Conf. Proc., BICS 2008: Proceedings of the 1st International Conference on Bio-Inspired Computational Methods Used for Solving Difficult Problems: Development of Intelligent and Complex Systems, Tg. Mureş, vol. 1117(1), pp. 3–11 (2008)
8. Logofătu, D.: Grundlegende Algorithmen mit Java, pp. 65–98. Vieweg, Wiesbaden (2008)
9. Logofătu, D., Dumitrescu, D.: Parallel Evolutionary Approach of Compaction Problem Using MapReduce. In: Schaefer, R., Cotta, C., Kołodziej, J., Rudolph, G. (eds.) PPSN XI. LNCS, vol. 6239, pp. 361–370. Springer, Heidelberg (2010)
10. De Micheli, G.: Synthesis and Optimisation of Digital Circuits. McGraw-Hill, Inc. (1994)
11. Murgai, R., Fujita, M., Krishnan, S.C.: Data Sequencing for Minimum-Transition Transmission. In: IFIP Int'l Conf. on VLSI (1997)
12. Murgai, R., Fujita, M., Oliveira, A.: Using Complementation and Resequencing to Minimize Transitions. In: Design Automation Conf., pp. 694–697 (1998)
13. Oliver, I.M., Smith, D.J., Holland, J.R.C.: A Study of Permutation Crossover Operators on the Traveling Salesman Problem. In: Int'l Conference on Genetic Algorithms, pp. 384–390 (1994)
14. Shen, W.-Z., Lin, J.-Y., Wang, F.-W.: Transistor reordering rules for power reduction in CMOS gates. In: ASP Design Automation Conf., pp. 1–6 (1995)
15. Sotiriadis, P.P., Chandrakasan, A.: Low Power Bus Coding Techniques Considering Interwire Capacitances. In: Proc. of IEEE Conf. on Custom Integrated Circuits (CICC 2000), pp. 507–510 (2000)
16. Stan, M.R., Burleson, W.P.: Bus Invert Coding for Low-Power I/O. IEEE Tr. on VLSI Systems 3(1), 49–58 (1995)
17. Stan, M.R., Burleson, W.P.: Low-Power encodings for Global Communication in CMOS VLSI. IEEE Tr. on VLSI Systems 5(4), 444–455 (1997)
18. Apache Hadoop, web-link (2009), http://hadoop.apache.org/

Application of Evolutionary Algorithms in Project Management

Christos Kyriklidis and Georgios Dounias

Management and Decision Engineering Laboratory,
Department of Financial and Management Engineering, University of the Aegean,
41 Kountouriotou Str. GR-82100, Greece
c.kiriklidis@gmail.com, g.dounias@aegean.gr

Abstract. The paper deals with "resource leveling optimization problems", a class of problems that are often met in modern project management. The problems of this kind refer to the optimal handling of available resources in a candidate project and have emerged, as the result of the even increasing needs of project managers in facing project complexity, controlling related budgeting and finances and managing the construction production line. For the effective resource leveling optimization in problem analysis, evolutionary intelligent methodologies are proposed. Traditional approaches, such as exhaustive or greedy search methodologies, often fail to provide near-optimum solutions in a short amount of time, whereas the proposed intelligent approaches manage to quickly reach high quality near-optimal solutions. In this paper, a new genetic algorithm is proposed for the investigation of the start time of the non-critical activities of a project, in order to optimally allocate its resources. Experiments with small and medium size benchmark problems taken from publicly available project data resources, produce highly accurate resource profiles. The proposed methodology proves capable of coping with larger size project management problems, where conventional techniques like complete enumeration is impossible, obtaining near-optimal solutions.

Keywords: Time Constraint Project Scheduling, Resource Levelling, Project Management, Genetic Algorithms.

1 Introduction

Operation research and artificial intelligence have evolved along parallel lines in the last three decades, often by attempting to competitively solve the same class of real-world optimization problems and in other circumstances to show how the integration of techniques from different fields can lead to interesting results on large and complex problems [1]. Computational intelligence methods and algorithms such as evolutionary or neural programming, and more recently also nature inspired intelligent techniques, have succeed to obtain high quality near-optimal solutions to hard optimization problems, thus becoming competitive approaches in OR fields. Nature inspired intelligent methods such as ant colony optimization, memetic algorithms and hybrid particle swarm optimization, have been successfully dealt with various other optimization problems such as the quadratic multiple container packing problem [2], the

L. Iliadis et al. (Eds.): AIAI 2014, IFIP AICT 436, pp. 335–343, 2014.

dynamic strategic planning for electric distribution systems [3], the bandwidth minimization problem [4], etc. The current work proposes the use of genetic algorithms for solving effectively the well-known optimization problem of resource leveling.

The Resource Leveling Problem (RLP), also known as time-constrained project scheduling problem (TCPSP) is an NP-hard optimization problem [5, 6]. The aim of this paper is to highlight the advantages of evolutionary intelligent algorithms in resource leveling problems. More specifically, an intelligent metaheuristic is proposed based in a new genetic algorithm, in order to produce an optimal resource management profile. An important contribution of this work is the way that the proposed approach investigates the various different feasible start-time values of non-critical activities of a project, thus leading to improved resource profiles and contributing to optimal resource allocation. The paper is organized as follows:

In Section 1, introductory comments are made. In Section 2, a literature review regarding both, the resource leveling problem and the proposed intelligent metaheuristic, is provided. In Section 3, the main methodological issues are presented in order to get a better understanding of the underlying mechanisms of the algorithm. In Section 4, the mathematical formulation of the resource leveling problem is analyzed, and complexity issues are discussed. Finally, the last Section concludes to remarks and interesting points which are summarized in an intuitive manner.

2   Literature Review

In the literature some selected studies regarding the application of several conventional and intelligent optimization algorithms and techniques on the resource leveling problem, are presented. Research work dealing with the resource leveling problem is quite extensive, as it is shown in Table 1. The given collection of papers contains a variety of methodologies ranging from dynamic programming and relaxation methods, to a number of intelligent approaches such as fuzzy optimal models, neural networks, genetic optimization based multicriteria techniques, Petri nets, ant colonies and memetic algorithms.

Analytical methods such as exhaustive search (complete enumeration approaches) [7], integer programming [8], dynamic programming [9] and branch-and-bound methods have been used to search for an accurate solution. However most of these methods have many limitations. Their major drawback is that they cannot be used to solve large and complex problems effectively [6, 9], as they are either computationally infeasible and lead to combinatorial explosion in cases of large and/or complicated projects [10].

For the effective management of large projects, researchers prefer to apply heuristic methods, which are successful in solving large and complex problems comparatively to analytical methods, although their effectiveness highly depends on the problem. Several efforts in the direction of developing heuristic rules for resource leveling, aiming at producing high-quality feasible solutions, have been made [11-17]. The models also presented by Neumann and Zimmermann [5,18] are indicative examples of a polynomial heuristic procedure for various types of resource leveling problems, where general temporal constraints are given by minimum and maximum time lags among activities.

In recent years, advanced computational methods like Genetic Algorithms (GAs) [6, 9, 19-26] and particle swarm optimization approaches [27, 28] have been used for solving resource leveling problems. Also, Kartam and Tongthong [29] introduced a methodology for solving resource leveling problems using Artificial Neural Networks (ANN). Also, Geng et al. [10] improved an ant colony optimization (ACO) approach, another nature-inspired algorithm, for non-linear resource leveling problems. However, potential drawbacks of these methods, such as premature convergence and poor exploitation, have attracted increasing attention from researchers and engineers, whose main aim is to upgrade their efficiency.

Finally, a few other interesting approaches to the resource leveling problem can be found, that do not fall in any of the above mentioned categories. Research by Jeetendra et al. [30] described the use of Petri nets (PNs), whereas Raja et al. [31] proposed the use of a Petri net in combination with a memetic algorithm.

Findings from the literature survey indicate the effectiveness of intelligent techniques such as genetic algorithms and neural networks, in solving resource leveling problems. Especially, in the case where new formulations of the problem at-hand are considered, these algorithms have proven their ability to yield high-quality, if not the optimal, solution. Literature shows the evolution of applied methodologies for the resource leveling problem, progressing from naïve techniques to more sophisticated ones. However, based on their weak points, ground for further development exists.

3   Resource Leveling

The project network production in the initial phase displays many peaks (maximum resource uses) and valleys (minimum resource uses) in resource profiling diagrams. This fact emanates from the non-uniform resource requirement when the project execution begins [32]. Two basic Resource Leveling versions are used in several studies. The fixed project duration restriction is common in both of them:

1. Reduction of the minimum and maximum resource uses, [20, 29, 32-35].
2. Reduction of resource use fluctuation from period t to period t+1 [10, 20, 24, 29, 30, 35-38].

A usual problem formulation is the following:

Let $A = \{1,2,..,n\}$ be the set of the project's activities to be scheduled, where activities 1 and n are dummy activities that represent the starting and ending phase, respectively. Duration and resource requirements for these activities equal to zero. The duration of each activity $i \in A$ is denoted by d_i. T denotes the total duration of the project. Also T is computed by CPM method, where the project network is designed as A.O.N. (Activity On Node). Precedence relations among the activities in A exist. These relations indicate which activities should be completed before a specific activity can start. The underlying assumption is that the type of relationships among activities is finish-to-start with zero lag.

Each activity i requires r_i units of k (k=1,...,K) resource types per time period. If a type of resource is used in resource leveling process, then $c_k=1$ else $c_k=0$.

$$r(i) = \sum_{k \in \{1,..,K\} i \in \{1,2,..n\}} c_k r_{ki} \tag{1}$$

The most often used multi-objective function in literature for the resource leveling problem uses one or blends more than one of the following seven objectives:

1. *min Gf* (minimization of the maximum resource usage for the project)
2. *min RLI* (difference between actual and desirable resource usage)
3. *min StD* (minimization of the standard deviation)
4. *min Step* (minimization of uniform resource use from period to period)
5. *min R^2* (minimization of the squared resource usage)
6. *min RIC* (relates variation of a selected resource histogram to an ideal one)
7. *min EV* (minimum entropy)

The present GA is able to use each of the above mathematical measurements for the resource profiles evaluation.

4   Algorithmic Framework

The paper applies an evolutionary algorithmic approach (GA) for solving the resource leveling problem. In particular, optimization problems can be tackled efficiently intelligent and evolutionary methodologies and algorithms. In the scope of this paper, the resource leveling problem corresponds to a typical optimization task.

The standard genetic algorithm was initially proposed by Holland [39]. The main characteristics of the algorithm lie in the concept of evolutionary process. Genetic algorithms apply the mechanisms of selection, crossover and mutation in order to evolve the members of a population, through a number of generations. The ultimate goal is to reach a population of good-quality solutions, approaching the 'optimum' region. In order to assess the quality of each member of the population, the concept of fitness value is introduced. In the case of the resource leveling problem, each solution represents a chromosome in the GA.

In Figure 1, the GA for the resource leveling problem is presented. Below, the main terms and processes mentioned in Figure 1 are explained.

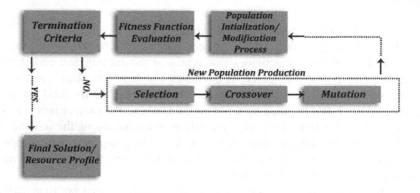

Fig. 1. GA operation

4.1   Construction of the Next Generation

The first generation is formed by randomly creating feasible resource profiles. To produce the subsequent generations, the top 10% (best chromosomes) of each generation is transported directly to the next generation, 70% of the descendant generation is produced by applying the two-point crossover operator on parents' chromosomes. Finally, the last 20% chromosomes of the offspring will occur from mutation generated in the same way by which the initial population was formed. Resource profiles are generated by taking the start-time of non-critical activities contained in each chromosome and the start-times of critical activities that always remain unchanged. Then, the fitness function (e.g. RLI) is used to evaluate each resource profile.

It is important to be mentioned that the crossover and mutation processes produce the sequence for the non-critical activities start-time production. The solution feasibility depends on the non-critical activity ES-LS as well as on the correlations that has each non-critical activity with the other non-critical activities. All possible different start-time values of all non-critical activities of a project are explored and are combined on the basis of the ES-LS value restrictions and on non-critical activities correlations. In other words only efficient solutions are produced and no efficient solution is excluded by the solution search process.

Let's denote as S: start-time (day), SP: time step (in days), SS_1: new early start value and SS_2: new late start value. After the initial population, a new restriction applies, which is a frame $\pm SP$ from the best chromosome start-times for the non-critical activities. GA applies local search around the previous generation best chromosome using this frame. SP also defines how close to the best chromosome the searching takes place, by receiving values between: $\{min\,SP = 1 : max\,SP = LS - (ES + 1)\}$.

When SP is defined by the decision maker, two new limits are set which satisfy the ES, LS restrictions:

$$for\ the\ SS_1: SS_1 \geq ES\ and\ for\ the\ SS_2: SS_2 \leq LS$$

These new bounds SS_1, SS_2 obtained from the above described process, as well as the correlations between non-critical activities, are the restrictions for the new start-time production taking place in Crossover process (Fig. 2).

The proposed technique for the the investigation of the start time of the non-critical activities of a project has the following three advantages which contribute to the GA effectiveness in optimal solution finding:

- Maintenance of the optimal solutions (TOP) from the previous generation in order to be compared with the newly produced solutions.
- The restriction $\pm SP$ from the optimal solution start-time in previous generation offers the possibility for faster convergence to an optimal solution.
- The BOT chromosomes receive randomly a start-time value according to a uniform distribution in corresponding interval ES - LS for each gene. This fact releases the GA algorithm from a premature convergence in a semi-optimal solution.

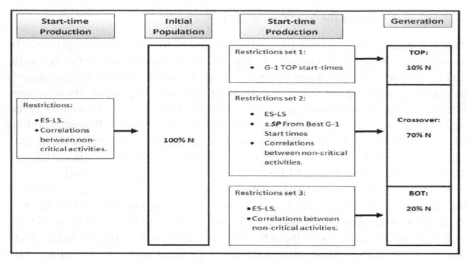

Fig. 2. New generation start-time production

4.2   Benchmark Experiments

The parameter settings of the GA approach used during the experimentation were (a) size of population set to 40, (b) number of generations set to 50, (c) best 10% chromosomes of each generation transferred directly to the next generation, (d) 70% of the current population subjected to crossover, (e) mutation in the remaining 20% of chromosomes, and (f) number of independent simulations set to 50 per criterion (parameters' values have been selected after extensive experimentation).

Firstly the proposed GA was compared with similar studies existing in literature. Exhaustive enumeration was also applied for small size problem and the proposed GA approach could identify the optimal solution for those problems. Specifically, 50 repeated runs of the algorithm were made, and in almost all cases the GA-approach provided excellent results identifying the global minimum of the benchmark projects (excluding the RLI criterion for the problem posed by [24], where the optimal solution was identified in 37 out of 50 repeated experiments, Table 1). Results in Table 1 show that in most cases the proposed approach is superior to the existing ones. The implementation of the algorithm was made in Matlab programming language. Experiments were conducted on a Pentium(R) Dual-Core CPU E5400 at 2,70 GHz and RAM   2,00 GB.

Table 1. Experimental results for benchmark problems existing in literature

Research Paper	Function	Optimal result	Result of Paper	GA best result	times found optimal in experiments
[10] Geng et al. (2011)	STD	**5,2034**	5,6151	**5,2034**	100%
[29] Katram & Tongthong (1998)	R^2	**885**	885	**885**	100%
[40] Kim et al (2005)	R^2	**448**	448	**448**	100%
[24] Leu et al (2000)	RLI	**25**	25	**25**	74%
[20] Leu and Hung (2002)	RLI	**11**	13,09	**11**	100%

Table 2. Results of experiments for 5 benchmark PSPlib projects

PROJECT		GA			PROJECT		GA		
		Gf	RLI	StD			Gf	RLI	StD
P 1	mean	65,26	829,2	15,34	P 4	mean	61,38	1084,8	17,05
	std	2,63	106,45	1,74		std	3,17	125,08	1,62
	min	59	657	12,78		min	54	839	14,15
P 2	mean	44,44	697,72	9,94	P 5	mean	50,32	973,12	13,50
	std	1,16	53,49	0,64		std	1,66	86,05	0,91
	min	42	552	7,89		min	46	765	11,75
P 3	mean	62,78	877,76	17,08					
	std	2,61	70,7	1,16					
	min	56	744	14,91					

5   Concluding Remarks

In this paper, the application of evolutionary algorithms for solving resource leveling problems was studied. The basic idea behind this work is the way that the proposed approach investigates the various different feasible start-time values of non-critical activities of a project, thus leading to improved resource profiles and contributing to optimal resource allocation. This mechanism (SP) assists the decision maker in searching efficiently the area surrounding the optimal star-time values. The proposed mutation process also assists in avoiding premature convergence in a semi-optimal solution. Experimental results are considered as near-optimal because mathematically the optimal solution cannot be found. With repeated runs one can conclude about the approximate area where the optimum lies.

In conclusion, evolutionary intelligence (GAs) proves capable of solving adequately complex optimization problems and can be proposed as an efficient and effective method for coping with large real-world projects. It is our firm belief that the optimal resource profile could be achieved through even more extensive experimentation and simulations, e.g. in a ship construction project consisting of more than 1000 activities, which is currently tested on its effectiveness. Research is also underway around the implementation of other competitive nature-inspired and hybrid intelligent methodologies on the same application domain.

References

1. Holsapple, C.W., Jacob, V.S., Whinston, A.B.: Operations Research and Artificial Intelligence. Ablex Publishing Corporation (1994)
2. Soak, S.-M., Lee, S.-W.: A memetic algorithm for the quadratic multiple container packing problem. Applied Intelligence 36, 119–135 (2012)

3. Ippolito, M.G., Morana, G., Riva Sanseverino, E., Vuinovich, F.: Ant Colony Search Algorithm for Optimal Strategical Planning of Electrical Distribution Systems Expansion. Applied Intelligence 23, 139–152 (2005)
4. Lim, A., Lin, J., Xiao, F.: Particle Swarm Optimization and Hill Climbing for the bandwidth minimization problem. Applied Intelligence 26, 175–182 (2007)
5. Neumann, K., Zimmermann, J.: Resource levelling for projects with schedule-dependent time windows. European Journal of Operational Research 117, 591–605 (1999)
6. Hegazy, T.: Optimization of Resource Allocation and Leveling using Genetic Algorithm. Journal of Construction Engineering and Management 125(3), 167–175 (1999)
7. Younis, M.A., Saad, B.: Optimal resource leveling of multi-resource projects. Computers and Industrial Engineering 31, 1–4 (1996)
8. Elwany, M.H., Korish, I.E., Barakat, M.A., Hafez, S.M.: Resource smoothening in repetitive projects. Computers and Industrial Engineering 35, 415–418 (1998)
9. Chan, W.T., Chua, D.K.H., Kannan, G.: Construction resource scheduling with genetic algorithms. Journal of Construction Engineering and Management 122, 125–132 (1996)
10. Geng, J.Q., Weng, L.P., Liu, S.H.: An improved ant colony optimization algorithm for nonlinear resource-leveling problems. Computers and Mathematics with Applications 61, 2300–2305 (2011)
11. Burges, A., Killebrew, J.: Variation in activity level in a cyclical arrow diagram. Journal of Industrial Engineering 13(2), 76–83 (1962)
12. Galbreath, R.: Computer program for leveling resource usage. J. Constr. Div., ASCE 91(1), 107–124 (1965)
13. Shaffer, L., Ritter, J., Mayer, W.: The critical path method. McGraw- Hill Book C, NY (1965)
14. Wiest, J.D., Levy, F.K.: A Management Guide to PERT/CPM. Prentice-Hall, NJ (1977)
15. Harris, R.B.: Packing method for resource leveling _PACK. Journal of Construction Engineering and Management, ASCE 116(2), 331–350 (1990)
16. Antill, J.M., Woodhead, R.W.: Critical Path Methods in Construction Practice, 3rd edn. Wiley (1982)
17. Moder, J.J., Philips, C.R., Davis, E.W.: Project Management with CPM, PERT and Precedence Diagramming, 3rd edn. Van Nostrand-Reinhold, New York (1983)
18. Neumann, K., Zimmermann, J.: Procedures for resource leveling and net present value problems in project scheduling with general temporal and resource constraints. European Journal of Operational Research 127, 425–443 (2000)
19. Leu, S.S., Yang, C.H.: GA-based multicriteria optimal model for construction scheduling. Journal of Construction Engineering & Management 125, 420–427 (1999)
20. Leu, S.S., Hung, T.H.: An optimal construction resource leveling scheduling simulation model. Canadian Journal of Civil Engineering 29, 267–275 (2002)
21. Huang, J.W., Wang, X.X., Chen, R.: Genetic algorithms for optimization of resource Allocation in Large Scale Construction Project Management. Journal of Computers 5, 1916–1924 (2010)
22. Li, J.H.: Combination of genetic & ant colony algorithms for multi-project resource leveling problem. Jisuanji Jicheng Zhizao Xitong/Computer Integrated Manufacturing Systems, CIMS 16, 643–649 (2010)
23. Zhou, L., Peng, W., Zhang, Z.: An ant colony system for solving resource leveling problem. In: International Conference on Intelligent Computation Technology & Automation (ICICTA), China, pp. 489–492 (2010)

24. Leu, S.S., Yang, C.H., Huang, J.C.: Resource leveling in construction by genetic algorithm-based optimization and its decision support system application. Automation in Construction 10, 27–41 (2000)
25. Liu, S.X., Wang, M.G.: Genetic algorithm for resource leveling problem in project scheduling. Xitong Gongcheng Lilun yu Shijian/System Engineering Theory and Practice 21, 24 (2001)
26. Senouci, A.B., Eldin, N.N.: Use of genetic algorithms in resource scheduling of construction projects. Journal of Construction Engineering and Management 130, 869–877 (2004)
27. Chen, Z.Y., Du, Z.D., Zhou, H.: Research on the unlimited resource leveling optimization with PSO. Tumu Gongcheng Xuebao/China Civil Engineering Journal 40, 93–96 (2007)
28. Guo, X., Li, N., Li, X.S.: Multi-resource leveling in multiple projects and vector evaluated particle swarm optimization based on Pareto. Kongzhi yu Juece/Control and Decision 25, 789–793 (2010)
29. Kartam, N., Tongthong, T.: An artificial neural network for resource leveling problems. Artificial Intelligence for Engineering Design, Analysis and Manufacturing: AIEDAM 12, 273–287 (1998)
30. Jeetendra, V.A., Krishnaiah, C.O.V., Prashanth, R.: Petri nets for project management and resource levelling. International Journal of Advanced Manufacturing Technology 16, 516–520 (2000)
31. Raja, K., Kumanan, S.: Resource leveling using Petrinet and memetic approach. American Journal of Applied Sciences 4, 317–322 (2007)
32. Pang, N., Shi, Y., You, Y.: Resource Leveling Optimization of Network Schedule Based on Particle Swarm Optimization with Constriction Factor. In: International Conference on Advanced Computer Theory and Engineering, pp. 652–656 (2008)
33. Leachman, R.C.: Multiple Resource Leveling in Construction Systems Through Variation of Activity Intensities. Naval Research Logistic Quarterly 30, 187–198 (1983)
34. Akpan, E.O.P.: Resource smoothing: A cost minimization approach. Production Planning and Control 11(8), 775–780 (2000)
35. Zhao, S.-L., Liu, Y., Zhao, H.-M., Zhou, R.-L.: GA based resource leveling optimization for construction project. In: Proceedings of the Fifth International Conference on Machine Learning and Cybernetics, Dalian, pp. 2363–2367 (2006)
36. Xiong, Y., Kuang, Y.P.: Ant colony optimization algorithm for resource leveling problem of contruction problem. In: The CRIOCM 2006 International Symposium on: "Advancement of Construction Management and Real Estate" (2006)
37. Roca, J., Pugnaghi, E., Libert, G.: Solving an Extended Resource Leveling Problem with Multiobjective Evolutionary Algorithms. International Journal of Computational Intelligence 4, 289–300 (2008)
38. Anagnostopoulos, K.P., Kouklinas, G.: A simulated annealing hyperheuristic for construction resource leveling. Construction Management and Economics 28, 163–175 (2010)
39. Holland, J.H.: Genetic Algorithms. Scientific American 267(1), 66–72 (1992)
40. Kim, J., Kim, K., Jee, N., Yoon, Y.: Enhanced Resource Leveling Technique for Resource Scheduling. Journal of Asian Architecture and Building Engineering 4(2), 461–466 (2005)

Inverse Reliability Task: Artificial Neural Networks and Reliability-Based Optimization Approaches

David Lehký, Ondřej Slowik, and Drahomír Novák

Brno University of Technology, Brno, Czech Republic
{lehky.d,slowik.o,novak.d}@fce.vutbr.cz

Abstract. The paper presents two alternative approaches to solve inverse reliability task – to determine the design parameters to achieve desired target reliabilities. The first approach is based on utilization of artificial neural networks and small-sample simulation Latin hypercube sampling. The second approach considers inverse reliability task as reliability-based optimization task using double-loop method and also small-sample simulation. Efficiency of both approaches is presented in numerical example, advantages and disadvantages are discussed.

Keywords: Inverse reliability, artificial neural network, reliability-based optimization, double-loop optimization, uncertainties, Latin hypercube sampling.

1 Introduction

To achieve desired level of reliability in limit state design is generally not an easy task. Uncertainties are involved in every part of structural system (e.g. material properties, geometrical imperfections, dead load, live load, wind, snow, corrosion rate, etc.). When performing either reliability assessment or advanced engineering design, it is certainly essential to take uncertainties into account using a probabilistic analysis. Reliability assessment requires forward reliability methods for estimating the reliability (usually theoretical failure probability and/or reliability index are determined). On the other hand, the engineering design requires an inverse reliability approach to determine the design parameters to achieve desired target reliabilities.

Some sophisticated approaches to determine design parameters (material properties, geometry, etc.) related to particular limit states have been proposed under the name "inverse reliability methods", e.g. a reliability contour method [1] and [2], iterative algorithm based on the modified Hasofer-Lind-Rackwitz-Fiessler scheme used in reliability analysis [3], Newton-Raphson iterative algorithm to find multiple design parameters [4] and [5], decomposition technique [6] or various implementation of artificial neural network (ANN) with other soft-computing techniques [7], [8] and [9].

The two methods proposed in this paper attempts to overcome the shortcomings of existing inverse reliability methods. The first one utilizes ANN too, but in a different way: Computational time is reduced by using a small-sample simulation technique called Latin hypercube sampling (LHS) in ANN based inverse problem proposed by Novák and Lehký in [10] and [11] first.

L. Iliadis et al. (Eds.): AIAI 2014, IFIP AICT 436, pp. 344–353, 2014.

The second one is double-loop reliability based optimization (RBO) approach. Classical optimization usually leads to solutions that lie at the boundary of the admissible domain, and that are consequently rather sensitive to uncertainty in the design parameters. In contrast, RBO aims at designing the system in a robust way by minimizing some objective function under reliability constraints. It provides the means for determining the optimal solution of a certain objective function, while ensuring a predefined small probability that a structure fails. Thus RBO methods have to mix optimization algorithms together with reliability calculations. The approach known as "double-loop" consists in nesting the computation of the failure probability with respect to the current design within the optimization loop (e.g. [12]). FORM-based double-loop approach has been proposed by Dubourg in [13, 14]. The authors developed a double-loop reliability-based optimization approach based on small-sample simulation and FORM [15, 16].

2   Inverse Reliability Task

The aim of classical (forward) reliability analysis is the estimation of unreliability using a probability measure called the theoretical failure probability, defined as:

$$p_f = P(Z \le 0), \tag{1}$$

where Z is a function of basic random variables $\mathbf{X} = X_1, X_2, \ldots, X_N$ called safety margin. This failure probability is calculated as a probabilistic integral:

$$p_f = \int_{D_f} f_{\mathbf{X}}(\mathbf{X}) \, d\mathbf{X} \tag{2}$$

where the domain of integration of the joint probability distribution function (PDF) above is limited to the failure domain D_f where $g(\mathbf{X}) \le 0$. The function $g(\mathbf{X})$, a computational model, is a function of random vector $\mathbf{X}$ (and also of other, deterministic quantities). Random vector $\mathbf{X}$ follows a joint PDF $f_{\mathbf{X}}(\mathbf{X})$ and, in general, its marginal variables can be statistically correlated. The explicit calculation of integral in (2) is generally impossible. Therefore a large number of efficient stochastic analysis methods have been developed during the last seven decades.

The inverse reliability task is the task to find design parameters corresponding to specified reliability levels expressed by reliability index or by theoretical failure probability. In general, an inverse problem involves finding either a single design parameter to achieve a given single reliability constraint or multiple design parameters to meet specified multiple reliability constraints. The design parameters can be deterministic or they can be associated with random variables described by statistical moments (mean value, standard deviation) and PDF. In case of mean value one need to choose if either standard deviation or coefficient of variation will be fixed.

2.1   Solution Based on Artificial Neural Networks

An efficient general approach of inverse reliability analysis is proposed to obtain design parameters of a computational model in order to achieve the prescribed reliability level. The inverse analysis is based on the coupling of a stochastic simulation of Monte Carlo type and an ANN. The design parameters (e.g. mean values or standard deviations of basic random variables) play the role of basic random variables with a scatter reflecting the physical range of design values. A novelty of the approach is the utilization of the efficient small-sample simulation method LHS used for the stochastic preparation of the training set utilized in training the ANN. The calculation of reliability is performed using the first order reliability method (FORM). Once the ANN has been trained, it represents an approximation consequently utilized in an opposite way: To provide the best possible set of design parameters corresponding to prescribed reliability.

The procedure of ANN based inverse reliability method is illustrated by a simple flow chart as shown in Figure 1 and is implemented as follows:

1. The design parameters are considered as random variables with selected (physically reasonable) appropriate scatter and probability distribution. Rectangular distribution is often used.
2. Random samples of design parameters (possibly correlated) are generated using LHS simulation method.
3. Stochastic model of analyzed problem is prepared including generated samples of design parameters.
4. Reliability analyses are performed repeatedly for individual samples of design parameters and set of reliability measures like failure probabilities or reliability indices are calculated.
5. Reliability measures obtained from simulations together with set of random design parameters serve as training set for ANN training. During training an error between simulated and desired outputs of ANN (here in form of MSE) is minimized using appropriate optimization technique (e.g. back propagation methods, evolutionary algorithms).
6. Desired reliability measures are used as an input signal which is distributed through ANN structure to its output where optimal design parameters are obtained.
7. Verification of the results by calculation of failure probabilities related to limit state functions using the optimal parameters is carried out. A comparison with target failure probabilities will show the extent to which the inverse analysis was successful.

In the case of inverse reliability analysis a double stochastic analysis is needed for the training set preparation for ANN (steps 2 and 4 of the procedure). In the outer loop random realizations of design parameters are generated using the LHS simulation technique. The inner loop represents the reliability calculation for one particular realization of design parameters. Here, the FORM approximation method is recommended due to computational demands. The number of simulations in outer loop is driven by ANN and only tens of simulations are usually needed.

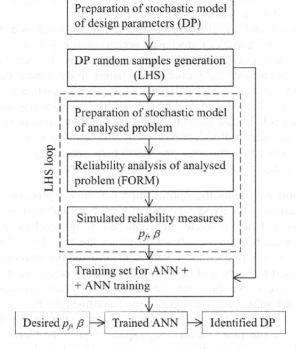

Fig. 1. A flow chart of proposed inverse reliability method

2.2   Solution by Small-Sample Double-Loop Reliability-Based Optimization

Typically, reliability-based optimization is formulated as:

$$\text{find } \mathbf{x}$$
$$\min f(\mathbf{x}) \tag{3}$$
$$\text{subject to: } P_f\left[g(\mathbf{x}, \mathbf{z}) \le 0\right] \le P_0, \quad \mathbf{l} \le \mathbf{x} \le \mathbf{u}$$

with P_f the probability of constraint satisfaction. The limit state $g = 0$ separates the region of failure $(g \le 0)$ and safe region $(g > 0)$ and is a function of the design variables $\mathbf{x}$ (and $\mathbf{l}$ and $\mathbf{u}$ are lower and upper bounds) and the uncertain variables $\mathbf{z}$. P_0 is the reliability level or performance requirement. The above inequality can be expressed by a failure probability multi-dimensional integral with the joint probability density function of probabilistic variables $\mathbf{z}$. Formulation based on reliability index instead of failure probability is popular especially in the context of FORM approximation.

From the theoretical point of view, RBO has been a well-established concept. However, computing failure probabilities appears as a technically involved problem analytically tractable for very simple cases only. This is because it is often a multi-dimensional integral equation for which the joint probability density function and/or limit state function g is unknown in explicit form, like FEM computational model. The same difficulty is with objective function $f(\mathbf{x})$ – it can be computationally demanding FEM analysis and the use of classical optimization technique is problematic or even impossible. Then an application of RBO for real-world problem is difficult.

Computational demands of reliability-based optimization are obvious from the formulation above. For the purposes of stochastic optimization it is necessary to repeatedly generate random realizations within the design space. It is also necessary for each of these realizations to calculate the probability of failure in the general case by computationally demanding (mostly numerical) integration of the equation (2). Therefore we suggest here an original small-sample double-loop RBO methodology where lower computational burden exists in case of both outer loop – minimization of objective function and inner loop – calculation of failure probability (or reliability index). A practical solution to the above-defined optimization problem is performed using the so-called double-loop approach. The algorithm is composed of two basic loops:

- **The outer loop** represents the optimization part of the process based on small-sample simulation Latin hypercube sampling. The simulation within the design space is performed in this cycle. For obtained design vectors of n-dimensional space $\mathbf{x}_i=(\mathbf{x}_1, \mathbf{x}_2,..., \mathbf{x}_n)$ objective function values are calculated. The best realization is then selected based on these values and utilized optimization method. Consequently the best realization of random vector $\mathbf{x}_{i,best}$ is compared with optimization constraints. These constraints may be formulated by any deterministic function which functional value can be compared with a defined interval of allowed values. Constraints are also possible to formulate as allowed interval of reliability index β for any limit state function (within design space of given problem). Calculations of reliability index of each generated random vectors $\mathbf{x}_i$ takes place in the inner loop. Note that it is necessary to use some of advanced meta-heuristic optimization techniques (e.g. simulated annealing or genetic algorithms) to avoid local minima.
- **The inner loop** is used to calculate reliability index (FORM-based) either for the need of checking of generated solutions – if they satisfy constraints, or to calculate the actual value of the objective function, if the target reliability index is set as goal of optimization process.

3   Numerical Example

Selected application originates from the civil engineering field of structural mechanics. The aim is to design the dimensions of a rectangular cross-section with width b and height h of a simply–supported beam made of timber (Figure 2). Both dimensions are considered as random variables with a variation of 5 %. The mean values of b and h are design parameters in the inverse reliability problem.

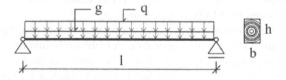

Fig. 2. Scheme of a simply–supported beam with a rectangular cross-section

The design is performed fully according to Eurocode 5. The ultimate limit state (ULS) as well as the serviceability limit state (SLS) is taken into account. Target reliability indices are $\beta_1 = 3.8$, $\beta_2 = 1.5$. The limit states are described by the following limit state functions g_1 and g_2:

$$g_1 = M_R - M_E$$
$$g_2 = u_{\lim, fin} - u_{net, fin}$$

(4)

where M_R is the bending moment of resistance, M_E is the bending moment of load action, $u_{\lim fin}$ is the final limit deflection and u_{netfin} is the final deflection caused by load action. Bending moments M_R and M_E are calculated as:

$$M_R = \theta_R \frac{1}{6} bh^2 k_{\mod} f_m$$
$$M_E = \theta_E \frac{1}{8} (g + q) l^2$$

(5)

where b and h are the width and height of rectangular cross-section, l is the length of the beam, f_m is flexural strength, k_{mod} is the modification factor taking into account the effect on the strength parameters of the duration of the load and the moisture content in the structure (value $k_{mod} = 0.8$ was considered), g is permanent load, q is variable load and θ_R and θ_E are the model uncertainties of resistance and load action. Deflections in the second limit state function g_2 are calculated as:

$$u_{\lim, fin} = \frac{l}{200}$$
$$u_{net, fin} = \theta_E \left(u_{1, fin} + u_{2, fin} \right)$$
$$u_{1, fin} = \frac{5}{384} \frac{g l^4}{E \frac{1}{12} bh^3} \left(1 + k_{1, def} \right)$$
$$u_{2, fin} = \frac{5}{384} \frac{q l^4}{E \frac{1}{12} bh^3} \left(1 + k_{2, def} \right)$$

(6)

where u_{1fin} and u_{2fin} are the final deflections caused by the permanent load and variable load, E is the modulus of elasticity of timber, $k_{1,def}$ is a factor which takes into account the increase in deflection with time due to the combined effect of creep and moisture and it belongs to permanent load and $k_{2,def}$ is the same factor but for variable load (values of $k_{1,def} = 0.8$ and $k_{2,def} = 0.25$ were used). Table 1 summarizes all random variables and their randomization. The values of the material parameters correspond to spruce timber. Randomization was carried out according to the recommendations of JCSS probabilistic model code [17]. Reliability analysis was carried out using the FORM method; the starting values were means; the tolerance for convergence was 10^{-4}. For both a training set preparation and purpose of optimization the design parameters were considered as random variables with rectangular distribution, see Table 2.

Table 1. Random variables and design parameters

Variable	Distribution	Mean	Std	COV
l [m]	Normal	3.5	0.175	0.05
b [m]	Normal	?	--	0.05
h [m]	Normal	?	--	0.05
E [GPa]	Lognormal (2 par)	10	1.3	0.13
f_m [MPa]	Lognormal (2 par)	34	8.5	0.25
g [kN/m]	Gumbel max EV 1	1.686	0.169	0.10
q [kN/m]	Gumbel max EV 1	2.565	0.770	0.30
θ_R [-]	Lognormal (2 par)	1	0.1	0.10
θ_R [-]	Lognormal (2 par)	1	0.1	0.10

Table 2. Randomization of design parameters for training set preparation and purpose of optimization

Variable	Distribution	Mean	Std	a	b
mean(b)	Rectangular	0.125	0.0144	0.10	0.15
mean(h)	Rectangular	0.225	0.0144	0.20	0.25

3.1   ANN Inverse Analysis

First, ANN inverse reliability analysis was carried out. The ANN (see Figure 3) consisted of one hidden layer having four nonlinear neurons (hyperbolic tangent transfer function) and an output layer having two output neurons (linear transfer function) which correspond to two design parameters – the mean values of width b and height h. The ANN has two inputs which correspond to two specified reliability indices, β_1 and β_2. For preparation of training set one hundred random samples of design parameters were generated using LHS method according to stochastic model in Table 2 and stochastic analyses were carried out to obtain corresponding reliability indices.

The resulting design parameter values are given in Table 3. To check their accuracy these values were used in equations (4) to (6) and reliability indices were calculated; see the comparison with the target reliability indices in Table 3. In the case of practical design the dimensions of cross-section should be selected from available set of dimensions. In our example, the resulting width and height would be $b = 140$ mm and $h = 220$ mm which gives the final reliability indices $\beta_{1,fin} = 4.068$ and $\beta_{2,fin} = 1.912$.

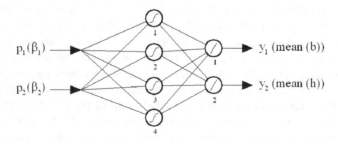

Fig. 3. A scheme of artificial neural network

3.2   Double-Loop Optimization Approach

If we define optimization problem according to definition at section 2.2 then inverse reliability problem can be also solved using double-loop reliability-based optimization approach. During the solution of the problem an option to determine a target value of reliability index for the selected limit state function was utilized. Therefore target reliability index for the limit state function g_2 was defined as $\beta_2 = 1.5$. As a constraint an interval (3.75, 3.85) of allowable values of reliability index β_1 for the limit state function g_1 was set.

During solution of the task using aimed multilevel sampling (AMS) optimization algorithm [15] the total number of 300 simulations was used. The result solution of the task is displayed in Table 3 and in Figure 4. Final solution corresponds well to the values obtained using ANN inverse analysis. The resulting cross-sectional area is $2.8302{\times}10^{-2}$ m^2 compared to $2.8385{\times}10^{-2}$ m^2 obtained from ANN inverse analysis. The graph in Figure 4 shows the gradual convergence of generated solutions toward the required values of reliability indices.

Table 3. Resulting values of design parameters and reliability indices β obtained by double-loop optimization approach

Approach	mean(b)	mean(h)	β_1	β_2	$\beta_{1,target}$	$\beta_{2,target}$
ANN inverse analysis	0.13244	0.21432	3.8001	1.5001	3.8	1.5
Double-loop optimization	0.1311555	0.215135	3.793	1.50009		

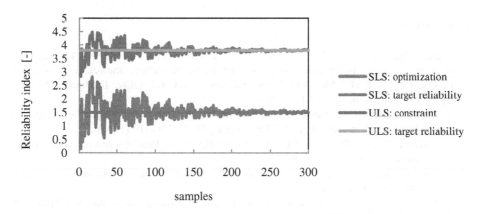

Fig. 4. Evolution of values of reliability indices during optimization

4   Conclusion

The paper presents two alternative approaches to solve inverse analysis task. Both approaches provide very good results, as is indicated in numerical example. Some advantages and disadvantages of the methods can be highlighted:

ANN inverse analysis will be probably more accurate and capable to solve also highly nonlinear problems. But more simulations are generally needed for good training of ANN (in case of very small numbers – training cannot be simply done at all). The step "training of ANN" requires deeper involvement of a user, which makes the usage of the approach difficult.

On the other hand the double-loop reliability-based optimization approach can solve problem satisfactorily using small number of simulations, but the lower accuracy can be expected. In case of highly nonlinear problems less efficiency can be expected comparing to ANN inverse analysis approach. The advantage of double-loop optimization approach is a transparency of solution and better understanding by general engineering practice.

The above mentioned summary is formulated based on testing approaches using limited number of numerical examples. The more systematic verification and testing are needed. Presents results indicate that both approaches have very good potential to solve inverse reliability task using small-sample simulation.

Acknowledgements. The authors give thanks for the support of the research project No. TA01011019 of Technology Agency of the Czech Republic (TACR) and the European Union's "Operational Programme Research and Development for Innovations", No. CZ.1.05/2.1.00/03.0097, as an activity of the regional Centre AdMaS "Advanced Materials, Structures and Technologies". In this undertaking, theoretical results gained in the project of the specific university research at Brno University of Technology, registered under the number FAST-J-14-2425 were exploited.

References

1. Winterstein, S.R., Ude, T.C., Cornell, C.A.: Environ-mental Parameters for Extreme Response: Inverse Form with Omission Factors. In: Schueller, Shinozuka, Yao (eds.) Proceedings of ICOSSAR 1993: Structural Safety and Reliability, Innsbruck, Austria, pp. 551–557. Balkema, Rotterdam (1994)
2. Maes, M.A., Huyse, L.: Developing Structural Design Criteria with Specifies Response Reliability. Canadian Journal of Civil Engineering 24(2), 201–210 (1997)
3. Der Kiureghian, A., Zhang, Y., Li, C.C.: Inverse Reliability Problem. Journal of Engineering mechanics 120(5), 1154–1159 (1994)
4. Li, H., Foschi, R.O.: An Inverse Reliability Method and its Application. Structural Safety 20, 257–270 (1998)
5. Sadovský, Z.: Discussion on: An Inverse Reliability Method and its Application. Structural Safety 22(1), 97–102 (2000)
6. Mínguez, R., Castillo, E., Hadi, A.S.: Solving the Inverse Reliability Problem Using Decomposition Techniques. Structural Safety 27, 1–23 (2005)
7. Shayanfar, M.A., Massah, S.R., Rahami, H.: An Inverse Reliability Method Using Networks and Genetic Algorithms. World Applied Sciences Journal 2(6), 594–601 (2007)
8. Cheng, J., Li, Q.S.: Application of the Response Surface Methods to Solve Inverse Reliability Problems with Implicit Response functions. Computational Mechanics 43(4), 451–459 (2009)

9. António, C., Hoffbauer, L.: Uncertainty Propagation in Inverse Reliability-based Design of Composite Structures. International Journal of Mechanics and Materials in Design 6(1), 89–102 (2010)
10. Novák, D., Lehký, D.: ANN Inverse Analysis Based on Stochastic Small-Sample Training Set Simulation. J. of Eng. Application of Artificial Intelligence 19, 731–740 (2006)
11. Lehký, D., Novák, D.: Solving Inverse Structural Reliability Problem Using Artificial Neural Networks and Small-Sample Simulation. Advances in Structural Engineering 15(11), 1911–1920 (2012)
12. Tsompanakis, Y., Lagaros, N., Papadrakis, M. (eds.): Structural Design Optimization Considering Uncertainties. Taylor & Francis (2008)
13. Dubourg, V., Noirfalise, C., Bourinet, J.-M.: Reliability-Based Design Optimization: An Application to the Buckling of Imperfect Shells. In: 4th ASRANet Colloquium, Athens, Greece (2008)
14. Dubourg, V., Bourinet, J.-M., Sudret, B.: A Hierarchical Surrogate-based Strategy for Reliability-based Design Optimization. In: Straub, D., Esteva, L., Faber, M. (eds.) Proc. 15th IFIP WG7.5 Conference on Reliability and Optimization of Structural Systems, Munich, Germany, pp. 53–60. Taylor & Francis (2010)
15. Slowik, O.: Reliability-based Structural Optimization. Master's thesis. Brno University of Technology. Supervisor: prof. Ing. Drahomír Novák, DrSc, Brno, Czech Republic (2014)
16. Slowik, O., Novák, D.: Algorithmization of Reliability-based Optimization. Transactions of the VŠB - Technical University of Ostrava. Civil Engineering Series (2014) ISSN: 1804-4824
17. JCSS.: JCSS Probabilistic Model Code, Joint Committee on Structural Safety, Technical University of Denmark, Kongens Lyngby, Denmark (2007)

Model-Based Generation of Realistic 3D Full Body Avatars from Uncalibrated Multi-view Photographs

Nicholas Michael and Andreas Lanitis

Visual Media Computing Lab, Dept. of Multimedia and Graphic Arts,
Cyprus University of Technology, 3036 Lemesos, Cyprus
{nicholas.michael,andreas.lanitis}@cut.ac.cy

Abstract. In today's world of rapid technological advancement, we find an increasing demand for low-cost systems that are capable of fast and easy generation of realistic avatars for use in Virtual Reality (VR) applications. For example, avatars can enhance the immersion experience of users in video games and facilitate education in virtual classrooms. Therefore, we present here a novel model-based technique that is capable of real-time generation of personalized full-body 3D avatars from orthogonal photographs. The proposed method utilizes a statistical model of human 3D shape and a multi-view statistical 2D shape model of its corresponding silhouettes. Our technique is automatic, requiring minimal user intervention, and does not need a calibrated camera. Each component of our proposed technique is extensively evaluated and validated.

Keywords: personalized avatars, 3D body shape modelling, multi-view ASM, image segmentation.

1 Introduction

In today's world of rapid technological advancement, we find an increasing demand for low-cost systems that are capable of fast and easy generation of realistic avatars with minimal user intervention. Such systems have applicability in many Virtual Reality (VR) applications. For example, they can be used in cultural heritage visualizations and to populate virtual worlds in multi-player computer games, enhancing a user's immersion experience. In addition, they can facilitate communication and education when incorporated in chat applications and in virtual classrooms. They can even assist users to make shopping decisions by allowing them to dress their avatars accordingly. Existing technologies make it possible to create 3D avatars, e.g., using 3D scanning devices such as Microsoft's Kinect, using 3D modelling software, etc. However, this kind of technologies tend to require dedicated hardware, calibration of imaging equipment, creative skills and considerable amount of post-processing manual intervention.

Motivated by this multitude of applications and the limitations of existing technologies, we develop a model-based method for the generation of realistic personalized full-body 3D avatars [1]. The novelty of our work is that it is automatic and it works using photographs taken by any uncalibrated low cost camera. Our method extends the work of Hilton et. al. [2], however ours does not require the user to mark the location of 3D landmarks in the 2D image, in order to extract model-silhouette correspondences, for

L. Iliadis et al. (Eds.): AIAI 2014, IFIP AICT 436, pp. 354–363, 2014.

the purpose of improving accuracy. Furthermore, the proposed method does not require the user to measure the camera's field of view and the subject's distance to the camera, which serve as camera calibration. Instead, and as our main contribution, we train and use a *multi-view* Active Shape Model (ASM) [3] of human silhouettes (as viewed from the front, left, right and back), in order to refine the extracted silhouettes obtained from the segmentation result and to register them to the projection of the 3D shape model in each of the orthogonal views. Our only assumption is that when taking the multi-view input images the user does not significantly change their depth relative to the camera. This means that the avatars can be generated quickly and easily, as no time is wasted for calibrating the camera nor for marking correspondences. Additionally, in order to increase the recognizability and hence the realism of the generated avatar, together with the full-body model we use a face-only 3D model, which has a significantly higher resolution aimed at capturing the more detailed facial characteristics.

An outline of our method is illustrated in Figure 1. First we train a statistical model of 3D human shape using a dataset of dense full-body 3D scans [4]. We project the 3D scans of the dataset to orthogonal views and extract the corresponding silhouettes, which we use to train our multi-view ASM [3], thus concluding the training phase. Once these two models are trained, we can deploy the proposed technique live as follows. Orthogonal input images are captured using an uncalibrated low cost camera and we extract 2D silhouettes using background subtraction and other image processing techniques. The extracted silhouettes are refined and registered to the 3D model's projection using the trained multi-view ASM. For improved accuracy around the face, we detect facial features (eyes, nose and mouth) using Viola/Jones detectors [5] and combine the detection result with the registered silhouettes, while performing an optimization over the space of permissible 3D shape parameter. Once the shape is reconstructed, the two 3D models (full-body and face-only) are aligned with each other, using a rigid transformation and then texture from input images is mapped to both models assuming a cylindrical model [2].

The remainder of our paper is organized as follows. Section 2 covers previous work on the problem of avatar generation. Section 3 discusses in detail each phase of our proposed method, such that Sect. 3.1 covers the components of the off-line model training phase and Sect. 3.2 describes the steps involved in the avatar generation phase. Section 4 presents experimentation for the evaluation of our work. We conclude with Sect. 5 where we also mention a few thoughts on possible future work.

2 Related Work

Previous efforts on the generation of realistic human-like avatars can be categorized into two groups. In one group are model-based methods (such as [2], [6], [7], [8], [9] and [10]) that rely on a model of human body shape to reconstruct the subject's geometry from a multi-view camera setup. In the other group are model-free methods (such as [11], [12], [13], [14] and [15]), which do not rely on a human body shape model and perform multi-view reconstruction, using for example, multi-view photometric stereo [12], [15] or even a setup of inexpensive range scanners like those found in the recently popularized Kinect device [11], [14], [16].

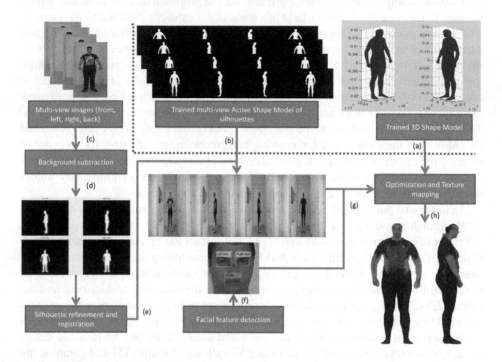

Fig. 1. Method overview (dotted line separates training and live phases): (a) A 3D shape model is trained by applying PCA on a dataset of dense full-body 3D scans, (b) A multi-view Active Shape Model is trained from the 2D silhouettes of the full-body 3D dataset, (c) Orthogonal input images are captured using an uncalibrated low cost camera, (d) Extract silhouettes using background subtraction, (e) Refine and register the silhouettes to the 3D model's projection using the multi-view ASM, (f) Detect facial features using Viola/Jones detectors, (g) Given the registered silhouettes and detected facial features optimize over the 3D shape parameters and map the texture from the input images, (h) Frontal and side views of generated 3D avatar

Fig. 2. (Left) Sample scans in used dataset [4]. Only 111 (in neutral pose) were used for training, (Right) The model can be used to generate random populations that follow the distribution of the training set by varying the shape parameters to vary e.g., height, weight, muscle/fat ratio, etc.

Model-based approaches use a prior model of the human body shape, which allows for consistent reconstruction across frames but the reconstructed shapes are limited by the utilized model, in that they may not accurately reconstruct a previously unseen human body shape that is not consistent with the training set. Model-free methods are more dynamic in nature and can reconstruct shape from any scene but because of their lack of a prior model, their reconstructed shapes may not be consistent even across neighboring frames and cannot handle ambiguities. An important advantage of model-based approaches is the uniformity of the avatars created in terms of the number and positioning of polygons, that facilitates the animation of the resulting avatars in virtual environments. In our approach we adopt a PCA model-based approach, since it naturally enforces consistency in the reconstructed shape across the four orthogonal views.

3   Methodology

The proposed avatar generation method involves two main phases: (i) the off-line model training phase and (ii) the live avatar generation phase. Figure 1 illustrates the overall work flow of our proposed technique. In the following subsections we describe the various steps in detail.

3.1   Off-line Model Training

In the off-line phase, we use a dataset of full-body range scans [4] (selecting only the 111 samples with a neutral pose – see Fig. 2 (right)) to train a 3D full-body shape model. Then the training samples are projected to four orthogonal views (front, left, right, back) to generate a training set of corresponding multi-view silhouettes, which we use to train a multi-view ASM model (see Fig. 1 (a-b)).

3D Model. In the training phase we utilize a Principal Component Analysis (PCA) model trained on range scans from a 3D body scan dataset [4] to extend the work of [2], [9] and [13]. The purpose of the PCA model is to learn the permissible modes of human body 3D shape variations reflected in the training set, so that the avatars we generate will have a realistic human shape. Each range scan is represented as a column vector, x, of vertex coordinates such that $x = [x_1, ..., x_N, y_1, ..., y_N, z_1, ..., z_N]^\top$, where N represents the number of vertices in the 3D mesh (in our case $N = 6449$). The range scans are aligned using Procrustes Alignment [17].

We apply eigen-decomposition on the covariance matrix of the aligned shapes, while keeping only the first m out of the resulting n eigenvectors, sorted in order of decreasing eigenvalue, λ_i, such that:

$$\arg \max_{1 \le m \le n} \left\{ \left(\sum_{i=1}^{m} \lambda_i \right) / \left(\sum_{i=1}^{n} \lambda_i \right) < v \right\} , \tag{1}$$

where $v \in [0, 1]$ is the amount of variance that we want the PCA model to capture (we have set $v = 0.98$ in our experiments, which resulted in $m = 13$). In this way, any new shape, x_i, can be represented as:

$$x_i \approx \bar{x} + C b_i , \tag{2}$$

where $\bar{x}$ is the mean shape of the model, C are the principal m eigenvectors of the learned shape manifold and b_i is a column vector of shape parameters, also known as the encoding of shape x_i. This encoding vector b is typically truncated, so that each element satisfies $b_i \in [-2\sqrt{\lambda_i}, +2\sqrt{\lambda_i}]$, where usually $k = 2$. This ensures that any shapes generated by (2) remain plausible with respect to the training set.

Silhouette Multi-view ASM. We project the 3D range scans of the training set to four orthogonal views (front, left, right and back) extracting the corresponding 2D silhouette contour in each view. Each silhouette contour is represented as a column vector, y_i, of 2D coordinates such that $y_i = [x_1, ..., x_C, y_1, ..., y_C]^\top$, where C represents the number of points in the 2D contour and $i \in \{$front, left, right, back$\}$. For each training range scan, the set of four silhouette vectors, representing its multi-view orthogonal 2D projection, is stacked vertically, resulting in the augmented column vector: $y = [y_{\text{front}}, y_{\text{left}}, y_{\text{right}}, y_{\text{back}}]^\top$.

The set of augmented column vectors is aligned using Procrustes Alignment and subsequently used to train by application of PCA a *multi-view* ASM model (we set $v = 0.98$, yielding $m = 33$), instead of training a separate ASM model per view. In this way, during the ASM search algorithm [3] the shape parameters are optimized across all four views simultaneously, instead of independently, providing robustness to outliers, as it naturally enforces shape consistency across all views, yielding a more accurate result. Figure 3 illustrates the modes of variation learned by the first few eigenvectors of the trained multi-view ASM.

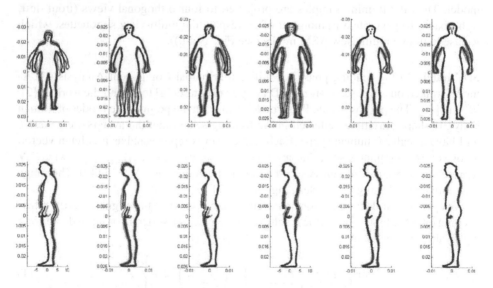

Fig. 3. Illustration of the modes of variation in the first 6 eigenvectors of our trained silhouette multi-view ASM for the frontal (top) and left-side (bottom) views. Blue plots represent the mean shape, while the green and red plots represent a variation from this mean by an amount of $-2\sqrt{\lambda_i}$ and $+2\sqrt{\lambda_i}$, respectively, where λ_i represents the ith eigenvalue and $i \in \{1, ...6\}$.

3.2   Avatar Generation

In the live phase, we extract the subject's silhouettes and then we use the trained models to reconstruct the 3D shape of the subject, register the face model and perform texture mapping (see Fig. 1 (c-h)).

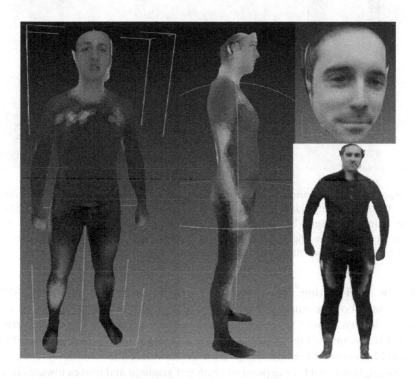

Fig. 4. Generated avatar samples

Image Acquisition. Image acquisition is done using a low cost camera. First we photograph the background and then the subject is photographed from the front, left, right and back, using a tripod to keep the camera stationary. We assume that the subject maintains a constant distance from the camera during the acquisition.

Segmentation. The segmentation step is needed to extract the silhouettes in each view. First the input images are converted to the CIELAB colorspace and we obtain their difference image with respect to the background. The difference image is then thresholded and we apply erosion and dilation to merge foreground regions, selecting the largest connected component as the foreground. We extract the contour by following 8-connected adjacent pixels on the silhouette boundary. Figure 5 illustrates the procedure.

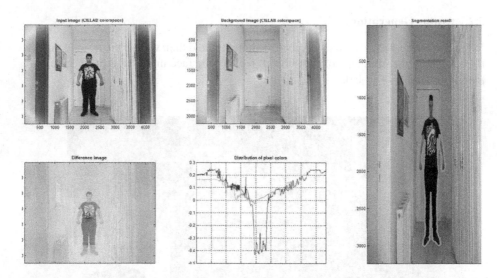

Fig. 5. Segmentation illustration: (Top left) Input image of subject converted to CIELAB colorspace, (Top middle) Input image of background converted to CIELAB colorspace, (Bottom left) Raw difference image, (Bottom middle) Average pixel values along y-axis (green graph) and average pixel values along x-axis (blue graph), which can be used to guide the selection of an appropriate segmentation threshold, (Right) Extracted silhouette

Multi-view ASM Fitting. The procedure for fitting the trained multi-view ASM to a set of four silhouette images follows the ASM search algorithm presented in [3]. For simplicity, we chose the search function in such a way that the model drives itself towards regions of high gradient, i.e. strong edges, which tend to coincide with foreground-background boundaries. In each iteration, each landmark searches a window in its neighborhood for the point of strongest gradient and moves towards it. Once all points have moved to new locations, the new shape vector is projected to the ASM shape manifold to regularize it and the resulting shape parameter vector is truncated to maintain shape plausibility. The ASM search terminates once the 2-norm of the shape parameter vector stabilizes within an ϵ-value between successive iterations (see Fig. 6).

3D Model Fitting. Silhouette points on the reconstructed 3D shape when projected to each of the four views, should minimize the RMS error with the corresponding silhouette points in the input images. Therefore, we seek to find the shape parameter vector, b^*, which minimizes the constrained objective given below:

$$b^* = \arg\min_{b} \sum_i \left(\frac{1}{N_i} \| P_i \left(\bar{x} + Cb \right) - Q_i Y_i \|_2 \right) , \tag{3}$$

$$\text{such that. } b_j \in [-2\sqrt{\lambda_j}, +2\sqrt{\lambda_j}] , \tag{4}$$

where $i \in \{\text{front}, \text{left}, \text{right}, \text{back}\}$, P_i is the projection matrix, Y_i is the matrix of 2D coordinates of the silhouette contour in the input image, Q_i is the alignment transformation matrix and N_i is the number of silhouette contour points in the ith view.

We solve the problem in (3) using CVX, a package for specifying and solving convex programs [18], [19].

Texture Mapping. Once we have the shape of the avatar we fill-in the realistic appearance by mapping texture information from the input images to the projected model vertices using the cylindrical model described in [2]. This involves converting the coordinates of each reconstructed 3D vertex to polar coordinates to determine which image needs to be looked up to get the texture. Then the 3D coordinates are projected to 2D image coordinates (using P_i from (3) above) to get the color value at the projected image pixel.

In order to obtain an avatar with a more realistic appearance, we then register the facial landmarks (nose, eyes, mouth) of the higher resolution face model (11655 vertices) to corresponding landmarks on the full-body 3D model. This allows us to determine the rigid transformation that aligns the two models. Hence, we transform the higher resolution face model and apply the texture mapping process again, transferring the realistic appearance to the high resolution face model (sample results shown in Fig. 4).

4 Evaluation

In order to evaluate the ability of the trained model to reconstruct a subject's 3D body shape when given four orthogonal images of the body's 2D projection, we conducted the following experiment. Using the trained PCA model of 3D body shape, we trained regression functions [4] for various physical shape-controlling parameters e.g., height. In this way, we were able to synthesize 100 new shape models (50 male, 50 female) having desired physical measurements:

- Weight: varied uniformly in the range [50kgs, 120kgs]
- Height: varied uniformly in the range [150cm, 200cm]
- % Muscle: varied uniformly in the range [0%, 100%]

From each synthetic shape we constructed four orthogonal images of its 2D projection (front, back, left, right). Landmark correspondences between image and model points were manually marked. The goal was then to recover the 3D synthetic shape when presented with only its four projection images and the manually marked landmark correspondences, and assess the 3D shape reconstruction accuracy. The shape parameters were estimated by minimizing a cost function of the average error between the image landmarks and the corresponding projected landmarks of the model. We performed the experiment twice, once using on average only 17 landmarks per view (selected similarly to the feature points in [2]) and once using 144 landmarks per view (sampled regularly around the silhouette outline). For each shape we then calculated the average relative point-to-point reconstruction error (calculated as the absolute difference between the reconstructed projection and the ground truth projection of each landmark, averaged over all 6449 landmarks; this was then divided by the maximum point to point distance between model vertices to yield the average relative point-to-point reconstruction error). The results were as follows: (i) 144 landmarks per view: $\mu = 0,81\%$, $\sigma = 0,1803$, (ii) 17 landmarks per view: $\mu = 0,90\%$, $\sigma = 0,2167$.

Finally, we calculated the sample correlations and p-values to determine correlations between the reconstruction error and the physical parameters. We found that the most significant correlations of reconstruction error were between weight and % muscle. More specifically, for weight we got $\rho = 0.2685$ and p-value=0.0069, indicating that error increases with increasing weight. For % muscle we got $\rho = 0.3503$ and p-value=0.0004, indicating that error increases with increasing % muscle because higher % muscle cause more bulging of the body thus greater deviation from the mean shape.

In another experiment, we evaluated the accuracy of the multi-view ASM against the accuracy of four separate single-view ASM models. The results, which are shown in Fig. 6 illustrate the superiority of the multi-view ASM. We are in the process of performing evaluation simulations of the proposed method against a large synthetic dataset as well as quantitative evaluation on real data. See Fig. 4 for a visual evaluation.

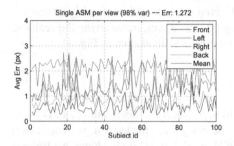 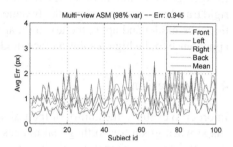

Fig. 6. Comparative evaluation of our silhouette registration method by a multi-view ASM on a synthetic dataset of 100 subjects. (Left) Plot of average relative silhouette registration error for each view, using single-view ASM models (98% variance). (Right) Corresponding plot evaluating silhouette registration error using a multi-view ASM model.

5   Conclusions and Future Work

We presented our ongoing work towards the generation of realistic animation ready avatars that will allow users and their friends to participate and collaborate in VR game adaptations. While this is work in progress and additional evaluation experiments are under way, early results obtained prove the potential of our overall approach. In the future we plan to increase the accuracy of model fitting and texture mapping so that the end result is cleaner and more realistic. Furthermore, we plan to add automatic rigging and animation on the personalized avatars, so that they can be easily incorporated in collaborative VR applications.

Acknowledgements. This work was partially supported by the Cyprus Research Promotion Foundation and the European Union Structural Funds (project VR-CAVE: IPE/NEKYP/0311/02). We would like to thank Dr Nils Hasler for providing the 3D human body database [4].

References

1. Michael, N., Kater, A.E., Lanitis., A.: Increasing user engagement in re-designed classic video games. In: Procs. of Joint Conference on Virtual Reality (2013)
2. Hilton, A., Beresford, D., Gentils, T., Smith, R., Sun, W., Illingworth, J.: Whole-body modelling of people from multiview images to populate virtual worlds. The Visual Computer 16(7), 411–436 (2000)
3. Cootes, T., Taylor, C., Cooper, D., Graham, J.: Active shape models-their training and application. Computer Vision and Image Understanding 61(1), 38–59 (1995)
4. Hasler, N., Stoll, C., Sunkel, M., Rosenhahn, B., Seidel, H.P.: A statistical model of human pose and body shape. Comput. Graph. Forum 28(2), 337–346 (2009)
5. Viola, P., Jones, M.: Robust real-time object detection. International Journal of Computer Vision (2001)
6. Anguelov, D., Srinivasan, P., Koller, D., Thrun, S., Rodgers, J., Davis, J.: Scape: Shape completion and animation of people. ACM Trans. Graph. 24(3), 408–416 (2005)
7. Lee, W., Gu, J., Magnenat-thalmann, N.: Generating animatable 3d virtual humans from photographs. In: Computer Graphics Forum., pp. 1–10 (2000)
8. Ahmed, N., de Aguiar, E., Theobalt, C., Magnor, M., Seidel, H.P.: Automatic generation of personalized human avatars from multi-view video. In: Proceedings of the ACM Symposium on Virtual Reality Software and Technology, VRST 2005, pp. 257–260. ACM, New York (2005)
9. Starck, J., Hilton, A.: Model-based multiple view reconstruction of people. In: ICCV, pp. 915–922. IEEE Computer Society (2003)
10. Weiss, A., Hirshberg, D.A., Black, M.J.: Home 3d body scans from noisy image and range data. In: Metaxas, D.N., Quan, L., Sanfeliu, A., Gool, L.J.V. (eds.) ICCV, pp. 1951–1958. IEEE (2011)
11. Cui, Y., Chang, W., Nöll, T., Stricker, D.: Kinectavatar: Fully automatic body capture using a single kinect. In: Proceedings of the 11th International Conference on Computer Vision, ACCV 2012, vol. 2, pp. 133–147. Springer, Heidelberg (2013)
12. Furukawa, Y., Ponce, J.: Accurate, dense, and robust multiview stereopsis. IEEE Transactions on Pattern Analysis and Machine Intelligence 32(8), 1362–1376 (2010)
13. Starck, J., Miller, G., Hilton, A.: Video-based character animation. In: Proceedings of the 2005 ACM SIGGRAPH/Eurographics Symposium on Computer Animation, SCA 2005, pp. 49–58. ACM, New York (2005)
14. Tong, J., Zhou, J., Liu, L., Pan, Z., Yan, H.: Scanning 3d full human bodies using kinects. IEEE Trans. Vis. Comput. Graph. 18(4), 643–650 (2012)
15. Vlasic, D., Peers, P., Baran, I., Debevec, P.E., Popovic, J., Rusinkiewicz, S., Matusik, W.: Dynamic shape capture using multi-view photometric stereo. ACM Trans. Graph. 28(5) (2009)
16. Mashalkar, J., Bagwe, N., Chaudhuri, P.: Personalized animatable avatars from depth data. In: Proceedings of the 5th Joint Virtual Reality Conference, JVRC 2013, pp. 25–32. Eurographics Association, Aire-la-Ville (2013)
17. Kendall, D.G.: A survey of the statistical theory of shape. Statistical Science 4(2), 87–99 (1989)
18. Grant, M., Boyd, S.: Graph implementations for nonsmooth convex programs. In: Blondel, V., Boyd, S., Kimura, H. (eds.) Recent Advances in Learning and Control. Lecture Notes in Control and Information Sciences, pp. 95–110. Springer (2008), http://stanford.edu/~boyd/graph_dcp.html
19. Grant, M., Boyd, S.: CVX: Matlab software for disciplined convex programming, version 2.1. (March 2014), http://cvxr.com/cvx

Data-Driven Motion Reconstruction Using Local Regression Models

Christos Mousas[1], Paul Newbury[1], and Christos-Nikolaos Anagnostopoulos[2]

[1] Department of Informatics
University of Sussex
Brighton BN1 9QJ, UK
{c.mousas,p.newbury}@sussex.ac.uk
[2] Department of Cultural Technology and Communication
University of the Aegean
Mytilene 81100, Greece
canag@ct.aegean.gr

Abstract. Reconstructing human motion data using a few input signals or trajectories is always challenging problem. This is due to the difficulty of reconstructing natural human motion since the low-dimensional control parameters cannot be directly used to reconstruct the high-dimensional human motion. Because of this limitation, a novel methodology is introduced in this paper that takes benefit of local dimensionality reduction techniques to reconstruct accurate and natural-looking full-body motion sequences using fewer number of input. In the proposed methodology, a group of local dynamic regression models is formed from pre-captured motion data to support the prior learning process that reconstructs the full-body motion of the character. The evaluation that held out has shown that such a methodology can reconstruct more accurate motion sequences than possible with other statistical models.

Keywords: character animation, local regressions, motion reconstruction.

1 Introduction

Full-body motion reconstruction is a process that is quite important in cases in which the ability to animate virtual characters while using a reduced number of sensors or user defined trajectories is necessary. Such techniques, especially those that are developed to reconstruct the motion of the character during the performance capture process can be quite beneficial in various areas that are related to virtual reality, such as rehabilitation and sports training, as well as in video games. Although various motion capture systems for capture of the user's performance, such as Vicon [1] and XSens [2], can provide desirable results, the basic limitation is the high cost for general family use. Recently, low-cost commercial products, such as those provided by Microsoft, Sony, and Nintendo, have developed next generation hardware devices to capture the online performances of individual players. However, a reduced number of input signals retrieved from those devices cause the motion reconstruction process to be challenging. The reason is that human motion has many degrees of freedom (DOF) and, therefore,

L. Iliadis et al. (Eds.): AIAI 2014, IFIP AICT 436, pp. 364–374, 2014.

motion models are required that can reconstruct the natural and realistic motion of a character while using few parameters.

In the proposed methodology, the motion reconstruction process is formulated in a maximum a posteriori (MAP) framework that is responsible for producing a natural-looking motion sequence that best matches the user-defined inputs. Specifically, the proposed methodology learns a group of local regression models in order to constrain the prior learning of the pre-captured motion data. Then, by searching within the motion database, by using K nearest motion examples that are similar to the previously reconstructed poses $q_{t-1}, ..., q_{t-m}$, and the motion sequences $q_{t_{k-1}}, ..., q_{t_{k-m}}$ along with their subsequent poses q_{t_k} for $k = 1, ..., K$ as the training data, it learns a predictive model for the reconstruction of the current character's posture q_t.

The proposed methodology can reconstruct a variety of motion sequences by using a reduced number of input trajectories, such as walking, running, jumping, and punching, as well as golf swings. Further, by evaluating the presented methodology with previous solutions for reconstructing the character's motion, the proposed approach can reconstruct motion sequences by reducing the reconstruction error. The remainder of this paper is organized as follows: Section 2 presents related work in data-driven character animation and motion reconstruction. Section 3 provides an overview of the proposed methodology. The proposed reconstruction methodology is explained in Section 4. Section 5, presents the results obtained from the evaluations of the proposed methodology versus those oft previously examined techniques. Finally, conclusions are drawn and the potential future work is discussed in Section 6.

2   Related Work

In data-driven motion synthesis techniques [3], the low-dimensional control signals that are obtained from the motion capture device are used to retrieve suitable motion sequences from a database that contains high-dimensional motion capture sequences. The reuse of pre-recorded human motion data requires efficient retrieval of similar motions from databases [4][5], as well as a good understanding of how motions must be parameterized in order to yield smooth transitions between several retrieved motion clips [6].

On the other hand, statistical motion models are often described as several mathematical functions that represent human motion by a set of parameters that are associated with probability distributions [7]. So far, using pre-captured motion data to learn statistical motion models have been used for full-body character control [8], in key frames interpolation [9], motion styles synthesis [10][11], facial animation [12] and speech-driven facial expressions [13][14], hands-over animation techniques [15][16], interactive creation of a character's pose [8][17] or control of human actions using vision-based tracking [18], real-time human motion control with inertial sensors [19] or accelerometer sensors [17], construction of physically-valid motion models for human motion synthesis [14] and so forth.

During the past years, a number of researchers have developed approaches that use sparse constraints provided by sensors to control high-dimensional human motions. A single depth camera to track and reconstruct various human motions, their approaches acquired no makers attached on user's body, however, no less than 15 control points needed to be used to segment the human body [20][21]. By combining inverse kinematics algorithms and a few constraints from eight magnetic sensors to provide an analytic solution for human motion control [22]. By using six to nine retro-reflective markers as the control points for online human motion reconstruction [18], and by using five inertial sensors for real-time upper-body control [23]. Recently, full-body human motion control was achieved using the positional and orientational constraints from six inertial sensors [19]. Finally, another solution [17] utilizing a few constraints provided by four accelerometer sensors achieved full-body motion control of the character.

3   Overview

In the proposed methodology a reduced number of input trajectories retrieved from a reference motion sequence are used to reconstruct the full-body motion of the character. Those input trajectories automatically transform the control inputs into realistic human motion by building a group of online local regression models during the runtime. An overview of the methodology appears in Figure 1.

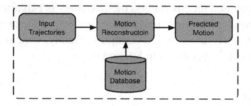

Fig. 1. An overview of the proposed system

For the motion reconstruction process, it is assumed that the character's actions can be represented as an $m - order$ Markov chain, such as the current posture of the character at the $t - th$ frame q_t can be considered to depend only on previous m reconstructed postures $Q_{t,m} = [q_{t-1}, ..., q_{t-m}]$. Thus, the probabilistic model should be able to fulfill $p(q_t|q_{t-1}, ..., q_1) = p(q_t|q_{t-1}, ..., q_{t-m})$. In the proposed local regression modelling methodology, the spatiotemporal directions to the models to constrain the transformation of the character's postures in the configuration space are added. This approach predicts how the humans move in each region and constrains the reconstructed motion to remain in the natural-looking space. For that reason, it was considered to be an online learned model to generate the desired posture q_t from various forms of kinematic constraints c_t that are specified by the input trajectories. Finally, the motion reconstruction process is optimized in a MAP framework by estimating the posture q_t that is satisfied by the input trajectories c_t along with the previous reconstructed postured $Q_{t,m}$ such as:

$$\arg\max_{q_t} p(q_t|c_t, Q_{t,m})$$
$$\propto \arg\max_{q_t} p(c_t|q_t)p(q_t|Q_{t,m}) \tag{1}$$

In this case, by applying the negative logarithm to the posteriori distribution function $p(q_t|c_t, Q_{t,m})$, the constrained MAP problem becomes an energy minimization problem, which is now represented as:

$$\arg\max_{q_t} \underbrace{-\ln p(c_t|q_t)}_{E_{likelihood}} + \underbrace{-\ln p(q_t|Q_{t,m})}_{E_{prior}} \tag{2}$$

where the likelihood term ($E_{likelihood}$) measures how well the reconstructed posture q_t fits the user defined input trajectories c_t, and the prior term (E_{prior}) measures the naturalness of the reconstructed posture. It should be noted that an optimal estimation of the reconstructed posture produced a natural motion that achieved the inputs that were specified by the user.

4 Motion Reconstruction

Using a reduced number of input trajectories to reconstruct the motion of the virtual character is quite challenging, since the control inputs cannot fully constrain the entire human motion to remain in the natural-looking space. Thus, in the proposed solution, a group of local regression models are used to solve this issue of motion reconstruction ambiguity. The methodology that used is presented in the following subsections.

4.1 Prior Motion Modelling

The prior motion modelling that was based on the local regression models is presented in this subsection. This model is responsible for adequately constraining the reconstructed posture of the character to remain in the natural-looking space. The novelty of the proposed model is that there is no need to find an appropriate structure for a global dynamic model, which would necessarily by high dimensional and non-linear. In the proposed methodology, a k-nearest neighbour ($k - NN$) searching algorithm is adopted to find the K motion examples that are contained in our motion capture database and are similar to the already reconstructed posture. The examples and their subsequent postures are used for our online learning process.

In order to estimate the current posture q_t at the $t - th$ frame, a searching of the database of the motion data is employed in the first steps. This searching process finds the motion segments that are most similar to the recently reconstructed motion segment $Q_{t,m} = [q_{t-1}, ..., q_{t-m}]$. Thus, the k nearest motion segments q_{t_k} for $k = 1, ..., K$ are chosen as training data to learn a predictive model by means of a statistical learning method of current posture q_t.

Now, assume a linear relationship between an input angle vector $x = [q_{t-1}, ..., q_{t-m}]$ and an output joint angle vector $y = q_t$. In this case, one should note that the predictive function for each DOF in the output q_t is learned separately. Then, by subtracting the means from the input and output training data, one assumes that the mean values of x and y are zeros. Therefore, the function of the proposed model is represented by using linear regression as follows:

$$y = a^T x + \beta_y \tag{3}$$

where the input joint angle x is an $m \times D$-dimensional vector. D represents the dimension of DOF for a virtual character and y is the joint angle value for the output motion. Vectors a and β_y are regression coefficients that represent a homoscedastic noise variable, which is independent of vector x. Moreover, given the K motion examples $(x_k; y_k)$ for $k = 1, ..., K$ that are similar to the current reconstructed poses, and by minimizing the expected squared error $E = \min \sum_{k=1}^{K} \|y_k - a^T x_k\|^2$, the coefficient a is obtained by the least squares solution that follows:

$$a = (X^T X)^{-1} X^T y \tag{4}$$

where the row of the matrix X stores the input joint angle vectors x_k for $k = 1, ..., K$, and K output joint angle values are stacked in vector y.

The proposed methodology calculates the projections of the highest correlation between the input joint angle matrix X and the output vector y. These projections can be obtained by maximizing the squared relationship as follows:

$$correlation^2(X_{u_j}, y) = \frac{(u_j^T X^T y)}{(u_j^T X^T X_{u_j})} \tag{5}$$

where u_j denotes the one of the projection's directions. It should be noted that since each of the projections X_{u_j} is orthogonal to others and its length is unit, it is possible to get $u_j^T X^T X_{u_j} = 1$. Hence, u_j is one column of the matrix U that includes the eigenvectors of the covariance matrix $C = (X^T X)^{-1} X^T yy^T X$. In the proposed model, X is projected onto U for only considering the projections. Thus, by minimizing $\|y - XU\gamma\|^2$ with respect to the reduced coefficient γ, one obtains:

$$a = U\gamma = U(U^T X^T XU)^{-1} U^T X^T y \tag{6}$$

Since each DOF is predicted separately in output q_t, the model has only one projection direction u for each time. The weight for each data point x_k can now be calculated by the Gaussian function, using its relative distance from the previously reconstructed postures $Q_{t,m} = [q_{t-1}, ..., q_{t-m}]$ as:

$$\omega_k = \exp\left(-\frac{1}{2}(x_k - Q_{t,m})^T W(x_k - Q_{t,m})\right) \tag{7}$$

where W denotes the diagonal matrix that contains the weights for each DOF. It should be noted that in the proposed implementation used an identity matrix to represent W. The eigenvectors are extracted from the matrix as:

$$C_\omega = (DX)^{-1} X^T Dyy^T DX \tag{8}$$

where D denotes a diagonal matrix that constrains ω_k along its diagonal. Moreover, the weighted regression coefficients can be represented as:

$$a_\omega = U(U^T X^T DXU)^{-1} U^T X^T Dy \tag{9}$$

In this case, assuming that there is a Gaussian distributed noise variable β_y, its standard deviation σ can be estimated by $y_k - \beta x_k$ for $k = 1, ..., K$. In our experiments, a predictive function for each DOF of the reconstructed posture is constructed. There-fore, to predict the $d-th$ DOF of the character's posture, the local regression model is described as:

$$q_{t,d} = a_{d,\omega}^T Q_{t,m} + N(0, \sigma_d) \tag{10}$$

where $q_{t,d}$ and σ_d are scalars, $q_{t,d}$ represents the $d-th$ DOF of the $t-th$ frame posture, and σ_d is the standard deviation of the $d-th$ predictive function. $a_{d,\omega}^T$ and $Q_{t,m}$ are vectors, where $a_{d,\omega}^T$ are the weighted regression coefficients for the $d-th$ DOF, and $Q_{t,m}$ is the reconstructed motion segment of the previous m postures of the character. The complexity of such a model for reconstructing the character's posture is $O(Km^2D^2)$, where K, m, and D represents the number of training data, the previous m postures, and the dimension of DOF for the virtual character repsectivelly.

4.2 Likelihood Estimation

The likelihood term ($E_{likelihood}$) of the MAP framework measures how well the corre-sponding joint in the reconstructed character's postures fits the user-defined constraints. Therefore, the likelihood term is formulated as:

$$\begin{aligned} E_{likelihood} &= -lnp(c_t|q_t) \\ &\propto \|f(q_t;s) - c_t\|^2 \end{aligned} \tag{11}$$

where q_t, c_t, and s are vectors. q_t denotes the joint angles of the reconstructed posture at frame t, s denotes the character's skeletal size, which is modelled according to the Acclaim Skeletal File format (ASF) as provided by [24]. Finally, c_t is the user-specified input trajectories that are retrieved from reference motion data. Finally, f denotes a forward kinematics function that calculates the global coordinates value of the current posture q_t of the character.

5 Implementation and Results

The following subsections briefly present the implementation and the results obtained from the evaluation process of the proposed methodology.

5.1 Implementation

For the implementation of the proposed methodology a gradient-based optimization was used using the Levenberg-Marquardt method [25] for the objective function that is defined in Equation (2). This method uses the most similar motion examples that are already in the motion database to initialize the optimization. The computational effi-ciency of the proposed motion reconstruction process relies on the scope of the search in the motion database. Thus, the process to find the K nearest neighbour is accelerated using the neighbour graph approach as presented in [18]. Finally, examples of posture reconstructed with the proposed methodology are shown in Figure 2.

Fig. 2. Examples of postures synthesized with the proposed methodology

5.2 Results

To evaluate the proposed methodology two different datasets were used. The first dataset contains 85,097 postures that are separated into five different actions that the character can perform: walking, running, jumping, punching, and swinging a golf club. The second dataset contains a total of 1.1 M poses that were downloaded from [24]. All of the motion sequences were recorded by use of a Vicon motion capture system that has a framerate of 120 fps. In the proposed implementation, the motion data were downsampled to 60fps in order to achieve more natural-looking motion for visualization. The effectiveness of the proposed approach was verified on various behaviours and the reconstruction error was evaluated against the ground truth data.

Comparing to Other methods: The proposed methodology was evaluated against three popular approaches for reconstructing the character's motion. Specifically, the methodology was evaluated against the Gaussian Process Latent Variable Model [17], the local PCA model [18] and the local PCR model [19], while reconstructing different actions that the character can perform. The results of this evaluation process appear in Figure 3. Specifically, the mean error of five different actions for all of the aforementioned techniques is presented. In the evaluation, we also adopt six constraint points that were used in [19]. The results show that our method achieved smaller mean errors than the other three techniques. In another aspect, while using only four control points (two wrists and two ankles), the three previous methods cannot reconstruct natural-looking human data. The results indicate that the proposed method is better than those of the two other local methods.

Using Different Number of Control Points: We tested a different number of control points for four methods. We chose from two to six positional control points. (1) Left wrist and right ankle; (2) left wrist and two ankles; (3) two wrists and two ankles; (4) root, two wrists and two ankles; (5) head, root, two wrists and two ankles. After testing with different motions, we concluded that the reconstruction errors usually decrease as the number of constraints increases. In addition, we also found that, in comparison to the six constraint points (head, center of torso, two wrists and two ankles), which was used in [19] to obtain a natural-looking, reconstructed human motion, we can use as few constraint points as possible (4 constraint points: two wrists and two ankles) to achieve a comparable result with the motion capture data. Table 1 is the average reconstruction error comparison for various numbers of control points. Despite the use of fewer constraint points, our model is more powerful for accurate motion reconstruction than the three previous methods.

Table 1. Reconstruction errors based on different numbers of control points for different algorithms

	2	3	4	5	6
GPLVM	56.76	41.88	16.13	9.45	5.92
LPCA	43.27	29.25	10.56	6.39	3.81
LPCR	38.71	23.61	7.33	4.73	3.22
Proposed Method	18.63	7.86	2.67	2.25	1.90

Using Different Datasets: Table 2 presents the average reconstruction errors of five different actions from the three aforementioned techniques, while using the different training database. The reconstruction errors were calculated using 3D positional constraints from six control points. We found that, for the GPLVM method, the reconstruction error is large when using a large and heterogeneous database. When using a small database, the reconstruction error is also larger than that of local modeling approaches. For the other local modeling methods, the reconstruction error decreased when the size

Table 2. Average reconstruction errors for four methods on different databases

	69888 poses	1.1 M poses
GPLVM	5.86	26.57
LPCA	3.92	3.02
LPCR	3.38	2.75
Proposed Method	2.01	1.56

of training database increased. In addition, our proposed model can achieve a smaller reconstruction error than others. By testing on different databases, we also verified the proposed model's power.

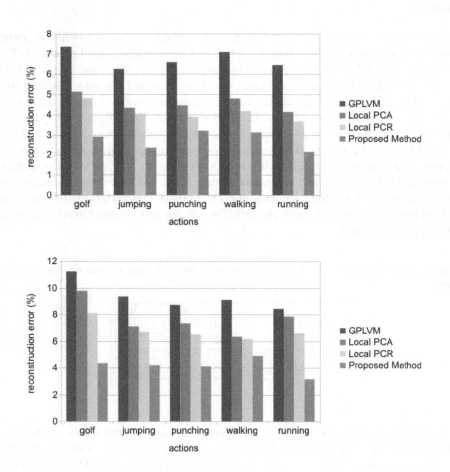

Fig. 3. A comparison of the proposed methodology to three popular algorithms: GPLVM, local PCA, and local PCR. The percentage error while using six (up), and four joints (down) for reconstruction of the motion of the character.

6   Conclusion

In this paper, a new local regression model was presented for reconstructing natural full-body human motion based on as few user-specified constraints as possible. The proposed method, which uses a data-driven approach, utilizes several nearest motion examples to construct a group of online local regression models for online motion reconstruction. However, based on the same defined constraints and motion database, the proposed method has a better force of constraint than the previous local models and thus can reconstruct more realistic human motions. Therefore, our proposed model is suitable for the next generation of hardware devices to exploit the motion capture system for a common use.

On the other hand, the proposed method has three limitations. Firstly, like other data-driven approaches, the database is crucial for the quality of reconstructed motion. The system will not produce a desired motion if the training data does not contain any desired motion patterns. For example, if the walking motion pattern is not included in the database, our system cannot reconstruct desired walking data. Secondly, user-specified constraints are also crucial for the final results. In fact, if user-specified constraints are not natural or self-conflicting, the reconstruction result will not be a realistic human motion that satisfies the user's constraints. Finally, the motion data must be previously arranged for online search. Like most local modeling approaches, a specific data arrangement structure is applied for motion data to accelerate the searching process.

References

1. Vicon Motion Capture Solution, from `http://www.vicon.com/` (accessed May 10, 2014)
2. Xsens Motion Capture Solution, from `http://www.xsens.com/` (accessed May 10, 2014)
3. Liu, H., He, F., Cai, X., Chen, X., Chen, Z.: Performance-Bbased Control Interfaces Using Mixture of Factor Analyzers. The Visual Computer 27(6-8), 595–603 (2011)
4. Keogh, E., Palpanas, T., Zordan, V.B., Gunopulos, D., Cardle, M.: Indexing Large Human-Motion Databases. In: 30th International Conference on Very Large Databases, pp. 780–791 (2004)
5. Müller, M., Röder, T., Clausen, M.: Efficient Content-Based Retrieval of Motion Capture Data. ACM Transactions on Graphics 24(3), 677–685 (2005)
6. Kovar, L., Gleicher, M.: Flexible Automatic Motion Blending with Registration Curves. In: ACM SIGGRAPH/Eurographics Symposium on Computer Animation, pp. 214–224. Eurographics Association (2003)
7. Mousas, C., Newbury, P., Anagnistopoulos, C.-N.: Evaluating the Covariance Matrix Constraints for Data-Driven Statistical Human Motion Reconstruction. In: Proc. of the 30th Spring Conference on Computer Graphics. ACM Press, New York (2014)
8. Wei, X.K., Chai, J.: Intuitive Interactive Human-Character Posing with Millions of Example Poses. IEEE Computer Graphics and Applications 31(4), 78–88 (2011)
9. Li, Y., Wang, T., Shum, H.Y.: Motion Texture: a Two-Level Statistical Model for Character Motion Synthesis. ACM Transactions on Graphics 21(3), 465–472 (2002)
10. Brand, M., Hertzmann, A.: Style Machines. In: 27th Annual Conference on Computer Graphics and Interactive Techniques, pp. 183–192. ACM Press, New York (2000)
11. Mousas, C., Newbury, P., Anagnistopoulos, C.-N.: Motion Style transfer in Correlated Motion Spaces. In: Proc. of the 12nd International Symposium on Smart Graphics. Springer (2014)
12. Weise, T., Bouaziz, S., Li, H., Pauly, M.: Realtime Performance-Based Facial Animation. ACM Transactions on Graphics 30(4), Article No.77 (2011)
13. Bregler, C., Covell, M., Slaney, M.: Video Rewrite: Driving Visual Speech with Audio. In: 24th Annual Conference on Computer Graphics and Interactive Techniques, pp. 353–360. ACM Press, New York (1997)
14. Brand, M.: Voice Puppetry. In: 26th Annual Conference on Computer Graphics and Interactive Techniques, pp. 21–28. ACM Press, New York (1999)

15. Mousas, C., Newbury, P., Anagnistopoulos, C.-N.: Efficient Hand-Over Motion Reconstruction. In: Proc. of the 22nd International Conference on Computer Graphics, Visualization and Computer Vision (2014)
16. Wheatland, N., Jörg, S., Zordan, V.: Automatic Hand-Over Animation Using Principle Component Analysis. In: Motion on Games, pp. 175–180. ACM Press, New York (2013)
17. Grochow, K., Martin, S.L., Hertzmann, A., Popović, Z.: Style-Based Inverse Kinematics. ACM Transactions on Graphics 23(3), 522–531 (2004)
18. Chai, J., Hodgins, J.K.: Performance Animation from Low-Dimensional Control Signals. ACM Transactions on Graphics 24(3), 686–696 (2005)
19. Liu, H., Wei, X., Chai, J., Ha, I., Rhee, T.: Realtime Human Motion Control with a Small Number of Inertial Sensors. In: Symposium on Interactive 3D Graphics and Games, pp. 133–140. ACM Press, New York (2011)
20. Shotton, J., Fitzgibbon, A., Cook, M., Sharp, T., Finocchio, M., Moore, R., Kipman, A., Blake, A.: Real-Time Human Pose Recognition in Parts from a Single Depth Image. In: IEEE Conference on Computer Vision and Pattern Recognition, pp. 1297–1304. IEEE Press (2011)
21. Wei, X., Zhang, P., Chai, J.: Accurate Realtime Full-Body Motion Capture Using a Single Depth Camera. ACM Transactions on Graphics 31(6), Article No. 188 (2012)
22. Semwal, S.K., Hightower, R., Stansfield, S.: Mapping Algorithms for Real-Time Control of an Avatar Using Eight Sensors. Presence: Teleoperators and Virtual Environments 7(1), 1–21 (1998)
23. Slyper, R., Hodgins, J.K.: Action Capture with Accelerometers. In: ACM SIGGRAPH/Eurographics Symposium on Computer Animation, pp. 193–199. Eurographics Association (2008)
24. Carnegie Mellon University, Motion Capture Database, from http://mocap.cs.cmu.edu/ (accessed May 10, 2014)
25. Lourakis, M.: Levmar: Levenberg-Marquardt Nonlinear Least Squares Algorithms in C/C++ (2014), http://www.ics.forth.gr/~lourakis/levmar (accessed May 10, 2014)

Calculation of Complex Zernike Moments with Geodesic Correction for Pose Recognition in Omni-directional Images

K.K. Delibasis[1], Spiros Georgakopoulos[1],
Vassilis Plagianakos[1], and Ilias Maglogiannis[2]

[1] University of Thessaly, Dept. of Computer Science and Biomedical Informatics,
Lamia, Greece
[2] University of Piraeus, Dept. of Digital Systems, Piraeus, Greece
kdelibasis@yahoo.com, {spyrosgeorg,vpp}@dib.uth.gr,
imaglo@unipi.gr

Abstract. A number of Computer Vision and Artificial Intelligence applications are based on descriptors that are extracted from imaged objects. One widely used class of such descriptors are the invariant moments, with Zernike moments being reported as some of the most efficient descriptors. The calculation of image moments requires the definition of distance and angle of any pixel from the centroid pixel. While this is straightforward in images acquired by projective cameras, it is complicated and time consuming for omni-directional images obtained by fish-eye cameras. In this work, we provide an efficient way of calculating moment invariants in time domain from omni-directional images, using the calibration of the acquiring camera. The proposed implementation of the descriptors is assessed in the case of indoor video in terms of classification accuracy of the segmented human silhouettes. Numerical results are presented for different poses of human silhouettes and comparisons between the traditional and the proposed implementation of the Zernike moments are presented. The computational complexity for the proposed implementation is also provided.

Keywords: Pattern recognition, computer vision, image descriptors, moment invariants, Omni-directional image/video, fish-eye camera, silhouette pose recognition.

1 Introduction

Pattern recognition in images is a very common task in artificial intelligence. Among other methods [1], invariant moments have been used extensively for providing regional descriptors to be used in pattern recognition problems [2] – [6]. The calculation of invariant moments of a region of an image requires a valid distance metric defined in the image domain (e.g. [3]). In omnidirectional images, the distance between two pixels cannot be defined in terms of their coordinates in the image frame, since the image is acquired not through a simple projection, but using a spherical element with 180 degrees field of view.

L. Iliadis et al. (Eds.): AIAI 2014, IFIP AICT 436, pp. 375–384, 2014.

This need has been recently addressed in a few publications. In [7] the implementation of the SIFT algorithm ([8]) is presented for omni-directional images, using the convolution operators on a sphere in the Fourier domain, whereas it has also been studied for wide angle images [9]. In other approaches, the SIFT algorithm has been applied to the unwrapped omni-directional images [10]. Pixel distance was redefined in conic sensor images [11].

The contribution of this work focuses on an efficient way of computing Zernike moment invariant (or any other invariant moments) for calibrated omni-directional images in the time domain. We propose the measurement of geodesic distance and angles between image pixels, based on the calibrated model of the acquiring fish-eye camera. The geodesically corrected implementation of Zernike moments presents reduced variability when applied to known geometric shapes and it achieves increased accuracy when applied to pose classification of segmented human silhouettes. The proposed implementation is computationally quite efficient, since it allows the processing of a high number of frames per second.

2   Proposed Methodology

2.1   Zernike Moment Invariant

Zernike moments of order m, n are defined by a set of radial polynomials $R_{nm}(r)$, which are orthogonal inside the unit circle:

$$R_{nm}(r) = \sum_{s=0}^{(n-|m|)/2} (-1)^s \frac{(n-s)!}{s!\left(\frac{n+|m|}{2}-s\right)!\left(\frac{n+|m|}{2}+s\right)!} r^{n-2s}, \tag{1}$$

where n positive integer and m integer such that n-|m| is even and |m|<n. Orthogonality is preserved since $\int_0^1 R_{nm}(r) R_{km}(r) dr = \frac{1}{2(n+1)} \delta_{nm}$.

Zernike polynomials are defined as

$$V_{nm}(r,\theta) = R_{nm}(r) e^{jm\theta} \tag{2}$$

The Zernike moments of a bivariate function $f(x,y)$ are defined as:

$$Z_{nm} = \frac{n+1}{\pi} \int_0^1 \int_0^{2\pi} f(r,\theta) V_{nm}(r,\theta) r dr d\theta \tag{3}$$

The above Eq. is easily adapted to be used in discrete images.

2.2   Omni-Directional Image Formation and Camera Calibration

The formation of omnidirectional image using a spherical camera, as presented in detail in [12], is shown in Fig. 1, using only two dimensions to facilitate understanding. In Fig. 2, neighbouring pixels are shown in different positions on the image sensor. It is clear that the distance of two pixels is different when measured on the sensor and on the position of the images points on the spherical optical element. Therefore, in order to produce accurate results, image processing algorithms that use pixel distances have to be re-implemented for omni-directional images. In this work, we utilize the calibration of the specific fish-eye camera, proposed in [12], according to which, for each pixel (i, j) of the image, the corresponding vector (θ, φ) on the spherical element is precalculated

$$(\theta, \varphi) = M(j, i) \tag{4}$$

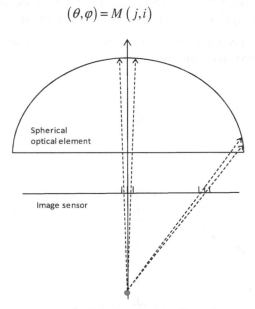

Fig. 1. Model of image formation for spherical omni-directional camera

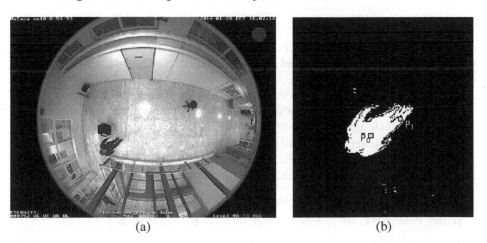

Fig. 2. A typical omni-directional image (a) and the segmented human silhouette (b). The strong deformation imposed by the fish-eye spherical lens is evident.

2.3   Calculation of Zernike Moment Invariants for Calibrated Omni-Directional Images – Geodesic Correction

The calculation of Zernike moments requires the distance and orientation with respect to the centre of mass for each pixel of the segmented object / pattern to be classified. Let us assume that we need to calculate Zernike moments in the case of the segmented human silhouette of Fig. 2(b). If P_0 is the centroid pixel, the distance and angle to any other pixel P_1 needs to be calculated. Let us use the calibration of the camera to obtain the spherical coordinates (θ_0, φ_0) and (θ_1, φ_1) on the unit sphere, of points P_0 and P_1, respectively. Now, the distance and relative angle of P_1 with respect to P_0 can be measured on the unit sphere.

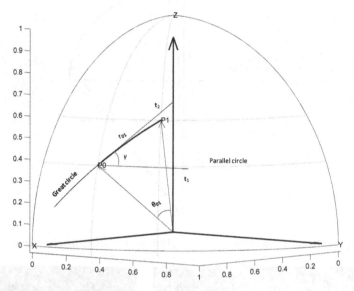

Fig. 3. The mapping of pixels P_0 and P_1 on the unit sphere and the calculation of their geodesic distance and angle. The great circle and the parallel circles through P_0 and P_1 are also shown.

It is well known that the geodesic curve of a sphere is a great circle. Thus, the distance between any two points on a sphere is the length of the arc that is defined by the two points and belongs to a circle that passes through the two points and has the same centre with the sphere. We will call this circle "great circle". Let v_0 and v_1 be the position vectors pointing to P_0 and P_1. The distance r_{01} of points P_0 and P_1 on the unit sphere is easily calculated, (assuming that $\cos^{-1}$ returns the result in radians):

$$\mathbf{v}_0 \cdot \mathbf{v}_1 = (\cos\theta_0 \cos\varphi_0, \sin\theta_0 \cos\varphi_0, \sin\varphi_0) \cdot (\cos\theta_1 \cos\varphi_1, \sin\theta_1 \cos\varphi_1, \sin\varphi_1) =$$
$$(\cos\theta_0 \cos\varphi_0 \cos\theta_1 \cos\varphi_1 + \sin\theta_0 \cos\varphi_0 \sin\theta_1 \cos\varphi_1 + \sin\varphi_0 \sin\varphi_1) \tag{5}$$

$$r_{01} = \cos^{-1}(\mathbf{v}_0 \cdot \mathbf{v}_1). \tag{6}$$

Zernike moments, like any other definition of moment invariants, require the distance between any two pixels as well as their angle γ. If we assume the parallel circle passing through P_0 and the great circle passing through P_0 and P_1, then angle γ is calculated as the angle between the tangent vectors t_0, t_1 of the two circles at P_0. The great circle through P_0, P_1 is calculated as following:

$$v = \lambda_0 v_0 + \lambda_1 v_1, \text{ with } |v| = 1. \tag{7}$$

After some algebraic operations we obtain:

$$\lambda_0^2 + 2\cos(v_0 \cdot v_1)\lambda_0\lambda_1 + \lambda_1^2 - 1 = 0. \tag{8}$$

After solving the above equation and requiring λ_0, λ_1 to be real number we obtain:

$$\lambda_0^2 \left(\cos^2(v_0 \cdot v_1) - 1\right) \geq 0,$$
$$\lambda_1 = -\cos^2(v_0 \cdot v_1)\lambda_0 \pm \sqrt{\left(1 - \cos^2(v_0 \cdot v_1)\right)\lambda_0^2 + 1} \tag{9}$$

Since execution speed is essential and taking into consideration that the circle is a very smooth curve (with constant curvature), the tangent vector is calculated by finite differences, as following. It is obvious that for $\lambda_0 = 1, \lambda_1 = 0$, $v = v_0$. We set $\lambda_0 = 0.9$ and use Eq. (9) to calculate λ_1. Then, $t_1 = 0.9v_0 + [\lambda_1]_{\lambda_0=0.9} v_1 - v_0$. The tangent vector of the parallel circle at P_0 is trivially calculated: $t_2 = (-P_{1y}, P_{1x}, 0)$. The angle γ is now easily obtained as

$$\gamma = \cos^{-1}\left(\frac{t_1 \cdot t_2}{|t_1||t_2|}\right). \tag{10}$$

The sign of angle γ will be set equal to the sign of $(\theta_1 - \theta_0)$. Thus, the distance and angle (r, θ) in Eq. (3) are replaced by r_{01} defined in (6) and γ defined in Eq. (10). This correction will call "geodesic" in the rest of the paper.

3   Experimental Results

3.1   Experiments with a Geometric Shape

A rectangular shape of known constant RGB values was captured in a video sequence of approximately 500 frames with different distances and positions from the fish-eye camera. Fig. 4(a) shows portions from 5 such frames of the original RGB video and the corresponding segmented ones are shown in Fig. 4(b). Segmentation was performed by thresholding, based on the known RGB values of the test pattern. The Zernike moments of the binary image were calculated, using the traditional definition and the geodesic corrections. Fig. 4(c) plots the Zernike for $n=2$, $m=0$. It is evident that the geodesic correction Zernike is less noisy than the traditional implementation. The negative peak is caused by a drastic change in scale. Fig. 4(d) and 4(e) show the fraction of Z_{60}/Z_{30} and the fraction Z_{40}/Z_{20} (see Eq. (3)). It can be observed that these fractions are immune to the sudden scale change. Furthermore, the fractions of the geodesically corrected Zernike are less noisy than the traditional ones.

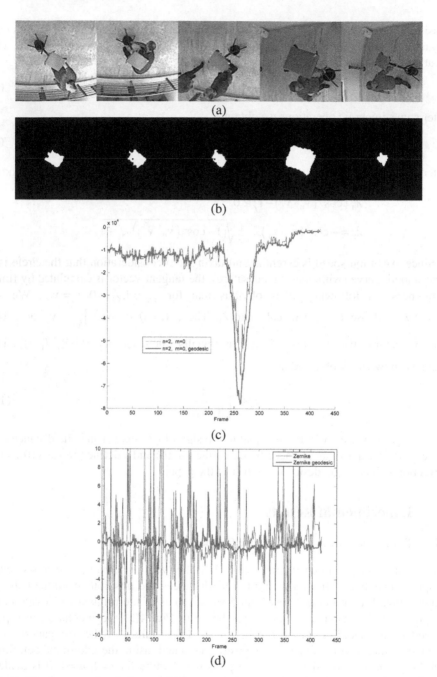

Fig. 4. Zernike moments before and after geodesic correction for the rectangular geometric shape (see text for details)

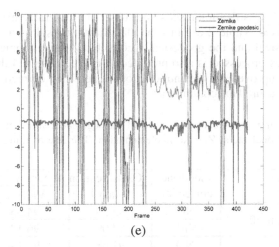

(e)

Fig. 4. (*Continued.*)

3.2   Classification of the Human Silhouette Pose

In this experiment, we acquired 2 different videos of a human with two different classes of poses: walking/standing and sitting. In each video the subjects sits in 2 or 3 different places. The human silhouette is segmented using an unsupervised segmentation algorithm presented in [12] and the Zernike moments are calculated twice: without and with the geodesic correction. The simple k-Nearest Neighbour (k-NN) classifier with k=3 was employed to classify each frame into the two classes, using the Euclidean distance metric between the Zernike features.

The classical Zernike moments have been used to classify the human action of each frame, utilizing moments of the first n orders, where $n = 3, 4, 5$. In Table 1 the results for each video are presented. The classification results for both videos are quite similar for moments of the same order. More specifically, in video 1 the classification accuracy increases for higher order Zernike moments. On the other hand, in the more complex video 2, the best results are obtained using N = 4. It is evident that higher order Zernike moments, results in an increase in computational complexity, as well as the CPU time needed to process each frame. The average number of frames per second (FPS) processed by applying the Zernike moments and the k-NN classifier is also reported.

Table 1. The resulting classification accuracy of the segmented silhouette pose for different orders of the traditional Zernike implementation. The frame processing rate is also provided.

Zernike order	Classification Accuracy (%)		FPS
	Video 1	Video 2	
n = 3, m=-1,+1	84.22	85.77	16 fps
n = 4, m=-2,0,+2	92.11	92.08	15 fps
n = 5, m=-3,-1,+1,+3	94.88	91.82	12 fps

To improve the classification results, the subset of the Zernike moments with radial symmetry were used. Table 2 exhibits the classification results for n up to 30, while $m=0$ for the classical and the geodesically corrected Zernike moments, respectively. The rate of frames processed is also provided. It can be observed that the classification results are improved by the introduction of geodesic correction, with minimal increase to the measured execution time.

Table 2. The resulting classification accuracy of the segmented silhouette pose for different orders of the traditional radial Zernike implementation. The rate of frame processing is also provided.

Zernike order	Geodesic correction	Classification Accuracy (%)		Execution speed
		Video 1	Video 2	
n = 2,4,6,8,10, m=0	NO	93.39	94.53	17 fps
	YES	94.09	96.13	17 fps
n = 2,4,6,8,…,20, m=0	NO	91.68	94.14	11 fps
	YES	94.24	95.94	12 fps
n = 2,4,6,8,…,30, m=0	NO	92.11	94.01	8 fps
	YES	92.24	95.75	9 fps

Fig. 5 graphically exhibits the classification results, using n≤20, m=0, without and with geodesic correction. Class 0 corresponds to sitting silhouette and class 1 corresponds to standing/walking silhouette. The algorithmic results are shown in blue and the ground truth in red. It can be seen that the geodesic correction improves the Zernike based classification accuracy.

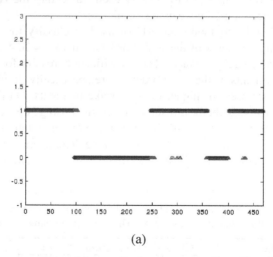

(a)

Fig. 5. Graphical representation of the results for pose estimation. Class 0: "sitting", Class 1: "standing/walking", using Central Zernike with n≤10, without geodesic correction (a) and with geodesic correction (b).

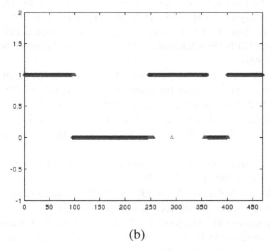

(b)

Fig. 5. (*Continued.*)

4   Discussion and Conclusions

As shown in the previous section the geodesic correction enhances the descriptive properties of Zernike moments, when applied to omni-directional images. The Zernike descriptors appear to be more stable (less variability for the same pattern). The accuracy of a classifier in terms of classifying two human postures was also increased when using Zernike moments with the proposed geodesic correction. Finally, the execution time was only marginally increased. The proposed methodology can significantly improve the accuracy of more complex activity recognition algorithms usually found in ambient assisted living environments

Acknowledgment. The authors would like to thank the European Union (European Social Fund ESF) and Greek national funds for financially supporting this work through the Operational Program "Education and Lifelong Learning" of the National Strategic Reference Framework (NSRF) - Research Funding Program: \Thalis \ Interdisciplinary Research in Affective Computing for Biological Activity Recognition in Assistive Environments

References

1. Mikolajczyk, K., Schmid, C.: Scale and affine invariant interest point detectors. International Journal of Computer Vision 1(60), 63–86 (2004)
2. Hu, M.K.: Visual pattern recognition by moment invariants. IRE Trans. Inform. Theory IT 8, 179–187 (1962)
3. Teague, M.: Image analysis via the general theory of moments. J. Opt. Soc. Am. 70(8), 920–930 (1980)

4. Abu-Mostafa, Y.S., Psaltis, D.: Recognitive Aspects of Moment Invariants. IEEE Transactions on Pattern Analysis and Machine Intelligence PAMI-6, 698–706 (1984)
5. Flusser, Zitova, B., Suk, T.: Invarant-based registration of rotated and blurred images. In: Proceedings of the IEEE 1999 International Geoscience and Remote Sensing Symposium, IGARSS 1999 (1999)
6. Schlemmer, M., Heringer, M., Morr, F., Hotz, I., Bertram, M.H., Garth, C., Kollmann, W., Hamann, B., Hagen, H.: Moment Invariants for the Analysis of 2D Flow Fields. IEEE Transactions on Visualization and Computer Graphics 13, 1743–1750 (2007)
7. Cruz-Mota, J., Bogdanova, I., Paquier, B., Bierlaire, M., Thiran, J.-P.: Scale Invariant Feature Transform on the Sphere: Theory and Applications. Int. J. Comput. Vis. 98, 217–241 (2012)
8. Lowe, D.G.: Distinctive image features from scale-invariant keypoints. International Journal of Computer Vision 60(2), 91–110 (2004)
9. Hansen, P., Corke, P., Boles, W., Daniilidis, K.: Scale invariant features on the sphere. In: International Conference on Computer Vision, pp. 1–8 (2007)
10. Tamimi, H., Andreasson, H., Treptow, A., Duckett, T., Zell, A.: Localization of mobile robots with omnidirectional vision using particle filter and iterative sift. Robotics and Autonomous Systems 54(9), 758–765 (2006)
11. Marhic, B., Mouaddib, E.M., Pegard, C.: A Localisation Method with an Omnidirectional Vision Sensor Using Projective Invariant. In: Proceedings of the 1998 IEEEJRSJ Inti. Conference on Intelligent Robots and Systems Victoria, B.C., Canada (October 1998)
12. Delibasis, K.K., Goudas, T., Plagianakos, V.P., Maglogiannis, I.: Fisheye Camera Modeling for Human Segmentation Refinement in Indoor Videos. In: PETRA 2013, Island of Rhodes, Greece, May 29-31 (2013), Copyright 2013 ACM 978-1-4503-1973-7/13/05

Fish-Eye Camera Video Processing and Trajectory Estimation Using 3D Human Models

Konstantina Kottari[1], Kostas Delibasis[1], Vassilis Plagianakos[1],
and Ilias Maglogiannis[2]

[1] University of Thessaly, Dept. of Computer Science and Biomedical Informatics,
Lamia, Greece
[2] University of Piraeus, Dept. of Digital Systems, Piraeus, Greece
kottarikonstantina@gmail.com, kdelibasis@yahoo.com,
vpp@dib.uth.gr, imaglo@unipi.gr

Abstract. Video processing and analysis applications are part of Artificial Intelligence. Frequently, silhouettes in video frames lack depth information, especially in case of a single camera. In this work, we utilize a three-dimensional human body model, combined with a calibrated fish-eye camera, to obtain three-dimensional (3D) clues. More specifically, a generic 3D human model in various poses is derived from a novel mathematical formalization of a well-known class of geometric primitives, namely the generalized cylinders, which exhibit advantages over the existing parametric definitions. The use of the fish-eye camera allows the generation of rendered silhouettes, using these 3D models. Moreover, we present a very efficient algorithm for matching that 3D model with a real human figure in order to recognize the posture of a monitored person. Firstly, the silhouette is segmented in each frame and the calculation of the real human position is calculated. Subsequently, an optimization process adjusts the parameters of the 3D human model in an attempt to match the pose (position and orientation relatively to the camera) of real human. The experimental results are promising, since the pose, the trajectory and the orientation of the human can be accurately estimated.

Keywords: fish-eye camera video processing, three-dimensional human modelling, posture recognition, minimization, generalized cylinders, and elliptical intersections.

1   Introduction

The field of automated human activity recognition utilizing fixed cameras of indoor environments has gained significant interest during the last years. It finds a variety of applications in diverse areas, such as assistive environments, smart homes, support for the elderly or the chronic ill, surveillance and security, traffic control, industrial processes, etc.

This work focuses on fish-eye camera video processing for pose estimation of sitting or standing/walking humans. Therefore, human silhouette segmentation of the video sequence is a prerequisite. Recognizing a human pattern is often possible via

L. Iliadis et al. (Eds.): AIAI 2014, IFIP AICT 436, pp. 385–394, 2014.

volume intersection [1] or a voxel-based approach [2,3]. Stereometry based models have also been constructed through calibrated camera pairs. Using triangulation, the depths of the points are calculated. This approach has been taken into account by Plänkers and Fua [4] and Haritaoglu et al. in [5]. Stereo vision is also used by Jojic et al. [6], with the optional aid of projected light patterns. The proposed algorithm is based on a parametric three-dimensional (3D) human model with limited degrees of freedom so that it allows efficient manipulation for standing/walking and sitting postures. Our aim is to estimate human position, trajectory and standing/sitting state, which would be useful towards human behavior recognition.

The first step in applications dealing with human activity recognition from video is the foreground segmentation. Most video segmentation algorithms are based on background subtraction. The background has to be modelled, since it may change due to a number of reasons, including: motion of background objects, changes in light conditions, or video compression artifacts. In this work, we employ the forward and inverse camera model that was proposed in [7]. We follow a "top-down" approach that matches the model rendered through the calibrated fish-eye camera, with the segmented frame of the video. Then, an optimization algorithm is utilised to find the model parameters and determine human orientation and pose. The rest of the paper is structured as follows: Section 2 discusses the technical details of the proposed algorithms, Section 3 presents some initial results, while Section 4 concludes the paper.

2   Proposed Methodology

2.1   Generalized   Cylinders

For the generation of the human model, we utilized the concept of generalized cylinders (GC), as proposed in [8]. More specifically, let C_1 be a piecewise smooth curve defined in a Cartesian coordinate system $OXYZ$, as:

$$r_1(t) = (x(t), y(t), z(t)), t \in [a,b] \subset R \tag{1}$$

and C_2 be a planar curve defined in an orthogonal local Cartesian system OXY. Letus now consider the surface S that is generated by moving the curve C_2 along C_1, so that its plane is perpendicular to the tangent vector of C_1, and the origin of OXY belongs to C_1.If we express the planar curve C_2 in polar coordinates $r_2 = r_2(u), u \in [0, 2\pi]$ and introduce a scale factor $s(t)$and a rotation factor $\phi(t)$ along the tangent vector of $C1$ as function of position along C_1, then the surface equation of the GC becomes:

$$x(t,u) = x(t) + \frac{s(t)y^{'}(t)r(u+\phi(t))\cos(u+\phi(t))}{P_2(t)}$$

$$+ \frac{s(t)x^{'}(t)z^{'}(t)r(u+\phi(t))\sin(u+\phi(t))}{P_1(t)P_2(t)} \tag{2}$$

$$y(t,u) = y(t) - \frac{s(t)x'(t)r(u+\phi(t))\cos(u+\phi(t))}{P_2(t)}$$

$$+ \frac{s(t)y'(t)z'(t)r(u+\phi(t))\sin(u+\phi(t))}{P_1(t)P_2(t)}$$ (3)

$$z(t,u) = z(t) - \frac{s(t)(x'(t))^2 + (y'(t))^2 r(u+\phi(t))\sin(u+\phi(t))}{P_1(t)P_2(t)}$$ (4)

where, $(t,u) \in [a,b] \times [0,2\pi]$, $P_1(t) = \sqrt{(x'(t))^2 + (y'(t))^2 + (z'(t))^2}$ and $P_2(t) = \sqrt{x'(t))^2 + y'(t))^2}$ .The complete proof is given in [8].

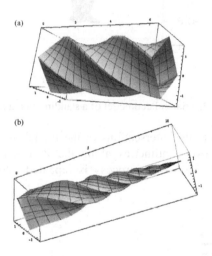

Fig. 1. Two examples of surfaces derived from equation (2) – (4) from [8]

2.2   3D Model Construction

In this work, a free triangulated model of a standing human (Fig.2) is utilised, defined by the Cartesian coordinates of approximately 27,000 vertices [9]. Since we are interested in simulating the rendering of the human model through the fish-eye camera in real time, we discard the triangle information of the model and we treat it as a cloud of points.

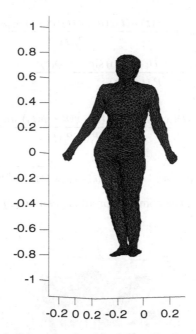

Fig. 2. Triangulatedmodel of a standing human [9]

Therefore, we compute the intersections of the model in a number of horizontal planes, in distance of two centimetres along the Z axis (feet – head direction) as shown in Fig.3 (a). The same process is repeated, along hands and legs – see Fig. 3(b).

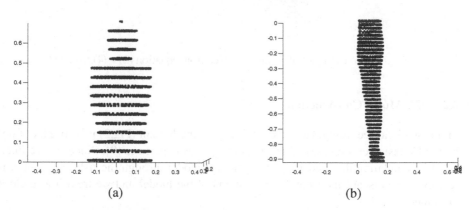

(a)   (b)

Fig. 3. Elliptical intersections of torso and leg

Each intersection is estimated for approximating an ellipse with its semi-axes a_{semi}, b_{semi} parallel to X and Y axis of coordinate system, as shown in Fig.4.

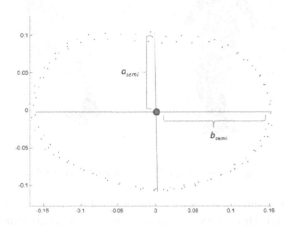

Fig. 4. Elliptical intersection of the human torsowith theXY plane

The parametrical equation of GC (4) can be simplified by using a piecewise straight line as curve C_1, each segment of which is defined by vector (a_0, b_0, c_0) and by assigning these ellipses as planar closed curveC_2as follows:

$$x = a_0 t + \frac{b_0 r \cos(u)}{P_2} + \frac{a_0 c_0 r \sin(u)}{P_2}$$

$$y = b_0 t - \frac{a_0 r \cos(u)}{P_2} + \frac{b_0 c_0 r \sin(u)}{P_2}$$

$$z = c_0 t - \frac{(a_0^2 + b_0^2) r \sin(u)}{P_2}$$

(5)

where $r = \dfrac{a_{semi} b_{semi}}{\sqrt{(b_{semi} \cos(u))^2 + (a_{semi} \sin(u))^2}}$ and a_0, b_0, and c_0 are determined by

the direction of the leading axis. In Fig.5 we see the result of elliptical patterning of model intersections, through the insertion of ellipses to the GC Eq. (5). Note that a torsional inconstancy at the knees section is being observed. This phenomenon has been explained in [8, section 4] and does not affect the optimization process that matches the 3D model to the segmented human silhouette.

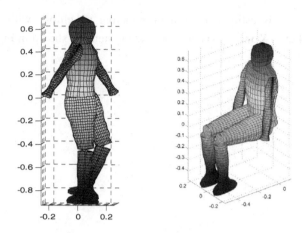

Fig. 5. D standing (left) and sitting human (right)

The estimation of human posture (sitting or standing) is based on the construction of the human model. Having the 3D standing human model constructed as described above, its transformation to match the sitting position can be easily performed by changing the model parameter (angles) at waist and knees. The result of that transformation is shown in Fig. 5 right, while Fig.6 depicts both models as they are utilised by the video-processing algorithm.

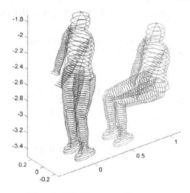

Fig. 6. 3D Human silhouettes

2.3   Video Processing Algorithm

In this work, we analyze videos captured by a fish-eye camera, fixed on the ceiling of a living environment. The recorded videos have been foreground segmented, while empty frames are being discarded. Then, the mask shown in Fig. 7(a) is applied to suppress noisy segmented pixels outside the field of view. The initial estimation of the real human position in the room coordinates is accomplished by the recently proposed algorithm in [7] based on the segmented frame pixels. For this purpose, we employ the calibration of the acquiring fish-eye camera that provides the spherical coordinates (θ, φ) for each pixel of the current frame, as well as the frame pixel that corresponds to any real world point (x,y,z), according to [7]:

$$(j,i) = M(x,y,z) \tag{6}$$

$$(\theta,\varphi) = M_1(i,j) \tag{7}$$

Let $PHI(i,j)$ hold the value of φ for pixel (i,j), as obtained by (7) and shown in Fig. 7(b). Thus, for any pose of the 3D parametric model, we can obtain the binary image I_M of the human model, rendered by the fish-eye model using (6). Fig.8illustrates image I_M combined for various standing and sitting models, as rendered by the calibrated fish-eye camera. Let I_S be segmented binary frame, after using the binary mask in Fig.7(a). The initial estimation of the person's position is obtained by locating the non-zero pixel (i_0,j_0) of I_S that holds the minimum value of angle φ. The objective function, which quantifies the match between the model and the segmented human silhouette as a function of its real world position (x,y) and its orientation θ_0, is defined as the intersection of the segmented silhouette I_S and the rendered human model I_M:

$$f(x,y,\theta_0) = \sum_{\substack{image \\ domain}} I_M \cap I_S \tag{8}$$

where I_M (defined above) is shown in red, I_S shown in green and their intersection $I_M \cap I_S$ is shown in yellow. Fig. 9 presents graphically the calculation of the objective function for one instance of each class (sitting and standing). Subsequently, the simplex [10] multidimensional unconstrained maximisation algorithm is utilised to optimise the objective function. The initial position (x,y) of the human is approximately computed from the first frame by finding the segmented human silhouette pixel (i,j) with maximum PHI, (as described in [7]) and used to initialize the simplex method. Thus, the maximisation algorithm computes the human model parameters that best match the segmented figure and returns the coordinates x and y, as well as the orientation θ_0 of model.

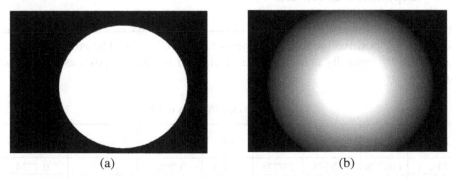

(a)   (b)

Fig. 7. (a) Binary mask used to exclude out of filed-of-view pixels in video frames (b) Visualization of the PHI angle for each frame pixel as resulted from the camera calibration [7]

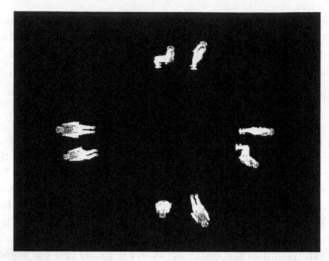

Fig. 8. Rendered standing and sitting human model in various angles and positions in the room

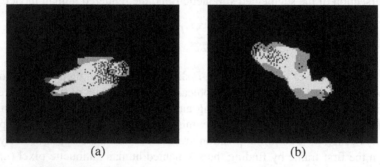

(a)   (b)

Fig. 9. Visualization of the calculation of the objective function (see text for details) for the standing (a) and sitting human (b)

3 Experimental Results

Classification results of two different videos are presented in Table 1, considering the sitting as "positive" and the standing as "negative" status. The ground truth was established by manually annotating the video with the human pose, as "sitting" or "standing".

Table 1. Pose classification results

	TP	TN	FN	FP	Accuracy	Sensitivity	Specificity
Video 1	110/278	112/278	23/278	33/278	0.7986	0.8271	0.7724
Video 2	98/164	54/164	5/164	7/164	0.9268	0.9515	0.8852

The proposed model-based algorithm is able to estimate the trajectory of the human silhouette, as well as its orientation. Fig. 10 shows the estimated positions and orientation of the human silhouette for video 1, as recovered by the optimization described above. The person moves from left to right at the bottom of the frame (A to B), then vice versa at the top of the frame and finally sits down at the point designated by E, where it rotates. Between points B and C the segmentation fails temporarily, however, the model-based tracking algorithm successfully recovers the silhouette's new position. The experimental results indicate that the simplex optimization method is robust and efficient. It needs approximately 30 iterations for each frame in order to converge. Each objective function evaluation requires approximately 30msec on an Intel i5 laptop with 4 GB Ram using the Matlab environment. Running time approaches the 900 milliseconds per frame.

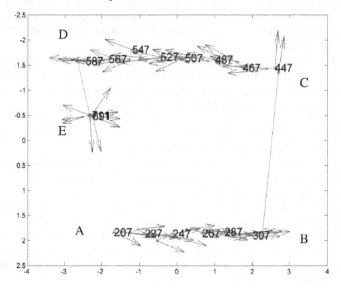

Fig. 10. Graphical representation of the results for model based trajectory and orientation estimation for video 1

4   Discussion and Conclusions

In this paper, an algorithm for estimating the trajectory of a human silhouette in indoor videos acquired by an omni-directional camera has been presented. The algorithm is based on a parametric 3D human model and it recovers the model parameters (translation and orientation) by optimizing a suitable defined objective function. Initial results show that the proposed algorithm can estimate the trajectory and orientation and discriminate between two different postures: sitting and standing. The proposed methodology may improve the accuracy of more complex activity recognition algorithms usually found in ambient assisted living environments. This methodology can be adopted for detecting higher level activity events and understand behavioural patterns.

Acknowledgment. The authors would like to thank the European Union (European Social Fund ESF) and Greek national funds for financially supporting this work through the Operational Program "Education and Lifelong Learning" of the National Strategic Reference Framework (NSRF) - Research Funding Program: \Thalis \ Interdisciplinary Research in Affective Computing for Biological Activity Recognition in Assistive Environments.

References

1. Bottino, A., Laurentini, A.: A silhouette-based technique for the reconstruction of human movement. Computer Vision and Image Understanding (CVIU) 83(1), 79–95 (2001)
2. Cheung, G.K.M., Baker, S., Kanade, T.: Shape-from silhouette of articulated objects and its use for human body kinematics estimation and motion capture. In: Proceedings of the Conference on Computer Vision and Pattern Recognition (CVPR 2003), Madison, WI, vol. 1, pp. 77–84 (2003)
3. Mikic, I., Trivedi, M., Hunter, E., Cosman, P.: Human body model acquisition and tracking using voxel data. International Journal of Computer Vision 53(3), 199–223 (2003)
4. Plänkers, R., Fua, P.: Tracking and modeling people in video sequences. Computer Vision and Image Understanding (CVIU) 81(3), 285–302 (2001)
5. Haritaoglu, I., Harwood, D., Davis, L.S.: W4S: A real-time system for detecting and tracking people in 2 1/2D. In: Burkhardt, H.-J., Neumann, B. (eds.) ECCV 1998. LNCS, vol. 1406, pp. 877–892. Springer, Heidelberg (1998)
6. Jojic, N., Gu, J., Shen, H.: S. Huang T.S: 3-Dreconstruction of multipart, self-occluding objects. In: Proceedingsof the Asian Conference on Computer Vision (ACCV 1998), HongKong, China, pp. 455–462 (1998)
7. Delibasis, K.K., Goudas, T., Plagianakos, V.P., Maglogiannis, I.: Fisheye Camera Modeling for Human Segmentation Refinement in Indoor Videos. In: PETRA 2013, Island of Rhodes, Greece, May 29-31 (2013), Copyright 2013 ACM 978-1-4503-1973-7/13/05
8. Delibasis, K.K., Kechriniotis, A., Maglogiannis, I.: A novel tool for segmenting 3D medical images based on generalized cylinders and active surfaces. Computer Methods and Programs in Biomedicine 111, 148–165 (2013)
9. http://www.3dmodelfree.com/models/20966-0.htm
10. Nelder, J.A., Mead, R.: A simplex method for function minimization. Computer Journal 7, 308–313 (1965)

Analysis of Relevant Features for Pollen Classification

Gildardo Lozano-Vega[1,2], Yannick Benezeth[2], Franck Marzani[2],
and Frank Boochs[1]

[1] i3mainz, Fachhochschule Mainz, Lucy-Hillebrand-Strasse 2, 55128 Mainz, Germany
gildardo.lozano@fh-mainz.de
[2] Le2i, Université de Bourgogne, B.P. 47870, 21078 Dijon Cedex, France

Abstract. The correct classification of airborne pollen is relevant for medical treatment of allergies, and the regular manual process is costly and time consuming. Aiming at automatic processing, we propose a set of relevant image-based features for the recognition of top allergenic pollen taxa. The foundation of our proposal is the testing and evaluation of features that can properly describe pollen in terms of shape, texture, size and apertures. In this regard, a new flexible aperture detector is incorporated to the tests. The selected set is demonstrated to overcome the intra-class variance and inter-class similarity in a SVM classification scheme with a performance comparable to the state of the art procedures.

Keywords: pattern recognition, feature extraction, feature evaluation, apertures, palynology.

1 Introduction

The correct estimation of airborne pollen concentration is important for the prevention and treatment of allergies. Traditional methods require manual and specialized labor, which is expensive, time consuming and susceptible to inconsistency. Commonly, collected airborne particles are analyzed manually under a brightfield microscope in order to count the frequency of different pollen taxa. With the introduction of computer vision techniques, pollen counting aspires to become automatic, faster and more accurate, which would enable a more frequent and broader analysis of samples.

Common strategies for image-based pollen classification follow a typical image classification process: image digitization, preprocessing, segmentation, pollen description, and pattern recognition. In all the cases, pollen description is based on different types of metrics and representations of the pollen image, known in the literature as *features*. Previous strategies are differentiated mainly by the type of employed features. Chen *et al.* found a relevant combination of seven general shape features, four aperture-colpus features, and a statistical gray-level feature to recognize *Birch, Grass* and *Mugwort* pollen with 97.2% of accuracy with a Linear Normal Classifier (LNC) and forward feature selection [1]. Unfortunately,

L. Iliadis et al. (Eds.): AIAI 2014, IFIP AICT 436, pp. 395–404, 2014.
© IFIP International Federation for Information Processing 2014

the aperture and colpus detectors are type-specific and are not easily usable for other taxa.

Additionally to general shape and aperture features, Boucher *et al.* used color features for the classification of 30 pollen taxa [2]. With the same deficiency as Chen *et al.* for the aperture detection, they achieved an accuracy of 77%. The average of 11.6 images/taxon in the dataset looks small for capturing all the variations. Interestingly, they used a knowledge-based classification instead of typical statistical methods. A good performance just employing five general shape features was achieved by Rodríguez Damián *et al.* in the classification of three types of *Urticaceae* pollen with 85.6% of accuracy with a minimum distance classifier [3]. Employing only Haralick measures, Li *et al.* classified four pollen taxa with 100% of accuracy using the Fisher's linear discriminant method [4]. However, the tested taxa does not appear to have strong inter-class similarity which leaves the performance unknown on typical allergenic pollen. The aforementioned results suggest that the synergic contribution of features from different descriptive foundations could be the key for a robust and accurate classification.

Atypically, Ranzato *et al.* used local invariant features without segmentation with accuracy of 78.2% on eight pollen taxa, and employed a Bayesian classifier with a Gaussian mixture probability density function [5]. Similarly, Ronnenberger *et al.* developed 3D features from the Haar integration framework, achieving 98.5% of precision with a Support Vector Machine (SVM) and the Radial Basis Function (RBF) as the transformation kernel, for 33 taxa after the rejection of uncertain results [6]. Although very impressive, unfortunately the method requires 3D volumetric scans of the particle, which looks impractical for real-time applications.

Our proposed solution takes advantage of the strength of diverse feature groups in order to provide robustness and accuracy. The rest of the paper is organized as follows: in section 2, the proposed strategy for description and classification is explained in detail. In section 3, experiments and results are presented. Finally, in section 4, conclusions of the proposed method and the following steps are stated.

2   Strategy for Pollen Description and Classification

2.1   Outline of the Proposed Solution

We propose the application of selected features for the classification of the five most important allergenic pollen taxa in Germany: *Alder, Birch, Hazel, Mugwort* and *Grass*. Starting with microscopic 2D images, the solution involves the segmentation of the pollen from the background and the description of the particle by different types of features. Our contribution is a robust set of descriptive features, which range from general to more specific, and it is able to overcome the most important problems in pollen recognition: the inter-class similarity and the intra-class variation. Robustness is provided by employing features related to different pollen characteristics such as shape, size, texture and apertures. Moreover, an unified flexible aperture detector is employed for first time in pollen classification instead of multiple fixed algorithms. Finally, a relevant set of features is systematically selected and applied to a support vector machine process for classification of the taxa. This classifier is chosen due to its well known performance in different tasks, particularly when high dimensional data are involved [7].

2.2 Segmentation of the Pollen Particles

Color images of magenta-dyed pollen are manually digitized from microscope slides and labelled by palynologists. The isolation of the particle from the background by a bounding contour is a necessary step to compute the set of features. For this purpose, the application of the active contour method is often found in previous works, for example in [3][6]. Although effective in filling gaps, this method requires a modelling stage for a good performance with the risk of loosing flexibility. In contrast, Chen *et al.* successfully used a simple two-step automatic thresholding and hole-filling for segmentation of three different pollen taxa [1].

Following this simple idea, our automatic method employs Otsu's thresholding because it is nonparametric, unsupervised and fast [8]. This method searches the threshold that minimizes the intra-class variance of the background and the particle pixels on gray level images. It was found experimentally that the method can determine the correct threshold even under variation of light intensity due to the scanning process, as long as a certain intensity difference between the background and the particle exists (*cf.* Fig. 1). After the binarization, it is easy to trace the contour using Suzuki's method [9]. Due to its high contrast, debris could affect the correct estimation of the contour. If not stuck to the pollen, debris is detected and rejected by its size and irregular shape.

| Alder | Birch | Hazel | Mugwort | Grass |

Fig. 1. Example of the five pollen taxa of interest and their segmentation using Otsu's automatic method. The detected contour in blue is enhanced for visualization.

2.3 Description of the Pollen Particles

According to their descriptive characteristics, features can be grouped as general shape features, elliptic Fourier descriptors, texture features, and apertures.

General Shape Features. As mentioned in section 1, shape features are proven to be suitable to characterize pollen for classification purposes [1][3]. The computation of these features is based on a few main variables. Additionally to *perimeter* and *area*, statistics of the distance between the centroid and the contour are used: *standard deviation, mean (Dmean), minimum (Dmin)* and *maximum (Dmax)*. The first part of Table 1 shows the ten classical features that are employed in the present work, and whose detailed equations are given in [10]. These features focus, besides on estimating the size, on quantizing statistically the complexity of the shape. However, there is no feature that relates to the outline of the pollen, which in general can be described as spherical, oblate or prolate. Because this property is pointed out by palynologists to be discriminative [11], we propose a point-by-point representation of the pollen outline by fitting an ellipse to the contour using the Joon Ahn *et al.* optimal fitting algorithm [12]. From this, useful rotation invariant features are added to the classical shape features: *ratio between major and minor axes length*, and *rms value, mean* and *standard deviation of the error of the fitting*.

Particularly, *perimeter, area* and *2eN* are features that are proportional to the shape size. If combined with the rest of the general shape features, these three features could introduce bias related to the pollen size differences in the evaluation of the ability of general shape features to discriminate shape. For this reason, they are not considered at the shape evaluation stage. However, size is an important discriminant of pollen taxa and these three features are included in a global evaluation together with other type of features.

Table 1. List of general shape features grouped in classical and ellipse-fitting-based

Classical shape features

Feature	Short name	Features	Short name
Perimeter	P	Radius dispersion	rdis
Area	A	Ratio Dmax/Dmin	ratio1
Roundness	R	Ratio Dmax/Dmean	ratio2
O'Higgins Undulation	U	Ratio Dmin/Dmean	ratio3
Complexity f	cf	2n Euclidean norm	2eN

Ellipse fitting features

Feature	Short name	Features	Short name
Axes Ratio	EF_ratio	Mean error	EF_mean
RMS error	EF_rms	Std Dev of the error	EF_std

Elliptic Fourier Descriptors (EFD). These descriptors are employed to represent shapes given their contour as point pairs (x, y). The method is based on treating each dimension, x and y, as separate functions, and computing the Fourier coefficients. The coefficients are then transformed into translation, rotation and scale invariant descriptors [13]. The EFD's contain all the information to reconstruct the original shape in an inverse operation. Moreover, the EFD's are ordered in relation to the frequency content of shape contour form low to high. Iwata *et al.* were able to describe subtle variations of similar root shapes belonging to the same taxa with principal component analysis on EFD's [14]. Due to the strong similarity among the classes, the root representation would have not been effective if general shape features had been employed instead. For our tests, 400 EFD's are computed, from which the first two descriptors are not considered since they are meaningless due to a normalization step, resulting in 398 EFD's.

Texture Features. This group of features is focused on recognizing the texture patterns that are formed by the contribution of the outer ornamentation and the inner mass of the pollen. Motivated by results of previous works, we are interested in testing the Haralick measures on the top five german allergenic taxa. Our approach employs eleven Haralick measures from the gray co-occurrence matrix(GLCM): *angular second moment, contrast, correlation, variance, inverse difference moment, sum average, sum variance, sum entropy, entropy, difference variance* and *difference entropy* [15]. 24 different pixel offsets are employed, resulting in a total of 264 features ranging from two to eleven pixels of separation distance (0.5 to 2.9 microns) in different orientations. Because of the compactness of the texture patterns, longer distances are not needed. Only the area inside the contour is considered in order to avoid influence from the particle border and the background.

Aperture Feature. Palynologists state that the type and amount of apertures is very discriminating of the taxa [11]. Apertures are morphological distinctive regions of the pollen wall, usually visible in typical microscopic observations. Due to their inter-class variety, previous approaches have struggled to detect this kind structures efficiently. Moreover, changes in the point of observation increase the variability of their appearance. Figure 2 shows examples of different aperture appearances. Previous recognition strategies have focused on just particular pollen types, which would require the development of new algorithms for other types. For example, Boucher *et al.* detected apertures of *Grass* by means of segmentation techniques, shape and color features [2]. Chen *et al.* detected the aperture and colpus of *Birch* and *Mugwort* based on the intensity profile of the image polar transformation, the Hough transform and a template matching technique [1].

The present strategy employs the proposed method by Lozano Vega *et al.* in which different aperture types can be detected with the same algorithm [16]. This flexibility is an advantage for the recognition of multiple taxa with multiple appearances. Moreover, the method was tested on *Birch* and *Alder* with recall above 80%. The representation of the apertures is through the histogram of primitive visual words contained in the image and densely extracted. These visual words capture most of the variance of the different patterns present in the pollen and are gathered in a codebook. The visual words are created from clusters based on the similarity of local feature vectors computed on small image patches. Local Binary Patterns (LBP) [17] is chosen due to its simple but efficient computation and successful results on the description and classification of complex image

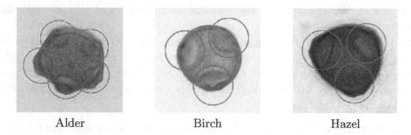

Alder   Birch   Hazel

Fig. 2. Different aperture appearances due to the change of the viewing angle and taxon variety. Aperture is highlighted by a green circle.

patterns such as faces [18]. The uniform (u2) normalized LBP version with a (8,1) neighborhood is employed as local feature to reduce the length of patterns and take advantage of the rotation invariance [19].

A SVM classifier is trained for the recognition of the aperture region histograms. SVM is a popular method that maps the feature vectors into a higher dimensional space in order to separate the positive and negative classes with a hyperplane with a maximum margin. The radial basis function is used as mapping kernel, and the penalty C and the kernel parameter γ are tuned with a grid search in a three-fold cross validation.

In order to evaluate an unknown pollen, multiple regions covering the particle are classified as *aperture* or *not an aperture* using the aforementioned method. A pixel-wise confidence map is created by the averaged votes of overlapping regions, which can be interpreted as the likelihood of the occurrence of apertures. Finally, apertures are found by estimating the local maxima of the map and the total amount is used as the aperture feature. Our implementation creates 20 visual words and square regions of 30 pixels by side, being able to detect apertures of *Alder, Birch,* and *Hazel* among five pollen types. An example of the confidence map and the detection of apertures is shown in Fig. 3.

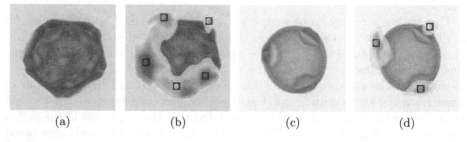

(a)   (b)   (c)   (d)

Fig. 3. Original images of *Alder* (a) and *Birch* (c) pollen and their respective confidence maps (b) and (d). Warmer colors (red-orange) on the maps indicate a high likelihood of detecting an aperture. Blue squares indicate the estimated location of the apertures after searching local maxima.

3 Experimental Results

We employed a statistical classification strategy for the evaluation of the feature groups on the discrimination of the five most important pollen taxa in Germany. The source microscope slides were prepared in a laboratory, stained with magenta, and scanned with a Keyence BZ-9000 microscope in brightfield mode with a magnification of 40x and a resolution of 0.26 microns/pixel. The dataset of individually cropped pollen images consisted of 100 labeled images/taxon, except for *Hazel*, for which 48 images were employed due to a defective staining. Automatic segmentation was performed according to the method described in section 2.2 and the complete set of features was computed according to section 2.3: 14 general shape features, 398 EFD's, 264 texture features, and the aperture feature.

Initially, each group of features was evaluated individually in order to evaluate the performance of the subsets and to be able to discover the most discriminant features. The aperture feature was not tested individually because results from a single-value feature are not relevant. Moreover, palynologists support firmly the discriminant importance of this characteristc [11]. Although SVM classification was selected especially for the high-dimensional EFD's and texture features, it was employed also for the rest of the experiments in order to keep comparability. The performance measure was the accuracy in a three-fold cross validation with stratified sampling and a training/testing ratio of 0.67/0.33. This method allows to verify the robustness to unseen data. We also tried five and ten folds in order to validate changes of the training/testing ratio (0.8/0.2 and 0.9/0.1 respectively). Since results were comparable, they are not shown here.

The optimal set of features can be found by examining all combinations, strategy called brute force feature selection. However, this solution is not computational efficient, particularly for high dimensional vectors, as in the case of EFD's and texture groups. For those cases, Sequential Forward Selection (SFS) is the method employed for discovering the most relevant features and for reducing the feature space dimensionality. SFS is a wrapper approach that searches the best set of features based on the classification performance. It begins with the empty subset of selected features $X = \{\emptyset\}$ and starts an iterative search of features $\{x_1, x_2, \ldots, x_n\}$ to grow X. In each iteration, all the combinations of X with a new feature is evaluated. The feature x_i that delivers the best performance is integrated permanently to X. The iteration is stopped until no better performance is obtained, and then the current X is regarded as the best subset. This method avoids evaluating all the combinations of features and it is especially fast when the optimal set is small. To keep consistency, the performance metric was a three-fold cross validation using a SVM classifier with RBF kernel.

For the individual experiments with the **general shape features**; *perimeter, area* and *2eN*; related directly to the pollen size; were excluded intentionally to avoid influence of the size differences in the evaluation of the shape recognition. Because of the reduced size of this group, brute force feature selection was feasible and employed in this feature group, with the advantage of finding the best subset of features.

The achieved accuracy was **83.49%** ±**2.19%** with *cf, rdis, ratio2, ratio3, EF_rms* and *EF_mean*. As expected, ellipse-fitting features came up as an important extension to classical features. Interestingly, neither R nor *EF_ratio*, which measure shape circularity, were selected. We consider that the intra-class variation of the shape of the pollen when projected into a 2D plane is strong enough to cause also a strong variation of the shape circularity, decreasing the discrimination power of this type of measures.

The accuracy with the **EFD's group** was **76.18%** ±**2.10%** with 34 descriptors using SFS, discarding 90% of the features. It is relevant to notice that 50% of the selected descriptors correspond to the first 90 positions, containing the lower frequencies of the contour.

The **texture group** reached an accuracy of **83.26%** ±**1.39%** with 42 features after applying SFS, discarding 84% of the features. The most discriminant Haralick measures were *angular second moment, contrast* and *sum average* with 21 features. The most frequent offsets belong to equivalent distances of two and ten pixels, which correspond together to 16 selected features.

Finally, we evaluated the performance of the classification using together **all the feature groups**. Only selected features from the individual tests were input in addition to *area, perimeter, 2eN* and the aperture feature for a total of 86 features. The classification accuracy increased up to **94.42%** ±**1.27%**, more than 10% above better than the individual groups. The confusion matrix is shown in Table 2 and the comparison with individual results in Fig. 4. Most errors are due to the mix up of *Alder and Birch*, which belong to the same family *betulaceae*, due to their mutual similarity in size, ornamentation and aperture number.

In summary, results show the importance of the contribution of different features to the classification accuracy. The combination of features of different nature allows to capture the diversity of visual information present in the pollen, which is linked to palynological properties. While the general shape features measure the shape complexity, EFD's can describe finely the pollen outline. Texture features capture appropriately pollen ornamentation patterns. Size and

Table 2. Confusion matrix of the classification of the five pollen taxa using all the feature groups together

		True				
		Alder	Birch	Hazel	Mugwort	Grass
Predicted	Alder	**91**	8	0	1	1
	Birch	8	**88**	1	1	0
	Hazel	1	2	**47**	0	0
	Mugwort	0	2	0	**98**	0
	Grass	0	0	0	0	**99**

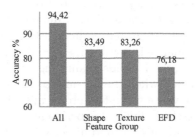

Fig. 4. Accuracy of the different tested groups. Using all groups together performed much better that the individual groups.

apertures are concepts which are statistically and semantically related to each taxon. The combination of the groups contributes to enhance the classification performance and the robustness against unknown data and errors on the contour.

4 Conclusions and Future Work

The performance of our proposal is among the best pollen recognition processes, comparable to that of Chen *et al.* with the addition of two more taxa and using a single method for aperture detection. More than 400 pollen samples from five pollen taxa and validation with different training and testing sets support the reliability of the results. Feature selection enables to identify relevant features and to reduce the feature space dimensionality. Assembling different feature groups facilitates to sum up strengths and to minify weaknesses, resulting in a more robust classification.

Focusing on the rich information (usually disregarded) in the palynology literature, we suspect that building a knowledge-structured classification system based on palynological properties could improve robustness while maintaining high accuracy. Those properties could be detected by taking advantage of the relevant features that are proposed in the present work.

Acknowledgements. The authors are grateful for their financial support to the German Bundesministerium für Wirtschaft und Technologie under the program Zentrales Innovationsprogramm Mittelstand ID KF2848901FR1, to the Conseil Regional de Bourgogne in France and to the Fond Europeen de Developpement Regional (FEDER). The authors are also grateful to Celeste Chudyk, Yann Ryann and Morad Larhriq for scanning and preparing the pollen datasets and to the Max Plank Institute for Chemistry for permitting the use of their facilities.

References

1. Chen, C., Hendriks, E.A., Duin, R.P., Reiber, J., Hiemstra, P., De Weger, L., Stoel, B.: Feasibility study on automated recognition of allergenic pollen: grass, birch and mugwort. Aerobiologia 22, 275–284 (2006)
2. Boucher, A., Hidalgo, P.J., Thonnat, M., Belmonte, J., Galan, C., Bonton, P., Tomczak, R.: Development of a semi-automatic system for pollen recognition. Aerobiologia 18(3), 195–201 (2002)
3. Rodríguez-Damián, M., Cernadas, E., Formella, A., González, A.: Automatic identification and classification of pollen of the urticaceae family. In: Proceedings of Advanced Concepts for Intelligent Vision Systems (ACIVS 2003), pp. 38–45 (2003)
4. Li, P., Treloar, W.J., Flenley, J.R., Empson, L.: Towards automation of palynology 2: the use of texture measures and neural network analysis for automated identification of optical images of pollen grains. Journal of Quaternary Science 19(8), 755–762 (2004)
5. Ranzato, M., Taylor, P.E., House, J.M., Flagan, R.C., LeCun, Y., Perona, P.: Automatic recognition of biological particles in microscopic images. Pattern Recognition Letters 28(1), 31–39 (2007)

6. Ronneberger, O., Wang, Q., Burkhardt, H.: 3D invariants with high robustness to local deformations for automated pollen recognition. In: Proceedings of the 29th DAGM Conference on Pattern Recognition, pp. 425–435 (2007)
7. Byun, H.-R., Lee, S.-W.: Applications of support vector machines for pattern recognition: A survey. In: Lee, S.-W., Verri, A. (eds.) SVM 2002. LNCS, vol. 2388, pp. 213–236. Springer, Heidelberg (2002)
8. Otsu, N.: A Threshold Selection Method from Gray-Level Histograms. IEEE Transactions on Systems, Man and Cybernetics 9, 62–66 (1979)
9. Suzuki, S., Abe, K.: Topological Structural Analysis of Digitized Binary Images by Border Following. Computer Vision, Graphics, and Image processing 30(1), 32–46 (1985)
10. Da Costa, L., Cesar Jr., R.: Shape analysis and classification: theory and practice. CRC Press Inc. (2001)
11. Erdtman, G.: An Introduction To Pollen Analysis. Chronica Botanica Company, U.S.A. (1943)
12. Joon Ahn, S., Rauh, W., Warnecke, H.: Least-squares orthogonal distances fitting of circle, sphere, ellipse, hyperbola, and parabola. Pattern Recognition 34(12), 2283–2303 (2001)
13. Nixon, M., Aguado, A.S.: Feature Extraction and Image Processing, 2nd edn. Academic Press (2008)
14. Iwata, H., Niikura, S., Matsuura, S., Takano, Y., Ukai, Y.: Evaluation of variation of root shape of Japanese radish (Raphanus sativus L.) based on image analysis using elliptic Fourier descriptors. J. Euphytica, 143–149 (1998)
15. Haralick, R.M., Shanmugam, K., Dinstein, I.: Textural Features for Image Classification. IEEE Transactions on Systems, Man and Cybernetics 3(6), 610–621 (1973)
16. Lozano-Vega, G., Benezeth, Y., Marzani, F., Boochs, F.: Classification of Pollen Apertures Using Bag of Words. In: Petrosino, A. (ed.) ICIAP 2013, Part I. LNCS, vol. 8156, pp. 712–721. Springer, Heidelberg (2013)
17. Ojala, T., Pietikäinen, M., Harwood, D.: A comparative study of texture measures with classification based on feature distributions. Pattern Recognition 29 (1996)
18. Huang, D., Shan, C., Ardabilian, M., Wang, Y., Chen, L.: Local Binary Patterns and Its Application to Facial Image Analysis: A Survey. IEEE Transactions on Systems, Man, and Cybernetics, Part C: Applications and Reviews 41(6), 765–781 (2011)
19. Ojala, T., Pietikäinen, M., Maenpaa, T.: Multiresolution gray-scale and rotation invariant texture classification with local binary patterns. IEEE Transactions on Pattern Analysis and Machine Intelligence 24, 971–987 (2002)

Identifying Features with Concept Drift in Multidimensional Data Using Statistical Tests

Piotr Sobolewski and Michał Woźniak

Wrocław University of Technology
ul. Wybrzeże Wyspiańskiego 27
50-370 Wrocław
{piotr.sobolewski,michal.wozniak}@pwr.edu.pl

Abstract. Concept drift is a common problem in the data streams, which makes the classifiers no longer valid. In the multidimensional data, this problem becomes difficult to tackle. This paper examines the possibilities of identifying the specific features, in which concept drift occurs. This allows to limit the scope of the necessary update in the classification system. As a tool, we select a popular Kolmogorov-Smirnov test statistic.

Keywords: Concept drift, detection, statistical test.

1 Introduction

Due to the evolution of the internet and the expansion of the decision making technology, the systems designed for classifying the data streams [2] recently became a popular area of research. In the field of machine learning, data streams are defined as sources of continuous data generation, examples of which can be found in real life e.g., shopping trends, stock market, weather control, surveillance systems or health care. Classification task in these areas is often hindered by various factors which cause undesirable changes in the data classification rules. Such phenomenon is called concept drift [6] and it is a major problem in the classification systems.

There are various methods described in the machine learning literature for defending against concept drift, mostly deploying one of the two popular strategies [4]:

- Adapting a learner at the regular intervals without considering whether the changes have really occurred or not,
- First detecting the concept changes and then adapting a learner to them.

The idea presented in this article has a potential of improving the classifier adaptation process as well as enhancing the efficiency of the concept drift detection algorithms.

L. Iliadis et al. (Eds.): AIAI 2014, IFIP AICT 436, pp. 405–413, 2014.
© IFIP International Federation for Information Processing 2014

2   Problem Description

We assume that in the multidimensional data, concept drift may influence only some specific features, leaving all other features in the same conceptual distribution model. Identifying these features may improve the adaptation of the classification systems, as well as provide useful information for the sophisticated concept drift detection algorithms, such as LDCnet [9].

Our previous experiments [8][10] have shown, that the popular test statistics, such as the Kolmogorov-Smirnov test [7], Wilcoxon rank sum [11] or Wald-Wolfowitz test [12] are capable of detecting concept drift with a similar efficiency as advanced methods, designed specifically for this purpose, such as the CNF test [3]. The most efficient in our experiments was the Kolomogorov-Smirnov test statistic, therefore we have selected it for further analysis in this article.

Kolomogorov-Smirnov test is a non-parametric statistic, as it makes no assumption about the distribution of data and therefore can be deployed on any data.

For the two-sample test, a Kolmogorov-Smirnov statistic is computed as

$$D_{n,m} = \sup_x |F_{1,n}(x) - F_{2,m}(x)| \tag{1}$$

where $F_{1,n}$ and $F_{2,m}$ are the empirical distribution functions of samples computed as:

$$F_n(t) = \frac{1}{n} \sum_{i=1}^{n} 1\{x_i \leq t\}, \tag{2}$$

where $(x_1, ..., x_n)$ are independent and identically distributed (i.i.d.) random variables laying in the real numbers domain with a common cumulative distribution function. The statistic is used to perform a KS-test to reject the null hypothesis at level α by computing:

$$\sqrt{\frac{nm}{n_m}} D_{n,m} > K_\alpha, \tag{3}$$

where K_α calculated from:

$$Pr(K \leq K_\alpha) = 1 - \alpha, \tag{4}$$

and K is a Kolmogorov distribution computed as:

$$K = \sup_{t \in [0,1]} |B(t)|, \tag{5}$$

$B(t)$ being the Brownian bridge [5].

In short, the Kolmogorov-Smirnov test compares the distributions of two samples by measuring a distance between the empirical distribution functions, taking into account both their location and shape.

In this paper we evaluate the possibilities of applying the Kolmogorov-Smirnov test statistic as a tool for identifying the fueatures, which are influenced by concept drift. For this purpose, the tool needs to accurately classify the true positives (sensitivity) as well as the true negatives (specificity).

3   Data

Due to limited availability of the real data with concept drift, the data used in experiments is taken from the UCI Repository of datasets [1] and concept drift is simulated by swapping the features with each other.

In mathematical notation, if a reference dataset DS is characterized by n features f then the concept drift is applied by swapping any two features i and j with each other. Data with swapped features forms a new dataset $DS_{i,j}$ and the role of the algorithm is to identify which features are influenced by concept drift (i.e. find the i and j).

Example swap of features 1 and 2:

$$DS = [f_1, f_2, ..., f_n]$$
$$DS_{1,2} = [f_2, f_1, ..., f_n] ,$$
(6)

The swaps are made for each possible pair of features, resulting in $\binom{n}{2}$ combinations, where n is the dimensionality of the dataset. Each of the dataset is described in general by the number of samples and number of features in Tab. 1.

Table 1. Datasets

Dataset	# of features	# of samples
breast	9	683
credit-australian	14	690
haberman	3	306
heart-c	13	297
heart-statlog	13	270
ionosphere	34	351
kr-vs-kp	36	3196
letter-recognition	16	20000
mfeat-mor	6	2000
nursery	8	12960
optdigits	64	3823
page-blocks	10	5473
pendigits	16	7494
pima-indians-diabetes	8	768
segmentation	19	210
tic-tac-toe	9	958
vehicle	18	846
vote	16	232
waveform	21	5000
yeast	8	1484

This method of simulating concept drift is relatively common in the machine learning literature [13].

4　Experiments

In the experiments, we use the original dataset D as the reference data and the drifted datasets $D_{i,j}$ (i and j indicate the features which are swapped), with the samples randomly drawn from the datasets $D_{i,j}$ and groupped into data windows DW of various sizes s.

The Kolomogorov-Smirnov statistic is evaluated on every feature f in the data window to reject the null hypothesis that the values arise from the same population as the values of features in the reference dataset D with confidence level of 5%. It means, that if the test returns the p-value lower than 0.05, then the analyzed feature is considered to be influenced by concept drift and the detection signal is noted and added to the scores.

In order to evaluate the specificity, i.e. the ability to identify the true negatives, the data windows which do not include any feature swap are evaluated and if in this test the statistic returns the p-value lower than 0.05, then the algorithm makes a mistake, as it results in a false positive concept drift detection.

A short description of the experimental process is described in the pseudocode in Fig. 1.

Algorithm 1. Pseudo-code of a single loop in experimental series

```
Notations:
  DS - original dataset with n features,
  DS_{i,j} - dataset with swapped features i and j,
  DW^s_{i,j} - data window of size s with features i and j swapped,
Single loop of experiment seriers:
  DW^s = draw s random data samples from DS
  For i = 1 to f
    For j = 1 to f
      DW^s_{i,j} = swap features i and j in DW^s
      For k = 1 to f
        Evaluate KS statistic on feature k of DW^s_{i,j} and feature k of D
        IF p-value < 0.05
          Note concept drift for feature k
        END IF
      END FOR
    END FOR
  END FOR
```

In the presented way, the sensitivity and specificity of the Kolmogorov-Smirnov test statistic are evaluated for every possible feature swap and for various sizes of the data windows.

5 Results

All presented values are averaged from the series of 1000 trials.

Tables 2 and 3 show the percentage of correctly detected concept drifts in certain features for the *breast* dataset (size of data window 20 and 50, respectively), where columns are the base features and rows are the features which swap them. The diagonals are the percentage of detected false positives, the lower the value the higher the specificity of the algorithm.

Tables 4 and 5 show how the window size influences the performance of algorithm for the *breast* and *credit − australian* datasets. The tables store the

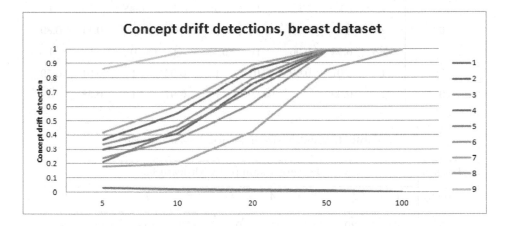

Fig. 1. Concept drift detection ratio for various window sizes

Table 2. Concept drift detection ratio in breast dataset, window size = 20

win size 20	Base feature								
Swap feature	1	2	3	4	5	6	7	8	9
1	0.02	0.7	0.67	0.86	0.82	0.89	0.27	0.91	1
2	0.65	0.02	0	0.02	0.93	0.01	0.5	0	0.47
3	0.59	0.01	0.01	0.03	0.93	0.03	0.43	0.07	0.63
4	0.84	0.02	0.02	0	0.95	0.01	0.71	0.03	0.34
5	0.68	1	0.99	0.99	0	1	0.03	1	1
6	0.73	0.02	0.08	0.02	1	0	0.8	0.04	0.33
7	0.39	0.56	0.62	0.63	0.17	0.78	0.03	0.85	1
8	0.89	0.03	0.03	0.01	1	0.02	0.83	0	0.16
9	1	0.6	0.58	0.27	1	0.42	0.99	0.13	0

Table 3. Concept drift detection ratio in breast dataset, window size = 20

win size 20	Base feature								
Swap feature	1	2	3	4	5	6	7	8	9
1	0.02	1	0.98	1	1	1	0.79	1	1
2	1	0	0	0.02	1	0.07	1	0.03	0.92
3	0.99	0.01	0	0.1	1	0.08	0.95	0.12	0.99
4	1	0.03	0.05	0.01	1	0.05	0.99	0.03	0.83
5	1	1	1	1	0	1	0.48	1	1
6	1	0.1	0.11	0.14	1	0	1	0.11	0.89
7	0.88	0.98	0.96	1	0.57	1	0	1	1
8	1	0.07	0.09	0.01	1	0.03	1	0.01	0.59
9	1	0.99	1	0.89	1	0.97	1	0.58	0

Table 4. Window size influence, *breast* dataset

breast	Feature 1 swapped with feature..								
win size	1	2	3	4	5	6	7	8	9
5	0.03	0.30	0.24	0.37	0.21	0.34	0.18	0.42	0.86
10	0.02	0.41	0.37	0.55	0.43	0.47	0.20	0.60	0.97
20	0.01	0.75	0.62	0.85	0.72	0.79	0.42	0.89	1
50	0.01	0.99	0.99	1.00	0.99	1.00	0.85	1.00	1
100	0.00	1.00	1.00	1.00	1.00	1.00	0.99	1.00	1

Table 5. Window size influence, *credit − australian* dataset

credit-aus	Feature 1 swapped with feature..													
win size	1	2	3	4	5	6	7	8	9	10	11	12	13	14
5	0.00	1.00	0.89	0.91	1.00	0.99	0.46	0.02	0.06	0.26	0.06	1.00	0.96	0.63
10	0.00	1.00	0.98	0.98	1.00	1.00	0.62	0.04	0.13	0.32	0.09	1.00	1.00	0.78
20	0.00	1.00	1.00	1.00	1.00	1.00	1.00	0.09	0.31	0.90	0.24	1.00	1.00	1.00
50	0.00	1.00	1.00	1.00	1.00	1.00	1.00	0.31	0.83	1.00	0.70	1.00	1.00	1.00
100	0.00	1.00	1.00	1.00	1.00	1.00	1.00	0.63	0.99	1.00	0.97	1.00	1.00	1.00

results obtained by swapping the first feature with other features. The results obtained for the *breast* dataset are also presented on the Diagram 1 for a more clear view of the efficiency trend in the domain of window size.

Finally, Tab. 6 shows the overall performance of the Kolmogorov-Smirnov statistic in identifying the features affected by concept drift, divided into the sensitivity and specificity scores for each of the datasets and for various window sizes. Specificity is presented only for window size 20, as the results were not significantly different for other window sizes.

Table 6. Overall sensitivity and specificity scores

	Sensitivity								Specificity	
win. size	20		50		100		200		20	
Dataset	avg	var	avg	var	avg	var	avg	var	avg	var
breast	0.53	0.15	0.70	0.17	0.77	0.14	0.82	0.11	0.99	0.00
credit-australian	0.92	0.06	0.95	0.04	0.96	0.03	0.97	0.02	0.99	0.00
haberman	0.92	0.06	0.95	0.04	0.96	0.03	0.97	0.02	0.99	0.00
heart-c	0.94	0.04	0.98	0.01	1.00	0.00	1.00	0.00	0.99	0.00
heart-statlog	0.95	0.03	0.98	0.01	1.00	0.00	1.00	0.00	0.99	0.00
ionosphere	0.60	0.16	0.70	0.15	0.76	0.14	0.81	0.13	0.97	0.00
kr-vs-kp	0.41	0.18	0.56	0.20	0.65	0.19	0.72	0.18	1.00	0.00
letter-recognition	0.48	0.18	0.62	0.19	0.72	0.17	0.78	0.15	0.99	0.00
mfeat-mor	0.48	0.19	0.62	0.19	0.72	0.17	0.78	0.15	0.99	0.00
nursery	0.47	0.18	0.62	0.19	0.71	0.18	0.77	0.16	0.99	0.00
optdigits	0.69	0.16	0.78	0.13	0.84	0.11	0.89	0.08	0.98	0.00
page-blocks	0.69	0.16	0.79	0.13	0.84	0.10	0.89	0.08	0.98	0.00
pendigits	0.68	0.16	0.79	0.13	0.84	0.10	0.89	0.08	0.97	0.00
pima-indians-diabetes	0.69	0.16	0.79	0.13	0.84	0.10	0.89	0.08	0.97	0.00
segmentation	0.70	0.16	0.79	0.13	0.84	0.10	0.89	0.08	0.98	0.00
tic-tac-toe	0.68	0.17	0.78	0.14	0.83	0.11	0.87	0.09	0.98	0.00
vehicle	0.71	0.16	0.80	0.13	0.85	0.10	0.89	0.08	0.98	0.00
vote	0.66	0.17	0.75	0.15	0.81	0.12	0.86	0.10	0.98	0.00
waveform	0.68	0.16	0.78	0.13	0.83	0.11	0.88	0.08	0.97	0.00
yeast	0.68	0.16	0.78	0.13	0.83	0.11	0.88	0.08	0.97	0.00

6 Discussion

In this paper we have proposed an unsupervised tool for enhancing the methods coping with concept drift. We have evaluated the efficiency of the

Kolmogorov-Smirnov test statistic in detecting the features affected by concept drift in the multidimensional data.

The most apparent conclusion is that the performance of algorithm depends on the data window size. Fig. 1 clearly shows this relation.

Regardless of the window size, algorithm achieves a very high specificity score, proving that the tool performs very well with true negatives, i.e. when there is no drift. It means, that the tool can be used for detecting features with concept drift without the need to worry about the false positive detections.

On the other hand, sensitivity i.e. the true positive detection rate, leaves a field for improvement. With increasing window size, sensitivity of the tool also increases, what suggests that the tool is more feasible for problems, which do not require a very limited window size.

Overall, the performance of the proposed tool is on a decent level, as e.g. in the optdigits dataset scenario, which has 2^{64} possible feature swap combinations, algorithm correctly identifies on average 88% of them with only 8% variance. Pairing it with the fact that the method does not require any supervision, the Kolmogorov-Smirnov test statistic can be considered an efficient tool for detecting the features with concept drift in multidimensional data. This functionality may be used for supporting the adaptation of classifiers as well as improving algorithms designed for detecting concept drift, such as LDCnet [9].

Further research aims on expanding the functionality of the mentioned LD-Cnet algorithm using the presented technique to battle concept drift in the multidimensional data.

References

1. Newman, D.J., Asuncion, A.: UCI machine learning repository (2007)
2. Babcock, B., Babu, S., Datar, M., Motwani, R., Widom, J.: Models and issues in data stream systems. In: Proceedings of the Twenty-First ACM SIGMOD-SIGACT-SIGART Symposium on Principles of Database Systems, PODS 2002, pp. 1–16. ACM, New York (2002)
3. Dries, A., Rückert, U.: Adaptive concept drift detection. Stat. Anal. Data Min. 2(5-6), 311–327 (2009)
4. Greiner, R., Grove, A.J., Roth, D.: Learning cost-sensitive active classifiers. Artif. Intell. 139(2), 137–174 (2002)
5. Revuz, D., Yor, M.: Continuous Martingales and Brownian Motion (Grundlehren der mathematischen Wissenschaften), 3rd edn. Springer (December 2004)
6. Schlimmer, J.C., Granger Jr., R.H.: Incremental learning from noisy data. Mach. Learn. 1(3), 317–354 (1986)
7. Smirnov, N.V.: Table for estimating the goodness of fit of empirical distributions. Ann. Math. Stat. 19, 279–281 (1948)
8. Sobolewski, P., Wozniak, M.: Sequential Tests of Statistical Hypotheses. The Annals of Mathematical Statistics 16(2), 117–186 (1945)
9. Sobolewski, P., Wozniak, M.: Ldcnet: minimizing the cost of supervision for various types of concept drift. In: Proceedings of the CIDUE 2013 - IEEE Symposium on Computational Intelligence in Dynamic and Uncertain Environments, CIDUE 2013, pp. 68–75 (2013)

10. Sobolewski, P., Woźniak, M.: Comparable study of statistical tests for virtual concept drift detection. In: Burduk, R., Jackowski, K., Kurzynski, M., Wozniak, M., Zolnierek, A. (eds.) CORES 2013. AISC, vol. 226, pp. 333–341. Springer, Heidelberg (2013)
11. Wilcoxon, F.: Individual Comparisons by Ranking Methods. Biometrics Bulletin 1(6), 80–83 (1945)
12. Wolfowitz, J.: On Wald's Proof of the Consistency of the Maximum Likelihood Estimate. The Annals of Mathematical Statistics 20, 601–602 (1949)
13. Zliobaite, I., Kuncheva, L.I.: Determining the training window for small sample size classification with concept drift. In: Proceedings of the 2009 IEEE International Conference on Data Mining Workshops, ICDMW 2009, pp. 447–452. IEEE Computer Society, Washington, DC (2009)

Features Extraction of Growth Trend in Social Websites Using Non-linear Genetic Programming

Umer Khayam, Durre Nayab, Gul Muhammad Khan, and S. Ali Mahmud

Center for Intelligent Systems and Network Research, UET Peshawar
{umerkhayam,nayaab_khan,gk502,sahibzada.mahmud}@nwfpuet.edu.pk

Abstract. Nonlinear Cartesian Genetic Programming is explored for extraction of features in the growth curve of social web portals and establishment of a prediction model. Daily hit rates of web portals provide the measure of the growth and social establishment behavior over time. Non-linear Cartesian Genetic Programming approach also termed as CGPANN has unique ability of dealing with the nonlinear data as it provides the flexibility in feature selection, network architecture, topology and other necessary parameters selection to establish the desired prediction model. A number of socially established web portals are used to evaluate the performance of the model over a span of two years. Efficient performance is shown by the system keeping the fact in consideration that only single independent web portal data is used for training the network and the same network was used for the other web portals for their performance evaluation. The system performance is significantly good as the system selects only the desired features from the features presented as input and achieves an optimal network and topology that produce the best possible results.

Keywords: Neuro Evolution, Artificial Neural Network, Cartesian Genetic Programming, Web Traffic Prediction, Web Portals, Future Demand.

1 Introduction

For the last few years a large number of web portals have been developed which intend to provide various kinds of valuable services to the internet users. Web portals such as social networks, online forums, job portals, blogging platforms, and news websites aim to provide uninterrupted free as well as paid services to their users. These web portals can also be used for business purposes or an advertising platform when its number of users increases and hit rate becomes high.

The forecast of internet traffic is an important issue [1] and accurate forecasting of web portals helps in determining the current growth and future prospect of the portals. It also provides the business department an idea to identify the features and patterns [3] that can lead to a higher hit rate in future, also an insight about the future demand and anomalies [4] [24] in the network can be detected. Software developers can get an idea about the types of browsers that are used to get access to the company website with the aid of forecasting.

The business departments, developers and the network administrators do the task of forecasting internet traffic intuitively with the help of market information about the

L. Iliadis et al. (Eds.): AIAI 2014, IFIP AICT 436, pp. 414–423, 2014.

usual behavior and future number of visitors [6]. This produces only a rough idea of what the future traffic will look like with a little day to day network administration. On the other hand, contributions from the areas of computational intelligence and artificial neural networks have replaced the intuition based forecasting methods [20] [21]. The forecasting results obtained by these neural networks are accurate, reliable and less time consuming compared to the former methods.

Different methods have been used to forecast the internet traffic and several experiments have been performed on real-world datasets using these methods in order to determine their accuracy [1] [2]. In addition, different time scales have been used for forecasting purposes [2] and some methods perform well in short term forecasting, while others perform better in large term forecasts. By accurate prediction and forecasting, the services of web portals can be extended, diminished or altered in order to maintain a high number of visitors.

The main intent of this work is to present the application of artificial neural networks to web traffic forecasting for web portals.

2   Literature Review

2.1   Traffic Forecasting of Web Portals

Forecasting of web traffic is one of a critical task [6] for researchers these days. There are several applications of internet traffic forecasting that aims for efficient traffic engineering, anomaly detections and business tools development [1]. Various techniques have been used to predict the internet traffic accurately. Paulo Cortez used a neural network ensemble approach and two important adapted time series methods such as ARIMA and Holt-Winters [2] in order to determine their accuracy for predicting traffic in TCP/IP based networks. G. Rutka inspected the performance of traffic models for forecasting the future traffic variations as precisely as possible using multilayer perceptron and radial basis function networks based on measured traffic history [7]. Machine learning methods have been used to analyze web traffic using decision trees [8]. A multi scale decomposition approach was used for real time traffic prediction that outperforms traffic prediction using neural network approach [9] and gives comparatively better results.

Many internet service providers use multi-protocol label switching to establish fill mesh of MPLS between pairs of the routers in a network in order to optimize the bandwidth resources in the network [10]. Even if MPLS is not used, with the knowledge of future demand of traffic matrix, the traditional allocation of the routing protocol weights can be done more efficiently [1]. The forecasted web traffic cab be used for anomaly detection [11] [12] in the network by comparing the actual traffic to the forecasted traffic in the network.

The forecasted web traffic can be used by the business departments [1] of the company to get an idea about the current popularity and future demand of their web services and applications. The statistical data about the visitors on a particular web portal contain patterns that are viable in determining the future demand. By analyzing statistical data, the factors effecting hit rate of the web portals can be identified, an insight about the future hit rate, types of visitors, and survival of the web portals can be predicted. Several factors have been identified in past that effect the number of visitors on these portals [1]. These factors include accessibility, browsers compatibility, trust level, flexibility, productive and organized content, and the targeted age, group and gender etc [2]. The business department and the developers of the company can take decisions based on the forecasted results in order to get increased traffic in the future.

Cartesian genetic programming and neural network models are used for forecasting purposes. Today the numbers of structures of the artificial neural networks and training algorithms that are applied in their learning process have evolved and these models are also used to forecast future network traffic. Similarly these models can be used to forecast the traffic patterns of web portals.

2.2   Cartesian Genetic Programming (CGP)

The idea of Cartesian Genetic Programming (CGP) was proposed in 1999 by Julian F Miller [22], that is a genetic programming approach with a two dimensional graphical arrangement of its functional units i.e. nodes. These programs are represented as directed acyclic structures operating in a feed forward fashion. CGP has two formats of network representation i.e. phenotype and genotype. The physical format of array of interconnected nodes represents the phenotypic structural space and its genotype can be represented as an array of integers i.e. genes. The nodes are the functional units that exhibit a node function, inputs, and outputs [23]. CGP provide a general platform for evolving the hybrid structure of any number of networks in any order. It provides a complete Cartesian architecture for their interconnectivity patterns producing less hybrid structures from a pole of networks. CGP is explored in a range of applications producing interesting results [23].

3   Cartesian Genetic Programming Evolved Artificial Neural Network

The strategy employed for the evolution of an ANN plays a vital role in ANNs and hence is a major concern for the researchers these days. The evolutionary strategy used for the evolution of ANN model proposed in this research is CGP. CGPANN is signal processing system that is based on some of the known organizing principles of the human brain [15] [16] [17]. CGP evolve the ANN with its unique architecture that makes it computationally cost effective and efficient. They are computational models that are capable of machine learning and pattern recognition. These systems are represented by a number of independent, simple processors called neurons which are interconnected with each other by weighted connections [18].   Figure 1 shows a typical CGPANN genotype and phenotype with inputs (I_0, I_1, I_2, ... I_9), outputs (O_0, O_1, O_2, ... O_9) , active nodes (0, 1, 2, 4, 5, 7, 8 & 9), inactive nodes (3 & 6) and weights (w_0, w_1, w_2......., w_{17}). The arity (number of inputs per node) of the network is 2 and the number of inputs to the system is 10.

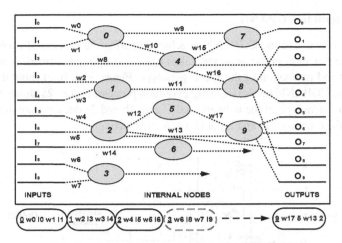

Fig. 1. A typical cgpann phenotype and genotype

The string of numbers underneath the nodal figure is the genotype arranged in boxes represent individual node in terms of its function, inputs and the weights associated with the inputs respectively. The box having dotted lines represents the inactive nodes.

The generic expressions for the CGPANN model can be given in the following equations. Equation (1) shows the system outputs (y_i) as the summation of system inputs (x_i) when the weights and node functions are not associated with them. Equation (2) and (3) shows the network outputs (y_i) with the weights (wi) and functions (f^1) are associated with the system inputs.

$$y_i = \sum_{i=1}^{N} x_i \tag{1}$$

$$y_i = \sum_{i=1}^{N} x_i w_i \tag{2}$$

$$y_i = f^i(y_i) = f^i\left(\sum_{i=1}^{N} x_i w_i\right) \tag{3}$$

Equation (4) and (5) shows the range (i) of the inputs, functions and weights respectively. As can be seen from Eq (4) the range (i) belongs to natural numbers (N) where N begins from 1 to NT. The weights belong to real numbers and are chosen in the range [-1, 1].

$$\{i | i \epsilon N 1 \leq i \leq N_T\} \tag{4}$$

$$w_i = \{w_i | w_i \epsilon R - 1 \leq w_i \leq 1\} \tag{5}$$

Equation (6) represent the single generation obtained from the network in terms of inputs (I), system outputs (y_i's) and ultimate output (O_p).

$$G_k = \{I, y_i, y_{i-1}, \dots \dots, y_1, O_p\} \tag{6}$$

Equation (7) shows the ultimate system output in terms of individual system outputs. Equation (8) shows the individual system output in terms of nodal outputs (y_i) and system inputs (I's) with weights (w_i's) associated with them.

$$O_p = \frac{1}{n} \sum_{i=1}^{N} Oi \tag{8}$$

$$Oi = \left(f\left(\sum(y_j w_j + y_{j-1} w_{j-1} + \dots \dots + y_1 w_1 + I w_k)\right)\right) \tag{9}$$

3.1   Mutation in CGPANN

The optimal network of the CGPANN is achieved during the process of evolution. During evolution mutation in the genotype takes place and optimal network is achieved by the selection and promotion of the fittest genotype. Figure 2 (a, b, c) show the mutation process of the CGPANN network during evolution process. Figure 2a shows the original phenotype and genotype of the network and Figure 2b and 2c demonstrate the process of mutation in the function (*f2*) and input (*I₃*) genes respectively.

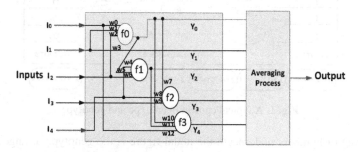

Fig. 2a. Original cgpann phenotype

Genotype1= $\underline{f0}$ $I_0 w_0 I_1 w_1 I_2 w_2$, $\underline{f1}$ $f_0 w_7 I_4 w_4 I_2 w_6$, $\underline{f2}$ $f_0 w_7 I_4 w_8 I_3 w_9$, $\underline{f3}$ $f_0 w_{10} f_1 w_{11} I_0 w_{12}$

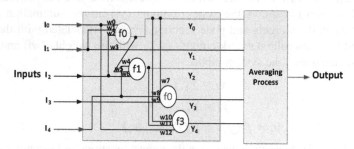

Fig. 2b. Mutation in the function of the network

Genotype2= $\underline{f0}$ $I_0 w_0 I_1 w_1 I_2 w_2$, $\underline{f1}$ $f_0 w_7 I_4 w_4 I_2 w6$, $\underline{f0}$ $f_0 w_7 I_4 w_8 I_3 w_9$ $\underline{f3}$ $f_0 w_{10} f_1 w_{11} I_0 w_{12}$

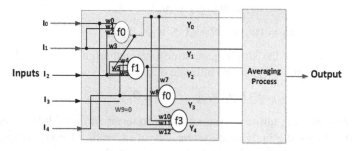

Fig. 2c. Mutation in the input to the node

Genotype3= $\underline{f0}$ $I_0 w_0 I_1 w_1 I_2 w2$, $\underline{f1}$ $f_0 w_7 I_4 w_4 I_2 w6$, $\underline{f0}$ $f_0 w_7 I_3 w9$, $\underline{f3}$ $f_0 w_{10} f_1 w_{11} I_0 w_{12}$

3.2   The Forecaster Model

The learning process of our forecaster model consists of adaptive modifying of the connection weights to improve the overall functionality of the neural network as the parallel signal processing system. The system learns the patterns and selects the optimal features in the historical data to predict the future values. The performance of the forecasting model is measured by Mean Absolute Percentage Error (MAPE) as it is a common metric in forecasting applications [6].

The forecaster model consists of single row and multiple columns of neurons trained by altering the values of the connections between the neurons. A nonlinear CGP is used for translating the network as it handles the nonlinear data such as that of web portals efficiently. The statistical data of six web portals for a period of one year is fed to the forecasting model and model is trained. The network is capable of feature selection i-e selecting optimal (not all) number of inputs and nodes for evolving the final network. The neural network model is trained on a one year hourly spaced datasets of Linkedin.com. The trained model is then used to forecast the internet traffic of web portals other web portals. The mean absolute percentage error is calculated between the forecasted values and the actual values.

4   Experimental Setup

A collection of time ordered dataset is taken for six web portals including Linkedin.com, Time.com, Tumblr.com, Answers.com, Hubpages.com and Collegehumor.com for a period of two years starting from October 2011 to October 2013. The data is obtained and verified by Quantcast.com that is a web portal used for monitoring the visitor hit rate on the portals available on the internet. The obtained data was the average daily hit rates of the six specified web portals for a period of two years.

During the training phase daily global hit rates of LinkedIn.com is used for training the system. Initially a random population of ten networks is produced for five independent seeds. The system parameters are defined and initialized. The number of offspring per generation is 9 under the $1+\lambda$ evolutionary strategy where λ is set to be 9. The initially generated genotype is mutated to produce nine offspring. The best genotype among these genotypes is selected and mutated again. The process is repeated unless the best network is achieved. The mutation rate (μr) is set to be 10%. The number of system inputs is set to be ten and that is the daily hit rates data and the arity of the system i.e. number of nodal inputs is set to be 5. Log sigmoid is taken as the activation function. These system parameters are chosen based on previous performance of CGPANN [5, 14] and evolutionary performance. The network is trained for variable number of inputs and outputs. The network takes seven days and fourteen days instances of the daily dataset as inputs to the network and forecasts the hit rate of the web portal for the next single day, seven days and fourteen days as outputs of the network.

Performance of the system is evaluated with MAPE (Mean Absolute Percentage Error) that is calculated by comparing the actual value of the hit rate with that estimated by the model.

The mathematical expression for MAPE is given below:

$$\text{MAPE} = \frac{1}{N}\sum i = (1 \text{ to N}) \left(\frac{|L_{Fi} - L_{Ai}|}{L_{Ai}}\right) \times 100$$

Where LAi is the actual value, LFi is the estimated value and N is the number of days.

5   Results and Analysis

During the training phase the system is been trained for one million generations and the best trained network is then tested for its performance. The system is trained on one web portal data i.e. LinkedIn.com and is tested on a different dataset of five web portals. The data with known hit rate values are estimated during the testing phase and compared with the actual values to evaluate the performance of the model.

The CGPANN forecaster model proposed in this work has been trained using daily averaged web traffic data of LinkedIn.com for one year, starting from October 2011 to October 2012. The results for the training session of each network are given in the following Table I. The output of the system is evaluated from the average of five independent evolutionary runs for each network with mentioned combination of inputs and outputs. The fittest network is achieved for the network with 7 inputs and 1 output during the training phase with the MAPE value of **1.1149.**

Table 1. Training results of CGPANN for web portal forecasting model

Nodes	7 in 1 out	7 In 7 out	14 In 7 out
50	0.11675	0.13302	0.132575616
100	0.11686	0.13457	0.132009816
150	**0.11493**	0.13291	0.13219517
200	0.11581	0.13402	0.132067988
250	0.11624	0.13377	0.133859813
300	0.11557	0.13290	0.132211834
350	0.11666	0.13318	0.132714417
400	0.11619	0.13349	0.133092481
450	0.11925	0.13347	0.13273484
500	0.11659	0.13286	0.13262616

For evaluation and testing, a new dataset is fed to the network for its validation. The new dataset is the statistical data of the visitor hit rate of Time.com, Tumblr.com, Answers.com, Hubpages.com and Collegehumor.com for a period of two years from ranging from October 2011 to October 2013. The testing phase results for the hit rates forecasting model are tabulated in Table II, III and IV for the mentioned combination of inputs and outputs. The proposed forecaster model takes seven days hit rates as input to the system and predicts the eighth day hit rate in the first case. In the second case it takes seven daily hit rates as input and predicts seven future daily hit rates as output of the system. In all the combinations of the inputs and outputs the forecaster model showed higher accuracy hence the model is proficient both in terms of single instant prediction and multiple instants predictions. The best MAPE values achieved for each web portal are highlighted in each table respectively.

The best MAPE value achieved by the model is 9.053% given in Table 2 where as other works done the web portal forecasting has achieved accuracies in the ranges such as 12-23% [1], 13–22% [2], 72.74 and 12.04 % [19].

Table 2. Testing results for one day hit rate as input with seven day's data as output

No. Of Nodes	Times	Tumblr	Answers	Hubpages	Collegehumor
50	0.1463	**0.09053**	0.14202	0.09647	0.1553
100	0.1433	0.09261	0.14445	0.09671	0.1541
150	0.1549	0.09426	0.14266	0.10058	0.1692
200	0.1537	0.09373	0.14592	0.10016	0.1652
250	0.1603	0.09414	0.14548	0.10192	0.1730
300	0.1430	0.09072	0.14254	0.09622	0.1507
350	0.1469	0.09361	0.14625	0.09828	0.1565
400	0.1545	0.09301	0.14100	0.09983	0.1677
450	0.1686	0.10202	0.15097	0.10875	0.1873
500	0.1424	0.09478	0.14661	0.09819	0.1542

Table 3. Testing results for seven days hit rates as input with seven days data as output

No. Of Nodes	Time	Tumblr	Answers	Hub pages	College humor
50	0.1442	0.1159	0.19272	0.10967	0.15613
100	0.1426	0.1152	0.19391	0.10884	0.15269
150	0.1270	0.1141	0.18955	0.10495	0.13731
200	0.1496	0.1191	0.19283	0.11337	0.16625
250	0.1571	0.1201	0.19474	0.11554	0.17485
300	0.1447	0.1165	0.19327	0.11035	0.15741
350	0.1530	0.1186	0.19458	0.11311	0.16570
400	0.1351	0.1151	0.19247	0.10721	0.14410
450	0.1112	0.1141	0.19209	**0.10121**	0.11393
500	0.1121	0.1134	0.19011	0.10173	0.11769

Table 4. Testing results for 14 days hit rates as input with 7 days data as output

No. Of Nodes	Time	Tumblr	Answers	Hub pages	College humor
50	0.12795	0.11923	0.19835	0.10806	0.14111
100	0.13847	0.11873	0.19897	0.10981	0.15333
150	0.15067	0.12162	0.20324	0.11404	0.16946
200	0.13021	0.11689	0.19627	0.10698	0.14178
250	0.09451	0.11605	0.19751	0.09905	**0.09592**
300	0.15046	0.12001	0.20089	0.11343	0.16898
350	0.14978	0.12036	0.20373	0.11343	0.16571
400	0.14785	0.12161	0.20692	0.11215	0.16292
450	0.15392	0.12095	0.20030	0.11530	0.17346
500	0.14222	0.11855	0.19732	0.11093	0.15814

6   Conclusion and Future Work

We have explored nonlinear CGP for the implementation of forecasting model for global web portals growth analysis and extracting the prominent features. A number of socially established web portals are analyzed for their growth process. The performance of the system revealed that the system is robust that learns the trends and extracts the optimal features responsible for the growth rate of these portals. The network is capable of feature selection i-e selecting optimal (not all) number of inputs and nodes for the evolution of the final network. The proposed system has the ability to obtain an optimal set of features, number of nodes and connections paradigm and morphology for the best possible prediction model for the task at hand. Further work can explore the social behavior on individual portals including: probing the posts and updates, advertisement on web portals for improving the news feeds, analyzing the intent and emotions in user updates, company survival capabilities analysis and business success or failure of start-up firm analysis.

References

1. Cortez, P., et al.: Internet traffic forecasting using neural networks. In: International Joint Conference on Neural Networks, IJCNN 2006. IEEE (2006)
2. Cortez, P., et al.: Multi Scale Internet traffic forecasting using neural networks and time series methods. Expert Systems 29(2), 143–155 (2012)
3. Fausett, L.: Fundamentals of Neural Networks: Architectures, Algorithms, and Applications. Pearson Education India (2006)
4. Krishnamurthy, B., et al.: Sketch-based change detection: methods, evaluation, and applications. In: Proceedings of the 3rd ACM SIGCOMM Conference on Internet Measurement, ACM (2003)

5. Nayab, D., Muhammad Khan, G., Mahmud, S.A.: Prediction of Foreign Currency Exchange Rates Using CGPANN. In: Iliadis, L., Papadopoulos, H., Jayne, C. (eds.) EANN 2013, Part I. CCIS, vol. 383, pp. 91–101. Springer, Heidelberg (2013)
6. Papagiannaki, K., et al.: Long-term forecasting of Internet backbone traffic. IEEE Transactions on Neural Networks 16(5), 1110–1124 (2005)
7. Rutka, G.: Neural network models for Internet traffic prediction. In: Proceedings of Electronics and Electrical Engineering, Lithuania, vol. 4(68), pp. 55–58 (2006)
8. Piramuthu, S.: On learning to predict web traffic. Decision Support Systems 35(2), 213–229 (2003)
9. Mao, G.: Real-time network traffic prediction based on a multiscale decomposition. In: Lorenz, P., Dini, P. (eds.) ICN 2005. LNCS, vol. 3420, pp. 492–499. Springer, Heidelberg (2005)
10. Davie, B.S., Rekhter, Y.: MPLS: technology and applications, vol. 1. Morgan Kaufmann Publishers, San Diego (2000)
11. Krishnamurthy, B., et al.: Sketch-based change detection: methods, evaluation, and applications. In: Proceedings of the 3rd ACM SIGCOMM Conference on Internet Measurement, ACM (2003)
12. Jiang, J., Papavassiliou, S.: Detecting network attacks in the internet via statistical network traffic normality prediction. Journal of Network and Systems Management 12(1), 51–72 (2004)
13. Khan, M.M., Khan, G.M., Miller, J.F.: Evolution of Optimal ANNs for Non-Linear Control Problems using Cartesian Genetic Programming. In: International Conference on Artificial Intelligence, ICAI, pp. 339–346 (2010)
14. Pacelli, V., Bavelacqua, V., Azzollini, M.: An Artificial Neural Network Model to Forecast Exchange Rates. Journal of Intelligent Learning Systems and Applications, JILSA 3(2A), 57–69 (2011)
15. Fausett, L.: Fundamentals of neural networks. Prentice Hall (1994)
16. Osowski, S.: Neural networks in algorithmic approach. WNT, Warsaw (1996) (in Polish)
17. Zurada, J.M.: Introduction to artificial neural systems. West Publishing Company (1992)
18. Ding, X., Canu, S., Denoeux, T.: Neural network based models for forecasting. In: Proceedings of Applied Decision Technologies Conf. (ADT 1995), Uxbridge, UK, pp. 243–252 (1995)
19. Gluszek, A., Kekez, M., Rudzinski, F.: Web traffic prediction with artificial neural networks. Wilga-DL Tentative. International Society for Optics and Photonics (2005)
20. Makridakis, S., Weelwright, S., Hyndman, R.: Forecasting: Methods and Applications. John Wiley & Sons, New York (1998)
21. Hanke, J., Reitsch, A.: Business Forecasting. Allyn and Bancon Publishing. Allyn and Bancon Publishing, Massachussetts (1989)
22. Rothermich, A.J., Miller, J.F.: Studying the emergence of multicellularity with cartesian genetic programming in artificial life. In: Proceedings of the 2002 UK Workshop on Computational Intelligence, pp. 397–403. GECCO (2002)
23. Miller, J.F.: "Cartesian Genetic Programming," Genetic Programming. Natural Computing Series. Springer, Heidelberg (2011)
24. Jiang, J., Papavassiliou, S.: Detecting network attacks in the internet via statistical network traffic normality prediction. Journal of Network and Systems Management 12(1), 51–72 (2004)

Comparison of Self Organizing Maps Clustering with Supervised Classification for Air Pollution Data Sets

Ilias Bougoudis[1], Lazaros Iliadis[1], and Stefanos Spartalis[2]

[1] Democritus University of Thrace,
Department of Forestry & Management of the Environment & Natural Resources,
193 Pandazidou st., 68200 N Orestiada, Greece
[2] Democritus University of Thrace,
Department of Production and Management Engineering, Xanthi, Greece
ibougoudis@yahoo.gr, liliadis@fmenr.duth.gr, sspart@pme.duth.gr

Abstract. Air pollution is a serious problem of modern urban centers. The objective of this research is to investigate the problem by using Machine Learning techniques. It comprises of two parts. Firstly, it applies a well established Unsupervised Machine Learning approach (UML) namely Self Organizing Maps (SOM) for the clustering of Attica air quality big data vectors. This is done by using the concentrations of air pollutants (specific for each area) for a period of 13-years (2000-2012). Secondly, it employs a Supervised Machine Learning methodology (SML) by using multi layer Artificial Neural Networks (ML-ANN) to classify the same cases. Actually, the ANN models are used to evaluate the SOM reliability. This is done, because there is no actual and well accepted clustering of the related data to compare with the outcome of the SOM and this adds innovation merit to this paper.

Keywords: Self Organizing Maps, Artificial Neural Networks, Classification, Air Pollution.

1   Introduction

Air pollution has been defined as the release of any substances affecting the normal cycle of any vital process or degrade infrastructure [10]. The target of this research effort is to reveal existing air pollution patterns by performing distinct clusterings for each selected measuring station of the Attica area. The clusterings are based on a vast volume of air quality data vectors related to specific air pollutants for each site. The motivation to use Self Organizing Maps was mainly related to their potential to produce a low-dimensional (typically two-dimensional) discretized representation of the input space of the training samples. Another reason was the unsupervised learning mode of SOM [6], [13]. It is well known, that the effectiveness of such an effort is obtained by estimating Sensitivity and Specificity indices as a result of the comparison between the actual clusters to the obtained ones [2]. However, in the case examined here, there is no well accepted clustering either from the literature or from a study of the civil protection authorities. For this reason, Multi Layer Feed Forward (MLFF) ANNs were developed for each measuring location which classify the

L. Iliadis et al. (Eds.): AIAI 2014, IFIP AICT 436, pp. 424–435, 2014.

available data of each site. The input vectors of the MLFF approach comprised of air pollutants, plus seven meteorological and five daily factors. The outputs of the MLFF ANNs were considered as a comparison metric for the validation of the SOM's efficiency. It is the first time that SML is employed to support the validation of the UML classification. Herein, we will designate by classification the process of supervised learning on labeled data and by clustering the process of unsupervised learning on unlabeled data.

1.1 Literature Review

There are several research efforts in the literature that use ANN to model pollution of the atmosphere [1], [5]. [11]. However, only recently some researchers have used SOM clustering to analyze air pollution of major cities. Patterns of air quality have been searched for Mexico City by Neme and Hernandez [10], whereas Karatzas and Voukantsis have done the same for the city of Thessaloniki [7]. Skön et al., have analyzed indoor air quality using SOM [12]. Li and Chou have investigated air pollution spatial variation with SOM [9]. Glorenec applied SOM to forecast Ozone peaks [3]. Unfortunately though, due to different pollution measurements and different approach methods, any comparison of these works is inaccurate. Moreover, due to station's malfunctions in our case, there were a lot of records missing, which made our effort even more challenging. Also, we had only daily measurements for PM pollutants, while daily for every other pollutant. To the best of our knowledge there is no similar research for the wider area of Attica.

2 Materials and Methods

2.1 Area of Research and Data

The area of research included four characteristic air pollution and meteorological stations. They were chosen as follows: "Athinas" is located exactly in the heart of the city centre. "Piraeus" has a unique position by the seaside, "Peristeri" is an urban site at Northwest and "Agia Paraskevi" is a suburb in the Northeastern part of Attica, close to a more mountainous area. So each measuring station is characterized by different topographic and geographic attributes. Figure 1 depicts the four locations under study. The available data sets that were used were related to the wider area of Athens, for a period of 13-years (2000-2012) and they varied from station to station. Overall, the pollutants of interest were hourly concentrations of monoxides plus dioxides like CO, NO, NO_2, SO_2, and daily concentrations of Particulate Matter like PM_{10}, ($\frac{\mu g}{m^3}$). The meteorological data were hourly values of air temperature (C°), solar radiation (Wm^{-2}), wind speed ($\frac{m}{sec}$) and direction (rad), pressure (mbar), Illumine and relative humidity. Finally, for every record we added its daily attributes; Year, Month, Day, Day_ID (1 for Monday, 7 for Sunday) and Hour. However not all stations measure the same pollutants. Thus, the inputs fed to both SOMs and MLFF ANNs were dependent on the sensors of each station. The meteorological parameters were the same everywhere and obtained from "Penteli" station, except from "Agia Paraskevi" station, where, because it is located between "Penteli" and "Thiseion" stations, we averaged the meteorological values for it.

Fig. 1. The four measuring stations

Data were obtained from the Greek ministry of Environment [4]. Totally 1,017,733 vectors without missing values were available, whereas an average as high as 18.82% of the data are missing with the station of Piraeus having the worst percentage equal to 33.67%.

The following table 1 presents the types of parameters measured in each station and percentage of missing values [14].

Table 1. Description of the stations employed for this research

ID	Station's name	Code	Missing values	Correct Data Vectors	Station's data
1	Ag. Paraskevi	AGP	12.32%	99,936	O_3, NO, NO_2, SO_2
2	Athinas	ATH	21.86%	89,058	O_3, NO, NO_2, CO, SO_2
3	Peristeri	PER	33.61%	75,668	O_3, NO, NO_2, CO, SO_2
4	Piraeus	PIR	33.67%	75,600	O_3, NO, NO_2, CO, SO_2
5	Penteli	PEN	3.66%	109806	Meteorological
6	Thiseion	THI	0.30%	113,632	Meteorological

2.2   Unsupervised Learning

In UML we let the algorithm decide how to group samples into classes that share common properties. Some examples of unsupervised learning are Kohonen's Self Organizing Maps, K-Means Clustering and Neural Gas Networks.

Self Organizing Maps (SOMs) are a well established unsupervised ML approach, based on competitive learning. Their main advantage is their ability to isolate clusters in high dimensional spaces [9]. The Self Organizing Map as all ANN consists of neurons (nodes). A weight vector is assigned to each neuron. This vector is of the same dimension as the input data vectors. The amount of nodes is the number of the clusters that will be used to group the input data. The obtained Map is a NxN space, where the data are scattered and arranged. The number of neurons is set as the square of the map. Their function can be summarized in four steps [8]:

Initialization: All of the connection weights of each cluster are initialized

Competition: In this stage for each input pattern, the neurons compete to each other in order to "win" this input. The neuron which adapts its value closest to the input "wins". We can define the discriminant function to be the squared Euclidean distance between the input vector x and the weight vector w_j for each neuron j as:

$$d_j(X) = \sum_{i=1}^{D} (X_i - W_{ji})^2 \qquad (1)$$

Cooperation: Here follows the creation of a neighbourhood located close to the previously winning neuron. In this way, the winning neuron creates a neighbourhood with other neurons, in order to cooperate with each other and win future inputs. If S_{ij} is the lateral distance between neurons i and j on the grid of neurons, we define a topological neighbourhood $T_{j,I(x)}$ where I(x) is the index of the winning neuron.

$$T_{j,I(X)} = \exp[-\frac{S_{j,I(X)}^2}{2\sigma^2}] \qquad (2)$$

Adaption: In this last stage, each neuron creates a neighbourhood or becomes a member of a neighbourhood and self - organizes, so that the feature map between inputs is formed. In practice, the appropriate weight update equation is: $\Delta W_{ji} = n(t)T_{j,I(X)}(t),(X_i - W_{ji})$ (3) [8].

In every step, all neurons adapt their weights to the current input, but not as much as the winner neuron and its neighbourhood [8]. In this way, each neighbourhood is suitable for certain values, and so the map is ordered and shaped.

3   SOM Clustering

In this study, the number of neurons was the only subject of experimentation, in the development of the feature maps. The main target was to isolate the extreme pollutant values of every station and then to record the meteorological and spatiotemporal characteristics for these particular cases. Through this research we have experimented with 2, 3 and 4 neurons. As a result, the obtained maps have 4, 9 and 16 clusters accordingly. Often, the selection of a small number of clusters may lead to loss of the pattern (if there is one). However in this effort, the numbers mentioned above have proven to be suitable. Larger numbers produced larger maps, separating our input data in a chaotic level. For every map produced, we kept the hits of each cluster, the cluster of each record, the neighbour weight distances and the weights of each input.

The following images show the results of the SOM clustering. They can represent quite efficiently almost every other case. The Neighbour Weight Distance Figure provides information about the neighbourhoods created; the darker colours represent larger distances, and the lighter colours represent smaller distances. In fact we have separated the neurons which were isolated and not part of any neighbourhood. These neurons had the extreme pollutant values and thus they correspond to very interesting cases.

The Input Weights Figure shows the weight assigned to each node from every input. It must be specified that input1 to input5 correspond to CO, NO, NO_2, O_3, SO_2 respectively. Lighter colour means larger value. Here, wherever we have bright colours (yellow or orange), we have the isolated neurons who are assigned the extreme values. Another clue for the most extreme neuron was that this neuron usually had less record than the others (Figure 4). The combination of these three figures helped us to figure out where the most hazardous values were. We are seeking yellow or orange from the Input's Weights Figure, while dark connections in the Distances Figure. The neuron in the bottom - left corner is the first, the one next to it is the second, while the last one is on the top - right corner.

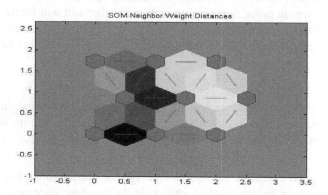

Fig. 2. Neighbor Distances for "Athinas" Hourly pollutants values, period 00-04 (the 1st neuron seems to be the extreme one)

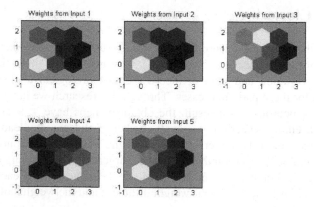

Fig. 3. Input weights for "Athinas" Hourly pollutants values, period 00-04 (the 1st neuron seems to be the extreme one for CO, NO, NO_2 and SO_2, while the third for O_3)

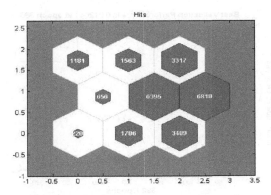

Fig. 4. Sample hits for "Athinas" Hourly pollutants values, period 00-04 (the first neuron has won 220 records, the forth 656, while the ninth 3317)

4   Pattern Recognition with MLFF ANNs

Classification is a form of Supervised Learning. After we obtained the Clusters for each station, we used the pattern recognition tool of MATLAB to classify the available data records by developing MLFF ANNs (three for each site plus a daily one for the stations were we had PM pollutants measures). The purpose was to figure out if obtained grouping was suitable. For the development of the optimal ANNs the data were divided in training, evaluation and testing sets by 70%-30% and 30% respectively. The evaluation metrics were mean square error and confusion matrices. The training function was trainscg (Scaled conjugate gradient backpropagation) and the number of hidden neurons 10, for all networks.

As it has already been mentioned, the input data for every station comprised of the measured pollutants plus seven meteorological parameters (as shown in table 1) and five daily parameters (year, month, day, day identification and hour). By this approach, evaluation became stricter, as we had 17 input parameters totally. As it is shown in the following Figures, the performance of the ANN is very accurate for the "Peristeri" station for the period 2000-2004.

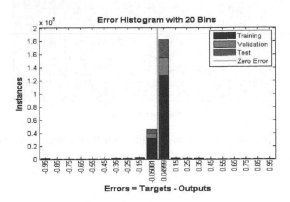

Fig. 5. Error Histogram for the MLFF ANN for the "Peristeri" station (00-04)

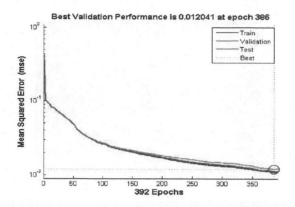

Fig. 6. Performance for the MLFF ANN for the "Peristeri" station (00-04)

If a pollutant was missing values (for example in Agia Paraskeui, the SO_2 pollutant has value only from 2000 to 2005), we created a separate SOM for it. Also, if 9 neurons could not group the extreme values of a pollutant, we developed a second SOM with more neurons (16), in order to manage to cluster the specific pollutant. Unfortunately, the confusion matrix of the networks with 16 clusters is not visible, as it contains 256 elements. For every network we obtained the confusion matrix and the percentages of correct and incorrect classifications. Figure 7 shows the confusion matrix which specifies the high compatibility level between clustering and classification for the "Peristeri" site. Unfortunately there is not enough space for all confusion matrices. However the high agreement of the two approaches is shown in the following tables.

All Confusion Matrix

	1	2	3	4	5	6	7	8	9	
1	0 0.0%	0 0.0%	0 0.0%	0 0.0%	0 0.0%	0 0.0%	0 0.0%	0 0.0%	0 0.0%	NaN% NaN%
2	0 0.0%	347 1.3%	5 0.0%	10 0.0%	28 0.1%	0 0.0%	81 0.3%	0 0.0%	0 0.0%	3.7% 96.3%
3	0 0.0%	0 0.0%	1022 3.8%	0 0.0%	6 0.0%	41 0.2%	0 0.0%	0 0.0%	0 0.0%	95.6% 4.4%
4	468 1.7%	47 0.2%	0 0.0%	1390 5.2%	0 0.0%	0 0.0%	20 0.1%	0 0.0%	0 0.0%	72.2% 27.8%
5	0 0.0%	89 0.3%	17 0.1%	0 0.0%	2001 7.4%	12 0.0%	3 0.0%	28 0.1%	34 0.1%	91.6% 8.4%
6	0 0.0%	0 0.0%	38 0.1%	0 0.0%	10 0.0%	5461 20.3%	0 0.0%	0 0.0%	58 0.2%	98.1% 1.9%
7	0 0.0%	60 0.2%	0 0.0%	68 0.3%	19 0.1%	0 0.0%	3777 14.0%	56 0.2%	0 0.0%	94.9% 5.1%
8	0 0.0%	0 0.0%	0 0.0%	0 0.0%	68 0.3%	0 0.0%	63 0.2%	5082 18.9%	89 0.3%	95.9% 4.1%
9	0 0.0%	0 0.0%	0 0.0%	0 0.0%	91 0.3%	4 0.0%	0 0.0%	44 0.2%	6318 23.4%	97.8% 2.2%
	0.0% 100%	93.9% 6.1%	94.5% 6.5%	94.7% 5.3%	90.0% 10.0%	99.0% 1.0%	95.8% 4.2%	97.5% 2.5%	97.2% 2.8%	94.2% 5.8%

Output Class (vertical) — Target Class (horizontal)

Fig. 7. Overall Confusion Matrix for the MLFF ANN for the "Peristeri" station (00-04)

5   Comparative Analysis between SOMs and MLFF ANNs

The tables below show the results from both the clustering and the classification procedures. Columns 2 to 4 are the average, minimum and maximum values for each pollutant. Columns 5 to 8 are the same values extracted from each cluster which contained the extreme values for each pollutant. The last column is the correct percentage of the classification process.

Table 2. Comparative analysis for "Agia Paraskevi" Station

AGP	ALL RECORDS			SOM			CLASS	
	AVG	MIN	MAX	AVG	MIN	MAX	RMSE	%
NO (00-04)	2,75	1	411	25,84	1	411		
NO_2 (00-04)	20,15	1	347	95,25	45	347	0,1	93,6
O_3 (00-04)	91,38	1	330	190,67	155	330		
SO_2 (00-04)	7,35	2	244	25,98	2	244		
NO (05-08)	2,19	1	141	47,96	22	141	0,14	82,9
NO_2 (05-08)	22,37	2	198	73,44	36	165		
O_3 (05-08)	76,34	1	279	166,9	143	279		
SO_2 (05-08)	6,02	2	100	32,2	24	100	0,06	98,1
NO (09-12)	2,46	1	111	11,71	1	111		
NO_2 (09-12)	13,23	1	287	49,94	22	287	0,1	92,5
O_3 (09-12)	86,91	2	251	146,01	129	251		
PM_{10}	26,13	6	292	127,75	91	292	0,19	87,9
$PM_{2,5}$	17,17	4	74	45,87	23	67		

Table 3. Comparative analysis for "Athinas" Station

ATH	ALL RECORDS			SOM			CLASS	
	AVG	MIN	MAX	AVG	MIN	MAX	RMSE	%
CO (00-04)	2,43	0,1	21,4	11,03	5	21,4	0,1	95,0
NO (00-04)	72,86	1	908	659,62	545	908		
NO_2 (00-04)	73,83	1	377	184,14	128	377		
O_3 (00-04)	33,22	1	253	124,51	93	253	0,1	82,5
SO_2 (00-04)	13,29	2	259	41,13	3	259		
CO (05-08)	1,78	0,2	11,8	7,93	4,1	11,8		
NO (05-08)	57,57	1	787	536	451	787	0,09	95,8
O_3 (05-08)	30,97	1	199	-	-	-		
SO_2 (05-08)	8,76	2	126	34,55	2	122		
NO_2 (05-08)	63,92	4	275	147,13	117	275	0,15	43,8
CO (09-12)	1,39	0,1	10,4	5,81	1,7	10,4		
NO (09-12)	50,72	1	678	431,76	362	678		
NO_2 (09-12)	55,35	3	323	94,4	39	263	0,09	97,5
O_3 (09-12)	34,33	1	186	-	-	-		
SO_2 (09-12)	7,63	2	86	19,85	8	60		

Table 4. Comparative analysis for "Piraeus" Station

PIR	ALL RECORDS			SOM			CLASS	
	AVG	MIN	MAX	AVG	MIN	MAX	RMSE	%
CO (00-04)	1,58	0,1	13,3	5,8	0,1	6,6		
NO (00-04)	55,81	1	590	343,09	246	590		
NO$_2$ (00-04)	65,41	1	243	108,09	47	224	0,12	90,3
O$_3$ (00-04)	35,05	1	217	114,17	81	217		
SO$_2$ (00-04)	23,45	2	293	93,57	37	293		
CO (05-08)	1,23	0,1	9,8	4,82	1,2	7,9		
NO (05-08)	50,66	1	902	377,05	300	902		
NO$_2$ (05-08)	66,26	1	296	129,25	12	198	0,16	73,6
O$_3$ (05-08)	36,2	1	190	118,54	95	190		
SO$_2$ (05-08)	18,07	2	275	124,23	57	275		
CO (09-12)	0,88	0,1	6,4	2,75	0,6	6,4		
NO (09-12)	34,67	1	504	279,9	210	504		
NO$_2$ (09-12)	49,77	1	277	118,97	61	277	0,14	81,6
O$_3$ (09-12)	42,23	1	192	109,32	92	192		
SO$_2$ (09-12)	10,36	2	279	50,4	2	207		
PM$_{10}$	41,89	11	185	104,93	72	185	0,22	91,6
PM$_{2.5}$	29,48	5	157	60,7	12	157		

Table 5. Comparative analysis for "Peristeri" Station

PER	ALL RECORDS			SOM			CLASS	
	AVG	MIN	MAX	AVG	MIN	MAX	RMSE	%
CO (00-04)	0,79	0,1	11,5	4,15	0,3	11,5		
NO (00-04)	17,72	1	427	198,97	133	427		
NO$_2$ (00-04)	42,74	1	289	106,38	22	289	0,1	94,2
O$_3$ (00-04)	55,93	1	257	146,76	110	257		
SO$_2$ (00-04)	14,14	2	272	86,83	6	272		
CO (05-08)	0,71	0,1	8,3	3,6	0,9	8,3		
NO (05-08)	16,27	1	447	202,81	137	447	0,2	60,6
NO$_2$ (05-08)	40,76	1	353	86,98	42	287		
SO$_2$ (05-08)	11,48	2	163	36,3	4	156		
O$_3$ (05-08)	52,98	1	284	224	194	284	0,09	2,3
CO (09-12)	0,52	0,1	8	2,55	0,7	8		
NO (09-12)	9,31	1	284	104,8	60	284		
NO$_2$ (09-12)	28,46	1	201	79,4	44	201	0,1	96,6
O$_3$ (09-12)	64,15	1	246	137,31	112	246		
SO$_2$ (09-12)	6,7	2	106	14,92	2	77		

6   Conclusions – Discussion

In almost every station we managed to isolate the extreme pollutants' values. From the clusterings we have obtained certain pollution patterns comprising of specific temporal, meteorological and air pollutant concentrations. While observing these conditions, we came to the conclusion that high values of CO and NO are present, when we have low temperature, high humidity, low solar radiation and low wind speed, between 8-12 AM or 6-10 PM. On the other hand, high levels of O3 appear under high temperature, low humidity and high solar radiation and between 12 - 6 PM.

In almost every case, the cluster which included the extreme values was in one corner of the map (in most cases, in the lower right cluster). This was because this cluster was isolated and not part of a neighbourhood.

Although most cases had their extreme values put together in the same cluster, the values of O3 were separated from the others. What is more, the cluster which contained the extreme values of O3 was across the one which contained the other pollutants.

Almost all percentages of correct classification were above 80%, showing that the SOM clustering can be considered reliable and compatible to the MLFF ANN classification.

In most cases, when we had two SOMS (with 9 and 16 clusters) the correct percentage of the clustering with 9 clusters is higher from the one with 16 clusters.

Most of the times, the network failed to assign the extreme records in the right neuron. Instead, it put every record in the closest cluster. Although this may sound like a malfunction of the classification, it may be used as a specific pattern recognition tool. This occurs because meteorological parameters act in a catalytic manner and they are not the primary source of pollutants.

This research is quite innovative and it has proven the ability of SOM to reveal clusters with specific attributes in real world problems and moreover their potential contribution for quality of life.

Future work will include the comparison and evaluation of SOM with other unsupervised methods (like fuzzy k- means and Neural Gas Networks), for the same case.

In the matrices below we have the daily attributes modes for every station's extreme pollutant values. Inside the parenthesis we have the counter for each mode (for example, most extreme records for NO2 appeared in 2009 511 times, while they occurred on Friday (DAY_ID is 5)). PM10 and PM2.5 have given the same results, so we combined them into one column.

Table 6. Daily attributes for extreme pollutant values for "Agia Paraskevi" Station

AGP	CO	NO	NO_2	O_3	SO_2	PM
YEAR	2009 (511)	2009 (511)	2009 (511)	2010 (453)	2003 (247)	2008 (5)
MONTH	1(253)	1(253)	1 (253)	8(499)	5(147)	4(5)
DAY	16(74)	16(74)	16(74)	6(81)	30(47)	21(2)
DAY ID	5(210)	5(210)	5(210)	6(277)	4(160)	3(4)
HOUR	8(159)	8(159)	8(159)	14(291)	10(127)	-

Table 7. Daily attributes for extreme pollutant values for "Athinas" Station

ATH	CO	NO	NO$_2$	O$_3$	SO$_2$	PM
YEAR	2009 (139)	2009 (139)	2009 (139)	2002 (444)	2009 (139)	-
MONTH	1(194)	1(194)	1(101)	7(363)	1(194)	-
DAY	3(28)	3(28)	3(28)	4(61)	3(28)	-
DAY_ID	5(72)	5(72)	5(72)	7(184)	5(72)	-
HOUR	9(298)	9(298)	11(165)	16(154)	11(90)	-

Table 8. Daily attributes for extreme pollutant values for "Piraeus" Station

PIR	CO	NO	NO$_2$	O$_3$	SO$_2$	PM
YEAR	2002 (145)	2002 (145)	2002 (294)	2002 (893)	2002 (294)	2007(11)
MONTH	12(76)	12(76)	3(151)	5(294)	3(151)	12(7)
DAY	30(27)	30(27)	26(36)	17(73)	26(36)	1(4)
DAY_ID	3(59)	3(59)	1(95)	6(500)	1(95)	2(9)
HOUR	9(85)	9(85)	10(136)	17(577)	10(136)	-

Table 9. Daily attributes for extreme pollutant values for "Peristeri" Station

PER	CO	NO	NO$_2$	O$_3$	SO$_2$	PM
YEAR	2011 (282)	2011 (282)	2011 (436)	2003 (479)	2011 (282)	-
MONTH	12(203)	12(203)	5(535)	7(579)	12(330)	-
DAY	2(43)	2(43)	13(71)	24(61)	20(44)	-
DAY_ID	5(113)	5(113)	5(231)	5(196)	5(113)	-
HOUR	9(210)	9(210)	8(177)	15(310)	11(103)	-

References

1. Alkasassbeh, M., Sheta, A.F., Faris, H., Turabieh, H.: Prediction of PM10 and TSP Air Pollution Parameters Using Artificial Neural Network Autoregressive, External Input Models: A Case Study in Salt, Jordan. Middle-East Journal of Scientific Research 14(7), 999–1009 (2013) ISSN 1990-9233
2. Fawcelt, T.: An Introduction to ROC Analysis. Pattern Recognition Letters 27(8), 861–874 (2006), doi:10.1016/j.patrec.2005.10.010.
3. Glorennec, P.Y.: Forecasting Ozone Peaks Using Self-organizing Maps and Fuzzy Logic. In: Air Pollution Modelling and Simulation, pp. 544–550. Springer (2002)
4. Iliadis, L., Spartalis, S., Paschalidou, A., Kassomenos, P.: Artificial Neural Network Modelling of the surface Ozone concentration. International Journal of Computational and Applied Mathematics 2(2), 125–138 (2007)
5. Jordan, M.I., Bishop, C.M.: "Neural Networks". In: Tucker, A.B. (ed.) Computer Science Handbook, 2nd edn. Section VII: Intelligent Systems. Chapman & Hall/CRC Press LLC, Boca Raton (2004) ISBN 1-58488-360-X

6. Karatzas, K., Voukantsis, D.: Studying and predicting quality of life atmospheric parameters with the aid of computational intelligence methods" iEMSs 2008: International Congress on Environmental Modelling and Software Integrating Sciences and Information Technology for Environmental Assessment and Decision Making 4th Biennial Meeting of iEMSs. In: Sànchez-Marrè, M., Béjar, J., Comas, J., Rizzoli, A., Guariso, G. (eds.) International Environmental Modelling and Software Society, iEMSs (2008)
7. Kirt, T., Vainik, E., Võhandu, L.: A method for comparing self-organizing maps: case studies of banking and linguistic data. In: Ioannidis, Y., Novikov, B., Rachev, B. (eds.) Proceedings of Eleventh East-European Conference on Advances in Databases and Information Systems, pp. 107–115. Technical University of Varna, Varna (2007)
8. Li, S.T., Cho, S.W.: Multi-Resolution Spatio-temporal Data Mining for the Study of Air Pollutant Regionalization. In: Proceedings of the 33rd Hawaii Conf. on System Sciences (2000)
9. Neme, A., Hernández, L.: Visualizing Patterns in the Air Quality in Mexico City with Self-Organizing Maps. In: Laaksonen, J., Honkela, T. (eds.) WSOM 2011. LNCS, vol. 6731, pp. 318–327. Springer, Heidelberg (2011)
10. Paschalidou, A., Iliadis, L., Kassomenos, P., Bezirtzoglou, C.: Neural Modeling of the Tropospheric Ozone concentrations in an Urban Site. In: 10th International Conference Engineering Applications of Neural Networks, pp. 436–445 (2007)
11. Skön, J.P., Johansson, M., Raatikainen, M., Haverinen-Shaughnessy, U., Pasanen, P., Leiviskä, K., Kolehmainen, M.: Analysing Events and Anomalies in Indoor Air Quality Using Self-Organizing Maps. International Journal of Artificial Intelligence 9(A12) (2012)
12. Yin, H.: Learning Nonlinear Principal Manifolds by Self-Organising Maps. In: Gorban, A.N., Kégl, B., Wunsch, D.C., Zinovyev, A. (eds.) Computing in Systems Described by Equations. Lecture Notes in Computer Science and Engineering (LNCSE), vol. 58, Springer, Berlin (1977)
13. Greek ministry of Environment, http://www.ypeka.gr/

Predicting Water Permeability of the Soil Based on Open Data

Jonne Pohjankukka, Paavo Nevalainen, Tapio Pahikkala, Eija Hyvönen,
Pekka Hänninen, Raimo Sutinen, Jari Ala-Ilomäki, and Jukka Heikkonen

University of Turku, Computer Science Dept.
{Jonne.Pohjankukka,ptneva,Tapio.Pahikkala}@utu.fi,
{Eija.Hyvonen,Pekka.Hanninen,Raimo.Sutinen}@gtk.fi,
Jukka.Heikkonen@utu.fi, jari.ala-ilomaki@metla.fi
http://www.utu.fi/en/units/sci/units/it/

Abstract. Water permeability is a key concept when estimating load bearing capacity, mobility and infrastructure potential of a terrain. Northern sub-arctic areas have rather similar dominant soil types and thus prediction methods successful at Northern Finland may generalize to other arctic areas. In this paper we have predicted water permeability using publicly available natural resource data with regression analysis. The data categories used for regression were: airborne electro-magnetic and radiation, topographic height, national forest inventory data, and peat bog thickness. Various additional features were derived from original data to enable better predictions. The regression performances indicate that the prediction capability exists up to 120 meters from the closest direct measurement points. The results were measured using leave-one-out cross-validation with a dead zone between the training and testing data sets.

Keywords: load bearing capacity of soil, water permeability, regression, k-nearest neighbor, mobility, sub-arctic infrastructure.

1 Introduction

This paper is about predicting the water permeability of the soil by regression analysis using publicly available multi-source data. Water permeability (also called hydraulic conductivity) is a central soil property related to soil type and soil texture. High permeability means that soil tends to stay dry and traversable most of the year, whereas low permeability creates a risk for mobility when precipitation is high. Mobility in arctic areas is of great interest to many different parties. E.g. the mining industry is interested about the mobility estimates when placing various facilities. The forest industry is interested on the load bearing capacity of the soil, since the route solutions can be adaptive to mobility predictions.

Our input data set consists of 44 features which are publicly available. The data is in raster format with grid resolution ranging from 10 meters to 50 meters.

L. Iliadis et al. (Eds.): AIAI 2014, IFIP AICT 436, pp. 436–446, 2014.

Water permeability of the soil has been measured in 1788 test spots at Northern Finland provided by the Geological Survey of Finland (GTK). It is an important attribute which, when combined with other features available, helps to determine the soil types. Related studies have been conducted in [1] where soil respiration rates are predicted from temperature, moisture content and soil type. Another related research was published in the paper of P. Scull, J. Franklin and O.A. Chadwick [2]. In their paper they use classification tree analysis for predicting the soil type in desert landscapes. R.P.O. Schulte et al.[3] focuses on soil moisture deficit, which is a related concept but not of concern in sub-polar areas, H. Gao et al. [4] and R.A. Chapuis [5] focus on water budget modeling, which was not yet attempted in our study. H.S. Mahmood et al. [6] uses on-site gamma-ray measurements for analysis of the farming soil. N.J. McKenzie [7] combines gamma-ray and digital elevation model to predict the chemical composition of the farming land. Closest to our paper is [8], where several data sources (topographic and remote sensing) are combined with 85 soil samples to assess the usability of the soil within and outside the sampled area. One can try to by-pass the water permeability estimation by directly learning the dynamic coupling between the precipitation and remotely observed soil moisture. This approach must include the digital terrain model (DTM) to estimate the water catchment. An example of this approach is [9].

The main novelty of this paper related to the previous studies is that the prediction is based on wide-area public data on a subpolar region. The features used in this paper are basically available through-out the arctic zone.

We use regression analysis to find a mapping between the publicly available data and water permeability of the soil. In the following, we present the regression methods in Ch. 2. Then we introduce the test area, the original data sets and derived features (Ch. 3) and describe the analysis process and results of the analysis (Ch. 4). The last part is for conclusions and future approaches (Ch. 5).

2 Regression Methods

Regularized least squares regression (RLS) is well known so we describe it mainly to introduce the variables and the notation used later in the paper. The explanatory variables $x_1, ..., x_p$ consist of given data and dependent variable y is the water permeability. We need to find a set of parameters $\mathbf{w} \in \mathbb{R}^p$ and $b \in \mathbb{R}$ such that the error function:

$$E(\mathbf{w}) = \frac{1}{n} \sum_{i=1}^{n} \left(y_i - \mathbf{w}^T \mathbf{x}_i - b \right)^2 + \frac{\lambda}{n} \mathbf{w}^T \mathbf{w} \qquad (1)$$

is minimized, where $\mathbf{x}_i \in \mathbb{R}^p$ is the input vector, $y_i \in \mathbb{R}$ is the response value, n is the number of observations and λ is the regularization parameter.

The k-nearest neighbors (k-NN) approach predicts the test sample by taking the average from k points nearest to it. Euclidean distance is the used metric in our analysis. Explicitly stated, if $y_1, ..., y_k$ are the response values of the k-nearest points to the test sample, then the response value for the test sample $\hat{y}_t$ is:

$$\hat{y}_t = \frac{1}{k} \sum_{i=1}^{k} y_i.$$

3   Test Area, Data Sets and Features

The research area is located in the northern part of the municipality of Sodankylä, which is a part of Finnish Lapland. The size of the target area is 18432 km^2. The center point of the rectangular target area is at ETRS-TM35FIN coordinates 7524 kmN, 488 kmE, zone 35.

The data set consists of aerial gamma-ray spectroscopy data (referred later as gamma-ray data, AGR) combined with electromagnetic (AEM), topographical (Z), peat bog mask (PBM) and The National Forest Inventory 2011 (VMI[1]) data when predicting the qualities and characteristics of the soil, namely its type and water permeability (WP). Gamma-ray data is inversely related on the amount of water on the soil, which can be used to predict the type of the soil. The forest inventory data describes the profile of tree species, their maturity and foresting state. Albeit this kind of data is not directly available elsewhere in northern sub-arctic areas (e.g. Russia, Canada), several studies are underway to predict the main characteristics of the forest by remote measurement methods [10]. These methods include LiDAR and various satellite measurements.

The data providers are:

Table 1. Data providers, data and the grid size

Provider	Data	Grid size
Geological Survey of Finland (GTK)	AGR, AEM WP	50 m
Finnish Forest Research Institute (Metla)	VMI, PBM	20 m
National Land Survey of Finland (NLS)	Z	10 m

When considering all the derived features used in the analysis we get a total of 96 data layers.

The test site has 1788 sample points, where many mechanical and electro-chemical properties of the soil were measured, see [11]. The water permeability is a theoretical value derived from the soil particle size distribution of the soil.

We now present our data sources and donors.

---

[1] VMI2011: http://www.metla.fi/ohjelma/vmi/vm11-info-en.html

3.1 Forest Inventory Data

The National Forest Inventory (VMI) holds the state of Finnish forests. The data is updated once in two years. The parameters are derived from various remote sensing sources, and several spot-wise verification and calibration methods are applied to it before publishing the data [12]. 44 numerical features include green mass, trunk dimensions and tree density per specie category. These multi-source features exhibit built-in dependencies, thus the final number of useful features is lower.

3.2 Aerial Gamma-Ray Data

The aerial gamma-ray data was provided by the Geological Survey of Finland (GTK). The raster data is based on gamma-ray flux from potassium, which is the decay process of the naturally occurring chemical element potassium (K). This data indicates many significant characteristics of the soil, including the tendency to stay moist after precipitation and tendency to frost heaving. Also the soil type, especially density, porosity, grain size and humidity of the soil have an effect to gamma-ray radiation. In Fig. 1 we present the gamma-ray data from Sodankylä target area. The bright end of the gray scale is for the high gamma radiation and hence less water in the locality of the pixel.

Fig. 1. Aerial data: gamma-ray (left) and electromagnetic data (right). Air-borne electromagnetic data is sensitive to geological properties to depth of hundreds of meters, but it also indicates some features of the top soil.

3.3 Electromagnetic Properties of Soil

The air-borne electromagnetic (AEM) data was provided by the Geological Survey of Finland (GTK). Primary AEM components, in-phase and quadrature, were transformed to apparent resistivity values by using a half-space model [13]. The apparent resistivity gives information on different kind of soil conductors. The apparent resistivity is governed by grain size distribution, water and electronic conductors content of soil and cumulative weathering.

3.4   Topographical Height Data

Topographical data provided by the National Land Survey of Finland (NLS) was included in the analysis. The data from NLS server is basically similar to aerial laser measurements (LiDAR) except LiDAR can reach denser grid. Instead of raw height alone we used local height difference, flow accumulation area, confluence and inclination described in [14]. These four derived features are more efficient for prediction than raw height data alone.

3.5   Peat Bog Mask

Peat bog mask is created from GTK aero-radiometric data and is courtesy of NLS and METLA. The grid size is 20 m and the value 1 indicates that peat thickness is over 60 cm. Value 0 indicates thickness less than 60 cm. The limit chosen is practical for mobility prediction.

3.6   Derived Features

The following features were derived from gamma-ray and electromagnetic data:

- Mean and variance over 3×3 window
- Mean and variance over Gabor filter with 8 orientations, see [15]
- Local Binary Pattern (LBP) with pixel radii $r \in \{1, 2\}$, see [16]

From topographical height we derived the following features: local height difference, ground inclination, convergence index and flow accumulation area. The definition of these features is at [17]).

There are several additional attributes possible to derive from topographical height data, and more geomorphological features will be employed in the future.

The regression methods use total of 44 original and 52 derived features, including the constant feature. The derived features are useful only if the original feature is continuous enough. E.g. the Forest Inventory data often has locally constant zones with abrupt changes and the derived features do not help much.

3.7   Water Permeability Exponent

This is the subject of prediction. Basically, the water permeability indicates the nominal vertical speed of water through the soil sample. The measurement of this quantity is indirect, based on soil particle size distribution, and the actual speed highly depends on the inhomogeneities (roots, rocks) and micro-cracks in the soil. This is why this quantity is descriptive and theoretical. In our analysis we are using a logarithmic quantity x_{wp} derived from water permeability speed v. For purposes of this presentation it is called as the water permeability exponent and defined as:

$$x_{wp} = -\log_{10} v, \ [v] = \frac{m}{sec}, \tag{2}$$

This formula has v as the vertical speed of water flow through the soil.

4   Analysis and Results

We are looking for methods which predict water permeability on areas, where there may not be direct water permeability measurements nearby. Therefore, we developed a modification of the leave-one-out cross-validation (LOOCV) for measuring the degree of spatial dependency from the nearby direct measurements, which we refer to as LOOCV with dead zone. Namely, the approach works on the measurement data just like an ordinary LOOCV in which each measurement at a time is omitted from the training set and used as a test point, except that we also remove from the training set all points that are within geographical distance r from the test point. This approach is illustrated in Fig. 2. By varying r, we can measure how far from the test area we assume the closest measurements to be at the very least. In addition, the results can be helpful in deciding how dense grid of direct measurement one should use in order to obtain a certain level of prediction performance.

We perform the regression of water permeability with the following three feature sets:
- location only
- features + location
- features only

where location refers to the geographical coordinates (e.g. latitude and longitude) and features to the ones described in Section 3. Note that one can not rely on the location information if there are no nearby direct measurements at all, and therefore we measure the prediction performance separately with these.

The prediction performances with the different feature sets as a function of the radius r of the dead zone are depicted in the two leftmost graphs in Fig. 3 on p. 443. The generic version based on feature data only gives weaker results, since the sample point arrangement at Sodankylä (see sample sets A and B in Fig. 2) and perhaps the phenomenon itself induce spatial dependency. No good generic regression method for this data set has been found, instead the problem is about how much additional samples are needed per target area to make the prediction useful.

The common k-NN method has one essential parameter, the number of neighbors k. The spatial dependency can be probed by adding the dead zone radius r to avoid the optimistic effect of the nearest neighbors. Fig. 2 depicts the modified leave-one-out arrangement, where k nearest points outside the dead zone of radius r are used for teaching. By varying r one gets a varied data set and a rough estimate on how dense it should be for it to predict well in new circumstances.

The same parameterized dead zone leave-one-out arrangement was used with regression, too.

4.1   Predicting Water Permeability

As mentioned in Sec. 3.7 before, the prediction subject is the water permeability exponent x_{wp} defined by Eq. 2. The values used for regularization parameter λ ranged from $2^{-15}, ..., 2^{15}$. k-NN parameter had $k \in \{1, 3, 6, 12, 22\}$. Two different

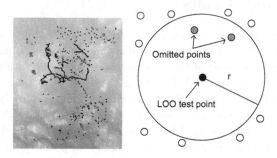

Fig. 2. *Left*: 1788 sample points. Set A (1187 points, marked with red circles, distance to the nearest neighbor $d_{NN} \leq 86m$) is tightly packed and set B is very sparse (601 points, marked with black dots, aver. $d_{NN} \approx 1.1\,km$). *Right*: the dead zone (with radius r) around the leave-one-out test point (black circle). The gray circles are omitted from the training set (white circles). Both the k-NN and RLS method address the training data only, e.g. the k nearest neighbours are selected from outside the circle.

error measures were used for estimating prediction performance: mean absolute error (MAE) and concordance index (CI) [18]. Explicitly, the error measures are:

$$MAE = \frac{1}{n} \sum_{i=1}^{n} \left| \frac{y_i - \hat{y}_i}{y_i} \right| \quad CI = \frac{1}{N} \sum_{y_i < y_j} h(\hat{y}_i - \hat{y}_j) \tag{3}$$

MAE prediction baseline $\tilde{y}$ is the best possible prediction under the assumption that the prediction will be constant, thus constraining all values $\hat{y}_j = \tilde{y}, j = 1..n$ in the error minimization process. MAE baseline becomes thus:

$$MAE_b = \arg\min_{\tilde{y}} \frac{1}{n} \sum_{i=1}^{n} |\frac{y_i - \tilde{y}}{y_i}|$$

The prediction performance should be better than this to be useful. The corresponding percentage values (MAPE and MAPE$_b$) have been used in the rest of the text.

Concordance index counts the occurrences when the prediction fails to be monotonical. In equation (3) we denote $N = | \{(i,j) \mid y_i > y_j\} |$ as the normalization constant which equals to the number of data pairs with different label values. $h(.)$ is Heaviside step function.

The values of the C-index range between 0.0 and 1.0, where 0.5 corresponds to a random predictor and 1.0 to the case where prediction is monotonically correct.

4.2   Results

The results for regression analysis can be seen in Fig. 3 on p. 443.

Both MAPE and C-index indicate rather good prediction performance to the distance of 120 m from the nearest soil sample point. This is seen both with k-NN and RLS methods. When MAPE is higher than the baseline, it is better to use baseline average than the prediction. MAPE baseline is the horizontal line in the lower figures in Fig. 3.

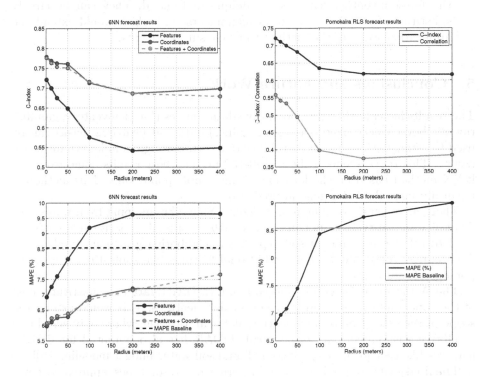

Fig. 3. *Left*: k-NN results with $k = 6$ and 3 different feature sets. *Right*: RLS results on features-only case. C-index and Pearson correlation at top and MAPE below. The prediction performance is adequate below 120 meters.

The dead zone radius $r > 0$ simulates a situation, where the test point is at least r distance away from the given training points. $r = 0$ is traditional LOO test arrangement and measures best the properties of the predicted value within the training set itself. It may be too optimistic, since we seek for generalization. A large radius $r \approx \infty$ is overly pessimistic, since it would use only tiny fragments of the training set and would completely distort the prediction.

The prediction performance near $r = 0$ seems to indicate rather good generalization ability, but the performance reduces drastically over the dead zone distance r. Further study, both theoretical and practical, must be done to properly address classical geoinformatics concepts such as spatial autocorrelation and spatial semivariance together with the general prediction ability. The problem is new, since spatial analysis in geosciences is usually applied in the sense of interpolation and extrapolation performance, and general prediction is usually analyzed in the terms of Machine Learning performance.

The feature selection was not attempted. There were two reasons for this:

- the number of features (96) remained modest.
- the forest inventory features are unique to Finland. They can be largely substituted by various remote measurements [12], which would extend the application scope of the method to whole sub-polar area.

5   Conclusions and Future Work

The results indicate that the chosen five data sources (forest inventory, gamma-ray, air-borne electromagnetic, topographical data and peat bog mask) can be used to estimate the water permeability to a certain range from known measurements. This range seems to be c. 120-150 m. The best results come from the k-NN method based on the location of the sample points only. This method is naturally unavailable for general prediction.

There are several possible improvements. Since the mapping from water permeability to soil types is not unique, see [19], a special majority rule could be used to select the dominant soil type from neighboring grid point predictions. Such expert rules would require additional features like sophisticated geomorphological categories.

The Aerial Light Detection and Ranging (LiDAR) data can substitute most of the topographical and forest inventory data features. This would extend the scope of the prediction to any location at the arctic zone, where only aerial and satellite measurements are economical. LiDAR has also potential for derived features like geological morphology [20] and soil water budget modeling [10].

The final goal is to predict the water permeability, soil types, approximate water budget and the load bearing capacity of the terrain in relation to the given weather forecast, while the model is based on remote measures and online learning based on measurements from the harvester fleet. The potential applications aim to wide-area routing and location planning. In this regard, even a modest prediction power of features-only prediction could yield a cumulative effect on route decisions.

Acknowledgements. This work is done as a part of ULJATH project, which is funded by the *Finnish Funding Agency for Technology and Innovation* (TEKES). ULJATH stands for *New Computational Methods For The Efficient Utilization of Public Data.*

The authors would like to thank the anonymous reviewers for valuable comments and suggestions.

References

1. Azzalini, A., Diggle, P.: Prediction of soil respiration rates from temperature, moisture and soil type. Journal of the Royal Statistical Society - Series C: Applied Statistics 43, 505–526 (1994)
2. Sculla, P., Franklin, J., Chadwick, O.: The application of classification tree analysis to soil type prediction in a desert landscape. Ecological Modelling 181, 1–15 (2005)
3. Schulte, R., Diamond, J., Finkele, K., Holden, N., Brereton, A.: Predicting the soil moisture conditions of irish grasslands. Irish Journal of Agricultural and Food Research 44, 95–110 (2005)
4. Gao, H., Tang, Q., Shi, X., Zhu, C., Bohn, T.J., Su, F., Sheffield, J., Pan, M., Lettenmaier, D.P., Wood, E.F.: Water budget record from variable infiltration capacity (vic) model. Algorithm Theoretical Basis Document for Terrestrial Water Cycle Data Records (2010)
5. Chapuis, R.: Predicting the saturated hydraulic conductivity of soils: a review. Bulletin of Engineering Geology and the Environment 71, 401–434 (2012)
6. Mahmood, H., Hoogmoed, W., van Henten, E.J.: Proximal gamma-ray spectroscopy to predict soil properties. Sensors 13, 16263–16280 (2013)
7. McKenzie, N., Ryan, P.: Spatial prediction of soil properties using environmental correlation. Geoderma 89, 67–94 (1999)
8. Emadi, M., Baghernejad, M., Pakparvar, M., Kowsar, S.: An approach for land suitability evaluation using geostatistics, remote sensing, and geographic information system in arid and semiarid ecosystems. Environ Monit Assess 164, 501–511 (2010)
9. J.P.W.P.R.H.: Soil moisture estimation using remote sensing. In: Proceedings of the 27th Hydrology and Water Resources Symposium, The Institute of Engineers Australia, Melbourne (2002)
10. Leutner, B., Müller, H., Wegmann, M., Beierkuhnlein, C.: Modelling biodiversity and forest structure using hyperspectral. In: 41st Annual Meeting of the Ecological Society of Germany (2011)
11. Hyvönen, E., Pänttäjä, M., Sutinen, M.L., Sutinen, R.: Assessing site suitability for scots pine using airborne and terrestrial gamma-ray measurements in finnish lapland. Canadian Journal of Forest Research 33-5(11), 796–806 (2003)
12. Tomppo, E., Katila, M., Mäkisara, K., Peräsaari, J.: Multi-source national forest inventory – methods and applications. Managing Forest Ecosystems, vol. 18. Springer (2008)
13. Hautaniemi, H., Kurimo, M., Multala, J., Leväniemi, H., Vironmäki, J.: The "three in one" aerogeophysical concept of gtk in 2004. Geological Survey of Finland, Special Paper 39, 21–74 (2005)
14. Schwanghart, W., Kuhn, N.: Topotoolbox: a set of matlab functions for topographic analysis. Environmental Modelling & Software 25, 770–781 (2010)
15. Weldon, T.P., Higgins, W.E., Dunn, D.F.: Efficient gabor filter design for texture segmentation. Pattern Recognition, 2005–2015 (1996), This could be removed, a more arcaic Gabor reference instead!
16. Ojala, T., Pietikäinen, M., Mäenpää, T.: Multiresolution gray-scale and rotation invariant texture. IEEE Transactions on Pattern Analysis and Machine Intelligence 24, 971–987 (2002)
17. Kiss, R.: Determination of drainage network in digital elevation models, utilities and limitations. Journal of Hungarian Geomathematics 2, 16–29 (2004)

18. Gönen, M., Heller, G.: Concordance probability and discriminatory power in proportional. Biometrika 92, 965–970 (2005)
19. Pohjankukka, J., Nevalainen, P., Pahikkala, T., Hyvönen, E., Sutinen, R., Hänninen, P., Heikkonen, J.: Arctic soil hydraulic conductivity and soil type recognition based on aerial gamma-ray spectroscopy and topographical data. In: Proceedings of the 22nd International Conference on, ICPR 2014 (to appear, 2014)
20. Sutinen, R., Hyvönen, E., Middleton, M., Ruskeeniemi, T.: Airborne lidar detection of postglacial faults and pulju moraine. Global and Planetary Change, 24–32 (2014)

Solar Radiation Time-Series Prediction Based on Empirical Mode Decomposition and Artificial Neural Networks

Petros-Fotios Alvanitopoulos[1], Ioannis Andreadis[1],
Nikolaos Georgoulas[1], Michalis Zervakis[2], and Nikolaos Nikolaidis[3]

[1] Department of Electrical and Computer Engineering,
School of Engineering, Democritus University of Thrace, Xanthi, Greece
[2] Laboratory of Digital Signal & Image Processing,
Department of Electronic and Computer Engineering,
Technical University of Crete (TUC), Chania, Greece
[3] Laboratory of Hydrochemical Engineering and Remediation of Soil (HERSLab),
Dept. of Environmental Engineering, Technical University of Crete (TUC), Chania, Greece
{palvanit,iandread,ngeorgou}@ee.duth.gr,
michalis@display.tuc.gr, nikolaos.nikolaidis@enveng.tuc.gr

Abstract. This paper presents a new model for daily solar radiation prediction. In order to capture the hidden knowledge of existing data, a time-frequency analysis on past measurements of the solar energy density is carried out. The Hilbert-Huang transform (HHT) is employed for the representation of the daily solar irradiance time series. A set of physical measurements and simulated signals are selected for the time series analysis. The empirical mode decomposition is applied and the adaptive basis of each raw signal is extracted. The decomposed narrow-band amplitude and frequency modulated signals are modelled by using dynamic artificial neural networks (ANNs). Nonlinear auto-regressive networks are trained with the average daily solar irradiance as exogenous (independent) input. The instantaneous value of solar radiation density is estimated based on previous values of the time series and previous values of the independent input. The results are promising and they reveal that the proposed system can be incorporated in intelligent systems for better load management in photovoltaic systems.

1 Introduction

Today there is an obvious need for eco-friendly and sustainable energy sources. It is well known that in developed world the daily life depends on the continuous supply of energy. Unfortunately, the exponential increase in energy demand compromises the ability of future generations to meet their needs. The provision of fossil fuel cannot be taken for granted. In recent years many European governments introduced subsidy programs in order to encourage the utilization of alternative energy resources with high efficiency and environmentally sound.

L. Iliadis et al. (Eds.): AIAI 2014, IFIP AICT 436, pp. 447–455, 2014.

Nowadays, there are many options for renewable energy sources. One of the most attractive and promising resource is the solar irradiance. Solar energy usage is small globally, but it is commonly accepted that it will increase substantially in the near future.

Solar energy is one of the most considerable design parameters in many research fields such as bioclimatic design of buildings, analysis of the performance of photovoltaic systems, agriculture science and meteorology. Moreover, in the case of autonomous energy systems with solar power source an automatic and efficient management of solar energy is required in order to minimize the strain of storage components and increase their lifetime.

Due to the widely use of solar energy there is an obvious need for a more precise modeling and close prediction of the solar irradiance future trend. Several studies for solar prediction exist in the literature. These studies involve empirical models [1, 2], statistical approaches [3, 4] and intelligent system approaches [5, 6].

This paper presents a new approach for solar radiation prediction by taking into account the past measurements of the series. The HHT operates on the original signal in order to extract the different spectral components. Afterwards, nonlinear autoregressive neural networks are trained with the average daily solar irradiance as exogenous input. The instantaneous values of solar radiation density are predicted for each of the partial component and are summarized in order to estimate the original signal's prediction.

The paper is organized as follows: Section 2 introduces the HHT. Section 3 refers to the nonlinear autoregressive networks, while section 4 deals with the proposed model implementation. Section 5 presents the experimental results.

2   The Hilbert Huang Transform

The HHT is an appropriate tool for analyzing non-stationary and nonlinear data. The HHT, a NASA's designated name, is proposed by Huang et al [7]. The essential difference between HHT and most currently data decomposition methods is that HHT provides an adaptive analysis of data. The term adaptivity refers to the requirement for an adaptive basis. There is no a priori reason to believe that an arbitrarily selected basis is able to describe the underlying processes of any given phenomenon. This algorithm comprises Empirical Mode Decomposition (EMD) and Hilbert Spectral Analysis (HSA). The combination of EMD and HSA provides a consistent and physically meaningful definition for instantaneous frequency and amplitude. First, the EMD process decomposes any complicated set of data into a finite and often small number of intrinsic mode function components. This decomposition is adaptive since it is based on local characteristics of the original signal and it can be applied to nonlinear and nonstationary signals. Second, the HSA is applied to the generated components in order to provide a meaningful time-frequency-energy distribution of time series.

2.1   Empirical Mode Decomposition (EMD)

The EMD step is based on the assumption that any time series consists of different, intrinsic modes of oscillation. Each mode of oscillation is derived from the data and fulfills the following conditions:

1   In the whole data set, the number of extrema and the number of zero-crossings must be equal or differ at most by one.
2   At any point, the mean value of the envelope defined by the local maxima and the envelope defined by the local minima is zero.

An intrinsic mode function (IMF) represents a simple oscillatory mode similar to a component in the Fourier-based simple harmonic function. IMFs constitute an a posteriori adaptive basis. The decomposition of any time series is described below.

First, all the local extrema of the series $x(t)$ are extracted. The local maxima are connected by a cubic spline to produce the upper envelope $u(t)$. Similarly the lower envelope $v(t)$ is created with the connection of the local minima. The mean of these two envelopes is designated as m_1, and the difference between $x(t)$ and m_1 is the first component h_1.

$$m_1 = \frac{u(t) + v(t)}{2} \tag{1}$$

$$h_1 = x(t) - m_1 \tag{2}$$

Usually, h_1 does not satisfy all the requirements of an IMF. Therefore, the previous process, entitled "sifting process", is repeated more times. In the subsequent sifting process, h_1 is treated as the data, then

$$h_{11} = h_1 - m_{11} \tag{3}$$

where m_{11} is the mean of the upper and lower envelopes of h_1. After repeated sifting, up to k times which is usually less than 10, h_{1k}, given by

$$h_{1k} = h_{1(k-1)} - m_{1k} \tag{4}$$

is designated as the first IMF component c_1 from the data, or

$$c_1 = h_{1k} \tag{5}$$

Typically, c_1 will contain the shortest-period component of the signal. The c_1 component is removed from the original signal in order to obtain the residue

$$r_1 = x(t) - c_1 \tag{6}$$

The residue r_1, which contains longer-period components, is treated as the new data and subjected to the same sifting process as it has been described above. This procedure can be repeated to obtain all the subsequent r_j functions as follows:

$$r_n = r_{n-1} - c_n \tag{7}$$

The sifting processs is terminated by either of the following predetermined criteria: either the component c_n or the residue r_n becomes so small that it is less than a predetermined value of consequence, or the residue, r_n, becomes a monotonic function, from which no more IMF can be extracted. Using the above equations the original data is the sum of the IMF components plus the final residue

$$x(t) = \sum_{i=1}^{n} c_i(t) + r_n(t) \tag{8}$$

2.2   Hilbert Spectrum Analysis (HSA)

In HSA the Hilbert transform (HT) is applied in each extracted IMF to obtain the representation of the original signal in time-frequency-energy domain. The application of HT in each IMF yields sharp frequency and time localization. For an arbitrary function, $x(t)$ its Hilbert transform, $y(t)$, is defined as

$$\text{(9)}$$

and the analytic signal z(t) is described by

$$\text{(10)}$$

where

$$\text{(11)}$$

the instantaneous frequency is given by

$$\text{(12)}$$

and the original signal can be expressed by

$$\text{(13)}$$

where (13) represents the general form of Fourier equation

$$\text{(14)}$$

3   Nonlinear Autoregressive Network with Exogenous Inputs

3.1   NARX Networks

Several processes in real world are described by time series. The ANNs have been widely used in the scientific field of time series prediction due to their inherent nonlinearity and high robustness in noise. It is well-known that ANN represents a nonparametric approach since there is no need to have any knowledge of the underlying processes that generate the original signal.

Typically, the challenge task of time series prediction can be expressed as finding the appropriate function l so as to acquire an estimate of the time series y at time $t+D$ ($D=1,2 ...$) given the past values of y up to time t, plus the values of exogenous input x:

$$\text{(15)}$$

where $y(t)$ and $x(t)$ represent the values of y and x in time t respectively. The variables d_y and d_x are the lags parameters of model and in case of D>1 we have the multi step ahead prediction of time series y.

NARX networks are recurrent neural architectures and it has been demonstrated that they are powerful in theory and at least equivalent to Turing machines. One can utilize those models rather than conventional recurrent networks without any computational loss [8]. The main advantages of NARX networks are: more effective learning of models and faster convergence than in other ANNs [9].

4   Proposed Method

4.1   Solar Radiation Dataset

In this paper, a study of forecasting half daily solar radiation is presented by taking into consideration past values of the solar radiation time series. The dataset contains solar radiation measurements provided by the Joint Research Centre (JRC) Renewable Energies unit of the European Commission [10]. Cubic spline interpolation is utilized in order to increase the size of data. Thus, from a 15-minute sample frequency is created a 1-minute sample frequency time series. Each daily artificial solar irradiance signal is a vector of length 840 which represents the daily 1-minute time interval solar irradiance samples. The geographic parameters latitude and longitude were set to 35°27'29" and 24°9'5", respectively. These parameters refer to Koiliaris river region in Crete, where the solar radiation prediction is studied to manage the energy demands of the CyberSensors monitoring system [11]. The daily solar power density samples, which are utilized in our analysis, are records that refer to the most adverse solar radiation conditions. This data is taken in December, which is the month that provides the lowest amount of solar energy.

4.2   Solar Radiation Analysis

In the present study the "clear-sky" values and the data for "real" conditions were taken into account. The former refers to values of irradiance where the sky is completely free of clouds. This signal is valuable since it provides the maximum values of solar irradiance that may appear at a given time in the examined month. The latter refers to values of solar irradiance with average cloud cover for the given month. In Fig.1 the "clear-sky" and the time series with average values of solar irradiance in December are presented for the examined region.

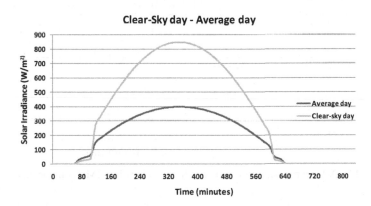

Fig. 1. Solar irradiance plot in December for the examined region with 25° angle

A typical solar radiation power density for a cloudy day is shown in Fig.2.

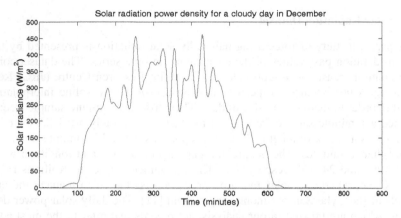

Fig. 2. Solar radiation power density for a cloudy day in December

The decomposition of each examined signal is implemented by means of EMD. An example of the EMD is presented in the following figure, where the solar power density signal from the previous figure is analyzed into a set of IMFs. It should be mentioned that the generated basis is an a posteriori basis created based on the local characteristics (local maxima and minima) of the time series. From Fig. 3 it can be noticed that the initial signal consist of five IMFs and a residue.

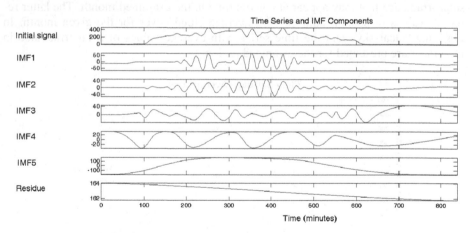

Fig. 3. Decomposition of a solar radiation signal into IMF components

4.3   The NARX Models

Nonlinear autoregressive with exogenous inputs networks were used for the solar radiation prediction. For each extracted intrinsic mode function a NARX network is trained and used for the prediction of the specific mode of oscillation. The NARX network is a two-layer feed forward artificial neural network. In the hidden layer a sigmoid transfer function is utilized and a linear transfer function is set to the output layer. In order to predict the future values of the time series $y(t)$, which in our case represents the solar radiation intrinsic mode function, the past values of time series $x(t)$ are also required. In the present study the clear-sky solar irradiance power density is treated as the exogenous input $x(t)$. Fig. 4 shows the topology of the NARX network, where the values d_y and d_x are the delays of $y(t)$ and $x(t)$ respectively.

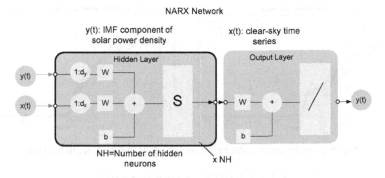

Fig. 4. Topology of the NARX network

5   Results

NARXs models are trained taking into account the solar radiation time series, which is derived from measurements of twenty five days. Five days records are utilized to assess the performance of the algorithm. The training function is set to Levenberg-Marquardt algorithm and the performance metric is the Mean Squared Error (MSE). The maximum horizon, which is taken, is 420 time steps ahead. The performance is estimated by taking into consideration the root-mean squared error (RMSE) and the maximum error. Table 1 shows the simulation results. The number of hidden neurons is set to 10 and the number of delays is 30. The validation and test data points are set to 15% of the original data, while the rest 70% is used for training.

Fig. 5 shows the simulation results of solar radiation for a half day horizon. As it can be seen from the below figures, solar radiation prediction is a complex task due to the extremely uncorrelated high frequency components. These high frequency components represent clouds and atmosphere attenuation, which are extremely difficult to predict using only past information. Thus, additional inputs are required in order to optimize the simulation results. Fig 6 shows an alternative view of the difference between the desired and predicted solar radiation values for the first day.

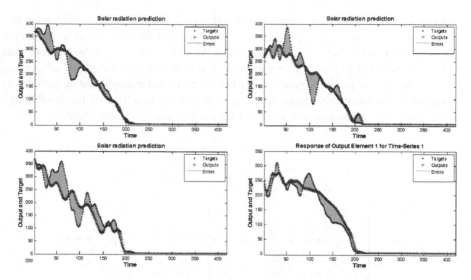

Fig. 5. Simulation of solar radiation for a half day maximum horizon

Table 1. RMSE from simulation results

	1st day	2nd day	3rd day	4th day	5th day
RMSE (W/m²)	26.891	21.732	31,009	39,658	35,729
MAX Error	97.847	60.130	120.510	129.815	128.858

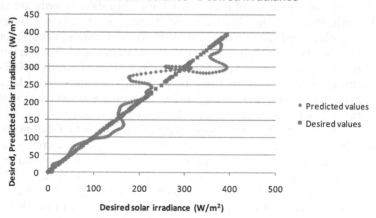

Fig. 6. Desired and predicted solar radiation values for the first day

6   Conclusions

In this paper the HHT is used for the decomposition of solar irradiance time series in order to predict solar radiation within a time horizon of half day ahead. From the results it can be seen that a more accurate prediction is required for the higher frequency components of the solar radiation time series. Taking into account extra data such as images from satellites, additional information regarding the impact of cloudiness on ground irradiance will be obtained. Moreover, deferencing of solar irradiance time series may generate stationary signals which are relatively easy to predict. In the nearby future, research will continue in this line by following the aforementioned approaches with ultimate goal to integrate this technique in an intelligent Energy Consumption Controller.

Acknowledgements. This work is elaborated through an on-going THALES project (CYBERSENSORS - High Frequency Monitoring System for Integrated Water Resources Management of Rivers). The Project has been co-financed by the European Union (European Social Fund – ESF) and Greek national funds through the Operational Program "Education and Lifelong Learning" of the National Strategic Reference Framework (NSRF) - Research Funding Program: Thales. Investing in knowledge society through the European Social Fund.

References

1. Batlles, F.J., Rubio, M.A., Tovar, J., Olmo, F.J., Alados-Arboledas, L.: Empirical modelling of hourly direct irradiance by means of hourly global irradiance. Energy 25, 675–688 (2000)
2. Loutzenhier, P.G., Manz, H., Felsmann, C., Strachan, P.A., Frank, T., Maxwell, G.M.: Empirical validation of models to compute solar irradiance on inclined surfaces for building energy simulation. Sol Energy 81, 254–267 (2007)
3. Safi, S., Zeroual, A., Hassani, M.: Prediction of global daily solar radiation using higher order statistics. Renew Energy 27, 647–666 (2002)
4. Kaplanis, S., Kaplani, E.: A model to predict expected mean and stochastic hourly global solar radiation I(h;nj) values. Renew Energy 32, 1414–1425 (2007)
5. Moustrisa, K., Paliatsos, A.G., Bloutsos, A., Nikolaidis, K., Koronaki, I., Kavadias, K.: Use of neural networks for the creation of hourly global and diffuse solar irradiance data at representative locations in Greece. Renew Energy 33, 928–932 (2008)
6. Rehman, S., Mohandes, M.: Artificial neural network estimation of global solar radiation using air temperature and relative humidity. Energy Policy 36, 571–576 (2008)
7. Huang, N.E., Shen, Z., Long, S.R., et al.: The empirical mode decomposition and the Hilbert spectrum for nonlinear and non-stationary time series analysis. In: Proceedings of the Royal Society, London, Series A, pp. 903–995 (1998)
8. Siegelmann, H.T., Horne, B.G., Giles, C.L.: Computational capabilities of recurrent narx neural networks. IEEE Trans. Syst., Man Cybern., pt. B 27, 208 (1997)
9. Gao, Y., Er, M.J.: NARMAX time series model prediction: feedforward and recurrent fuzzy neural network approaches. Fuzzy Sets and Systems 150, 331–350 (2005)
10. http://re.jrc.ec.europa.eu/pvgis/apps4/pvest.php
11. Moirogiorgou, K., Efstathiou, D., Zervakis, M., Nikolaidis, N.P., Stamatellos, G., Andreadis, I., Georgoulas, N., Savakis, A.E.: High Frequency Monitoring System for Integrated Water Resources Management of Rivers. In: 1st EWaS-MED International Conference: Improving Efficiency of Water Systems in a Changing Natural and Financial Environment, Thessaloniki, Greece, pp. 1–6 (April 2013)

Extracting Trends Ensembles in Solar Irradiance for Green Energy Generation Using Neuro-evolution

Mehreen Rehman, Jawad Ali, Gul Muhammad Khan,
and Sahibzada Ali Mahmud

Department of Electrical Engineering,
University of Engineering and Technology Peshawar, Pakistan 25000
{mehreen,jawad.ali,gk502,sahibzada.mahmud}@nwfpuet.edu.pk

Abstract. Globally, there are variations in climate, there is fossil fuel depletion, rising fossil fuel prices, increasing concern regarding energy security, and awareness about the environmental impacts of burning fossil fuels. These factors lead to a growing interest around the world in green and renewable energy resources, solar energy being a common one. A Neuro-evolutionary approach is explored to extract the trend ensembles in the solar irradiance patterns for renewable electric power generation, using the data taken from stations in Al-Ahsa, Kingdom of Saudi Arabia. The algorithm, based on Cartesian Genetic Programming Evolved Artificial Neural Network (CGPANN) was developed and trained for hourly and 24-hourly prediction, using the solar irradiance value as the input parameter. It was tested to predict solar irradiance on hourly, daily, and weekly basis. The proposed technique is **95.48%** accurate in solar irradiance prediction.

Keywords: Solar irradiance Forecasting, Neural Networks, Cartesian Genetic Programming (CGP), Neuro-evolution, Time Series Prediction.

1 Introduction

Most places on the earths surface receive clean and abundant solar energy, free of cost, throughout the year. Owing to the rising fossil fuel costs and the degradation of atmosphere by these fossil fuels, there is a dire need for economical and effective harnessing and utilization of solar energy. Because the world has to pay the cost of losing its energy resources at an alarmingly fast pace, there is an increasing reliance of power generation on renewables, such as solar energy. Worldwide efforts are being taken to increase the capacity of renewable power production. However, the intermittent nature of renewables makes it a challenging task to integrate them into any system. Accurate solar irradiance is therefore essential in improving the efficiencies of solar and wind energy based applications, obtaining optimized power production [1], overcoming the challenges of integrating them in the power grid and for the efficient management and operation of solar thermal energy plants. Solar radiation data gives us information

L. Iliadis et al. (Eds.): AIAI 2014, IFIP AICT 436, pp. 456–465, 2014.

about the amount of energy that hits the earths surface during a particular time frame and is required for research on effective solar energy utilization. Because this data is not readily available, and because the intensity and availability of solar radiation is dependent upon several other environmental factors, alternate ways of generating the data must be developed [2].

Artificial Neural Networks is an emerging technology for handling complex, practical problems, such as forecasting. Neural networks are capable of making computationally efficient, faster and more practical predictions than any of the conventional methods.

Our work mainly focuses on introducing a new and efficient solution based on the Neuro-evolutionary technique termed as the Cartesian Genetic Programming Evolved Artificial Neural Network (CGPANN) [3] to extract the trend ensembles in the solar irradiance patterns for renewable power generation. Neuro-Evolution is the phenomenon that involves the artificial evolution of the entities of an Artificial Neural Network (ANN). In CGPANN, Neuro-Evolution exploits the powerful structural properties of Cartesian Genetic Programming [4] and the functional properties of an ANN. This leads to the evolution of all the network parameters; node weights, system inputs, node inputs, node functions, network topology as well as the system outputs [3]. The research solution discussed here is specific only to CGPANN. The proposed technique uses past solar irradiance data to estimate the hourly, 12-, 24-, 48- and 168-hourly solar radiation patterns.

2   Literature Review

2.1   Computational Intelligence Techniques for Solar Irradiance Forecasting

Estimation methods used in the literature for forecasting solar radiation include stochastic, analytic [5], and artificial neural networks [6].The analysis by Mubiru and Banda [7] shows that owing to the nonlinear, non-stationary nature of solar radiation; ANNs are thought to be superior to empirical methods in estimating the global solar radiation.

Autoregressive (AR) and autoregressive moving average (ARMA) are models frequently used for the estimation of solar irradiance [8, 9]. Angstrom [10] was the pioneer to apply Regression models based on sunshine duration. Ahhi et al. [11] have used various input parameters of an ANN to predict solar radiation.

Kalogirou Model applies the standard back propagation learning algorithm. The model uses the hourly records for one year to predict hourly solar radiation. Other ANN models are; The Mohandes Model [6]; The Kemmoku Model [1]; The Reddy Model [13]; The Sozen Models [14] that has explored Scaled conjugate gradient (SCG), Pola-Ribiere conjugate gradient (CGP), and Levenberg-Marquardt (LM) learning algorithms and logistic sigmoid transfer function; The Cao Model [15] that establishes a recurrent BP network for forecasting of solar irradiance; The Soares Model [16] that develops a neural network exploring feature determination and pattern selection techniques. Linares-Rodriguez et. al. [17] explores an articial neural network ensemble model for estimating global solar radiation from Meteosat satellite images. This model uses clear-sky estimates and satellite images as input variables. Tymvios et al., Angela et. Al [20] have introduced a single-parameter model for solar irradiance prediction. Lorenz et al., [21] use numerical weather predictions (NWPs) as input parameters to NNs to estimate global irradiance. Hocaoglu et al., [22] study and compare feed-forward NNs with seasonal AR models. Cao and Lin, [23] combined NNs with wavelets to predict next day hourly values of global irradiance, incorporating various meteorological parameters as inputs to the models. Jain and Goel, [24] developed a multilayer feed forward neural network model for solar radiation forecasting in India. For the input to the network, the geographical, solar and meteorological parameters were taken from different areas of India. Azadeh et. al. [25] proposed an ANN model based on multilayer perceptron approach, incorporating all climatological and meteorological factors as input variables for prediction of solar global radiation. The results of the integrated ANN-MLP model have been compared to those of angstroms model [10]. There is a notable decrease in the MAPE value from 14.78% by the Angstrom model to 7.5% by the integrated ANN model.

2.2 Cartesian Genetic Programming evolved Artificial Neural Network (CGPANN)

CGPANN is a Neuro-Evolutionary technique based on Cartesian Genetic Programming where the nodes are replaced by articial neurons, the basic elements of CGPANN. A CGPANN node or neuron is composed of inputs, weighted connections, non-linear activation functions and outputs. The genotype is generated by encoding all these attributes of the neural networks [3]. The parent genotype undergoes mutation, producing the offspring. The network and its parameters continue to evolve until the best possible solution is obtained [4]. In this paper, CGPANN is used as the principle estimator in the prediction system. For the production of the next generation population in CGPANN, (1+) evolutionary strategy is recommended. The 10 % mutation rate used implies that only 10% of the genes take part in mutation to give the offspring. During the process of evolution some neurons, whose outputs remain unconnected (Junk Nodes) may not participate at all in generating the system output whereas others (Active Nodes) may play an active role in producing the system output. The weights are generated randomly (ranging from -1 to 1). The output can either be the system inputs or the output of any of the nodes. The neurons in the network are not fully connected. The systems inputs are not connected with all of the neurons in the input layer. The fundamental entities of a typical node in a CGPANN can be observed in Figure 1. The selection of the inputs is from the input array I_p, such that

$$I_p = [i_p1, i_p2, i_p3, ..., i_pn]$$ (1)

The weighing matrix [array of weights] W_g,

$$W_g = w_{g1}, w_{g2}, w_{g3}, ..., w_{gn}$$ (2)

for a particular genotype g, consists of random weights ranging from +1 and -1. The output of a junction in ANN that takes x_i as junction input, where the input is multiplied by a randomly alloted weight, w_{gi} can be represented by

$$O_{n'} = \sum_{i=1}^{N} x_i w_{gi} \tag{3}$$

If the inputs to a node are N, output $O_{nj'}$ is

$$O_{nj} = \xi^j(O_{n'}) = \xi^j \left(\sum_{i=1}^{N} x_i w_{gi} \right) \tag{4}$$

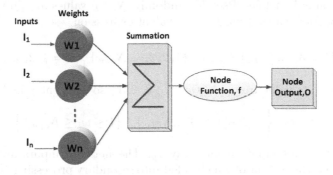

Fig. 1. The Nth node of CGPANN

here, ξ is the node function. Also, j is defined by

$$\{j | j\epsilon N \wedge N_T \geq j \geq 1\} \tag{5}$$

If I_g is the input set to G_k, where each input i has unique value:

$$i\epsilon R \wedge 1 \geq i \geq 0 \tag{6}$$

then, network G_k consists of randomly chosen inputs I_g, node outputs, O_{nj} , for a single output O_p such that

$$O_p = \frac{1}{n} \sum_{i=1}^{N} O_i \tag{7}$$

where

$$O_i = \xi \left(\sum (O_{nj} W_{gj} + O_{nj-1} W_{gj-1} + \ldots + O_{n1} W_{g1} + I W_{gk}) \right) \tag{8}$$

Here, W_{gj} , W_{gj-1}, ... W_{g1} are Random subsets of W such that W_{gk} is a subset W_{gj} and

$$W_{gj} = \{w_{gk}|w_{gk}\epsilon R \wedge +1 \geq w_k \geq -1\} \tag{9}$$

now

$$G_k = \{I, O_{nj}, O_{nj-1}, ..., O_{n1}, O_p\} \tag{10}$$

if we have two successive genotypes G_k and G_l. G_l is the product of G_k by mutating $\mu\%$ characteristics of the genotypes. Let the total entries in G_k be N_{cfw} then

$$N'_{cfw} = \mu \times N_{cfw} \tag{11}$$

N_cfw represents genotype entries that are to be mutated to get G_l. The set ζ has ω as a unique entry such that

$$\{\omega|\omega\epsilon\zeta\forall(W_g, I, O_{nj})\} \tag{12}$$

Each value in ω is defined Pseudo-randomly. N_cfw values are provided by the available N_cfw entities using the equivalent expression

$$\left\{\omega_i|\omega_i\epsilon\{1,2,3,...,N_{cfw}\} \wedge \omega_i \subset \{1,2,3,...,N_{cfw}\}\right\}, i = 1,2,3,...N'_{cfw} \tag{13}$$

Where ω_i can its value from $\{N_cfw, ... 3, 2, 1\}$ so each entry in ζ is

$$\gamma(i) = \left\{\omega_i|\omega_i\epsilon\{1,2,3,...N_{cfw}\} \wedge \{1 \leq \omega_i \leq N_{cfw}\}\right\} \tag{14}$$

The network takes inputs from array I_p. The network inputs are fed to the primary layer, the output of which is fed into secondary processing nodes, known as intermediate layer nodes. Activation function ξ, that is employed for this particular CGPANN is Log-sigmoid function, given by Eq. 16.

$$\xi(x) = \frac{1}{1 + e^{-x}} \tag{15}$$

Here x is input to the activation function $\xi(x)$. x is defined by the Eq. 17.

$$O(c, g, j) = \sum_{i=1}^{N} w_g(c, i).I(g, c, i) \tag{16}$$

Whereby N defines the quantity of inputs to the given node j for n = 1, 2, 3... N numbered inputs. K are the network inputs in set $k = 1, 2, 3...N...K$. g represents a particular genotype in a population and c is the node taking part in the operation. $w(i, c)$ is the randomly assigned weight to each input and $I(c, g, i)$ is defined as a randomly selected input to the node j.

$$I(g, c, i) = PRG([I(g, c, 1), I(g, c, 2), ..., I(g, c, N)...$$
$$..., I(g, c, k)] : [O(g, c, 1), O(g, c, 2), ...O(g, c, j1)]) \tag{17}$$

A typical CGPANN genotype and phenotype is shown in Fig. 2. The figure shows a single row, four nodes, with three inputs per nodes, five inputs and outputs network.

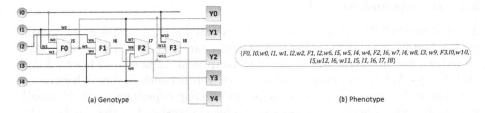

Fig. 2. A typical CGPANN genotype and phenotype

3   Application of CGPANN to Solar Irradiance Forecasting

3.1   Experimental Setup

The solar irradiance data of 1997, collected from Saudi Arabian stations in Al-Ahsa [26], were used to train the network for hourly and 24-hourly prediction and subsequently testing was done on the data acquired from 1997 as well as 1998, for hourly, 12-, 24-, 48- hourly and 7 days prediction capability of the model. The inputs to the network are the solar irradiance values, where irradiance, given in W/m2 , is the intensity of electromagnetic radiation incident per unit area. Al-Ahsa lies at an altitude of 178 m, 25.3°N latitude and 49.48°N longitude. The average temperature is 304 K, relative humidity 23% and the daily average solar global radiation is 21.6MJ/m². Although the network has been tested and trained using the data from 1997, the datasets for both belong to different months which mean that the data used for training is not being used for testing. The Mean Absolute Percentage Error (MAPE) value is used as the performance criterion to compare the offspring to the fittest network (the parent) and decide which of these are fit enough to be promoted to the next generation. The parent network further produces nine more networks by mutation. The process continues until the desired fitness is achieved or the maximum number of generations is completed. For training the network, all experiments are run for one million generations. Because a mutation rate of 10% is thought to produce more satisfactory results in comparison to other rates, it has been used in our experiment (Kadilar & Alada, 2009; Huang et al., 2006; Chen et al., 2008).

MAPE is given by:

$$MAPE = \frac{1}{N} \sum_{i=1}^{N} \left(\left| \frac{L_F - L_A}{L_A} \right| \right) \times 100\% \tag{18}$$

$$Fitness = 100\% - MAPE \tag{19}$$

Where L_F is the predicted value, L_A is the actual value and n is the number of days.

3.2 Results and Analysis

The proposed CGPANN model has been evaluated using 24 hours input network for the hourly, daily and weekly prediction of solar irradiance. The network takes 24 entries for the prediction of the next hour, next 12 hours, 24 hours, 2 days and 7 days. While the number of nodes on the network is varied between 50 and 500 with a step size of 50, training and testing experiments are repeatedly performed. Subsequently, a phenotype was being translated by the genotype that showed the best fitness level and then testing was carried on hourly, daily, and weekly basis. All the network nodes however may not be utilized by the final phenotype. Usually 5 to 10 percent of nodes participate in the production of the phenotype.

During testing, an independent dataset was given to the network for its validation. Table 1 contains the MAPE values calculated using 24 hours input network for the prediction of the next one hour, 12hrs, and 1 day. The network used here has been trained for hourly prediction. The values change with the changing number of nodes. MAPE values increase as the network is being used for long-term forecasting. The best result has been for the hourly prediction for 500 nodes i.e. a MAPE of only 4.52%.

Table 2 shows the testing results of predictions on 12-hourly, daily, and weekly basis, using the network model that has been trained for 24-hourly prediction. A lowest MAPE value of 5.36 has been achieved for the 12-hourly prediction. Hence, with the increase in time span, the forecasting of the model is affected

Table 1. Testing Results for Different Number of Nodes using 24 Hours Input Network for Prediction of the Next One Hour, 12 hours, and 1 Day, for Hourly Prediction Training

Prediction	Year	50	100	150	200	250	300	350	400	450	500
Hourly	1998	6.65	4.83	4.61	4.79	4.65	4.77	5.05	5.15	4.94	4.52
12-hourly	1997	9.88	7.48	7.03	7.06	6.98	6.64	7.44	7.57	7.22	6.78
	1998	11.01	8.37	7.97	7.96	7.87	7.55	8.36	8.49	8.14	7.68
24-hourly	1997	10.48	8.42	7.99	8.20	8.09	7.55	8.73	8.94	8.31	7.83
	1998	11.72	9.50	9.10	9.27	9.13	8.62	9.83	10.03	9.41	8.88

Table 2. Testing Results for various Number of Nodes for 24 Hours Input Network and Prediction of the Next 12 hours, 1 day, 2 Days, and 1 Week, for 24-hourly Prediction Training

Prediction	Year	50	100	150	200	50	300	350	400	450	500
12-hourly	1997	6.78	5.41	5.37	5.36	5.39	5.42	5.41	5.38	5.42	5.37
	1998	8.15	6.67	6.64	6.63	6.65	6.67	6.67	6.65	6.68	6.64
24-hourly	1998	8.08	6.68	6.65	6.64	6.66	6.70	6.68	6.66	6.69	6.65
48-hourly	1997	7.80	5.98	5.92	5.90	5.95	6.03	5.99	5.93	6.00	5.92
	1998	9.41	7.45	7.41	7.41	7.43	7.50	7.45	7.42	7.46	7.42
7 days	1997	8.30	7.43	7.28	7.17	7.39	7.52	7.56	7.26	7.59	7.31
	1998	12.41	8.52	8.37	8.29	8.48	8.66	8.62	8.36	8.66	8.40

owing to environmental, technical and climatic issues. The difference in values of Tables 1 and 2 explain the fact that no matter what the number of nodes, the CGPANN model exhibits better results when forecasting over a shorter period. However, the model that is trained to predict a single instance is also capable of predicting up to 24 hours without much error (see table 1). And the model that is trained to predict 24 hours can predict up to 7 days (168 hours) in advance with more than 90% accuracy (see table 2), indicating the robustness of the model. Another fact revealed is that the duration for which the network is trained is directly proportional to the efficiency of the model. This is indicated by the lower MAPE values in Table 2 corresponding to 24-hourly prediction training compared to those in Table 1 that correspond to hourly prediction training.

Figure 3 represents a graph that displays the deviation between the actual and the estimated data in terms of the normalized values of irradiance. It is obvious that the estimated data is following the frequent fluctuations in the actual data throughout the time frame i.e. 300 hours. The graph indicates only a minor difference between the two sets of values, confirming the accuracy and robustness of the CGPANN model.

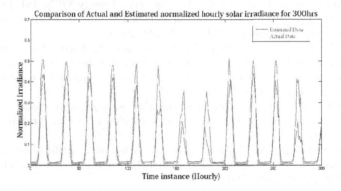

Fig. 3. The actual and the estimated normalized solar irradiance for 300 hours

4   Conclusion

The use of renewable energy necessitates exact estimation of renewable potential throughout the globe. Our research is an effort to introduce a technique for solar irradiance forecasting based on the Neuro-evolutionary technique called Cartesian Genetic Programming evolved Artificial Neural Network (CGPANN). The use of this technique could be very useful for improving efficiencies of solar based applications, the integration of renewable energy resources or the management of solar energy systems. The experiments display promising results, yielding an accuracy of 95.48 % and MAPE of 4.52% for hourly prediction. As the only input to the learning model is the solar radiation patterns, in the future, other parameters such as environmental conditions can be incorporated as inputs to produce even more accurate results.

464   M. Rehman et al.

References

[1] Koeppel, G., Korpas, M.: Using storage devices for compensating uncertainties caused by non-dispatchable generators. In: 2006 International Conference on Probabilistic Methods Applied to Power Systems, pp. 1–8 (2006)

[2] Dorvlo, A.S., Jervase, J.A., Al-Lawati, A.: Solar radiation estimation using artificial neural networks. Applied Energy 71(4), 307–319 (2002)

[3] Khan, M.M., Khan, G.M., Miller, J.F.: Efficient representation of recurrent neural networks for markovian/non-markovian non-linear control problems. In: 2010 10th International Conference on Intelligent Systems Design and Applications (ISDA), pp. 615–620. IEEE (2010)

[4] Miller, J.F.: Cartesian genetic programming. Springer (2011)

[5] Dagestad, K.: Estimating global radiation at ground level from satellite images, Ph.D. thesis, Department of Meteorology,University of Bergen, Bergen, Norway (2005)

[6] Mohandes, M., Balghonaim, A., Kassa, M., Rehman, S., Halawani, T.: Use of radial basis functions for estimation monthly mean daily solar radiation. Solar Energy 2000 68(2), 161–168 (2000)

[7] Mubiru, J., Banda, E.J.K.B.: Estimation of monthly average daily global solar irradiation using artificial neural networks. Solar Energy 82(2), 181–187 (2008)

[8] Reikard, G.: Predicting solar radiation at high resolutions: a comparison of time series forecasts. Solar Energy 83(3), 342–349 (2009)

[9] Torres, J.L., Garcia, A., De Blas, M., De Francisco, A.: Forecast of hourly average wind speed with ARMA models in Navarre (Spain). Solar Energy 79, 65–77 (2005)

[10] Angstrom, A.: On the computation of global solar radiation from records of sunshine. Arkiv Geophysik 3(23), 551 (1956)

[11] Ahhi, M.S., Jama, M.: Prediction of monthly average daily global radiation in Al Ain city- UAE using artificial neural networks. In: Proc. 4th RES 2010, pp. 109–113 (2010)

[12] Reddy, K.E., Ranjan, M.: Solar resource estimation using artificial neural networks and comparison with other correlation models. Energy Conversion and Management 44(15), 2519–2530 (2003)

[13] Sozen, A., Arcaklioglu, E., Ozalp, M., Kanit, E.G.: Solar-energy potential in Turkey. Applied Energy 80(4), 367–381 (2005)

[14] Cao, S., Cao, J.: Forecast of solar irradiance using recurrent neural networks combined with wave analysis. Applied Thermal Engineering 25(2-3), 161–172 (2005)

[15] Soares, J., Oliveira, A.P., Boznar, M.Z., Mlakar, P., Escobedo, J.F., Machado, A.J.: Model-ling hourly diffuse solar radiation in the city of Sao Paulo using artificial neural network technique. Applied Energy 79(2), 201–214 (2004)

[16] Linares-Rodriguez, A., Ruiz-Arias, J.A., Pozo-Vazquez, D., Tovar-Pescador, J.: An artificial neural network ensemble model for estimating global solar radiation from Meteosat satellite images. Energy 61, 636–645 (2013)

[17] Bacher, P., Madsen, H., Nielsen, H.A.: Online short-term solar power forecasting. Solar Energy 83(10), 1772–1783 (2009)

[18] Angela, K., Taddeo, S., James, M.: Predicting Global Solar Radiation Using an Artificial Neural Network Single-Parameter Model. Advances in Artificial Neural Systems (2011)

[19] Lorenz, E., Heinemann, D., Wickramarathne, H., Beyer, H., Bofinger, S.: Forecast of ensemble power production by grid-connected PV systems. In: Proceedings of the 20th European PV Conference, Milano (2007)

[20] Hocaoglu, F.O., Gerek, O.N., Kurban, M.: Hourly solar radiation forecasting using optimal coefficient 2-D linear filters and feed-forward neural networks. Solar Energy 82(8), 714–726 (2008)

[21] Cao, J., Lin, X.: Study of hourly and daily solar irradiation forecast using diagonal recurrent wavelet neural networks. Energy Conversion and Management 49(6), 1396–1406 (2008)

[22] Jain, R., Goel, B.: Prediction of Global Solar Radiation Using Artificial Neural Network. International Journal of Advanced Research in Electrical, Electronics and Instrumentation Engineering 2(10) (2013)

[23] Azadeh, A., Maghsoudi, A., Sohrabkhani, S.: An integrated artificial neural networks approach for predicting global radiation. Energy Conversion and Management 50(6), 1497–1505 (2009)

[24] Yang, K., Koike, T., Ye, B.: Improving estimation of hourly, daily, and monthly solar radia-tion by importing global data sets. Agricultural and Forest Meteorology 137(1), 43–55 (2006)

Optimal Control Using Feedback Linearization for a Generalized T-S Model

Agustín Jiménez[1], Basil Mohammed Al-Hadithi[1], Juan Pérez-Oria[2], and Luciano Alonso[2]

[1] Intelligent Control Group, University Politecnica de Madrid, Centre for Automation and Robotics UPM - CSIC, Spain. C/ J. Gutierrez Abascal, 2. 28006-Madrid, Spain
www.disam.upm.es/control
[2] University of Cantabria Santander, Spain
http://www.teisa.unican.es/gic/index.php

Abstract. In this paper, a fuzzy feedback linearization is used to control nonlinear systems described by Takagi-Suengo (T-S) fuzzy systems. In this work, an optimal controller is designed using the linear quadratic regulator (LQR). The well known weighting parameters approach is applied to optimize local and global approximation and modelling capability of T-S fuzzy model to improve the choice of the performance index and minimize it. The approach used here can be considered as a generalized version of T-S method. Simulation results indicate the potential, simplicity and generality of the estimation method and the robustness of the proposed optimal LQR algorithm.

1   Introduction

Feedback linearization has been used successfully to address some practical problems. These include the control of helicopters, high performance aircraft, industrial robots and biomedical devices. More applications of the methodology are being developed in industry [8].

It is well known that a robust a effective controller requires an accurate and efficient model. In [9], an interesting method is developed to identify nonlinear systems using input-output data. The main problem encountered is that T-S identification method can not be applied when the MFs are overlapped by pairs.

Nonlinear control systems based on the T-S model have attracted lots of attention during the last twenty years (e.g., see [12], [5] and [6]). It provides a powerful solution for development of function approximation, systematic techniques to stability and design of fuzzy control systems in view of fruitful conventional control theory. They also allow relatively easy application of powerful learning techniques for their identification from data.

T-S fuzzy models are proved to be universal function approximators [11]. But it was clearly shown that the number of fuzzy rules increases as the approximation error tends to zero. So it becomes difficult to make use of the universal approximation property of T-S fuzzy modelling for practical purposes. Moreover, if the number of rules is bounded, the resulting set of functions is nowhere dense

L. Iliadis et al. (Eds.): AIAI 2014, IFIP AICT 436, pp. 466–475, 2014.

in the space of approximated functions [10]. These conflicting objectives have motivated researchers to find a balance between the specified accuracy and the computational complexity of the resulting fuzzy model.

In [1] and [7] new and efficient approaches are presented to improve the local and global estimation of T-S model. The first approach uses the minimum norm method to search for an exact optimum solution at the expense of increasing complexity and computational cost. The second one is a simple and less computational method, based on weighting of parameters.

The rest of the paper is organized as follows. A description of feedback linearization methodology is presented in section 2. In section 3, the estimation of T-S fuzzy model is discussed. In section 4, the design of an optimal LQR controller is developed. In section 5, an example of an inverted pendulum is illustrated to demonstrate the validity of the proposed approach. The conclusions of the validity of the proposed estimation approach and the robustness of the LQR are explained in section 6.

2   Feedback Linearization

Feedback linearization is an approach to nonlinear control design which has attracted a great deal of research interest. The central ideal is to algebraically transform a nonlinear system dynamics into (fully or partly) linear one so that linear control techniqus can be applied. This differes entirely from conventional linearization (i.e. Jacobian linearization) in that feedback linearization is achieved by exact state transformation and feedback, rather than by linear approximations of the dynamics. Supposing that the system can be modeled as follows:

$$y^{(n)}(t) = f_n(y(t), \dot{y}(t), \cdots, y^{(n-1)}(t)) + b_n(y(t), \dot{y}(t), \cdots, y^{(n-1)}(t))u(t) \quad (1)$$

Let us define the following vector:

$$x = \begin{bmatrix} y(t) \ \dot{y}(t) \cdots y^{(n-1)}(t) \end{bmatrix}^T \in \Re^n$$

The system can be modeled as follows:

$$\dot{x}_n(t) = f_n(x(t)) + b_n(x(t))u(t) \quad (2)$$

For systems which can be expressed in the controllable and canonical form, the control action can be described as:

$$u(x) = \frac{1}{b_n(x)}(s(x) - f_n(x)) \quad (3)$$

Substituting (3) in (2) to obtain the feedback system:

$$\dot{x}_n(t) = f_n(x) + b_n(x)u(t) = f_n(x) + b_n(x)\frac{1}{b_n(x)}(s(x) - f_n(x)) = s(x) \quad (4)$$

Let us suppose the following control law:

$$s(x) = Sx = [-s_0 \ -s_1 \cdots - s_{n-2} \ -s_{n-1}]x \quad (5)$$

The feedback system becomes a linear one with a characteristic polynomial $p(\lambda)$ where the s_i are chosen so that the the characteristic polynomial:

$$p(\lambda) = \lambda^n + s_{n-1}\lambda^{n-1} + \cdots + s_1\lambda + s_0$$

The first inconvenient is the need to know explicitly the nonlinear system model, which in general can not be obtained because of the difficulty of identifying their parameters. Secondly, an adequate choice of the characteristic polynomial should be made. A possible solution for the first problem is by using the T-S fuzzy model for estimating nonlinear systems.

$$S^{(i_1 \cdots i_n)} \ : \ If \ z_1 \ is \ M_1^{i_1} \ and \dots z_f \ is \ M_n^{i_n} \ then$$
$$\dot{x} = a_0^{(i_1 \cdots i_n)} + A_n^{(i_1 \cdots i_n)}x + B_n^{(i_1 \cdots i_n)}u$$

where

$$A_n^{(i_1 \cdots\cdots i_n)} = [a_{n1}(x) \ \ a_{n2}(x) \ \ \cdots \ \ a_{nn}(x)]$$

Applying this model, the identification of its parameters can be realized using the proposed estimation methods which will be described in section 3. We suppose that we have a first estimation of the T-S model parameters. In order to obtain such estimation, the classical least square method can be used around the equilibrium point. The objective is to obtain a local approximation of the system.

$$\dot{x}_n = a_o^0 + a_{n1}^0 x_1(x) + a_{n2}^0 x_2(x) + \cdots + a_{nn}^0 x_n(x) + b_n^0 u \qquad (6)$$

A design of an optimal controller is carried out to show the effectiveness of the proposed estimation methods. The well known LQR controller is applied to solve the second problem proposed in this work which is how to choose the optimal characteristic polynomial $p(\lambda)$ (see section 4).

3   Estimation of Fuzzy T-S Model's Parameters

An interesting method of identification is presented in [9]. The idea is based on estimating the nonlinear system parameters minimizing a quadratic performance index. It is based on the identification of functions of the following form:

$$f : \Re^n \longrightarrow \Re$$

$$y = f(v_1, v_1, \dots, v_n)$$

Each IF-THEN rule for an nth order system can be written as follows:

$$S^{(i_1 \cdots i_n)} : If \ v_1 \ is \ M_1^{i_1} \ and \dots v_n \ is \ M_n^{i_n} \ then \qquad (7)$$
$$\hat{y} = p_0^{(i_1 \cdots i_n)} + p_1^{(i_1 \cdots i_n)}v_1 + p_2^{(i_1 \cdots i_n)}v_2 + \dots + p_n^{(i_1 \cdots i_n)}v_n$$

where the fuzzy estimation of the output is:

$$\hat{y} = \frac{\sum_{i_1=1}^{r_1} \cdots \sum_{i_n=1}^{r_n} w^{(i_1 \cdots i_n)}(\mathbf{v}) \left[p_0^{(i_1 \cdots i_n)} + p_1^{(i_1 \cdots i_n)}v_1 + \dots + p_n^{(i_1 \cdots i_n)}v_n \right]}{\sum_{i_1=1}^{r_1} \cdots \sum_{i_n=1}^{r_n} w^{(i_1 \cdots i_n)}(\mathbf{v})} \qquad (8)$$

where,

$$w^{(i_1\ldots i_n)}(\mathbf{v}) = \prod_{l=1}^{n} \mu_{l i_l}(v_l)$$

being $\mu_{j i_j}(v_j)$ the membership function that corresponds to the fuzzy set $M_j^{i_j}$.

Let $\{v_{1k}, v_{2k}, \ldots, v_{nk}, y_k\}$ be a set of input/output system samples. The parameters of the fuzzy system can be calculated as a result of minimizing a quadratic performance index:

$$J = \sum_{k=1}^{m}(y_k - \hat{y}_k)^2 = \|Y - XP\|^2 \tag{9}$$

where

$$Y = \begin{bmatrix} y_1 \ y_2 \ \ldots \ y_m \end{bmatrix}^T$$

$$P = \begin{bmatrix} p_0^{(1\ldots1)} \ p_1^{(1\ldots1)} \ p_2^{(1\ldots1)} \ \ldots \ p_n^{(1\ldots1)} \ \ldots p_0^{(r_1\ldots r_n)} \ \ldots \ p_n^{(r_1\ldots r_n)} \end{bmatrix}^T \tag{10}$$

$$X = \begin{bmatrix} \beta_1^{(1\ldots1)} \ \beta_1^{(1\ldots1)} v_{11} \ldots \beta_1^{(1\ldots1)} v_{n1} \ldots \beta_1^{(r_1\ldots r_n)} \ \ldots \ \beta_1^{(r_1\ldots r_n)} v_{n1} \\ \beta_m^{1\ldots1} \ \beta_m^{(1\ldots1)} v_{1m} \ldots \beta_m^{(1\ldots1)} v_{nm} \ldots \beta_m^{(r_1\ldots r_n)} \ \ldots \ \beta_m^{(r_1\ldots r_n)} v_{nm} \end{bmatrix}$$

and

$$\beta_k^{(i_1\ldots i_n)} = \frac{w^{(i_1\ldots i_n)}(\mathbf{v}_k)}{\sum_{i_1=1}^{r_1} \cdots \sum_{i_n=1}^{r_n} w^{(i_1\ldots i_n)}(\mathbf{v}_k)} \tag{11}$$

If X is a matrix of full rank, the solution is obtained as follows:

$$J = \|Y - XP\|^2 = (Y - XP)^T(Y - XP)$$
$$\nabla J = X^T(Y - XP) = X^T Y - X^T XP = 0 \tag{12}$$
$$P = (X^T X)^{-1} X^T Y$$

In the case when the matrix X is not of full rank, an effective approach with few computational effort, based on the well known parameters' weighting method [2], [1] and [7]. This method can also be used for parameters tuning of T-S model from local parameters obtained through the identification of a system in an operating region or from any physical input/output data. We suppose that we have a first estimation

$$P_0 = [p_0^0 \ p_1^0 \ p_2^0 \ldots p_n^0]^T$$

of the T-S model parameters. In order to obtain such an estimation, the classical least square method can be used around the equilibrium point. This first approximation can be utilized as reference parameters for all the subsystems. Then, the parameters' vector of the fuzzy model can be obtained minimizing:

$$J = \sum_{k=1}^{m}(y_k - \hat{y}_k)^2 + \gamma^2 \sum_{i_1=1}^{r_1} \cdots \sum_{i_n=1}^{r_n} \sum_{j=0}^{n}(p_j^0 - p_j^{(i_1\ldots i_n)})^2$$

$$= \|Y - Xp\|^2 + \gamma^2 \|P_0 - p\|^2 = \left\| \begin{bmatrix} Y \\ \gamma P_0 \end{bmatrix} - \begin{bmatrix} V \\ \gamma I \end{bmatrix} p \right\|^2 = \|Y_a - V_a p\|^2 \tag{13}$$

where

$$p_0 = \underbrace{[P_0 \ P_0 \ldots P_0]^T}_{r_1 . r_2 \ldots r_n}$$

In this case, the factor γ represents the degree of confidence of the parameters initially estimated [2], [3], [4] and [7].

4 Design of an Optimal Controller

A design of an optimal controller based LQR controller is applied to show the validity of the proposed estimation methods. The characteristic polynomial of the feedback system is calculated for this system so that finally, the parameters of this polynomial is used for the definition of $s(x)$. In order to calculate the control coefficients, any control design methodology through state feedback control can be applied.

Optimal selection of closed loop poles will lead to a trade-off between speed of dynamic response and control effort. Together with the proposed estimation method, LQR might be an appropriate choice. With this regulator we guarantee the stability and gaining a balance between static and dynamic behavior of the system with admissible control actions. Selection of values of Q and R matrices determines the dynamic speed of the controller as well as amplitudes of state variables and control signals.

$$\dot{x} = Ax + Bu$$

$$x \in \Re^n, u \in \Re^m, A \in \Re^{n \times n}, B \in \Re^{n \times m}$$

The objective is to find the control action u(t) to transfer the system from any initial state $x(t_0)$ to some final state $x(\infty) = 0$ in an infinite time interval, minimizing a quadratic performance index of the form:

$$J = \int_{t_0}^{\infty} (x^T Q x + u^T R u) dt$$

where $Q \in \Re^{n \times n}$ is a symmetric matrix, at least positive semidefinite one and $R \in \Re^{m \times m}$ is also a symetric positive definite matrix. The optimal control law is then computed as follows:

$$u(t) = -Kx(t) \tag{14}$$

$$K = -R^{-1}B^T L \tag{15}$$

where the matrix $L \in \Re^{n \times n}$ is a solution of the Riccati equation:

$$0 = -Q + LBR^{-1}B^T L - LA - A^T L$$

The feedback system becomes:

$$\dot{x} = Ax + Bu = Ax - BKu = (A - BK)x$$

and the characteristic polynomial is:

$$p(\lambda) = det(\lambda I - A + BK)$$

It is aimed in this work to design an optimal LQR controller for the local approximation in (6). In order to obtain such estimation, the classical least square method can be used around the equilibrium point.

4.1   Application of the Proposed Control Algorithm for T-S Systems

The new proposed control algorithm can be considered as a two level one: the first level includes the calculation of the resulting matrices $a_0(x)$, $A(x)$ and $B(x)$ from the above identified fuzzy system and the second one is obtaining the control action mentioned above in (3). The control law becomes:

$$u(x) = \frac{1}{b_n(x)}(SX - a_{0n}(x) - A_n(x)x) = -\frac{1}{b_n(x)}(a_{0n}(x) + (A_n(x) - V)x) \quad (16)$$

5   Illustrative Example

Consider the problem of estimating an inverted pendulum using the above mentioned estimation methods. The inverted pendulum can be represented as follows:

$$\ddot{\theta} = \frac{gsin\theta - cos\theta(\frac{u+ml\dot{\theta}^2 sin\theta}{M+m})}{l(\frac{4}{3} - \frac{mcos^2\theta}{M+m})} \quad (17)$$

Where θ denotes the angular position (in radians) deviated from the equilibrium position (vertical axis) of the pendulum and $\dot{\theta}$ is the angular velocity, g(gravity acceleration), M(mass) of the cart=1 kg, m(mass) of the pole=0.1 kg, l is the distance from the center of the mass (m) of the pole to the cart=0.5 m. Assuming that $x_1 = \theta$ and $x_2 = \dot{\theta}$, then (17) can be rewritten in state space form as follows:

$$x_1 = \theta$$
$$\dot{x}_1 = x_2$$
$$x_2 = \dot{\theta}$$
$$\dot{x}_2 = \frac{gsin(x_1) - cos(x_1)(\frac{u+mlx_2^2 sin(x_1)}{M+m})}{l(\frac{4}{3} - \frac{mcos^2(x_1)}{M+m})} \quad (18)$$

Firstly, the model of the inverted pendulum is estimated in three operation points for both the angle and its derivative. The universe of discourse of the angle is $[\frac{-\pi}{4}, \frac{\pi}{4}]$ rad. and the one of the angular velocity is $[-5, 5]$ $\frac{rad}{seg}$. Both MFs for the angle x and its derivative $\dot{x}$ are shown in figures 1 y 2 respectively.

If we apply the T-S method directly to this example, then the condition number of the matrix X is 3.4148e +015, which shows clearly a non reliable result.

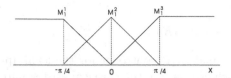   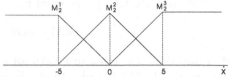

Fig. 1. Membership functions for the angle x of the inverted pendulum   **Fig. 2.** Membership functions for the angular velocity of the inverted pendulum

Using the parameters' weighting method with weighting factor $\gamma = 0.01$, the inverted pendulum fuzzy model can be represented as follows:

$$S^{11} : If \ (x_1 \ is \ M_1^1) \ and \ (x_2 \ is \ M_2^1) \ then$$
$$\dot{x}_2 = -8.2354 + 3.3866x_1 - 0.2040x_2 - 1.0443u$$
$$S^{12} : If \ (x_1 \ is \ M_1^1) \ and \ (x_2 \ is \ M_2^2) \ then$$
$$\dot{x}_2 = -8.3313 + 3.0983x_1 + 0.0000x_2 - 1.0443u$$
$$\vdots$$
$$S^{33} : If \ (x_1 \ is \ M_1^3) \ and \ (x_2 \ is \ M_2^3) \ then$$
$$\dot{x}_2 = 8.0784 + 3.8914x_1 - 0.2297x_2 - 1.0903 \tag{19}$$

By using the identification with the classical minimum square method in an interval close to the equilibrium point

$$x_1 \in [\frac{-\pi}{4}, \frac{\pi}{4}] \quad x_2 \in [-2.5, 2.5]$$

The linear model of the system in this interval is:

$$\dot{x}_2 = 0.0092 + 15.2665x_1 - 0.0000x_2 - 1.4187u$$

As it has been mentioned before, the inestable equilibrium point which will be also the objective of control is $x_1 = x_2 = u = 0$. In order that this point becomes an equilibrium one in the fuzzy model, a_0^{22} should be zero. The resultant mean square error from this approximation is 0.0013. In this case, the condition number of the extended matrix X_a becomes 1.4569e +004, thus improving the reliability of results.

As it can be observed, the results obtained through the parameters weighting method are always better than with the original T-S method. In fact, the proposed method correspond to the best possible result using TS method when the interval covering tends to totality, but this limit is not achievable since the problem would not be solvable any more.

Consider the problem of stabilizing and balancing of swing up of an inverted pendulum. The control of this system is a widely used performance measure of a controller, since this system is unstable and highly nonlinear. The objective is to maintain the inverted pendulum upright with θ despite small disturbances due to wind or system noises. The proposed optimum LQR is applied to calculate the desired parameters of the characteristic polynomial $p(\lambda)$ defined in the previous rules,

$$\begin{bmatrix} \dot{x}_1 \\ \dot{x}_2 \end{bmatrix} = \begin{bmatrix} 0 & 1 \\ 15.2665 & 0 \end{bmatrix} \begin{bmatrix} x_1 \\ x_2 \end{bmatrix} + \begin{bmatrix} 0 \\ -1.4187 \end{bmatrix} u$$

minimizing the following performance index:

$$J = \int_{t_0}^{\infty} (100x_1^2 + 10x_2^2 + u^2)dt$$

and the computed control becomes

$$u(t) = -Kx(t) = \begin{bmatrix} 25.4509 & 6.7734 \end{bmatrix} x(t)$$

The matrix of the feedback system becomes:

$$A_R = A - BK = \begin{bmatrix} 0 & 1 \\ -20.8408 & -9.6095 \end{bmatrix}$$

and its characteristic polynomial is:

$$p(\lambda) = \lambda^2 + 9.6095\lambda + 20.8408$$

and the eigen values are:

$$\lambda_1 = -6.3029 \quad \lambda_2 = -3.3065$$

Thus, the control becomes:

$$u(x) = \frac{1}{b_n(x)}(a_{0n}(x) + (A_n(x) - S)x) \tag{20}$$

where:

$$S = \begin{bmatrix} 20.8408 & -9.6095 \end{bmatrix}$$

Figure 3 shows the transient response of the inverted pendulum controlled by the proposed LQR subjected to an initial condition. The results obtained show that the system is stabilized by applying the proposed LQR.

Moreover, we have proven that the LQR is robust and invariant against measurement noise. If we suppose that the angle sensor introduces a noise of $\sigma = 1°$, it can be observed that the output is only affected by $\sigma = 0.04°$ as shown in figure 4. Figure 5 shows several trajectories in state space form of the system for several initial conditions.

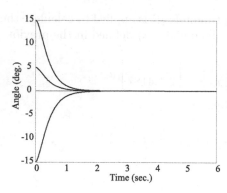

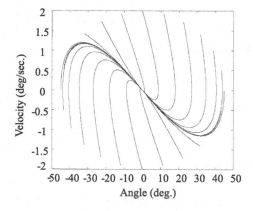

Fig. 3. Transient response of the inverted pendulum by the fuzzy LQR

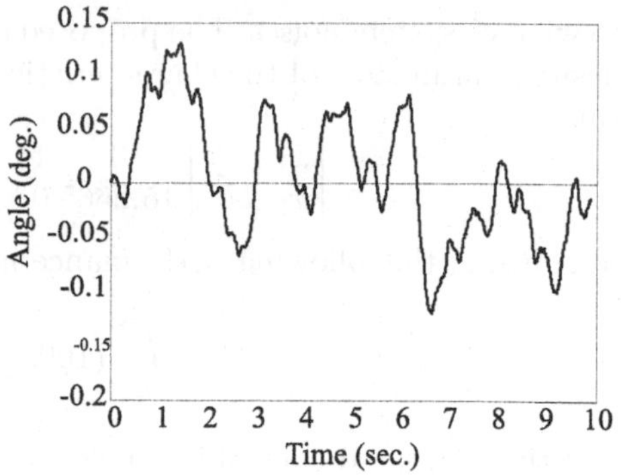

Fig. 4. Transient response of the inverted pendulum by the fuzzy LQR subjected to measurement noise

Fig. 5. Trajectories in state space form of the system for several initial conditions

6 Conclusions

A feedback linearization is used to nonlinear control design for fuzzy systems described by T-S. The main objective is to obtain an improved performance of highly non-linear unstable systems. In this work, the well known weighting parameters approach is proposed to optimize local and global approximation and modelling capability of T-S fuzzy model to improve the choice of the performance index and minimize it. An inverted pendulum mounted on a cart is chosen to evaluate the robustness, effectiveness, accuracy and remarkable performance of proposed estimation approach. The results show that the proposed approach is less conservative than those based on (standard) T-S model. Simulation results indicate the potential, simplicity and generality of the estimation method and the robustness of the LQR algorithm. We prove that the proposed estimation algorithm converge very fast, thereby making them very practical to use.

Acknowledgements. This work is funded by Spanish Ministry of innovation and Science (ARABOT project DPI2010-21247-C02-01).

References

1. Al-Hadithi, B.M., Jiménez, A., Matía, F.: New Methods for the Estimation of T-S Model Based Extended Kalman Filter and its Applications to Optimal Control for Nonlinear Systems. Optimal Control, Applications and Methods 33(5), 552–575 (2012)
2. Al-Hadithi, B.M., Jiménez, A., Matía, F.: Variable Structure Control with Chattering Reduction of a Generalized T-S Model. Asian Journal of Control 15(1), 1–14 (2013)
3. Al-Hadithi, B.M., Barragán, A.J., Andújar, J.M., Jiménez, A.: Variable Structure Control with chattering elimination and guaranteed stability for a generalized T-S model. Applied Soft Computing 13, 4802–4812 (2013)
4. Al-Hadithi, B.M., Jiménez, A., Matía, F.: A new approach to fuzzy estimation of Takagi-Sugeno model and its applications to optimal control for nonlinear systems. Applied Soft Computing 12, 280–290 (2012)
5. Gang, F.: A survey on analysis and design of model-based fuzzy control systems. IEEE Trans. Fuzzy Syst. 14(5), 676–697 (2006)
6. Hseng, T., Li, S., Tsai, S.-H.: Fuzzy Bilinear Model and Fuzzy Controller Design for a Class of Nonlinear Systems. IEEE Trans. Fuzzy Syst. 15(3), 494–506 (2007)
7. Jiménez, A., Al-Hadithi, B.M., Matía, F.: Improvement of Takagi-Sugeno Fuzzy Model for the Estimation of Nonlinear Functions. Asian Journal of Control 14(2), 320–334 (2012)
8. Slotine, J.-J.E., Li, W.: Applied nonlinear control. Prentice Hall (1991) ISBN: 0-13-040890-5
9. Takagi, T., Sugeno, M.: Fuzzy identification of systems and its applications to modeling and control. IEEE Trans. Syst., Man, Cybern. SMC-15(1), 116–132 (1985)
10. Tikk, D., Baranyi, P., Patton, R.J.: Polytopic and TS Models are Nowere Dense in the Approximation Model Space. In: Proc. IEEE Int. Conf. Syst. Man Cybern. Methodology,
11. Zeng, K., Zhang, N.Y., Xu, W.L.: A comparative study on sufficient conditions for Takagi-Sugeno fuzzy systems as universal approximators. IEEE Trans. Fuzzy Syst. 8(6), 773–780 (2000)
12. Barragán, A.J., Al-Hadithi, B.M., Jiménez, A., Andújar, J.M.: A general methodology for online TS fuzzy modeling by the extended Kalman filter. Applied Soft Computing 18, 277–289 (2014)

HHT-Based Artificial Seismic Accelerograms Generation

Eleni Vrochidou[1], Petros Alvanitopoulos[1], Ioannis Andreadis[1],
Anaxagoras Elenas[2], and Katerina Mallousi[1]

[1] Department of Electrical and Computer Engineering, Democritus University of Thrace,
GR-67100 Xanthi, Greece
[2] Department of Civil Engineering, Democritus University of Thrace,
GR-67100 Xanthi, Greece

Abstract. A new efficient approach for generating spectrum-compatible seismic accelerograms is proposed. It is based on the Hilbert-Huang Transform (HHT); one natural seismic accelerogram is decomposed into frequency and amplitude components. The components are appropriately modified to synthesize the artificial seismic accelerogram that appears to have compatible acceleration spectrum with the natural seismic accelerogram. The HHT is an adaptive signal processing technique for analyzing nonlinear and non-stationary data such as seismic accelerograms. With HHT a seismic accelerogram is decomposed into a finite and small set of components. These components have well defined instantaneous frequencies, estimated by the first derivative of the phase of the analytic signal. The method is tested using twenty natural seismic records and a comparison with two established methodologies is provided.

Keywords: Hilbert-Huang Transform, artificial spectrum-compatible seismic accelerograms, frequency components.

1   Introduction

Earthquake engineering is the scientific field concerned with the study of the behavior of structures subject to seismic loading. Earthquake accelerograms (seismic signals) are required in order to simulate the response of structures. Disaster scenarios need an extensive set of seismic signals and although a large database of recordings of seismic excitations exists, for many regions there is lack of actual acceleration time-history records [1]. Artificial spectrum-compatible accelerograms are widely used in the dynamic analysis of structures. The proposed method is based on HHT for generating spectrum-compatible earthquake accelerograms. It is well-known that the HHT is an appropriate signal processing technique for analyzing non-stationary and nonlinear signals such as seismic signals [2, 3].

The HHT decomposes the earthquake accelerogram into a finite number of Intrinsic Mode Functions (IMFs), and provides an energy-frequency-time distribution, the Hilbert spectrum. The extraction of the IMFs is based on the local characteristics of the seismic signal and not on a priori assumptions. Unlike traditional signal processing techniques, in HHT the frequency is defined through differentiation and there is no uncertainty principle limitation on time or frequency resolution from the

L. Iliadis et al. (Eds.): AIAI 2014, IFIP AICT 436, pp. 476–486, 2014.
© IFIP International Federation for Information Processing 2014

convolution pairs based on a priori bases. Thus, HHT provides sharper energy-frequency-time distribution [2]. Moreover, the finite number of the extracted IMFs is adequate to describe any seismic signal and therefore reduces the computational burden of the algorithm.

Previously reported methods in the literature employ HHT to generate spectrum-compatible accelerograms [4, 5]. In method [4] six natural records are required to describe the high and low frequency areas of the target design spectrum. The artificial spectrum-compatible seismic signal is obtained by solving an optimization problem. The difference between the target spectrum and the response spectrum of the artificially generated seismic signal is treated as the cost function. In contrast to the aforementioned method, in the proposed technique only one initial earthquake signal is required. Moreover, it can be of any intensity and thus no limitations are imposed. Another method is proposed in [5].Artificial seismic signals are generated and submitted to a correlation study. The correlation study demonstrates that the artificially generated seismic signals share similar seismic parameter values to those of natural accelerograms.

The proposed methodology is applied to twenty natural seismic accelerograms. In order to reveal the effectiveness of the proposed method, a comparison with two well-established algorithms for generating spectrum-compatible accelerograms is presented. The first is Gasparini and Vanmarcke algorithm [6] implemented by SIMQKE software [7]. The second is Hallodorson and Papageorgiou algorithm [8] implemented by SeismoArtif artificial earthquake accelerograms generator [9]. Results demonstrate that the proposed method is as reliable as the methodologies implemented by the two established Computer-Aided Design (CAD) tools.

2   The Hilbert-Huang Transform

The HHT was introduced by Huang et al [2] and comprises two parts: the Empirical Mode Decomposition (EMD) and the Hilbert Spectral Analysis (HSA).

2.1   Empirical Mode Decomposition (EMD)

The EMD can decompose complex data into a finite and small number of IMFs. The method is considered to be valid as it is an adaptive method and since the

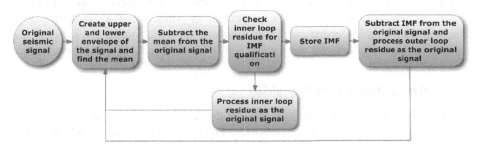

Fig. 1. Flow chart of the EMD

decomposition is based on the local characteristics of the signal, it is appropriate for non-stationary and non-linear signals [2, 3]. The decomposition process for a seismic signal is described in Figure 1 [2, 3].

2.2   Hilbert Spectral Analysis (HSA)

The HSA applies the Hilbert Transform to each IMF so as to obtain the instantaneous frequency data and construct the energy-frequency-time distribution. For all $c_j(t)$ components, the Hilbert Transform $y_j(t)$ is:

$$y_j(t) = \frac{1}{\pi} P \int_{-\infty}^{+\infty} \frac{c_j(\tau)}{t - \tau} d\tau \tag{1}$$

where P denotes the Cauchy Principal Value. The IMF components are defined by the equation:

$$c_j(t) = Re(a_j(t)e^{i\theta_j(t)}) = a_j(t)\cos\theta_j(t) \tag{2}$$

where $Re()$ is the real part. The initial signal is then written as:

$$X(t) = Re(\sum_{j=1}^{n} a_j(t)\cos(\int 2\pi f_j(t)dt)) \tag{3}$$

The residue $r_n(t)$ is not included to the above equation because it is either a mono-tonic function or a constant. Equation (3) shows that the amplitude and frequency can be expressed as functions of time and can strengthen the time-frequency distribution of the amplitude. This distribution $H(\omega, t)$ is called the Hilbert spectrum.

3   Response Spectrum and Strong Motion Duration

3.1   Response Spectrum

The response spectrum is an essential tool in earthquake engineering and is the plot of the peak response acceleration, displacement or velocity of a single-degree-of-freedom oscillator in dependence of its fundamental period under the same seismic excitation. Figure 2 demonstrates the concept of the earthquake response spectrum.

3.2   Strong Motion Duration (SMD)

A major issue of the spline fitting arises at the boundaries since cubic splines have wide swings if they are left unattended [2]. To eliminate end effects, only the strong motion duration of the seismic excitation is considered in the proposed methodology. The strong motion duration (SMD) is the time of the earthquake where the most seismic energy is released. The SMD after Trifunac/Brady [10] is defined as time elapsed between 5% and 95% of the Husid diagram [11] and is defined as follows:

$$T_{0.90} = T_{0.95} - T_{0.05} \tag{4}$$

where $T_{0.90}$ is the SMD and $T_{0.95}$ and $T_{0.05}$ are the time elapsed at the 95% and 5% of the Husid diagram, respectively.

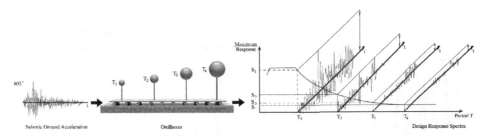

Fig. 2. Concept of the earthquake response spectrum

4   Overview of the Proposed Method

According to the proposed methodology, the original signal is subjected to the HHT so as to obtain the n amplitude and frequency components. The frequency components of the SMD are submitted to a smoothing procedure and addition of random Gaussian noise. The amplitude components of the SMD are also modified. The artificially generated accelerogram $AGA(x, t)$ is derived by reconstructing the seismic signal from the new modified instantaneous frequency ω_{jNew} and amplitude a_{jNew} components:

$$AGA(x,t) = Re \sum_{j=1}^{n} (a_{jNew}(t)) e^{j \int \omega_{jNew}(t)dt} \tag{5}$$

Each new amplitude component is given by:

$$a_{jNew}(t) = x_k a_{jOld}(t), \quad k = 1, 2, ..., n \tag{6}$$

where, n is the number of extracted IMFs of the initial seismic signal, x is a vector containing n scaling parameters of values randomly distributed in the interval [0.8-1.3] and $a_{jOld}(t)$ is the initial amplitude value at time t.

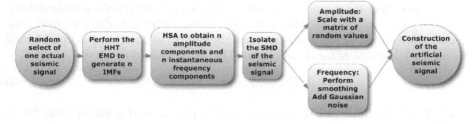

Fig. 3. Block diagram of the proposed method

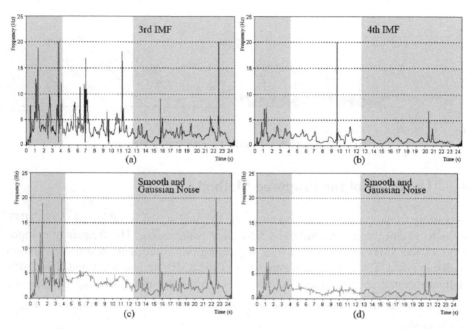

Fig. 4. (a) The 3rd and (b) 4th IMF of Victoria, Mexico time history record after HSA and (c), (d) after smoothing and addition of Gaussian noise

The new frequency content is obtained after a two step procedure performed to the frequency components of the natural seismic signal extracted from the EMD. This procedure is applied only to the IMFs that have in the SMD, mean frequency value lower than 5 Hz. This usually occurs after the third IMF and above. The rest of the IMFs remain unaffected. The response spectrum acceleration is determined in period range 0.2 s to 4 s (0.25 Hz to 5 Hz). This range covers the majority of low-, mid- and high-buildings [12]. The first three IMFs are of high frequency, thus of low eigenperiod, so there is no meaning of changing them as they do not contribute to the approximation of the spectrum at the eigenperiods of our interest.

The first step of the procedure employs a smoothing process to the SMD of the appropriate frequency IMFs by utilizing a moving average filter. At a second stage, Gaussian noise is added to create new frequency components that could resemble to the frequency components that were derived from the initial signal. Figure 3 shows the block diagram of the proposed method.

Figures 4(a) and 4(b) show two frequency components of the earthquake in Victoria Mexico. The smoothing and addition of Gaussian noise are performed to the SMD of the earthquake signal and the new components are generated and depicted in Figures 4(c) and 4(d). For the specific earthquake signal the total duration is 24.44 s while the strong motion duration is between 4.15 s and 12.67 s.

5   Empirical Results

The experimental results presented in this section include a comparison of the proposed method with similar state of the art techniques that are widely used for generating artificial spectrum-compatible accelerograms. Database, numerical results and system setup used in the validation process are described in the following sections.

5.1   Experimental Setup

In order to assess the performance of the proposed method, an experimental comparison is performed with respect to two other supervised approaches, invoked by the programs SIMQKE [7] and SeismoArtif [9]. SIMQKE is based on Gasparini and Vanmarche algorithm [6]. It estimates a power spectral density function from a specific response spectrum and generates the amplitudes of sinusoidal signals that have phase angles randomly distributed between 0 and 2π. The SeismoArtif, provides the Synthetic Accelerogram Generation and Adjustment method, based on Hallodorson and Papageorgiou algorithm [8]. From a synthetic accelerogram simulated by the user, the spectrum-compatible artificial accelerogram is defined by adapting its frequency content using the Fourier Transform.

The Root Mean Square Error (RMSE) between the response spectrum of the generated accelerogram and the spectrum of the natural accelerogram is calculated in all experiments:

$$ RMSE = \sqrt{\frac{1}{N} \sum_{j=1}^{n} (SA_{generated} - SA_{target})^2} \tag{7} $$

where, N is the number of samples. The calculation of the RMSE is utilized for the numerical validation of the compatibility between the response spectra of the original and generated seismic accelerograms.

5.2   Performance Evaluation

Table 1 includes all information regarding the seismic events [13], that are selected to test the method, and the values of RMSE for the proposed method (PM), the SIMQKE (M1) and the SeismoArtif (M2).

Table 1. Numerical results of the method for 20 seismic events

Seismic event	Country	Date	Station / Component	RMSE PM	RMSE M1	RMSE M2
Mammoth Lakes	U.S.A.	05/27/80	FIS / 000	$1.80 \cdot 10^{-2}$	$1.46 \cdot 10^{-2}$	$5.09 \cdot 10^{-2}$
Dinar	Turkey	01/10/95	Dinar / 090	$4.94 \cdot 10^{-2}$	$2.45 \cdot 10^{-2}$	$4.89 \cdot 10^{-2}$
Victoria	Mexico	06/09/80	Station 6604 / 045	$2.04 \cdot 10^{-2}$	$3.08 \cdot 10^{-2}$	$3.43 \cdot 10^{-2}$
Gazli	Uzbekistan	05/17/76	Karakyr / 000	$3.79 \cdot 10^{-2}$	$1.38 \cdot 10^{-2}$	$3.60 \cdot 10^{-2}$
Tabas	Iran	09/16/78	Tabas / LN	$9.12 \cdot 10^{-2}$	$1.17 \cdot 10^{-1}$	$7.81 \cdot 10^{-2}$
Spitak	Armenia	12/07/88	Gukasian / 000	$1.06 \cdot 10^{-2}$	$1.31 \cdot 10^{-2}$	$4.69 \cdot 10^{-2}$
Morgan Hill	U.S.A.	04/24/84	Station 1180 / 000	$8.86 \cdot 10^{-3}$	$3.84 \cdot 10^{-3}$	$1.52 \cdot 10^{-2}$
Patra	Greece	07/14/93	Station 00001 / NS	$5.05 \cdot 10^{-2}$	$1.31 \cdot 10^{-1}$	$5.83 \cdot 10^{-2}$
Pyrgos	Greece	03/26/93	Station 00001 / NS	$4.07 \cdot 10^{-2}$	$5.33 \cdot 10^{-2}$	$7.28 \cdot 10^{-2}$
Loma Prieta	U.S.A.	10/18/89	Santa Cruz / EW	$2.60 \cdot 10^{-2}$	$4.96 \cdot 10^{-2}$	$3.46 \cdot 10^{-2}$
Big Bear	U.S.A.	06/28/92	Lake-Civic Center / 270	$2.94 \cdot 10^{-2}$	$2.55 \cdot 10^{-2}$	$5.29 \cdot 10^{-2}$
Dinar	Turkey	01/10/95	Balikesir / 090	$2.29 \cdot 10^{-4}$	$3.10 \cdot 10^{-4}$	$1.49 \cdot 10^{-2}$
Erzican	Turkey	03/13/92	Erzican / NS	$1.57 \cdot 10^{-2}$	$2.66 \cdot 10^{-2}$	$3.53 \cdot 10^{-2}$
Friuli	Italy	05/06/76	Barcis / 270	$1.43 \cdot 10^{-3}$	$1.51 \cdot 10^{-3}$	$1.10 \cdot 10^{-2}$
Imperial Valley	U.S.A.	10/15/79	Bonds Corner / 230	$5.68 \cdot 10^{-3}$	$4.56 \cdot 10^{-3}$	$3.25 \cdot 10^{-2}$
Kobe	Japan	01/16/95	Fuk / 090	$7.17 \cdot 10^{-3}$	$3.31 \cdot 10^{-3}$	$3.16 \cdot 10^{-2}$
Loma Prieta	U.S.A.	10/18/89	Apeel 10 Skyline / 090	$9.47 \cdot 10^{-3}$	$2.52 \cdot 10^{-2}$	$7.48 \cdot 10^{-2}$
Mammoth Lakes	U.S.A.	05/27/80	Convict Creek / 090	$1.20 \cdot 10^{-2}$	$9.88 \cdot 10^{-3}$	$2.24 \cdot 10^{-2}$
Northridge	U.S.A.	03/20/94	Anacapa Island / 270	$1.24 \cdot 10^{-3}$	$3.97 \cdot 10^{-4}$	$2.53 \cdot 10^{-2}$
San Fransisco	U.S.A.	03/22/57	Golden Gate / 100	$4.52 \cdot 10^{-3}$	$8.96 \cdot 10^{-3}$	$2.07 \cdot 10^{-2}$

It can be observed that all RMSE values are very close in all methods. However, the PM leads to spectrum compatible artificial seismic accelerograms with lower RMSE in most experiments. The PM can be directly compared with the two well-known procedures and be advantageous since is not an iterative method. The RMSE calculated for the two standard methods is the average of ten measurements derived from different number of iteration (1, 3, 5, 7, 9, 11, 13, 15, 20 and 25) while the RMSE for the PM is the average of ten measurements of only one iteration each. In Table 2 it is recorded the RMSE for six experiments with the PM, and the number of iterations that need both M1 and M2 in order to obtain approximately the same range of values for the RMSE. The PM performs one iteration to achieve that range of RMSE, while at least three iterations are required for M1 and over five for M2.

Figure 5 shows the natural seismic signal of Victoria, and the generated seismic signal with the PM, M1 and M2. The artificial signals generated by the two programs and the PM are of the same duration and compatible with the same target spectrum. SIMQKE generates artificial signals by varying the random phase angles of the sinu-soidal signals and all generated signals are restricted to have a uniform shape that is

Table 2. RMSE toward number of iterations for the proposed method, SIMQKE and SeismoArtif for six seismic events

Seismic event (Station)	Method	RMSE	Number of iterations
Mammoth Lakes (FIS)	PM	$1.80 \cdot 10^{-2}$	1
	M1	$5.76 \cdot 10^{-2}$	11
	M2	$1.74 \cdot 10^{-2}$	19
Dinar (Dinar)	PM	$4.94 \cdot 10^{-2}$	1
	M1	$3.60 \cdot 10^{-2}$	3
	M2	$4.58 \cdot 10^{-2}$	7
Victoria (Station 6604)	PM	$2.04 \cdot 10^{-2}$	1
	M1	$2.25 \cdot 10^{-2}$	19
	M2	$2.77 \cdot 10^{-2}$	25
Gazli (Karakyr)	P.M	$3.79 \cdot 10^{-2}$	1
	M1	$4.12 \cdot 10^{-2}$	3
	M2	$3.87 \cdot 10^{-2}$	11
Tabas (Tabas)	PM	$9.12 \cdot 10^{-2}$	1
	M1	$9.21 \cdot 10^{-2}$	7
	M2	$9.27 \cdot 10^{-2}$	5
Spitak (Gukasian)	PM	$1.06 \cdot 10^{-2}$	1
	M1	$1.04 \cdot 10^{-2}$	5
	M2	$1.08 \cdot 10^{-2}$	25

provided by a predetermined envelop. In SeismoArtif, the generation of the synthetic accelerogram starts from a Gaussian white noise that is multiplied by the envelop shape suggested by Saragoni and Hart [14]. Both methods derive accelerograms that follow predetermined envelop shapes, thus, accelerograms generated with either of the two methods tend to resemble among them.

In order to demonstrate the ability of the PM to generate multiple different accelerograms, the standard deviation (SD) is used as a metric. Figure 6 demonstrates the SD between ten artificially generated seismic signals with the PM and the mean of them, starting from the natural record of Victoria, in contrast to the SD of ten artificially generated seismic signals with M1 and M2 and the mean of them. During the SMD of the signal, the SD is greater with the PM rather than with M1 and M2.

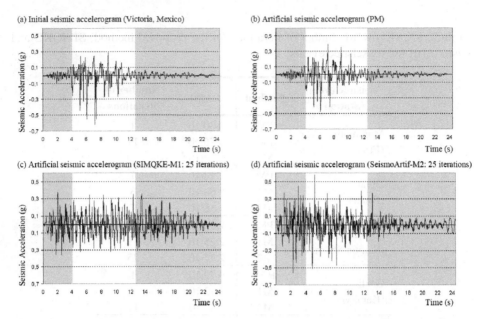

Fig. 5. (a) Victoria, Mexico time history record (b) artificially generated accelerogram with the PM (c) artificial seismic signal derived from M1 (d) artificial seismic signal derived from M2

Figure 7(a) shows the target spectrum and the mean spectra of the proposed and the two established methods evaluated with the programs SIMQKE and SeismoArtif. All spectra are compatible with the target spectrum and have slight differences (Table 1).

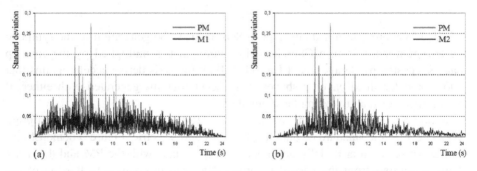

Fig. 6. Standard deviation between ten artificially generated signals (compatible with the Victoria spectrum) and the mean of them, using (a) the PM and SIMQKE (M1) (b) the PM and SeismoArtif (M2)

The mean spectrum is the average of ten artificially generated signals. The mean spectrum of the PM is evaluated from ten measurements with different amount of smoothing by varying the span of the moving average filter and adding more noise to the components. The mean spectrum of M1 and M2 stem from ten measurements with different number of iterations.

Finally, all generated artificial seismic accelerograms with the PM follow the restrictions of Eurocode 8 [15]. According to Eurocode 8, in the range of periods between $0.2T_1$ and $2T_1$ (T_1 is the fundamental period of the structure) no value of the mean 5% damping elastic spectrum, calculated from all time histories, should be less than 90% of the corresponding value of the 5% damping elastic response spectrum. In Figure 7(b) are demonstrated the 90% of the target response spectrum and the mean spectra of four artificially generated seismic signals with the PM.

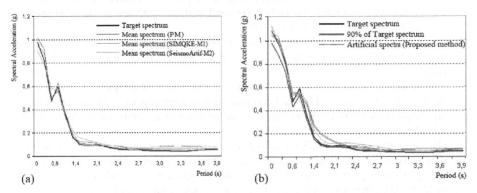

Fig. 7. (a) Target spectrum and mean spectra of the PM, M1 and M2 (b) target spectrum, spectra of four artificial generated signals with the PM and 90% of the target spectrum

6   Conclusions

Artificial spectrum-compatible accelerograms are used instead of natural acceleration records for the dynamic analysis of many critical structures for sites with no strong ground motion data. A new method for generating spectrum-compatible seismic accelerograms based on the HHT is proposed. One randomly selected seismic event is decomposed in amplitude and frequency components. The amplitude components of the SMD are scaled and the appropriate frequency components of the SMD are subjected to smoothing and addition of Gaussian noise.

The PM is flexible, since it requires only one seismic event, that can be from any region globally and of any intensity. This option makes the algorithm faster without affecting its accuracy by choosing seismic accelerograms with a definite number of points. Moreover, in order to analyze the non-stationary data only a small number of IMFs is required.

Comparison with two other state of art methods, one based on Gasparini and Vanmarcke algorithm (calculated by SIMQKE) and the other based on Hallodorson and Papageorgiou algorithm (calculated by SeismoArtif), has shown the satisfactory performance of the PM. The introduction of SD revealed that the PM can derive multiple different accelerograms. The proposed method is not an iterative method and all generated signals follow the restrictions of the Eurocode 8.

References

1. Carballo, J.E.: Probabilistic seismic demand analysis: spectrum matching and design. Doctoral dissertation, Stanford University, Stanford, California, USA (2000)
2. Huang, N.E., Shen, Z., Long, S.R., et al.: The empirical mode decomposition and the Hilbert spectrum for nonlinear and non-stationary time series analysis. Proceedings of the Royal Society A 454, 903–995 (1998)
3. Alvanitopoulos, P., Andreadis, I., Elenas, A.: Interdependence between seismic parameters and damage indicies based on Hilbert-Huang Transform. IOP Journal of Measurements Science and Technology 21, 1–14 (2010)
4. Ni, S., Xie, W., Pandey, M.: Application of Hilbert-Huang transform in generating spectrum-compatible earthquake time histories, ISRN Signal Processing, Article ID 563678 (2011)
5. Vrochidou, E., Alvanitopoulos, P., Andreadis, I., Elenas, A.: Correlation between seismic intensity parameters of HHT-based synthetic seismic accelerograms and damage indices of buildings. In: 8th IFIP Int. Conf. on Artificial Intelligence & Applications (AIAI), Chalkidiki Greece, September, pp. 1–6 (2012)
6. Gasparini, D., Vanmarcke, E.H.: SIMQKE – A program for artificial motion generation, user's manual and documentation. Publication No R76-4, Massachusetts Institute of Technology, Civil Engineering Department, Cambridge Massachusetts (1976)
7. Gelfi, P.: SIMQKE_GR version 2.7. University of Bescia Italy (2012), http://dicata.ing.unibs.it/gelfi/software/simqke/simqke_gr.htm
8. Halldorsson, B., Papageorgiou, A.S.: Calibration of the specific barrier model to earthquake of different tectonic regions. Bulletin of the Seismological Society of America 95, 1276–1300 (2005)
9. Computer program SeismoArtif, version 1.0.0.: 'SeismoArtif's help system-© 2002-2012', Seismosoft Ltd.
10. Trifunac, M., Brady, A.G.: A study on the duration of strong earthquake ground motion. Bulletin of the Seismological Society of America 65, 581–626 (1975)
11. Husid, R.: Analisis de terremoros: analisis general. Rev. IDIEM 8, 21–42 (1969)
12. Bommer, J.J., Elnashai, A.S.: Displacement spectra for seismic design. Journal of Earthquake Engineering 3, 1–32 (1999)
13. PEER Ground Motion Database, Pacific Earthquake Engineering Research Center, http://peer.berkeley.edu/peer_ground_motion_database
14. Saragoni, G.R., Hart, G.C.: Simulation of artificial earthquakes. Earthquake Engineering and Structural Dynamics, 219–267 (1974)
15. EC8, Eurocode 8: Design of Structures for Earthquake Resistance- Part 1: General Rules, Seismic Actions, and Rules for Buildings. European Committee for Standardization, Brussels, Belgium (2004)

Peak Observer Based Self-tuning
of Type-2 Fuzzy PID Controllers

Engin Yesil, Tufan Kumbasar, M. Furkan Dodurka, and Ahmet Sakalli

Istanbul Technical University, Faculty of Electrical and Electronics Engineering,
Control and Automation Engineering Department, Maslak, TR-34469, Istanbul, Turkey
{yesileng,kumbasart,dodurkam,sakallia}@itu.edu.tr

Abstract. Fuzzy PID (proportional-integral-derivative) controllers are commonly used as an alternative to the conventional PID controllers. In order to improve the control system performance of these controllers many self-tuning methods are already studied. It is mostly observed that the self-tuning mechanism should tune the scaling factors of the fuzzy controller to enhance the transient system performance. On the other hand, these studies only focus on the ordinary (Type-1) Fuzzy PID controllers. In this study, Type-2 Fuzzy PID controllers are studied and a peak observer based self-tuning method is proposed for these controllers. In order to show the benefit of the proposed approach, several Matlab simulations are performed where different type of fuzzy control structures are compared. The results obtained from the simulation studies clearly show the advantage of the proposed approach.

Keywords: Type-2 Fuzzy PID controllers, self-tuning, peak observer.

1   Introduction

Ordinary (type-1) Fuzzy PID (proportional-integral-derivative) controllers (FPID) are accepted as an alternative to conventional PID controllers since they are analogous to the PID controllers from the input-output relationship point of view [1-3]. Numerous techniques have been developed in the literature for analyzing and designing a wide variety of Type-1 Fuzzy PID (T1-FPID) control systems [3-6]. The design parameters of the T1-FPID controllers can be summarized within two groups, structural parameters and tuning parameters [6]. The structural parameters include input/output variables to fuzzy inference such as fuzzy sets, Membership Functions (MFs) shapes, rules and inference mechanism. Tuning parameters include input/output Scaling Factors (SFs) and the parameters of the MFs. Starting with the pioneering study by Qiao and Mizumoto [3], various online tuning mechanisms have been presented to improve the control performance of the fuzzy control system in presence of parameter variations and nonlinearities [7-10]. Thus, Self-Tuning Type-1 Fuzzy PID (STT1-FPID) controllers have been proposed where the SFs or the parameters of MFs have been adjusted in an online manner [7-12]. However, the main research focus was on the tuning of the SFs since their effect on the system's response can easily be

L. Iliadis et al. (Eds.): AIAI 2014, IFIP AICT 436, pp. 487–497, 2014.

observed [8-9]. The presented self-tuning mechanisms for the SFs provide extra degrees of freedom to the fuzzy control structure, and also enhance new tuning structures which have to be determined as updating functions or extra fuzzy inference mechanisms.

Recently, researchers began investigating Interval Type-2 Fuzzy Logic Controllers (IT2-FLCs) which have demonstrated significant control performance improvements in comparison to its type-1 counterpart [13-15]. It has been shown in various works that the T1-FPID controllers using Type-1 Fuzzy Sets (T1-FSs) might not be able to fully handle the high levels of uncertainties associated with control applications while the Interval Type-2 Fuzzy PID (IT2-FPID) controller using Interval Type-2 Fuzzy Sets (IT2-FSs) might be able to handle such uncertainties to produce a better control performance [15-17]. It has been shown that IT2-FPIDs achieve better control performances because of the additional degree of freedom provided by the Footprint of Uncertainty (FOU) in their antecedent MFs [14-15, 18]. In literature, the design of the IT2-FPID controllers is usually solved by extending MFs of an existing T1-FPID controller or by employing optimization algorithms [19-21].

In this study, the peak observer (PO) based self-tuning method, which was first proposed for T1-FPID controllers, is extended for IT2-FPID controllers. Till now, there is a very limited research on the self-tuning methods for the Type-2 Fuzzy Logic Controllers. The benefit of this approach is examined via various simulation studies performed on First Order plus Dead Time (FOPDT) systems. The simulation results clearly show that IT2-FPID controller achieve a better performance than T1-FPID and PO based T1-FPID controllers. When the proposed self-tuning approach is employed to the IT2-FPID controller the control system performance increases significantly.

2   The Design of Fuzzy PID Controllers

Interval Type-2 Fuzzy PID controllers are generalized forms of Type-1 Fuzzy PID controllers, which are mostly mentioned as an alternative to conventional PID controllers since they are analogous to the PID controllers from the input-output relationship point of view [1-3]. PID controllers are widely used in many control areas such as process control, adaptive control, robust control, nonlinear control, automation and robotics and they are still the most used controller type in the industry [22]. The design strategies of conventional PID controllers depends on the mathematical model of the system, however conventional PID controllers may not provide satisfactory results in case of imperfect model. Conventional PID controllers provide a linear transformation or mapping between their inputs and outputs while Fuzzy PID controllers provide a nonlinear transformation or mapping between their inputs and outputs.

In literature, various fuzzy controller structures are proposed including PD-type (Proportional-Derivative), PI-type (Proportional-Integral) and PID-type [4, 9]. In this context, fuzzy controllers are constructed by choosing the inputs to the error (e) and derivative of the error ($\dot{e}$) and the output as the control signal (u). The inputs of Fuzzy Logic Controllers (FLCs) (regardless to the type of the controller) is as follow

$$e = r - y \qquad (1)$$

$$\dot{e} = \frac{de}{dt} = \frac{d(r - y)}{dt} \qquad (2)$$

where r is the reference signal or setpoint and y is the system output. In fuzzy controller structures (regardless to the type of the controller), input SFs normalize the inputs to the universe of discourse, particularly, K_e normalizes the error signal and K_d normalize the derivative of error signal as follow

$$E = K_e e, \quad E \in [-1,1] \qquad (2)$$

$$\dot{E} = K_d \dot{e}, \quad \dot{E} \in [-1,1] \qquad (3)$$

where E is the normalized error signal of the FLC and $\dot{E}$ is the normalized derivative of error signal of the FLC. Here, the universe of discourses of the inputs (E and $\dot{E}$) is defined in the interval $[-1, 1]$. The FPID controller structure including inputs, SFs and the FLC (T1-FLC or IT2-FLC) is illustrated as in Fig. 1. K_a is a constant gain, and K_b is the integral gain of the output of the FLC. Here also notice that, output SFs make the control signal applicable for a particular problem or a real life problem because the output of the FLC could only take values in the interval $[-1,1]$ due to its universe of discourse of the output MFs. Then the generalized formulation of the output of a FLC that used in this study is as follows:

$$u = K_a U + K_b \int U \, dt, \quad U \in [-1,1] \qquad (4)$$

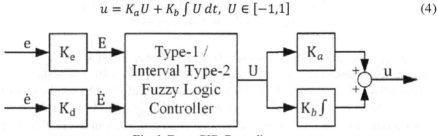

Fig. 1. Fuzzy PID Controller

2.1   The Type-1 Fuzzy PID Controller Structure

In this subsection, the internal structure and the design strategies of the employed T1-FPID controllers are presented. The rule structure is as follows:

$$R_q: \text{ If } E \text{ is } A_{1j} \text{ and } \dot{E} \text{ is } A_{2j} \quad \text{THEN} \quad U \text{ is } C_q,$$
$$i, j = 1,2,3; \quad q = 1, \dots, Q = 9 \qquad (5)$$

where A_{1j} is and A_{2j} are the antecedent MFs for the inputs E and $\dot{E}$, respectively, C_q is the consequent crisp set and Q is the number of rules.   In the handled T1-FPID structure, a symmetrical 3x3 rule base is used as shown in Fig. 2a. Here, three Type-1 Fuzzy Sets (T1-FSs) for each input domain (E and $\dot{E}$) are used and they are denoted as N (Negative), Z (Zero) and P (Positive). The T1-FSs of the T1-FLC are defined in [-1 1] interval as shown in Fig. 2b. The consequent part is defined with five singleton consequents, as shown in Fig. 2c. The implemented T1-FLCs use the product implication and the weighted average defuzzification method.

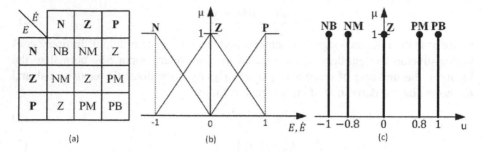

(a)   (b)   (c)

Fig. 2. (a) The fuzzy rule base, (b) the antecedent MFs, (c) the consequent MFs used for the FPID controllers

2.2   The Interval Type-2 Fuzzy PID Controller Structure

In this subsection, the general structure of the employed IT2-FPID controller is presented. In the design of the IT2-FPID, the same symmetrical 3x3 rule base given in Fig. 2a and consequent MFs of T1-FPID illustrated in Fig. 2c are used. Thus, the rule structure of the IT2-FLC is as follows:

$$R_q: \text{ IF } E \text{ is } \tilde{A}_{1i} \text{ and } \dot{E} \text{ is } \tilde{A}_{2j} \text{ THEN } U \text{ is } C_q,$$
$$i, j = 1,2,3; \quad q = 1, \dots, Q = 9 \tag{6}$$

where C_q is the consequent crisp set, Q is the number of rules and $\tilde{A}_{1i}$ and $\tilde{A}_{2j}$ are the antecedent Interval Type-2 Fuzzy Sets (IT2-FSs). We will denote the IT2-FSs as N (Negative), Z (Zero) and P (Positive). The antecedent IT2-FSs can be described in terms of upper MFs ($\overline{\mu}_{\tilde{A}_{1i}}$ and $\overline{\mu}_{\tilde{A}_{2i}}$) and lower MFs ($\underline{\mu}_{\tilde{A}_{1i}}$ and $\underline{\mu}_{\tilde{A}_{2i}}$) to form the Foot of Uncertainty (FOU), which provides extra degree of freedom in IT2-FSs. The antecedent IT2-FSs are defined again in [-1 1] interval. Thus, the FOU will be created by the heights of the lower MFs (m_1 and m_2) of the IT2-FLC. The implemented IT2-FLC uses the center of sets type reduction/ defuzzification method [23].

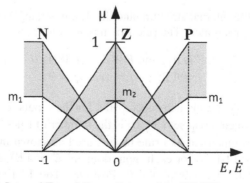

Fig. 3. Interval Type-2 Fuzzy Sets used for IT2-PID controller

3   The Peak Observer Based Self-tuning Method

As shown in Fig. 1, the actual control signal of the closed loop system is a function of scaling factors and the output of the FLC as $u = f(U, K_a, K_b)$ or $u = f(U, \int U, K_a, K_b)$. The FLC performs a nonlinear mapping between its inputs and output, so the output of the FLC can also be expressed as a function of its inputs namely, error and derivative of the error as $U = f(E, \dot{E})$ or $U = f(e, \dot{e}, K_e, K_d)$. Note that $f(.)$ represents a functional mapping. Consequently, the FPID controller structure becomes analogous to conventional PID structure. This analysis was examined in detail by considering the input space into several subspaces and mathematical expression of MFs [3] and the control signal can be written as

$$u = K_a K_0 + K_b K_0 t + (K_a K_e K_1 + K_b K_d K_2)e + (K_b K_e K_1) \int e \, dt + (K_a K_d K_2)\dot{e} \quad (7)$$

where K_0, K_1 and K_2 are the constants which are calculated with respect to antecendent and consequent membership functions. They are calculated with respect to values of consequent sets and apexes of the antecedent membership functions for corresponding inputs and subspaces. Here, K_0 is a nonlinear term, which converges to zero by time. So one can say that a FPID controller is similar to a conventional PID controller with variable gains. Then after neglecting the nonlinear term for the steady state, the PID equivalent control components of FPID can be achieved as follows

$$K_p = K_a K_e K_1 + K_b K_d K_2 \quad (8)$$

$$K_I = K_b K_e K_1 \quad (9)$$

$$K_D = K_a K_d K_2 \quad (10)$$

where K_p is the proportional gain, K_I is the integral gain and the K_D is the derivative gain of the fuzzy controller.

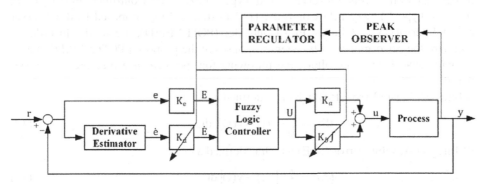

Fig. 4. Peak observer based self-tuning for Fuzzy PID controllers

The peak observer based self-tuning method for T1-FPID controllers has been proposed in [3]. The block diagram of the proposed method is shown in Fig. 4. The peak observer observes the output of the system and transmits a signal at each peak time

and measures the absolute peak value. The parameter regulator tunes the controller
parameters K_d and K_b simultaneously at each peak time according to the peak value.
The algorithm for tuning these parameters is as follows:

$$K_d = \frac{K_{d_s}}{\delta_k}, K_b = \delta_k K_{b_s}$$ (11)

where K_{d_s} and K_{b_s} are the initial values of K_d and K_b, δ_k are the peak values. It can
easily be deduced from the relation given in (9) that if in the meanwhile of
ing K_b, K_d is increased in the same rate as K_b is decreased, the equivalent proportional control strength will remain unchanged. Then, the system can always keep
quick reaction against the error under this condition. This is achieved by updating the
integral coefficient as the reciprocal of the derivative coefficient. As it is seen from
(12), the integral factor is decreased in a direct proportion to the peak error value.

4   Simulation Studies

In this section, the performance of the proposed self-tuning method for IT2-FPID
controllers is examined via various cooperative simulations. For this purpose, different processes are used for simulations but as it is well-known a large number of industrial plants can be modeled by a first-order plus dead time (FOPDT) transfer function.
The wide use of FOPDT is due both to its simplicity as well as its ability to capture
the essential dynamics of several industrial processes. Therefore, the results obtained
only for FOPDT process are presented here. The transfer function of FOPDT is as
follows:

$$G(s) = \frac{K}{Ts+1} e^{-Ls}$$ (12)

where K is the process gain, T is the time constant and finally L is the time delay.

In the simulations, four different control schemes, namely Type-1 Fuzzy PID Controller (T1-FPID), Peak Observer based Type-1 Fuzzy PID Controller (POT1-FPID),
Interval Type-2 Fuzzy PID Controller (IT2-FPID), and the proposed Peak Observer
based Interval Type-2 Fuzzy PID Controller (POIT2-FPID), are used. In order to
compare the performance of transient responses of the proposed POIT2-FPID controller with other fuzzy controllers, the following four performance measures are considered:

i) Integral Square Error (ISE) which is defined as

$$\text{ISE} = \int_0^\infty (r(t) - y(t))^2 \, dt$$ (13)

ii) Integral Absolute Error (IAE) which is defined as

$$\text{IAE} = \int_0^\infty |r(t) - y(t)| \, dt$$ (14)

iii) Integral Time Absolute Error (ITAE) which is defined as

$$\text{ITAE} = \int_0^\infty t |r(t) - y(t)| \, dt$$ (15)

iv) Total Variation (TV) [11] of the control input u(t), which is defined as

$$TV = \sum_{i=1}^{\infty} |u_{i+1} - u_i| \qquad (16)$$

As the first step, the nominal process parameters are set to K=1, T=1 and L=0.2. The sampling time of the simulation is set to 0.05 s and noise is added to the process output to make the simulations more realistic. Then the input SFs of the T1-FPID controller are designed in order to have a response without an overshoot but small rise time and settling time. For this purpose, first the input scaling factor of the error (K_e) is calculated as

$$K_e = \frac{1}{r(t_f) - y(t_f)} \qquad (17)$$

where $r(t_f)$ and $y(t_f)$ are the values of the reference and system output at the time of the reference variation. Since the unit step changes are studied for the simulations K_e is set to 1, then the other scaling factors are determined as $K_d = 0.5141$, $K_a = 0.077$, $K_b = 7.336$ after optimization. As the second step, the IT2-FPID controller is design using the same scaling factors obtained for T1-FPID controller. For the design procedure only the height of the lower membership functions (m_1 and m_2) are designed. After the optimization procedure, m_1 and m_2 are set to 0.3 and 0.9, respectively. The step responses of the process for four different control schemes are illustrated in Fig. 5. As clearly seen from Fig. 5, the set-point following performance of all fuzzy controllers are very close to each other for the nominal process. Since there is no overshoot, the self-tuning mechanism did not tune the SFs.

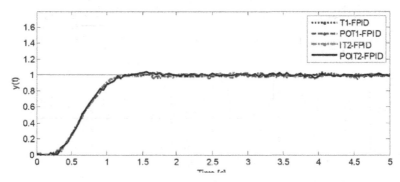

Fig. 5. The set-point following performance of the fuzzy controllers for the nominal process

In order to show the benefit of using IT2-PID controllers, and the proposed PO based Type-2 Fuzzy PID controller the system parameters of the process are perturbed. The step responses of four controller for the first perturbed process, which has the parameters as K=1.3, T=1.9, L=0.4, is given in Fig. 6a. For this case, IT2-FPID controller gives a better set-point following performance than the T1-FPID controller because of its extra degree of freedom. As it is seen in Fig. 6b and Fig. 6c, the two scaling factors (K_d and K_b) of the fuzzy controllers are tuned as the overshoots occurs, therefore the performance of the fuzzy controllers is improved drastically. The control signals illustrated in Fig. 6d, shows that IT2-FPID controller with peak-observer based self-tuning mechanism has a smoother control action in comparison to the other fuzzy controllers.

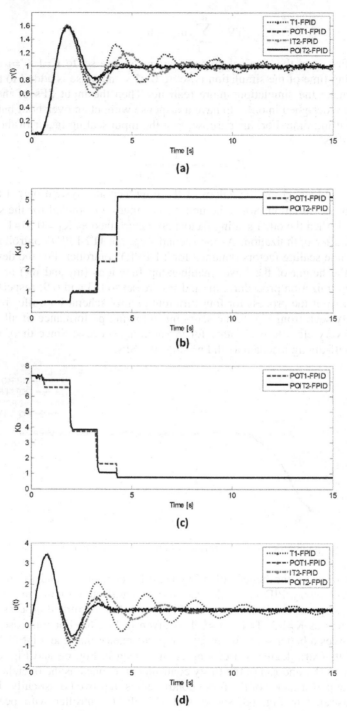

Fig. 6. The first the perturbed process (K=1.3, T=1.9, L=0.4): (a) The process output; (b) The change of Kb; (c) The change of Kd; (d) The control signals.

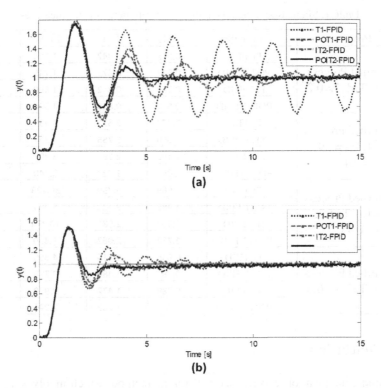

Fig. 7. The set-point following performance of the fuzzy controllers for the perturbed processes: (a) K=1.1, T=1.3, L=0.45; (b) K=3, T=3, L=0.3

The simulation results obtained for the first perturbed process show that IT2-FPID controller has a better set-point following performance than the T1-FPID controller because of FOU. Moreover, the peak observer based self-tuning mechanism improves the performance of the fuzzy controllers. The best performance from the point of transient response is obtained by the proposed IT2-FPID controller with the peak observer mechanism. Moreover, two more simulations are performed for two other perturbed processes and the control performances are given in Fig. 7. Parallel to the first simulation results, IT2-FPID controller always outperforms the T1-FPID controller. The Peak Observer based T1-FPID is more robust against parametric uncertainties; consequently the process output inherits smaller oscillations. When the proposed Peak Observer self-tuning method is employed to IT2-FPID controller, its performance again increases drastically.

The performance results of the fuzzy controllers are tabulated in Table 1. As it is seen from Table 1, the IT2-FPID controller is always better than the T1-FPID controller because of the additional degree of freedom provided by the FOU in their antecedent MFs. Also, the peak observer self-tuning methods improve the fuzzy control system performance. On the other hand, since the peak observer changes the scaling factors of the fuzzy controller, the control signal varies and thus results with higher TV values compared to the counterparts without a self-tuning mechanism.

Table 1. Simulation comparison using different performance measures

Process	Controller	ISE	IAE	ITAE	TV
Nominal Process K=1, T=1, L=0.2	T1-FPID	0.552	0.842	1.564	0.409
	POT1-FPID	0.547	0.872	1.745	0.970
	IT2-FPID	0.550	0.842	1.590	0.364
	POIT2-FPID	0.550	0.838	1.508	0.926
Perturbed Process K=1.3, T=1.9, L=0.4	T1-FPID	1.272	2.677	9.430	1.589
	POT1-FPID	1.050	1.755	3.357	1.401
	IT2-FPID	1.160	2.181	5.497	1.000
	POIT2-FPID	1.055	1.733	3.240	1.275
Perturbed Process K=1.1, T=1.3, L=0.45	T1-FPID	2.969	5.739	36.223	3.756
	POT1-FPID	1.457	2.423	5.659	1.690
	IT2-FPID	1.503	2.981	10.394	1.476
	POIT2-FPID	1.252	2.062	4.310	1.461
Perturbed Process K=3, T=3, L=0.3	T1-FPID	0.836	1.710	4.355	1.166
	POT1-FPID	0.789	1.488	3.278	1.290
	IT2-FPID	0.789	1.472	2.979	0.768
	POIT2-FPID	0.763	1.382	2.940	1.173

5   Conclusion

In this study, the peak observer based self-tuning method, which involves a peak observer and parameter regulator mechanisms, is extended to IT2-FPID Controllers. The advantage of the proposed approach is tested via Matlab simulations and the control system performance of the proposed method is compared with T1-PID controllers with and without a self-tuning mechanism, and also with an IT2-FPID controller without a self-tuning method. The obtained result clearly shows that peak-observer based self-tuning approach improved the performance in comparison to the IT2-FPID controller. As a future work, this approach will be implemented in a PLC and real-time experiments will be performed.

Acknowledgments. This research is supported by the Scientific and Technological Research Council of Turkey (TUBITAK) under the project (113E206). All of these supports are appreciated.

References

1. Galichet, S., Foulloy, L.: Fuzzy controllers: synthesis and equivalences. IEEE Transactions on Fuzzy Systems 3(2), 140–148 (1995)
2. Huang, T.T., Chung, H.Y., Lin, J.J.: A fuzzy PID controller being like parameter varying PID. In: IEEE International Fuzzy Systems Conference Proceeding, vol. 1, pp. 269–275 (1999)

3. Qiao, W.Z., Mizumoto, M.: PID type fuzzy controller and parameters adaptive method. Fuzzy Sets and Systems 78(1), 23–35 (1996)
4. Li, H.X., Gatland, H.B.: Conventional fuzzy control and its enhancement. IEEE Transactions on Systems, Man and Cybernetics Part B 26(5), 791–797 (1996)
5. Duan, X.G., Li, H.X., Deng, H.: Effective tuning method for fuzzy PID with internal model control. Industrial & Engineering Chemistry Research 47, 8317–8323 (2008)
6. Hu, B., Mann, G.K.I., Gasine, R.G.: New methodology for analytical and optimal design of fuzzy PID controllers. IEEE Transactions on Fuzzy Systems 7(5), 521–539 (1999)
7. Mudi, R.K., Pal, N.R.: A robust self-tuning scheme for PI- and PD-type fuzzy controllers. IEEE Transactions on Fuzzy Systems 7(1), 2–16 (1999)
8. Woo, Z.W., Chung, H.Y., Lin, J.J.: A PID-type fuzzy controller with self-tuning scaling factors. Fuzzy Sets Systems 115, 321–326 (2000)
9. Guzelkaya, M., Eksin, I., Yesil, E.: Self-tuning of PID-type fuzzy logic controller coefficients via relative rate observer. Engineering Applications of Artificial Intelligence 16, 227–236 (2003)
10. Ahn, K.K., Truong, D.Q.: Online tuning fuzzy PID controller using robust extended Kalman filter. Journal of Process Control 19, 1011–1023 (2009)
11. Karasakal, O., Guzelkaya, M., Eksin, I., Yesil, E.: An error-based on-line rule weight adjustment method for fuzzy PID controllers. Expert Systems and Applications 38(8), 10124–10132 (2011)
12. Karasakal, O., Guzelkaya, M., Eksin, I., Yesil, E., Kumbasar, T.: Online tuning of fuzzy PID controllers via rule weighting based on normalized acceleration. Engineering Applications of Artificial Intelligence 26(1), 184–197 (2013)
13. Wu, D., Tan, W.W.: Interval Type-2 Fuzzy PI Controllers: Why They Are More Robust. In: IEEE International Conference on Granular Computing, pp. 802–807 (2010)
14. Wu, D.: On the Fundamental Differences between Type-1 and Interval Type-2 Fuzzy Logic Controllers. Transactions on Fuzzy Systems 20(5), 832–848 (2012)
15. Hagras, H.: A Hierarchical Type-2 Fuzzy Logic Control Architecture for Autonomous Mobile Robots. IEEE Transactions on Fuzzy Systems 12(4), 524–539 (2004)
16. Castillo, O., Melin, P.: Type-2 Fuzzy Logic Theory and Applications. Springer, Berlin (2008)
17. Wu, D., Tan, W.W.: A simplified type-2 fuzzy logic controller for real-time control. ISA Transactions 45(4), 503–516 (2006)
18. Kumbasar, T., Eksin, I., Guzelkaya, M., Yesil, E.: Type-2 fuzzy model based controller design for neutralization processes. ISA Transactions 51(2), 277–287 (2012)
19. Kumbasar, H.: Hagras: Big Bang–Big Crunch optimization based interval type-2 fuzzy PID cascade controller design strategy. Inform. Sci. (2014), http://dx.doi.org/10.1016/j.ins.2014.06.005
20. Kumbasar, T.: A simple design method for interval type-2 fuzzy PID controllers. Soft Computing, 1–12 (2013)
21. Yesil, E.: Interval type-2 fuzzy PID load frequency controller using Big Bang-Big Crunch optimization. Applied Soft Computing 15, 100–112 (2014)
22. Astrom, K.J., Hagglund, T.: Advanced PID Control. ISA (2005)
23. Liang, Q., Mendel, J.M.: Interval type-2 fuzzy logic systems: theory and design. IEEE Transactions on Fuzzy Systems 8(5), 535–550 (2000)

TSK Fuzzy Modeling with Nonlinear Consequences

Jacek Kabziński and Jarosław Kacerka

Institute of Automatic Control, Lodz University of Technology
{jacek.kabzinski,jaroslaw.kacerka}@p.lodz.pl

Abstract. We propose to generalize TSK fuzzy model applying nonlinear functions in the rule consequences. We provide the model description and parameterization and discus the problem of model training and we recommend PSO for tuning parameters in membership functions and in nonlinear part of a rule consequence. We also propose some more or less formalized approach to nonlinear consequence selection and construction. Several examples demonstrate the main features of the proposed fuzzy models. The proposed approach reduces the average obtained model Root Mean Square Error (RMSE) with regard to the linear fuzzy model, as well that it allows to reduce the model complexity preserving the desired accuracy.

Keywords: fuzzy modeling, TSK fuzzy model, fuzzy model training.

1 Introduction

Among various fuzzy modeling techniques the approach proposed by Takagi, Sugeno and Kang [1, 2], so called TSK model, remains one of the most popular and effective. The standard TSK model employs affine functions in consequences of fuzzy rules. So the idea of modeling is based on local linear models dominating if the rule activation strength is high. The model possesses the universal approximation property [3] but in many practical problems the number of rules to achieve the desired accuracy is really high. Motivated by many experiences from modeling mechatronic systems (see example 2 below) we claim that it is possible to propose local nonlinear models in many practical applications. Having some knowledge on the nature of the modeled phenomenon we may foresee nonlinear functions that should appear in the model, even if we are not sure about the exact parameters of these functions.

In this contribution we propose to generalize TSK model by accepting nonlinear functions of model inputs in fuzzy rule consequences. We propose a formal model description and parameterization. Next we consider the problem of model training from the numerical data, which is more difficult than in the standard case. We also propose some formal methods to develop nonlinear consequences. The aim of the proposed modification is to reduce the model complexity preserving the desired accuracy, and we demonstrate on several examples that that it really happens.

The idea of using nonlinear functions in fuzzy rule consequences was already mentioned in literature, but only particular cases were considered. In [4] switched Takagi-Sugeno models with an affine nonlinear consequent part are used to control

L. Iliadis et al. (Eds.): AIAI 2014, IFIP AICT 436, pp. 498–507, 2014.
© IFIP International Federation for Information Processing 2014

switched nonlinear dynamical systems and in [5] it is proved that under some special assumptions such fuzzy model can approximate a particular class of nonlinear functions, nonlinear dynamic systems and nonlinear control systems.

2   Fuzzy TSK System with Nonlinear Consequences

We consider a fuzzy, single output, multi input model given by the rules

$$R_i: IF \ (x_1 is \ \mu_{i1}) and \ \dots \ and \ (x_n is \ \mu_{in}) \ THEN \ y \ is \ y_i = f_i(x_1, \dots, x_n), \qquad (1)$$

where $x = (x_1, x_2, \dots, x_n)^T$ is the vector of inputs, $i = 1, \dots, R$ is the rule number, μ_{ij} are membership functions, each defined by three parameters $\bar{\mu}_{ij}, \bar{\bar{\mu}}_{ij}, \breve{\mu}_{ij}$, and y_i are consequences

$$f_i(x) = f_i(x_1, x_2, \dots, x_n) = a_{i,1}\varphi_{i,1}(x, p_{i,1}) + a_{i,2}\varphi_{i,2}(x, p_{i,2}) + \cdots + \\ a_{i,m_i}\varphi_{i,m_i}(x, p_{i,m_i}). \qquad (2)$$

Parameters of the consequences are organized as follows: $p_{i,j}$ is a $s_{i,j}$ –dimensional vector parameterizing nonlinear function $\varphi_{i,j}$ while $a_{i,j}$ are scalars. For each rule we define its activation level as

$$\alpha_i(x) = \prod_{j=1}^{n} \mu_{ij}(x_j) \qquad (3)$$

and denote

$$\sigma(x) = \sum_{i=1}^{R} \alpha_i(x) \qquad (4)$$

The model output is calculated from

$$y(x) = \frac{1}{\sum_{i=1}^{R} \alpha_i(x)} \sum_{i=1}^{R} [\alpha_i(x) f_i(x)]. \qquad (5)$$

If we represent the vector of linear combination parameters as

$$a = \left[a_{1,1}, \dots, a_{1,m_1}, a_{2,1}, \dots, a_{2,m_2}, \dots, a_{R,1}, \dots, a_{R,m_R} \right]^T \qquad (6)$$

or

$$\bar{a}_i = \left[a_{i,1}, \dots, a_{i,m_i} \right]^T, \ a = \begin{bmatrix} \bar{a}_1 \\ \vdots \\ \bar{a}_R \end{bmatrix} \qquad (7)$$

and we refer to the vector of activated consequences:

$$v_i(x) = \alpha_i(x) \left[\varphi_{i,1}(x, p_{i,1}), \dots, \varphi_{i,m_i}(x, p_{i,m_i}) \right], v(x) = [v_1(x), \dots, v_R(x)], \qquad (8)$$

we get a short description of the system output:

$$y(x) = \frac{1}{\sigma(x)} v(x) a, \qquad (9)$$

which stresses that the model is linearly parameterized by a. This important feature will be utilized during model training and makes such model attractive in adaptive control applications.

If the same set of functions appears in each of the consequences, i.e.

$$m_1 = m_2 = \cdots = m_R = m, \varphi_{i,j}(x, p_{i,j}) = \varphi_j(x, p_j), j = 1,2, \dots, m \qquad (10)$$

we may describe the model output as

$$y(x) = \frac{1}{\sigma(x)} [\varphi_1(x, p_1), \quad \dots \quad , \varphi_m(x, p_m)][\bar{a}_1, \quad \dots \quad , \bar{a}_m] \begin{bmatrix} \alpha_1(x) \\ \vdots \\ \alpha_R(x) \end{bmatrix} \qquad (11)$$

and after accepting notation

$$z^T = [\varphi_1(x, p_1) \quad \varphi_2(x, p_2) \quad \cdots \quad \varphi_m(x, p_m)], \hat{A} = \begin{bmatrix} a_{1,1} & \cdots & a_{R,1} \\ \vdots & \ddots & \vdots \\ a_{1,m} & \cdots & a_{R,m} \end{bmatrix}$$

$$\xi = \begin{bmatrix} \alpha_1(x) \\ \vdots \\ \alpha_R(x) \end{bmatrix} \frac{1}{\sigma(x)} \qquad (12)$$

we have another short output description

$$y(x) = z^T \hat{A} \xi. \qquad (13)$$

Similarly to the classical TSK model with linear consequences our model may be represented as a multi-layer neural network. In the first layer actual values of the membership functions are calculated, in the second the activation level for the each rule is calculated, in the third layer normalized activation strength $\bar{\alpha}_i(x) = \frac{\alpha_i(x)}{\sigma(x)}$ of each rule is computed, in the fourth products $\bar{\alpha}_i f_i$ are obtained and the final node performs the summation $y = \sum_i \bar{\alpha}_i f_i$.

3   Model Training

The task to construct a TSK fuzzy model can be divided into two steps: first we have to design the model structure – recognize relevant inputs, decide about membership functions, plan the rules and the consequences etc. While the structure is established we optimize the system parameters based on some numerical data representing the desired behavior of the model – this stage of modeling is usually called model training or tuning.

In case of the proposed model with nonlinear consequences the model structure selection may be performed using any standard approach as placing membership functions on a grid, by clustering, or others [6]. The problem of nonlinear consequences selection is discussed separately. Here we concentrate on the model training methods.

The proposed model parameters may be divided into three sets: parameters of membership functions (parameters μ), parameters that appear nonlinearly in the consequences (parameters p) and parameters that are coefficients of linear combination in consequences (parameters a). The training data is collected in a M-dimensional target output vector $\bar{Y}$ and each entry $\bar{y}^i$ corresponds to an input vector x^i while the model output for the same input is y^i. The aim of training is to find model parameters minimizing the RMSE

$$J = \sqrt{\frac{1}{M}\sum_{i=1}^{M}(\bar{y}^i - y^i)^2} \to min. \tag{14}$$

Calculation of optimal parameters a, for given parameters μ and p, is trivial since according to (9) a is the solution in the least squares sense to system of equations

$$\bar{Y} = \Phi a, \quad \Phi = \begin{bmatrix} \frac{1}{\sigma(x^1)}v(x^1) \\ \vdots \\ \frac{1}{\sigma(x^M)}v(x^{M1}) \end{bmatrix}, \quad a = \Phi \backslash \bar{Y}. \tag{15}$$

Unfortunately the target function J in (14) is heavily nonlinear function of parameters μ and p. It may be nonsmooth and possess many local minima. So we propose to apply evolutionary optimization to find optimal model parameters. We have experimented with many evolutional optimization algorithms and finally we recommend PSO as well suitable for the problem. We use a well-known version as described in [8]. Parameters $[\mu, p]$ are coded as particles X. Each calculation of the fitting function for the given $[\mu, p]$ is done as follows: calculate Φ and a from (15), next calculate J from (14).

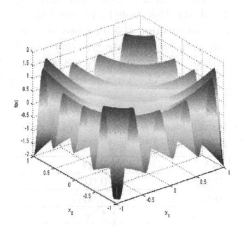
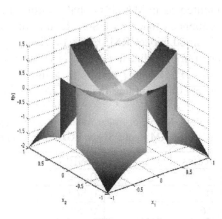

Fig. 1. Example function $f_1(x_1, x_2)$ **Fig. 2.** Example function $f_2(x_1, x_2)$

4   Examples

The proposed model training approach was tested on several modeling problems.
Example 1. Let us consider the following nonlinear functions defined for $x_1, x_2 \in$ [−1,1] and plotted in fig. 1 and 2

$$f_1(x) = (-x_1^2 - x_2^2) \cdot \cos(13 \cdot (x_1 + x_2)),   \tag{16}$$

$$f_2(x) = (-x_1^2 - x_2^2) \cdot sign(|x_1| - 0.2) \cdot sign(|x_2| - 0.2)   \tag{17}$$

and assume that we know the general form of these functions as

$$f_1(x) = \alpha(x_1, x_2) \cdot \cos(p \cdot (x_1 + x_2)),   \tag{18}$$

$$f_2(x) = \alpha(x_1, x_2) \cdot sign(|x_1| - p_1) \cdot sign(|x_2| - p_2).   \tag{19}$$

This assumption motivates the following choice of consequences in rule number i

$$y_i = a_{i0} + a_{i1} \cdot x_1 + a_{i2} \cdot x_2 + a_{i3} \cdot \cos(p \cdot (x_1 + x_2))   \tag{20}$$

$$y_i = a_{i0} + a_{i1} \cdot x_1 + a_{i2} \cdot x_2 + a_{i3} \cdot sign(|x_1| - p_1) \cdot sign(|x_2| - p_2)   \tag{21}$$

The fuzzy model structure was designed by selecting membership functions

$$\mu_{ij}(x_j) = \frac{1}{1 + \left(\frac{x_j - c_{ij}}{a_{ij}}\right)^{2b_{ij}}}   \tag{22}$$

initialized on an uniform grid. We have tested the performance of modeling by: standard TSK model with linear consequences trained by ANFIS (adaptive-network-based fuzzy inference system [9], the widely accepted standard for TSK systems training), TSK model with nonlinear consequences tuned by genetic algorithm (GA implemented as in Matlab Global Optimization Toolbox), TSK model with nonlinear consequences tuned by PSO. Results of modeling of function (14) in terms of average obtained model RMSE are presented in table 1. It may be observed, that modeling error for nonlinear models is lower regardless of model complexity, i.e. number of rules. PSO tuning results in a much more precise model – error is lower by 1 to 3 orders of magnitude than for GA tuned model.

Table 1. RMSE of linear model tuned by ANFIS and nonlinear models tuned by GA and PSO. Presented RMSE of PSO and GA tuned models is an average of 10 algorithm executions.

Rules	ANFIS	GA	PSO
4	0.5643	0.5369	0.0271
9	0.5611	0.5149	0.0017
16	0.5281	0.4962	5.5e-4
25	0.4845	0.1165	2.0e-5
36	0.4071	0.0523	3.7e-5
49	0.3266	0.0345	6.3e-5

Additional experiments demenstrated that both GA and PSO benefit from estimating the value of nonlinear p parameters prior to optimizations. Such estimation was performed by minimizing the value of the same RMSE objective with constant parameters of membership functions, i.e. membership functions were initialized as a uniform grid and remained unchanged – only p parameters were tuned by GA or PSO for a low number of iterations. During the main optimization the search space for p was limited to the neighborhood of the estimated value. Such two stage optimization results in further model improvement, e.g. for a 4 rule model the average RMSE is 0.029 for GA tuning and 0.006 for PSO.

Example 2. The data presented in fig. 3 describes the q-axis flux component of a permanent magnet synchronous motor as function of current and rotor position. Acceptable fuzzy model with affine consequences requires at least N=11 rules. It is visible that for currents bigger than 1A we observe flux oscillations with rotor position. The frequency of these oscillations is 6 times bigger than the rotor angular speed. Hence we propose nonlinear consequences

$$y_i = a_{i1}x_1 + a_{i2}x_2 + a_{i0} + a_{i3}sin(6x_1) + a_{i4}cos(6x_1), i = 1 \dots N \qquad (23)$$

where x_1 is a rotor position and x_2 is q-axis current. With this it was possible to reduce the number of rules and the model with 4 rules only was equivalent to 11 rules model with linear consequences, as it is demonstrated in table 2. Presented error is calculated as RMSE for the whole set of collected data (80 000 points), while model training was based on points equal to local average of measurements over a grid of 900 points.

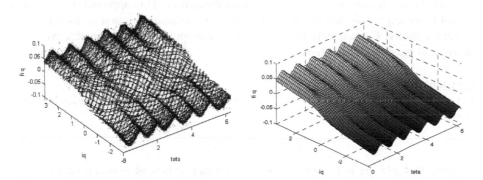

Fig. 3. The q-axis motor flux (Ψ_q [Wb]) as function of motor current (i_q [A]) and rotor position (θ [rad]). The data collected from the flux observer.

Fig. 4. Fuzzy model output corresponding to the data from fig. 3. The model with 4 rules and consequences (23) trained by PSO.

Table 2. Comparison of fuzzy models. The training was based on 900 local average values and test error was calculated using all 80 000 points from fig. 3.

	11 rules with affine consequences trained by ANFIS	4 rules with nonlinear consequences (23) trained by PSO
Training error	0.0056	0.0060
Test error	0.0079	0.0070

5   Nonlinear Consequences Construction

The obvious problem we have to solve constructing a TSK model with nonlinear consequences is how to select and parameterize nonlinear functions that appear in the consequences. Usually, in practical problems, the expert knowledge is sufficient to propose the consequences, as it was demonstrated in example 2. We may also propose a more formal approach.

Additive correction approach
If we may assume (from an expert knowledge for example) what kind of nonlinearities may be present in the system model we add these nonlinearities to the consequences. Finally we get the consequence for rule number i

$$y_i = a_{i0} + a_{i1}x_1 + \cdots + a_{in}x_n + a_i g(x, p). \tag{24}$$

The first part of (24) is a standard affine function as in classical TSK model and the last one is selected to cope with expected nonlinearities. This approach is especially useful if we expect discontinuities in the data – for example friction is discontinuous at zero velocity. Consequents similar to (24) were applied in all examples presented above and prove to be effective.

Constant coefficient approach
Let us consider a function $f: R^n \rightarrow R$ to be modelled and assume that it may be represented on a compact set A as

$$f(x) = \alpha(x)g(x, p). \tag{25}$$

Assume that $g(x, p)$ is known, although the actual values of constant parameters p may be unknown. Assume that A is covered by a grid defined by selection of points for each axis:

$$x_{k,0} < x_{k,1} < \cdots < x_{k,n_k}, k = 1, \dots, n. \tag{26}$$

We denote the middle points of the grid by

$$\bar{x}_{k,j} = \frac{x_{k,j} - x_{k,j-1}}{2}, \quad j = 1, \dots, n_k, \quad k = 1, \dots, n \tag{27}$$

and define the rules

$$R_{j_1,\dots,j_n}: IF \left(x_1 \text{ is near } \bar{x}_{1,j_1}\right) and \; \dots and \left(x_n \text{ is near } \bar{x}_{n,j_n}\right) THEN \; y \text{ is}$$

$$y_{j_1,\dots,j_n} = a_{j_1,\dots,j_n} g(x,p). \tag{28}$$

Of course the initial guess for $a_{j_1,\dots,j_n}$ is $\left(\bar{x}_{j_1,\dots,j_n}\right)$, $\bar{x}_{j_1,\dots,j_n} = \left(\bar{x}_{1,j_1}, \dots, \bar{x}_{n,j_n}\right)$, or the average value of the data representing $\alpha(x)$ in the part of the grid around $\bar{x}_{j_1,\dots,j_n}$, if such information is available.

<u>Linear approximation approach</u>
Assuming that $\alpha(x)$ is sufficiently smooth, it may be approximated by

$$\alpha(x) \approx \alpha\left(\bar{x}_{j_1,\dots,j_n}\right) + \sum_{i=1}^{n} \left.\frac{\partial \alpha}{\partial x_i}\right|_{\bar{x}_{i,j_i}} (x_i - \bar{x}_{i,j_i})$$

$$= \alpha\left(\bar{x}_{j_1,\dots,j_n}\right) - \sum_{i=1}^{n} \left.\frac{\partial \alpha}{\partial x_i}\right|_{\bar{x}_{i,j_i}} \bar{x}_{i,j_i} + \sum_{i=1}^{n} \left.\frac{\partial \alpha}{\partial x_i}\right|_{\bar{x}_{i,j_i}} x_i \tag{29}$$

what motivates the rules

$$R_{j_1,\dots,j_n}: IF \left(x_1 \text{ is near } \bar{x}_{1,j_1}\right) and \; \dots and \left(x_n \text{ is near } \bar{x}_{n,j_n}\right) THEN \; y \text{ is}$$

$$y_{j_1,\dots,j_n} = a_{0,j_1,\dots,j_n} + a_{1,j_1,\dots,j_n} x_1 g(x,p) + \dots + a_{n,j_1,\dots,j_n} x_n g(x,p) \tag{30}$$

Example 3. Proposed methods of constructing rule consequences were applied to modeling of example function (16). The tested model consisted of 4 or 9 rules with consequences defined as:

- standard affine:

$$y_i = a_{i0} + a_{i1}x_1 + a_{i2}x_2 \tag{31}$$

- nonlinear - consisting of linear part and added nonlinear function:

$$y_i = a_{i0} + a_{i1}x_1 + a_{i2}x_2 + a_{i3}\cos\left(p \cdot (x_1 + x_2)\right) \tag{32}$$

- nonlinear with constant coefficients:

$$y_i = a_i \cos\left(p \cdot (x_1 + x_2)\right) \tag{33}$$

- nonlinear constructed according to the linear approximation approach:

$$y_i = a_{i0} + a_{i1}x_1 \cos\left(p \cdot (x_1 + x_2)\right) + a_{i2}x_2 \cos\left(p \cdot (x_1 + x_2)\right) \tag{34}$$

Model with linear consequences was trained by ANFIS, while nonlinear models were trained by PSO. The objective of training was to minimize the RMSE model error. Nonlinear models - gained much better accuracy than the linear model as is demonstrated in table 3 and shown in figures 5 and 6 presenting modeling error of a sample linear and nonlinear model.

To illustrate the complexity of linear model required to match the accuracy of nonlinear models we continue constructing of linear models for as many as 144 rules trained by ANFIS. The resulting model's RMSE is similar to error of a 4-rule nonlinear model.

Table 3. RMSE of linear (ANFIS) and nonlinear models: PSO additive (eq. (32)), PSO constant (33), PSO lin.apx. (34) for 4 and 9 rules. Higher number of rules modeled for comparison for linear model only. Presented RMSE of PSO tuned models is an average of 10 algorithm executions.

Rules	ANFIS	PSO additive	PSO constant	PSO lin.apx.
4	0.5643	0.0271	0.0461	0.0024
9	0.5611	0.0017	0.0014	0.0003
81	0.1921	-	-	
100	0.0693	-	-	-
144	0.0254	-	-	-

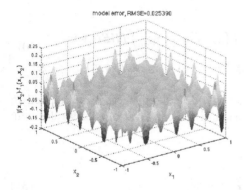 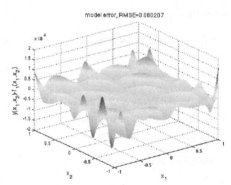

Fig. 5. Modeling error of function (16) for a 144-rule ANFIS trained model with linear consequents

Fig. 6. Modeling error of function (16) for a 9-rule PSO trained model with nonlinear consequents given by eq. (34)

6   Conclusions

We propose to generalize TSK fuzzy model applying nonlinear functions in the rule consequences. We provide the model description and parameterization and discus the problem of model training and we recommend PSO for tuning parameters in membership functions and in nonlinear part of a rule consequence. We also propose some more or less formalized approach to nonlinear consequence selection and construction. Several examples demonstrate the main features of the proposed fuzzy models.

The proposed approach allows to reduce the model complexity preserving the desired accuracy. A model with smaller number of rules is simpler for real time application and easier interpretable. Nonlinear consequences enable modeling of function imperfectly approximated by standard TSK models, for example discontinuous functions [10]. As the constructed models are linear in some of parameters they may be efficiently utilized in adaptive control of dynamical systems, as it was proposed for example in [11]. It is also important that the proposed model may be described as a neural network and so NN-training techniques may be utilized for the model tuning.

References

1. Takagi, T., Sugeno, M.: Fuzzy Identification on Systems and its Applications to Modeling and Control. IEEE Trans. on Systems, Man, and Cybern. 15, 116–132 (1985)
2. Sugeno, M., Kang, G.: Structure Identification of Fuzzy Model. Fuzzy Sets and Systems 28, 329–346 (1998)
3. Nguyen, H.T., Sugeno, M.: Fuzzy Systems: Modeling and Control. Springer (1998)
4. Chen, S.-S., Chang, Y.-C., Wu, J.-L., Cheng, W.-C., Su, S.-F.: Fuzzy Control Design for Switched Nonlinear Systems. In: Proc. of SICE Annual Conference, pp. 352–357 (2008)
5. Rajeshm, R., Kaimal, M.R.: T–S fuzzy model with nonlinear consequence and PDC controller for a class of nonlinear control systems. Applied Soft Computing 7, 772–782 (2007)
6. Kabziński, J.: One-Dimensional Linear Local Prototypes for Effective Selection of Neuro-Fuzzy Sugeno Model Initial Structure. In: Papadopoulos, H., Andreou, A.S., Bramer, M. (eds.) AIAI 2010. IFIP AICT, vol. 339, pp. 62–69. Springer, Heidelberg (2010)
7. Parsopoulos, K.E., Vrahatis, M.N.: UPSO: A Unified Particle Swarm Optimization Scheme. In: Proceedings of the International Conference of "Computational Methods in Sciences and Engineering" (ICCMSE 2004). Lecture Series on Computer and Computational Sciences, vol. 1, pp. 868–873. VSP International Science Publishers, Zeist (2004)
8. Shi, Y., Eberhart, R.: A modified particle swarm optimizer. In: IEEE World Congress on Computational Intelligence, The 1998 IEEE International Conference on Evolutionary Computation Proceedings, May 4-9, pp. 69–73 (1998)
9. Jang, J.R.: ANFIS: Adaptive-network-based fuzzy inference system. IEEE Trans. Syst. Man Cybern. 23, 665–684 (1993)
10. Jastrzebski, M.: A generalized Takagi-Sugeno-Kang model designed for approximation of discontinuous dependeces, Zeszyty Naukowe. Elektryka - Politechnika Łódzka, pp. 15–24 (2005)
11. Kabziński, J.: Fuzzy Friction Modeling for Adaptive Control of Mechatronic Systems. In: Iliadis, L., Maglogiannis, I., Papadopoulos, H. (eds.) Artificial Intelligence Applications and Innovations. IFIP AICT, vol. 381, pp. 185–195. Springer, Heidelberg (2012)

Novel Techniques for Text Annotation
with Wikipedia Entities

Christos Makris and Michael Angelos Simos

Computer Engineering and Informatics Department,
University of Patras, Greece
{makri,asimos}@ceid.upatras.gr

Abstract. Text annotation is the procedure of identifying the semantically dominant words of a text segment and attaching them with conceptual content information in their context. In this paper, we propose novel methods for automatic annotation of text fragments with entities of Wikipedia, the largest knowledge base online, a process commonly known as *Wikification* aiming at resolving the semantics of synonymous and polysemous terms accurately. The cornerstone of our contribution is a novel iterative Wikification approach, converging at optimal annotations while balancing high accuracy with performance. Our first two methods can be fine-tuned through a machine-learning technique over large homogenous data sets. Our experimental evaluation resulted in remarkable improvement over state-of-the-art Wikification approaches.

Keywords: Text semantic annotation, Wikipedia entities, Semantic Data Linking on Web, Data Mining, Information Retrieval, Ontologies.

1 Introduction

Wikipedia is one of the largest online knowledge repositories, consisting of more than 4 million entities (also called lemmas or articles), built over more than with over 20 million contributing users. The online "crowdsourcing" nature of Wikipedia enables the creation and maintenance of knowledge entities in a quite diverse manner resulting in widespread and commonly accepted textual descriptions, through the consensus of a large number of people. Wikipedia inherits the main principles of web, elegantly combining textual content and cross-references with hyperlinks providing plenteous semantic information that can be exploited for representing the described entities.

Text annotation is the process of identifying the semantically dominant words (called spots or anchors) in a text segment and attaching them with additional information expressing conceptual content in their current context. Recent research works [1, 3, 4, 5, 6, 9, 12, 13, 15] explored various approaches for efficient text annotation with Wikipedia entities. Text annotation is a fundamental preprocessing step of many information retrieval activities like semantic indexing, cross-referencing content on web, clustering, classification, for efficient rankings, text summarization, natural

L. Iliadis et al. (Eds.): AIAI 2014, IFIP AICT 436, pp. 508–518, 2014.

language generation, exploiting conceptual similarity of documents, and sentiment analysis.

Text annotation techniques face the entity disambiguation problem, where multiple candidate Wikipedia article annotations exist for a specific term, deriving by polysemy and synonymy, common properties of natural languages. Disambiguating to Wikipedia entities is similar with classic Word Sense Disambiguation tasks [14], but also requires compatibility with additional information provided by the link structure of Wikipedia. The core of most existing methods involves the textual context and the collective agreement of other identified spots with every candidate disambiguation of a specific spot.

The aim of current work is the introduction of simple and accurate disambiguation methods for small texts, using Wikipedia as the underlying catalogue, thus focusing on the same kind of texts as in TAGME system [3] and the approaches introduced in [11, 13].

2   Literature Overview

Wikipedia disambiguation techniques can be distinguished into two main classes: local and global [15]. Local approaches exploit context information concerning only the specific spot, while global approaches rely on a collective agreement of all spots in a text segment and their disambiguation. Moreover, some approaches involve collection-level methods, performing a collective disambiguation of all entity references across a given document collection.

Disambiguation based on Wikipedia entities has attracted many researchers [1, 9, 11, 12, 13, 15], with most prominent the approach proposed in [3], where the authors focused on real time annotation of short texts existing on tweets, web search engine results etc, where textual context is quite limited. The TAGME system was extended in [10] by using information from knowledge resources such as WordNet [2], and exploiting the PageRank values inherent in the Wikipedia pages. The system can be extended to handle other ontologies [7] as well.

3   Proposed Methods

The touchstone of our proposed techniques is TAGME's voting scheme that assigns to each candidate sense of a spot a value, based on the collective agreement of this annotation from the other spots in the same text. Our first approach investigates some minor changes over TAGME's scheme. The second method proposes a more analytical model for the problem. Finally our iterative method, introduces a novel approach.

Spots Extraction
We use a simple yet very effective method for preprocessing the input text fragments for the extraction of candidate Wikipedia entities. Firstly, every text fragment is separated into tokens (words, punctuation, symbols, etc) to be grouped later, forming

phrases when possible. The grouping is performed by matching as many adjacent tokens as possible with phrases (token groups) extracted from Wikipedia's links anchors. Complex rules have been applied allowing loose matching, handling minor details and further improving our results. The final step of spots extraction process is the removal of words not containing semantic information, (stopwords), symbols, punctuation etc. A more complex morphosyntactic analysis model is to be developed as future work.

3.1 Method 1

Let a_i be an anchor to be annotated, and $Pg(a_i)$ the list of candidate Wikipedia articles for this annotation, derived from occurrences of a_i in Wikipedia hyperlinks as anchor to the above articles. A vote provided by a spot c to the annotation $a_i \rightarrow p_j$ with p_j in $Pg(a_i)$ is evaluated as:

$$vote_c(p_j) = \sum_{p_l \in p_g(c)} srel(p_j, p_l) * P(p_l|c) / |Pg(c)|$$

where $P(p_l/c)$ is the prior probability of c is pointing to p_l (also known as commonness). Commonness is pre-calculated by parsing the Wikipedia dataset and for each observed anchor-Wikipedia pair dividing the number of appearances that the anchor points to the corresponding article by the total appearances of the anchor. The quantity $srel(p_j, p_l)$ is a measure devised by us to depict the relatedness between two Wikipedia pages that is computed by taking into account the commonality of the incoming Wikipedia links to the two pages, as the cardinality of intersection divided by the maximum cardinality set, thus:

$$srel(p_j, p_l) = \frac{|in(p_j) \cap in(p_l)|}{max\left(|in(p_j)|, |in(p_l)|\right)}$$

where in(p) as in [2, 4] denotes the set of Wikipedia pages pointing to page p.
The disambiguation weight of a probable annotation $a_i \rightarrow p_j$ is defined as:

$$score_{a_i}(p_j) = \sum_{a_t \neq a_i} vote_{a_t}(p_j)$$

The final annotation is selected based on maximum commonness score after a filtering step among Wikipedia articles within a voting score distance threshold τ from the maximum voting score.

We have tested relatedness formulas like [5, 13], but the highest accuracy is achieved on this schema by our simple formula (srel), evaluating the incoming link intersection of two articles. A common observation throughout our experiments is that the heterogeneity of compared articles in size, results in imbalances of relatedness score, when using approaches like [5] and [13]. A better formula is to be developed as future work.

3.2 Method 2

Let $Pg(a_i)$ be the set of candidate Wikipedia articles $(p_{a_i}j)$ annotations for each spot as anchor a_i. The anchor-Wikipedia pairs that appear less than three times, or in less than 0.1% of total occurrences of the anchor are filtered. Let a text fragment contain k spots a_i where $i \in \{1, 2, ..., k\}$.

The core logic of this method involves computing a global score for every probable combination of anchor senses represented as possible Wikipedia article annotations (p_{a_i}). There are $\prod_1^k \binom{|Pg(a_i)|}{1} = \prod_1^k |Pg(a_i)|$ probable combinations, that can be reduced by filtering more anchor-Wikipedia low probability pairs. Let *Gscore* be the global evaluation score of an annotation combination, as described above:

$$Gscore\left(p_{a_1}, p_{a_2} \cdots p_{a_k}\right) = \sum_{i=1}^{k-1} \sum_{w=i+1}^{k} Bscore(p_{a_i}, p_{a_w})$$

At first in "Method 2 (*rel*)" we evaluated a relatedness measure between two Wikipedia Articles:

$$rel(p_j, p_l) = 1 - \frac{log\left(max\left(|in(p_j)|, |in(p_l)|\right)\right) - log\left(|in(p_j) \cap in(p_l)|\right)}{log(W) - log(min(|in(p_j)|, |in(p_l)|))}$$

proposed by Milne & Witten [13] as the *Bscore* formula, where W denotes the size of Wikipedia. Still our simpler relatedness approach (*srel*) outperforms this commonly utilized relatedness measure yielding more accurate results, for most entities pairs having small incoming links intersection sets.

Finally, in "Method 2 comrel" we tested a more complex formula for *Bscore*:

$$Bscore\left(p_{a_i}, p_{a_w}\right) = comrel\left(p_{a_i}, p_{a_w}\right) = srel\left(p_{a_i}, p_{a_w}\right)P(p_{a_i}|a_i)P(p_{a_w}|a_w)$$

where $P(p_{ai}/a_i)$ is the Commonness of the anchor a_i pointing the article p_{ai}. This way *BScore* scales respectively with the commonality of the two articles combination.

As a final step, again we perform filtering by keeping all candidate combinations achieving a *Gscore* up a threshold distance τ from the maximum *Gscore*. Among the filtering results, we select the combination maximizing the total commonness $\sum_{i=0}^{K} P(p_{a_i}, a_i)$. For very small threshold values (t), the selection of annotations yielding the highest Bscore is voted as the dominant. If the threshold is large, then a selection with the highest total commonness score is selected.

Threshold Optimization

The most important fine tuning in our methods schemas involves the optimal selection of τ. We propose a method for optimizing τ threshold, given a Wikipedia entity manually annotated training set, having homogenous statistics with the to-be-annotated dataset. This technique can also be used to determine experimentally the upper bound of performance for our methods (by training with our experimental dataset).

Based on this optimization technique we explore the threshold value intervals for which a selection with the optimal annotations passes through the final filtering step. Then we just find the values interval for τ, that belongs to the maximum number of each selection interval, for the selections that perform a correct annotation.

Thus, the problem of optimizing τ value can be deducted to a simple optimization problem of selecting a value that belongs to as more intervals as possible, by simply checking the values on intervals edges.

3.3   A Novel Approach on Text Annotation

Various model schemas have been introduced for the text annotation problem in previous works with most common the voting scheme and its many variations, to which we contributed during the previous part of this paper. One of the most important breakthroughs of the current work is the introduction of a novel method for semantic anchor annotation with Wikipedia entities, through an iterative approach, aiming at converging to an optimal candidate for each anchor as a result of constant improvement of the approximate solutions of each iteration. The principal intuition of this method, lies on the analysis of human semantic interpretation process of text segments containing polysemous terms. Such a process would involve an iterative evaluation procedure, until some termination criteria are met. For the most of the common text segments disambiguation, those criteria may be met before iterations are required, but in complex polysemous context, an evaluation of each candidate annotation may be necessary, consisting an iterative procedure until some criteria for a decision with a degree of certainty are met. The main logic difference of this method and method 2, which evaluates every possible annotation combination among anchors of a text segment, is the targeted evaluation of semantically meaningful combinations (as evaluated by some commonness and relatedness formulas), resulting a vast complexity reduction.

A candidate model of such a process may be epitomized by the following iterative method:

Let $s_0, s_1, s_2, \ldots s_n$ be the spots of a text fragment, and $P_{si}1 \ldots P_{si}m \in Pg(s_i)$, the m candidate Wikipedia entity annotations of spot s_i, where $i \in \{0,1,2,3\ldots n\}$. Let $Pg(s_i)$ the set of candidate Wikipedia entity annotations of spot s_i. Each state is constituted of lists of candidate Wikipedia entity annotation of each spot sorted by a ranking criterion. At initial state, the candidate annotation lists are sorted by commonness in Wikipedia where $p_{si}0$, the most common entity of each spot s_i).

s_0	s_1	s_2	$\ldots$	s_n
$p_{s0}0$	$p_{s1}0$	$p_{s2}0$		$p_{sn}0$
$p_{s0}1$	$p_{s1}1$	$p_{s2}1$		$p_{sn}1$
$p_{s0}2$	$p_{s1}2$	$p_{s2}2$		$p_{sn}2$
$\ldots$	$\ldots$	$\ldots$	$\ldots$	$\ldots$
$p_{s0}m_0$	$p_{s1}m_1$	$p_{s2}m_2$		$p_{sn}m_n$

So the disambiguation approximation during the initial step of the iteration is $p_{s0}0$ $p_{s1}0$ $p_{s2}0 \ldots p_{sn}0$, equivalent with the commonness annotation approach. The iterative step involves sorting each of the spots list in descending order of its elements R metric values. Every iteration results in approximation improvement of the optimal annotation entities, by exploiting as initial seed the commonness of the candidate entities, and combining relatedness at each iteration step until convergence. The evaluated formula for the R metric is:

$$R(p_{s_i}x) = \sum_{w=0}^{n} srel(p_{a_i}, p_{a_w}) * P(p_{a_i}|a_i) * P(p_{a_w}|a_w)$$

$$Where: srel(p_i, p_l) = \frac{|in(p_j) \cap in(p_l)|}{max(|in(p_j)|, |in(p_l)|)}$$

The iterative algorithm is terminated when one of the following convergence criteria is met:

- A maximum number of iterations is reached. This is more of a security criterion, for handling the case of infinite iterations, due to other convergence criteria not being met. It is rarely activated, since in most cases, the rest of the convergence conditions are already met. The enforcement of large values results in more accurate results, relying more and more on relatedness for the cases where one of the other termination conditions is not yet satisfied. The application of smaller values, weights commonness more over relatedness, thus results in faster overall and worst case execution time, with an expense in accuracy.
- There are zero disturbances during the last k iterations. This means that a convergence is due to happen, since no changes on the list rankings are observed.
- The top m elements of each spots list maintain their position for k iterations. Thus the probability of disturbances is decreased.
- Small difference of each of the spot lists elements values is observed between iteration steps, stabilizing the ranking and leading to convergence.

The convergence speed of the above criteria can be enhanced by filtering the lower k percentile elements of each lists spots, after m iterative steps, since our evaluation revealed a fairly decreased probability of the lowest k elements occupying top positions on the lists rankings, before some of the convergence criteria is satisfied. This behavior is intuitively explained by the fact that for the vast majority of cases, the ranking results tend to converge at optimal values through each iteration.

The real advantages of the third method, involve remarkable improvements compared with our two first contributions. Its iterative nature, allows balancing between speed and accuracy, with the appropriate fine tuning of the convergence conditions parameters. Flaccid convergence criteria lead to more mature results in terms of relatedness, approximating at most cases the accuracy of our second method. Tighter convergence criteria imply improved average and worst case time complexity, without significant sacrifices in accuracy. Finally we should note that convergence occurs within the first iterations in most cases with high accuracy, a fact that allows us conclude correlation between convergence speed (in iterations) and the degree of certainty of the results accuracy, as is intuitively expected.

We exploited the above correlation among convergence speed and accuracy, by composing a post processing pruning step and evaluating the experimental results during our evaluation.

4   Experimental Evaluation

For the experimental evaluation of the methods we introduced above, we used the datasets of TAGME [3], available online by the authors. Our principal focus involved

the Wiki-Disamb30 data file consisting of 1.4M short texts randomly selected from
Wikipedia pages, each one containing approximately 30 words and at least one ambiguous anchor. Common precision metrics were used as in [13, 9, 2], in order to evaluate in practice the performance of our algorithms.

In table 1 and figure 1, we present the precision of methods 1 and 2, after optimization of their parameters after training. The optimal result column of table 1, presents
the maximum possible accuracy of the method, using optimal parameters, calculated
through training with the entire evaluated dataset. Figure 2, shows the precision of
method 3, by varying the number of max iterations parameter. Figure 3 presents the
evaluation of method 3 pruning step (precision/recall of pruning). The evaluation of
method 3 is presented separately from methods 1 and 2, since it doesn't involve a
machine learning procedure.

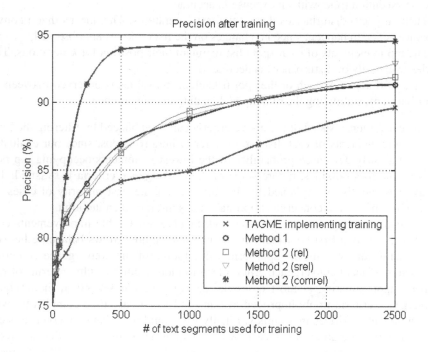

Fig. 1.

We used an Ubuntu Server 12.04LTS on a quad core 2.9 GHz 64 bit PC utilizing
8GBytes of main memory. The proposed methods were implemented in Python 2.7.6
over the Wikipedia dump files instance enwiki-20130503, loaded on a PostgreSQL
database.

We ran our methods on randomly selected subsets of Disamb30 dataset, since the
time complexity of method 2 was beyond our time resources for a full scale experiment, having in all cases similar results. The results, provided below are over a subset
of more than 10K disambiguated spots. We also performed a simplified implementa-

tion of TAGME disambiguation algorithm, as described in [3] as performance baseline, applying a similar threshold optimization technique of its ϵ value.

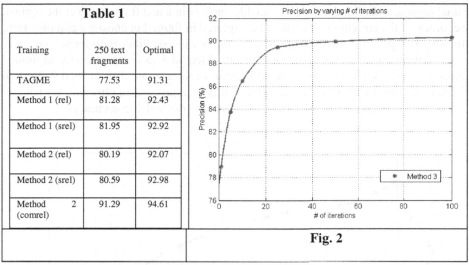

| Table 1 | | | Fig. 2 |

Training	250 text fragments	Optimal
TAGME	77.53	91.31
Method 1 (rel)	81.28	92.43
Method 1 (srel)	81.95	92.92
Method 2 (rel)	80.19	92.07
Method 2 (srel)	80.59	92.98
Method 2 (comrel)	91.29	94.61

Table 1 and figure 1 present our detailed results on the precision of the two first
methods proposed, expressing their capacity to give the same senses/Wikipedia articles initially attached to the text fragments anchors. Throughout our experiments, we
investigated the upper bound of our methods precision while annotating every available anchor of the dataset.

Our first method outperforms TAGME, yielding more accurate optimal results up
to 92.43%. This improvement is an outcome of the use of *srel* formula, depicting the
necessity of a better relatedness measure between two Wikipedia entities for similar
annotation algorithm schemas. The behavior of this method under our τ-*optimization*
technique was also improved comparatively with TAGME, converging faster near
optimal results.

The second method reached optimal accuracy of 94.61%, but the time complexity
of this method was by far larger than both our first method and TAGME. The rel and
srel formulas were not as effective, yielding respectively 92.07% and 92.98%, optimal
accuracy. But when using *comrel* formula, the improvement in both accuracy and
training set size for fast convergence to optimal results was remarkable. This fact was
analyzed in depth, concluding that the analytic model of Method 2 creates a high
ranking quality pre-filtering list, with the correct article of an annotation in the top of
that list in most cases. That leads to very small optimal threshold values (τ) selection,
thus fast convergence to the optimal τ. In conclusion, we note the necessity of weighting commonness with relatedness by multiplying as our *comrel* formula has proven.

Finally we evaluated our iterative method's precision, (figure 2), varying the maximum iterations parameter. The optimal results for this method's accuracy exceeded
90.71%. We have not developed an optimization technique for the current method due
to its numerous parameters, thus its complexity. We applied common precision and

recall metrics for the overall evaluation including the pruning step (figure 3), yielding exceptional conclusions for the quality of pruning.

To conclude, method 3 yields less accurate results from the previous methods optimal results, when annotating all available anchors in a text fragment. Yet the optimal accuracy of the current method, is to be explored in future but consist a complex task. In very low pruning recall conditions, accuracy exceeds 98%, with ambiguous anchors contributing as well in these high results. Method 3 accuracy outperforms limited training versions of methods 1, but still remains less accurate than method 2. Method 3 balances speed with accuracy, with the ability to adjust this equilibrium, yet with no expectancy of outperforming method 1 in speed, or method 2 in accuracy.

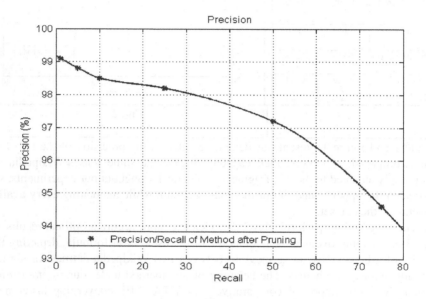

Fig. 2.

5   Conclusion

In this paper we propose three novel methods. The first method, revealed the potentiality for further improvement of the relatedness measures between Wikipedia entities. Our second method yields very accurate results, yet requires substantial resources and time. The τ-*optimization* algorithm we developed for training our two first methods, allowed exploring the optimal upper bound of their performance, and the behavior after training with various training set sizes.

The cornerstone of this contribution is a novel iterative approach of the Wikification task, not explored by previous work. This technique, achieves convergence through a series of iterative steps, each of which aims at improving approximation of the optimal annotations.

The development of a better relatedness measure for Wikipedia entities, based on both incoming links, and contextual content is included among our future plans. We also aim at further improving, fine tuning our methods performance and engineering their large scale deployment as an online service, after reserving the necessary resources.

Acknowledgments. This research has been co-financed by the European Union (European Social Fund – ESF) and Greek national funds through the Operational Program "Education and Lifelong Learning" of the National Strategic Reference Framework (NSRF) - Research Funding Program: Thales. Investing in knowledge society through the European Social Fund"

References

1. Cucerzan, S.: Large-Scale Named Entity Disambiguation Based on Wikipedia Data. In: Proceedings of EMNLP-CoNLL 2007, pp. 708–716 (2007)
2. Fellbaum, C. (ed.): WordNet, an electronic lexical database. The MIT Press (1998)
3. Ferragina, P., Scaiella, U.: TAGME: on-the-fly annotation of short text fragments (by wikipedia entities). In: Proceedings of the 19th ACM International Conference on Information and Knowledge Management (CIKM 2010), pp. 1625–1628. ACM, New York (2010)
4. Gabrilovich, E., Markovitch, S.: Computing semantic relatedness using Wikipedia-based explicit semantic analysis. In: Sangal, R., Mehta, H., Bagga, R.K. (eds.) Proceedings of the 20th International Joint Conference on Artifical Intelligence (IJCAI 2007), pp. 1606–1611. Morgan Kaufmann Publishers Inc., San Francisco (2007)
5. Han, X., Sun, L., Zhao, J.: Collective entity linking in web text: a graph-based method. In: Proceedings of the 34th International ACM SIGIR Conference on Research and Development in Information Retrieval (SIGIR 2011), pp. 765–774. ACM, New York (2011)
6. Han, X., Zhao, J.: Named entity disambiguation by leveraging wikipedia semantic knowledge. In: Proceedings of the 18th ACM Conference on Information and Knowledge Management (CIKM 2009), pp. 215–224. ACM, New York (2009)
7. Hoffart, J., Suchanek, F.M., Berberich, K., Lewis-Kelham, E., Melo, G., Weikum, G.: YAGO2: exploring and querying world knowledge in time, space, context, and many languages. In: Proceedings of the 20th International Conference Companion on World Wide Web (WWW 2011), pp. 229–232. ACM, New York (2011)
8. Hoffart, J., Yosef, M.A., Bordino, I., Fürstenau, H., Pinkal, M., Spaniol, M., Taneva, B., Thater, S., Weikum, S.: Robust disambiguation of named entities in text. In: Proceedings of the Conference on Empirical Methods in Natural Language Processing (EMNLP 2011), pp. 782–792. Association for Computational Linguistics, Stroudsburg (2011)
9. Kulkarni, S., Singh, A., Ramakrishnan, G.: and. Chakrabarti, S. Collective annotation of Wikipedia entities in web text. In: Proceedings of the 15th ACM SIGKDD International Conference on Knowledge Discovery and Data Mining, pp. 457–466. ACM, New York (2009)
10. Makris, C., Plegas, Y., Theodoridis, E.: Improved Text Annotation with Wikipedia Entities. In: Proceedings of the 28th Annual ACM Symposium on Applied Computing, pp. 288–295. ACM, New York (2013)

11. Meij, E., Weerkamp, W., Rijke, M.: Adding semantics to microblog posts. In: Proceedings of the 5th ACM International Conference on Web Search and Data Mining (WSDM 2012), pp. 536–572. ACM, New York (2012)
12. Mihalcea, R., Csomai, A.: Wikify!: linking documents to encyclopedic knowledge. In: Proceedings of the 16th ACM International Conference on Information and Knowledge Management (CIKM 2007), pp. 233–242. ACM, New York (2007)
13. Milne, D., Witten, I.H.: Learning to link with Wikipedia. In: Proceedings of the 17th ACM Conference on Information and Knowledge Management (CIKM 2008), pp. 509–518. ACM, New York (2008)
14. Navigli, R.: Word Sense Disambiguation. ACM Computing Surveys 41(2), 10:1–10:69 (2003)
15. Ratinov, L., Roth, D., Downey, D., Anderson, M.: Local and global algorithms for disambiguation to Wikipedia. In: HLT 2011, vol. 1 (2011)

An Avatar-Based Weather Forecast Sign Language System for the Hearing-Impaired

Juhyun Oh[1], Seonggyu Jeon[1], Minho Kim[2], Hyukchul Kwon[2], and Iktae Kim[3]

[1] Technical Research Institute, Korean Broadcasting System, Seoul, Korea
{jhoh,cheez}@kbs.co.kr
[2] Dept. of Computer Science, Pusan National University, Busan, Korea
{karma,hckwon}@pnu.ac.kr
[3] Bluenine Inc., Seoul, Korea
blue9naa@naver.com

Abstract. In this paper, we describe a text-to-animation framework for TV weather forecast sign language presentation. To this end, we analyzed the last three years' weather forecast scripts to obtain the frequency of each word and determine the order of motion capture. About 500 sign language words were chosen and motion-captured for the weather forecast purpose, in addition to the existing 2,700 motions prebuilt for daily life. Words that are absent in the sign language dictionary are replaced with synonyms registered in *KorLex*, the Korean Wordnet, to improve the translation performance. The weather forecast with sign language is serviced via the Internet in an on-demand manner and can be viewed by PC or mobile devices.

Keywords: sign language, avatar animation, closed caption, machine translation, wordnet.

1 Introduction

Closed caption and sign language broadcasts are provided with the terrestrial digital television (DTV) services in Korea for the hearing-impaired people. Even though closed captions are provided for almost the whole broadcast time, sign language broadcasting covers only about 5% of it. We propose a system that translates the closed captions of weather forecast programs into Korean sign language (KSL) and present it with three-dimensional (3D) avatar animation. The translated sign language data are sent via the Internet for the receivers such as personal computer (PC) and mobile devices to show the corresponding sign language animation. The system consists of the Korean-KSL translator, the sign language avatar animation system, and the server system that provides the closed caption and video for the most recent weather forecast in an on-demand manner.

A similar system for Japanese sign language (JSL) was proposed to use TV program making language (TVML) [1,2] for high-quality computer animation. However the translation is not automatic in their work and the user should input the sign language words in the order of JSL. For Brazilian sign language (LIBRAS), automatic

L. Iliadis et al. (Eds.): AIAI 2014, IFIP AICT 436, pp. 519–527, 2014.

translation and middleware structure for DTV was also proposed in [3,4]. Similarly to the existing studies, the purpose of the proposed system in this paper is to provide information when human interpreters are not available and not to replace human interpreters.

This paper is organized as follows. In Section 2, the Korean-KSL translation method is described. In Section 3, the 3D avatar animation scheme is introduced. In Section 4, the system to provide the weather forecast sign language service is described. The experimental results and conclusion are given in Section 5 and 6 respectively.

2   Korean-KSL Translation

Table 1 shows an example of Korean-KSL translation in a couple of weather forecast sentences by a professional sign language interpreter. Our goal is to implement an automatic translator that outputs similar KSL results when the same Korean weather forecast input is given.

Table 1. An example of manual Korean-KSL translation in weather forecast scripts, expressed in English

Korean	Recently it is cold at every weekend.
KSL	week + end + cold
Korean	It is raining from the early morning in the southern area and Jeju Island.
KSL	warm + place + and + Jeju Island + place + morning + from + rain

2.1   Korean-KSL Dictionary

There are about 10,000 words in KSL Dictionary [5]. It is very difficult to build the Korean-KSL dictionary and capture the motions for all these 10,000 words. Thus we investigated the weather forecast scripts for the past three years, from Korean Broadcasting System (KBS) and a few other sources. After some preprocessing of the weather forecast scripts, the scripts are divided by part of speech (POS). For this purpose we use the POS tagger of Pusan National University with about 1 million registered words [6]. The accuracy of the POS tagger is about 98%. Since the basic word orders of Korean and KSL are similar as subject-object-verb (S-O-V), direct word-to-word translation rule is applied in the proposed system.

Table 2 shows ten most frequent words for noun, adjective, verb, and adverb in the analyzed scripts. Notice that the POS in Korean is different from that of English for some words. In Table 2, the most frequently used noun and adjective are 'temperature' and 'high' respectively. The cumulative frequency in Table 2 is important because it shows the translation capability when that word and the above ones in the table were registered in the dictionary. For example, if only ten most frequently used nouns were registered in the dictionary and motion-captured, the system can probabilistically process 33% of the nouns in the script.

Table 2. Ten most frequently used noun, adjective, verb, adverb words (expressed in English) in the weather forecast scripts. C.F. means the cumulative frequency of each POS.

Noun			Adjective			Verb			Adverb		
Word (Kor/Eng)	Frequency (times)	C.F. (%)	Word	Freq.	C.F.	Word	Freq.	C.F.	Word	Freq.	C.F.
temperature	4,180	6	high	1,277	14	go down	1,451	11	a little	742	11
tomorrow	3,295	11	many	1,173	27	rise	854	18	gradually	702	21
day	2,664	15	exist	1,125	40	come	844	24	mostly	628	30
today	2,602	19	clear	841	49	rise	821	31	again	576	38
rain	2,092	22	low	798	58	continue	737	36	somewhat	470	45
Seoul	1,747	24	strong	609	65	hang	712	42	more	417	51
morning	1,611	27	similar	480	70	take	628	46	occasionally	284	55
region	1,585	29	large	376	74	pass by	493	50	again	269	59
central area	1,426	31	dark	343	78	fall	440	54	especially	204	61
whole country	1,293	33	cold	313	81	be seen	430	57	but	185	64

2.2 KSL Synonym Dictionary

Synonyms are translated into same sign language words. For example, 'house', 'housing' and 'abode' are synonyms and all translated into one sign language word. The synonym dictionary is built based on KorLex [7], as shown in Fig.1. Using this synonym dictionary, a word that is absent in the KSL dictionary can be translated to a synonym, increasing the translation success rate. Without the synonym dictionary, non-registered words would have been represented with finger spelling, decreasing the usefulness of the system.

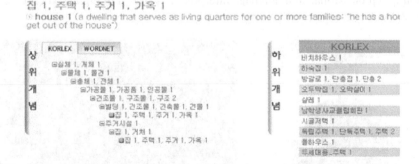

Fig. 1. The word '집(house)' found in KorLex

2.3 Word Sense Disambiguation with KorLex

In Korean, the word '눈' (pronounced as 'nun') can mean 'eye' or 'snow'. This kind of ambiguity should be resolved in the context. A simple rule to determine the word

'눈' as 'snow' is shown in Table 3. If the context words (CW) exists around the ambiguous word, then it is translated to 'snow'.

Table 3. A simple rule for the ambiguous word '눈'

눈 CW → snow	
(R1)	CW = [내리다(fall)	날리다(fly)	흩날리다(flutter)]

This kind of rules suppose the exact match to the CWs, therefore its recall is low. We replace the CWs with the synonym set number in KorLex, to increase the recall. Case particle restriction (R3) and CW verb conjugation restriction (R4) are also applied, as shown in Table 4.

Table 4. Modified rule for the ambiguous word '눈'

눈+P CW → snow	
(R1)	Context = B1
(R2)	CW = [02674938	02041026	02197925]
(R3)	P = [subjective, complement, objective, auxiliary particle]
(R4)	Conjugation = 1001 + 2001

3   Sign Language Avatar Animation

We use about 2,700 sign language motions for daily life owned by Primpo Inc [8]. Additionally, we captured 507 words dedicated for weather forecast at a studio. 15 Vicon motion capture cameras [9] and 41 infra-red markers were used to capture the motion of a professional sign language interpreter, as shown in Fig.2. Cyberglove [10] was used to capture the hand motion.

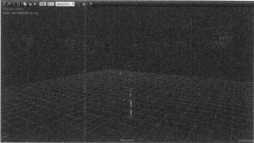

Fig. 2. The sign language interpreter (left) and the corresponding captured motion (right)

A 3D female avatar model was built for sign language animation. Fig.3 shows the avatar model and bones associated to it.

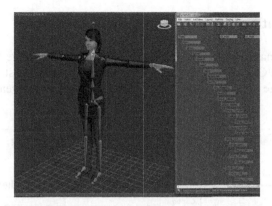

Fig. 3. 3D avatar and bones.

When animating a sign language sentence, two contiguous sign words are over-lapped to each other similarly in human sign language. 15 frames at the end of each sign word are blended with those of the following word, using a linear motion blend-ing equation [11]. Let two sign word motions to blend be M_i and M_j, and the k-th joint of concern J_{ik} and J_{jk}. If the two motions are blended with the weights w_i and w_j respectively, the blended position of the k-th joint is given as

$$J_{k_{blended}} = \frac{J_{i_k} \cdot w_i + J_{j_k} \cdot w_j}{w_i + w_j} \tag{1}$$

where $0 \leq w_i \leq 1$ and $w_j = 1 - w_i$ in this case. Similarly the blended quaternion $q_{blended}$ is given as

$$q_{blended} = q_i \left(q_i^{-1} q_j \right)^{w_j} \tag{2}$$

where q_i and q_j represent the quaternions to be blended [12]. Fig.4 shows a blended motion frame using the original motions of two different sign words.

Fig. 4. Motion blending

4   The Weather Forecast Sign Language System

In the service scenario, when a hearing-impaired user launches the 'weather forecast' application on a PC or mobile devices, the latest weather forecast video is provided with sign language animation. Fig.5 shows the system for the weather forecast sign language service.

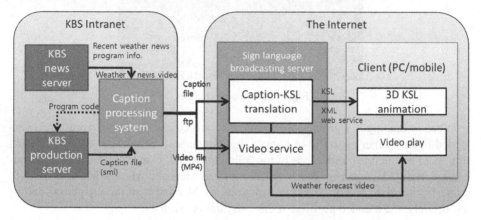

Fig. 5. Weather forecast sign language system

```xml
<?xml version="1.0" encoding="UTF-8"?>
- <KoreanSignLanguage>
   - <Description>
       <!--내일에 대한 정보들-->
       <SourceFile>20130824.Weather.txt</SourceFile>
     </Description>
   - <Translation>
       <!--번역 정보들-->
     - <Sentence startTime="00:00" id="1">
         <Original>오늘 반가운 봄비가 내리기 시작했습니다.</Original>
       - <MorphemeAnalysis>
           <!--형태소 분석 정보들-->
         - <Token id="1">
             <Word>오늘</Word>
             <Morpheme>오늘/일반명사</Morpheme>
           </Token>
         + <Token id="2">
         + <Token id="3">
         + <Token id="4">
         - <Token id="5">
             <Word>시작했습니다.</Word>
             <Morpheme>시작/동작성명사</Morpheme>
             <Morpheme>하다/동사화접미사</Morpheme>
             <Morpheme>었/선어말어미</Morpheme>
             <Morpheme>습니다/종결어미</Morpheme>
             <Morpheme>./온점</Morpheme>
           </Token>
         </MorphemeAnalysis>
       - <TranslationResult>
           <!--한국어 수화 번역쌍-->
         - <Token id="1">
             <Korean>오늘</Korean>
             <Sign>PA019033</Sign>
           </Token>
         + <Token id="2">
         + <Token id="3">
         - <Token id="4">
             <Korean>내리다</Korean>
             <Sign>PA008016</Sign>
           </Token>
         </TranslationResult>
       </Sentence>
     </Translation>
   </KoreanSignLanguage>
```

Fig. 6. Simplified XML schema for the KSL translation result

The 'caption processing system' asks the news server for the latest weather forecast information. If there is a new weather forecast, it inquires the corresponding closed caption from the production server, using the program code of the weather forecast. The weather forecast video in Flash Video (FLV) format originally targeted for real-time message protocol (RTMP) streaming is transcoded to H.264 (MP4). The closed caption is translated into KSL with the method described in section 2 in the sign language broadcasting server. The closed caption in KSL is distributed to clients using an XML web service. In a client device, weather forecast video and sign language animation are composited together and presented to the hearing-impaired user. Fig.6 shows the simplified XML schema for the KSL translation result.

5   Experimental Results

Fig.7 shows the result screens on a PC and an Android phone. We use Unity3D engine [13] to develop an application for both PC and mobile devices. The avatar can be moved to left or right, and can be enlarged or reduced, as the hearing-impaired users want the sign language broadcasting to be.

Fig. 7. Sign language animation on a PC (top) and an Android (bottom) phone

To our best knowledge, there has been no similar research that translates weather forecast texts into KSL to generate sign language animation. Hence we quantitatively compare the translation results in Table 5, with and without the Korlex-based synonym dictionary and word sense disambiguation.

Table 5. Quantatative comparison of the word translation rate

	Corpus size (words)	Word translation rate	
		without KorLex	with KorLex
Internal corpus	82,303	95.87%	96.17%
External corpus	1,448	88.60%	90.68%

In Table 5 the word translation rate is given as the ratio of the number of the words for which correct translation words were chosen, to the number of total words in the corpus. By applying KorLex the translation rates were increased, especially in terms of the translation failures. For example, the translation failure rate for the external corpus was reduced from 11.4% to 9.32% by applying KorLex. Notice that this is only a simplified measure for the evaluation purpose, because the characteristics of sign language such as non-manual signals and spatial representation are difficult to include in the evaluation, and were not considered in this paper.

6   Conclusion

The system proposed in this paper generates 3D sign language animation by translating the closed captions in DTV, for the hearing-impaired people to view the weather forecast with sign language. In order to find the frequency of each word, we analyzed the last three years' weather forecast scripts from several sources. We built sign language synonym dictionary using KorLex, to improve the translation performance. KorLex was also used for the word sense disambiguation process. We captured the motions of the additional 500 words for the weather forecast purpose. The motions were applied to a 3D avatar with motion blending. The servers and applications were developed to show weather forecast video and avatar animation in an on-demand manner.

Although the proposed system is targeted only for KSL users, we believe that it will be helpful to the other similar researches, especially if the source language and target sign language have a similar word order as in Korean and KSL.

In the future work, we will expand the translator and motion database for all kind of TV programs other than the weather forecast, and develop a real-time sign language broadcasting system. Constructing a sign language corpus using daily TV programs can be another future work.

References

1. Kaneko, H., Hamaguchi, N., Doke, M., Inoue, S.: Sign language animation using TVML. In: Proceedings of the 9th ACM SIGGRAPH Conference on Virtual-Reality Continuum and its Applications in Industry. ACM (2010)
2. Kato, N., Kaneko, H., Inoue, S., Shimizu, T., Hiruma, N.: Machine translation to Sign Language with CG-animation. ABU Technical Review 245 (2011)

3. Araújo, T., Ferreira, F., Silva, D., Lemos, F., Neto, G., Omaia, D., Filho, G., Tavares, T.: Automatic generation of Brazilian sign language windows for digital TV systems. Journal of the Brazilian Computer Society 19(2), 107–125 (2013)
4. Ferreira, F., Lemos, F., Neto, G., Araújo, T., Filho, G.: Providing support for sign language in middlewares compliant with ITU J.202. In: IEEE International Symposium on Multimedia (2011)
5. http://www.korean.go.kr/ (2014)
6. Kang, M.: Variable-Length Category-Based N-Gram Part-of-Speech Tagging Model Considering Morph-Syntactic Characteristics of Korean, Doctoral thesis, Pusan National University (2007) (in Korean)
7. Yoon, A., Hwang, S., Lee, E., Kwon, H.: Construction of Korean Wordnet KorLex 1.5. Journal of KIISE: Software and Applications 36(1), 92–108 (2009) (in Korean)
8. http://www.primpo.co.kr/ (2014)
9. http://www.xsens.com/products/ (2014)
10. CyberGlove Systems. CyberGlove II brochure (2009), http://www.cyberglovesystems.com/products/cyberglove-II/overview
11. Slot, K.: Motion Blending. Copenhagen University. Department of Computer Science (2007)
12. Shoemake, K.: Animating rotation with quaternion curves. In: Proceedings of ACM SIGGRAPH 1985. Annual Conference Series, pp. 245–254. ACM Press (1985)
13. https://unity3d.com/unity/ (2014)

Interpretation of Possessive and Reflexive-Possessive Pronouns of Bulgarian Language in DATR

Velislava Stoykova

Institute for Bulgarian Language, Bulgarian Academy of Sciences,
52, Shipchensky proh. str., bl. 17, 1113 Sofia, Bulgaria
vstoykova@yahoo.com

Abstract. The paper presents a formal interpretation of possessive and reflexive-possessive pronouns in Bulgarian language using non-monotonic approach to formal logical representation. It offers a semantic network to model inflectional morphology of definite article for related pronouns using DATR language for lexical knowledge representation. The proposed encoding combines non-monotonic orthogonal semantic networks hierarchical representation and relational database approach. Finally, some conclusions about presenting inflectional morphology using non-monotonic semantic networks approach are presented.

Keywords: Non-monotonic Reasoning, Knowledge Acquisition and Representation, Natural Language Processing.

1 Introduction

Non-monotonic approach has been widely used for formal representation in various Natural Language Processing tasks. Its main advantage is that it allows formal account for both regular and irregular cases and is suitable for modeling any type of grammar feature. Also, some knowledge representation languages which allow non-monotonic interpretation have been successfully used to model various types of language features (like inflection, syntax, etc.) for related and non-related natural languages. Further, we are going to present a formal interpretation of inflectional morphology of possessive and reflexive-possessive pronouns in Bulgarian language by using DATR language for lexical knowledge representation as a suitable formal framework.

2 The DATR Language

The DATR language for lexical knowledge representation is a non-monotonic language for defining the inheritance networks through path/value equations [4]. It has both an explicit declarative semantics and an explicit theory of inference allowing efficient implementation, and at the same time, it has necessary expressive power to encode lexical entries presupposed by the work in the unification grammar tradition [2,3].

In DATR information is organized as a network of nodes, where a node is a collection of related information. Each node has associated with it a set of equations that define

L. Iliadis et al. (Eds.): AIAI 2014, IFIP AICT 436, pp. 528–536, 2014.
© IFIP International Federation for Information Processing 2014

partial functions from paths to values where paths and values are both sequences of atoms. Atoms in paths are sometimes referred to as attributes.

DATR is functional, it defines mapping which assigns unique values to node attribute-path pair, and the recovery of these values is deterministic. With respect to its universality, DATR's formal properties and techniques underlay both rule-based inference and non-monotonic inference by default, and allow to account for language phenomena such as regularity, irregularity, and subregularity by using deterministic parsing.

The semantics of DATR uses non-monotonic inference and default inheritance, and allows generalization-capturing representation of inflectional morphology. DATR has expressive power which is capable to encode and process both syntactic and morphological rules and it allows representation of grammar knowledge by using semantic networks.

The DATR language for lexical knowledge representation offers syntagmatic operators (which can be used to define concatenation procedures) and paradigmatic operators (which can be used to define specific structure of the inflecting rules, and for further development of part-of-speech interpretations).

The DATR language has a lot of implementations, however, our application was made by using QDATR 2.0 (consult URL http://www.cogs.susx.ac.uk/lab//nlp/datr/datrnode49.html for a related file bul_det.dtr). This PROLOG encoding uses Sussex DATR notation [13].

DATR allows construction of various types of language models (language theories), however, the analysed interpretations underlay architecture of a rule-based formal grammar and a lexical database. The particular queries to be evaluated are all related inflected word forms, and the implementation allows to process words in Cyrillic alphabet.

The DATR language has been applied for developing inflectional morphology for a lot of languages including Russian [1] and Polish [5]. The ideas used for Russian nominal inflection interpretation underlay that of a paradigm and the encoding presents a resolving of a tabular conceptualization encoding task.

3 Bulgarian Possessive Pronouns in DATR

The semantics of possessive pronouns in Bulgarian language includes various types of relation like: possession (depending whether it is an object or a subject of possession), part-of-whole, relational, etc. The main semantic relationship of possession varies depending whether it is referred to the possessor or to the thing being possessed [7]. Only the full forms of possessive pronouns have inflection. They are: 'moj' (my), 'tvoj' (your), 'negov' (his), 'nein' (her), 'nash' (our), 'vash' (your), 'tehen' (their).

They have grammar features of person, number, gender and definiteness. The grammar feature of person is not inflectional and expresses information both at the level of syntax and at the hypertext level through agreement. The full forms imply information both about the possessor and the object being possessed using agreement in number and gender.

The grammar feature of definiteness implies information about the possession at the syntactic level using agreement and is expressed by a formal morphological marker which is an ending morpheme [6]. It is different for genders however, for masculine gender two types of definite morphemes exist – to determine a defined in a different way entities, which have two phonetic alternations, respectively. For feminine and for neuter gender only one definite morpheme exists, respectively. For plural, two definite morphemes are used depending on the ending vocal of the main plural form.

possessed		POSSESSOR								
		singular					plural			definiteness
		person								
number	gender	1st pers.	2nd pers.	3rd person			1st pers.	2nd pers.	3rd pers.	
				gender						
				male	female	neuter				
sg.	male	moj	tvoj	negov	nein	negov	nash	vash	tehen	undefined
sg.	female	moja	tvoja	negova	nejna	negova	nasha	vasha	tjahna	
sg.	neuter	moe	tvoe	negovo	nejno	negovo	nashe	vashe	tjahno	
pl.		moi	tvoi	negovi	nejni	negovi	nashi	vashi	tehni	
sg.	male	moja	tvoja	negovija	nejnija	negovija	nashija	vashija	tehnija	defined
sg.	male	mojat	tvojat	negovijat	nejnijat	negovijat	nashijat	vashijat	tehnijat	
sg.	female	mojata	tvojata	negovata	nejnata	negovata	nashata	vashata	tjahnata	
sg.	neuter	moeto	tvoeto	negovoto	nejnoto	negovoto	nasheto	vasheto	tjahnoto	
pl.		moite	tvoite	negovite	nejnite	negovite	nashite	vashite	tehnite	

Fig. 1. The Bulgarian possessive pronouns inflection table representation

The features of gender and number of definite article are different from the gender and number features of possessive pronouns, themselves. The former are inflectional whereas the later are not inflectional, even both they can express agreement and both they are significant for part-of-speech parsing. The relation between them is important starting point for our interpretation. The inflectional forms of possessive pronouns are presented at Fig 1.

Our approach to inflectional morphology of possessive pronouns is based on the encoding already published in [9]. It presents an inheritance semantic network consisting of different inflectional hierarchical type nodes of rule-based formal grammar and a lexical database (the pronouns). The particular queries to be evaluated are related inflected forms. It also offers an account of sound alternations.

The interpretation is based on adjectives' encoding [8] and takes as a starting point linguistic motivation, in particular, the priority of one or another grammar feature. Thus, it uses definite article's grammar features of gender and number to triger the value of related inflected forms. It also is similar to the approach presented in [10,11] to possessive pronouns formal representation using frameworks of Universal Networking Language [12] which present another type of semantic network.

Further, we are going to extend that approach by relating inflectional rules and inflectional type hierarchy to grammar features of possessive pronouns themselves.

3.1 The Inflectional Type Hierarchy

The DATR analysis of possessive pronouns starts with node DET which defines all inflecting morphemes for definite article as follows:[1]

```
DET:
    <sing undef>          ==
    <sing def_2 masc>     == _ja
    <sing def_2 masc_1>   == _a
    <sing def_1 masc>     == _jat
    <sing def_1 masc_1>   == _ut
    <sing def_1 femn>     == _ta
    <sing def_1 neut>     == _to
    <plur undef>          ==
    <plur def_1>          == _te.
```

The basic node AdjG defines grammar rules for generating all inflected forms for the feature of gender, number, and definiteness and is exactly the same as for the adjectives.

```
AdjG:
    <sing undef masc> == "<root>"
    <sing undef femn> == "<root gend>" _a
    <sing undef neut> == "<root gend>" _o
    <sing def_2 masc> == "<plur undef masc>" DET
    <sing def_1 masc> == "<plur undef masc>" DET
    <sing def_1>      == "<sing undef>" DET
    <plur undef>      == "<root gend>" _i
    <plur def_1>      == "<plur undef>" DET.
```

Node Adj inherits all grammar rules from node AdjG but employs grammar rules for generating inflected forms for the feature of comparison of degree. It also is the same as for the adjectives. We employ it to make encoding as short as possible regardless the fact that possessive pronouns generates compared inflected forms very rare. The pronouns 'negov' (his) and 'nein' (her) uses this type of inflectional rules.

```
Adj:
    <> == AdjG
    <compar> == po_  "<>"
    <superl> == naj_ "<>".
```

Node Adj_2 is defined as it is employed for the adjectives and it presents inflectional rules of possessive pronoun 'tehen' (their), which realises two types of phonetic alternation during the process of inflection. At this node an additional inflectional base form <root plur> is introduced to account for the complexity.

```
Adj_2:
    <> == Adj
    <plur undef> == "<root plur>" _i.
```

---

[1] Here and elsewhere in the description we use Latin alphabet to present morphemes instead of Cyrillic. Because of the mismatching between both some typically Bulgarian phonological alternations are marked by two letters, whereas in Cyrillic alphabet they are marked by one.

Node `Adj_4` is defined also for adjectives. It encodes inflectional rules of possessive pronouns 'nash' (our) and 'vash' (your) and inherits all grammar rules of node `Adj`. The new employed grammar rule changes inflectional morpheme for neuter gender into -e.

```
Adj_4:
    <> == Adj
    <sing undef neut> == "<root gend>" _e.
```

The new node `Adj_5` is employed and it consists of inflectional rules for possessive pronouns 'moj' (my) and 'tvoj' (your). It defines feminine and neuter gender forms by using palatal morphemes, and generates defined inflected forms of masculine gender by using the `<root gend>` base.

```
Adj_5:
    <> == Adj
    <sing undef femn> == "<root gend>" _ja
    <sing undef neut> == "<root gend>" _e
    <sing def_1 masc> == "<root gend>" DET
    <sing def_2 masc> == "<root gend>" DET.
```

The grammar features of gender and number of possessive pronouns themselves can be presented as relational information introducing node `PossPron` which consists of rules that relate lexical database information presented at the nodes of possessive pronouns themselves ('Moj', 'Tvoj', 'Negov', 'Nein', 'Nash', 'Vash' and 'Tehen') to inflectional grammar rules nodes `AdjG`, `Adj_2`, `Adj_4` and `Adj_5`. The node is as follows:

```
PossPron:
<first sing_p sing undef masc> == <sing undef masc> Moj
<first sing_p sing undef femn> == <sing undef femn> Moj
<first sing_p sing undef neut> == <sing undef neut> Moj
<first sing_p sing def_2 masc> == <sing def_2 masc> Moj
<first sing_p sing def_1 masc> == <sing def_1 masc> Moj
<first sing_p sing def_1>      == <sing def_1> Moj
<first sing_p plur undef>      == <plur undef> Moj
<first sing_p plur def_1>      == <plur def_1> Moj
<secound sing_p sing undef masc> == <sing undef masc> Tvoj
<secound sing_p sing undef femn> == <sing undef femn> Tvoj
<secound sing_p sing undef neut> == <sing undef neut> Tvoj
<secound sing_p sing def_2 masc> == <sing def_2 masc> Tvoj
<secound sing_p sing def_1 masc> == <sing def_1 masc> Tvoj
<secound sing_p sing def_1>      == <sing def_1> Tvoj
<secound sing_p plur undef>      == <plur undef> Tvoj
<secound sing_p plur def_1>      == <plur def_1> Tvoj
<third sing_p masc_p sing undef masc> == <sing undef masc> Negov
<third sing_p masc_p sing undef femn> == <sing undef femn> Negov
<third sing_p masc_p sing undef neut> == <sing undef neut> Negov
<third sing_p masc_p sing def_2 masc> == <sing def_2 masc> Negov
<third sing_p masc_p sing def_1 masc> == <sing def_1 masc> Negov
<third sing_p masc_p sing def_1>      == <sing def_1> Negov
<third sing_p masc_p plur undef>      == <plur undef> Negov
<third sing_p masc_p plur def_1>      == <plur def_1> Negov
<third sing_p femn_p sing undef masc> == <sing undef masc> Nein
```

```
<third sing_p femn_p sing undef femn> == <sing undef femn> Nein
<third sing_p femn_p sing undef neut> == <sing undef neut> Nein
<third sing_p femn_p sing def_2 masc> == <sing def_2 masc> Nein
<third sing_p femn_p sing def_1 masc> == <sing def_1 masc> Nein
<third sing_p femn_p sing def_1>          == <sing def_1> Nein
<third sing_p femn_p plur undef>          == <plur undef> Nein
<third sing_p femn_p plur def_1>          == <plur def_1> Nein
<first plur_p sing undef masc> == <sing undef masc> Nash
<first plur_p sing undef femn> == <sing undef femn> Nash
<first plur_p sing undef neut> == <sing undef neut> Nash
<first plur_p sing def_2 masc> == <sing def_2 masc> Nash
<first plur_p sing def_1 masc> == <sing def_1 masc> Nash
<first plur_p sing def_1>      == <sing def_1> Nash
<first plur_p plur undef>      == <plur undef> Nash
<first plur_p plur def_1>      == <plur def_1> Nash
<second plur_p sing undef masc> == <sing undef masc> Vash
<second plur_p sing undef femn> == <sing undef femn> Vash
<second plur_p sing undef neut> == <sing undef neut> Vash
<second plur_p sing def_2 masc> == <sing def_2 masc> Vash
<second plur_p sing def_1 masc> == <sing def_1 masc> Vash
<second plur_p sing def_1>      == <sing def_1> Vash
<second plur_p plur undef>      == <plur undef> Vash
<second plur_p plur def_1>      == <plur def_1> Vash
<third plur_p sing undef masc> == <sing undef masc> Tehen
<third plur_p sing undef femn> == <sing undef femn> Tehen
<third plur_p sing undef neut> == <sing undef neut> Tehen
<third plur_p sing def_2 masc> == <sing def_2 masc> Tehen
<third plur_p sing def_1 masc> == <sing def_1 masc> Tehen
<third plur_p sing def_1>      == <sing def_1> Tehen
<third plur_p plur undef>      == <plur undef> Tehen
<third plur_p plur def_1>      == <plur def_1> Tehen.
```

The new introduced paths <first>, <second> and <third> refer to the feature of person, paths <masc_p> and <femn_p> refer to the feature of gender, and paths <sing_p> and <plur_p> refer to the feature of number of possessive pronouns themselves. The generated inflected forms of possessive pronoun 'Negov' are given at the Appendix.

4   Bulgarian Reflexive-Possessive Pronoun in DATR

The semantics of reflexive-possessive pronoun in Bulgarian language combines semantics of possession relationship and that of reflexivity. It expresses possession relationship between the possessor (defined by the subject in the sentence, and agreed with it in gender and number) and the thing being possessed (to which the pronoun is referred to, and agrees in gender and number). The reflexive-possessive pronoun is one and it has full and short form, and both they can be used with respect to agreement. However, only its full form 'svoj' (-self) has inflectional grammar features of person, gender, number, and definiteness, which are similar to that of adjectives and of possessive pronouns. Its inflected forms are given at Fig. 2.

534   V. Stoykova

number	singular			plural
gender	male	female	neuter	
undefined	svoj	svoja	svoe	svoi
defined	svoja(t)	svojata	svoeto	svoite
	svojat			

Fig. 2. The Bulgarian reflexive-possessive pronoun's inflection table representation

4.1 The Inflectional Type Hierarchy

The DATR formal account of reflexive-possessive pronoun inflectional morphology [9] uses inflectional rules already defined at node Adj_5 and is part of it. It also uses the same principle as for possessive pronouns. The entire interpretation is extended whith rules which relate reflexive-possessive pronoun to its grammar features of person and gender (the person and gender of the subject in the sentense which are reflexive) and the gender and number grammar features of its definite article (the gender and number of the complement in the sentence). The node RefPoss is introduced which is as follows:

```
RefPoss:
      <sing undef masc> == <sing undef masc> Svoj
      <sing undef femn> == <sing undef femn> Svoj
      <sing undef neut> == <sing undef neut> Svoj
      <sing def_2 masc> == <sing def_2 masc> Svoj
      <sing def_1 masc> == <sing def_1 masc> Svoj
      <sing def_1>      == <sing def_1> Svoj
      <plur undef>      == <plur undef> Svoj
      <plur def_1>      == <plur def_1> Svoj.
```

The encoding differ with that for possessive pronouns 'moj' and 'tvoj' because the features of person and gender of pronoun 'svoj' are reflexive and should be defined by the person and gender of the subject in the sentense. Thus, entire interpretation gives a relation also to syntactic interpretation for part-of-speech analysis. The related generated inflected forms of reflexive-possessive pronoun 'svoj' are given at the Appendix.

5   Conclusion

The entire formal semantic network interpretation of inflectional morphology of possessive and reflexive-possessive pronouns in Bulgarian language uses non-monotonic approach to account for both regular and irregular word forms by introducing different nodes with related inflectional rules. It also uses the features of gender and number as a trigger to change the values of related inflected forms and a relational database approach to relate the grammar features of definite article to grammar features of pronouns themselves allowing further syntactic interpretation.

Thus, it is good as a future works to extend that approach by enlarging the application offering syntactic rules.

References

1. Corbett, G., Fraser, N.: Network Morphology: a DATR account of Russian nominal inflection. Journal of Linguistics 29, 113–142 (1993)
2. Evans, R., Gazdar, G.: Inference in DATR. In: Fourth Conference of the European Chapter of the Association for Computational Linguistics, pp. 66–71 (1989a)
3. Evans, R., Gazdar, G.: The semantics of DATR. In: Cohn, A.G. (ed.) Proceedings of the Seventh Conference of the Society for the Study of Artificial Intelligence and Simulation of Behaviour, pp. 79–87. Pitman/Morgan Kaufmann, London (1989b)
4. Evans, R., Gazdar, G.: DATR: A language for lexical knowledge representation. Computational Linguistics 22(2), 167–216 (1996)
5. Czuba, K.: The DATR Web Pages at Sussex (1994), http://www.cogs.susx.ac.uk/lab/nlp/datr/datrnode49,filepolish_n.dtr
6. Gramatika na suvremennia bulgarski knizoven ezik. Morphologia, tom. 2 (1983) (in Bulgarian)
7. Nicolova, R.: The Bulgarian Pronouns Nauka i izkustvo. Sofia (1986) (in Bulgarian)
8. Stoykova, V.: The definite article of bulgarian adjectives and numerals in DATR. In: Bussler, C.J., Fensel, D. (eds.) AIMSA 2004. LNCS (LNAI), vol. 3192, pp. 256–266. Springer, Heidelberg (2004)
9. Stoykova, V.: Bulgarian possessive and reflexive-possessive pronouns in DATR. In: Trappl, R. (ed.) Cybernetics and Systems 2010, pp. 426–432. Austrian Society for Cybernetic Studies, Vienna (2010)
10. Stoykova, V.: The Inflectional Morphology of Bulgarian Possessive and Reflexive-possessive Pronouns in Universal Networking Language. Procedia Technology 1, 400–406 (2012)
11. Stoykova, V.: Representation of Possessive Pronouns in Universal Networking Language. In: Iliadis, L., Papadopoulos, H., Jayne, C. (eds.) EANN 2013, Part II. CCIS, vol. 384, pp. 129–137. Springer, Heidelberg (2013)
12. Uchida, H.: Universal Networking Language. UNDL Foundation (2005)
13. The DATR Web Pages at Sussex (1997), http://www.cogs.susx.ac.uk/lab/nlp/datr/

Appendix

```
Negov:
        <> == Adj
        <root> == negov
        <root_gend> == negov.


PossPron: <third sing_p masc_p sing undef masc> == negov.
PossPron: <third sing_p masc_p sing undef femn> == negova.
PossPron: <third sing_p masc_p sing undef neut> == negovo.
PossPron: <third sing_p masc_p sing def_2 masc> == negovija.
PossPron: <third sing_p masc_p sing def_1 masc> == negovijat.
PossPron: <third sing_p masc_p sing def_1 femn> == negovata.
PossPron: <third sing_p masc_p sing def_1 neut> == negovoto.
PossPron: <third sing_p masc_p plur undef>       == negovi.
PossPron: <third sing_p masc_p plur def_1>       == negovite.
```

```
Svoj: <> == Adj_5
      <root> == svoj
      <root gend> == svo.

RefPoss: <sing undef masc> == svoj.
RefPoss: <sing undef femn> == svoja.
RefPoss: <sing undef neut> == svoe.
RefPoss: <sing def_2 masc> == svoja.
RefPoss: <sing def_1 masc> == svojat.
RefPoss: <sing def_1 femn> == svojata.
RefPoss: <sing def_1 neut> == svoeto.
RefPoss: <plur undef>      == svoi.
RefPoss: <plur def_1>      == svoite.
```

Forecasting Algorithm Adaptive Automatically to Time Series Length

Kolyo Onkov[1] and Georgios Tegos[2]

[1] Department of Mathematics, Computer Science and Physics, Agricultural University,
12 Mendeleev, 4000 Plovdiv, Bulgaria
kolonk@au-plovdiv.bg
[2] Department of Information Technology,
Alexander Technological Educational Institute of Thessaloniki, P.O. Box 14561, 54101, Greece
gtegos@gen.teithe.gr

Abstract. The developed forecasting algorithm creates trend models based on varying length time series by eliminating its oldest member. The constructed criterion evaluates the attained models through estimating the ratio between the average of the stochastic errors for the forecasted period and the average of real values. The best model and forecasting are automatically achieved in contrast to statistical software systems SPSS, STATISTICA, etc. where this process is accomplished progressively by the user. Therefore, this forecasting algorithm is adaptive to the length of time series. Component oriented approach has been used for software implementation. Simulation experiments have been carried out to test the forecasting algorithm using the multidimensional time series database on fishery in Greece. This algorithm is more efficient in case forecasting is applied on large number of time series because it saves time and efforts.

Keywords: varying length time series, automatic model fitting, criterion, adaptive algorithm.

1 Introduction

Time series forecasting refers to the process of identifying past relationships and trends in historical data for predicting future values [1]. Forecasting is important in time series analysis because it plays a central role in management since it precedes decision making [2]. There are many time series analysis techniques related to forecasting [3]. Trend modelling can be used for forecasting if it is assumed that the studied event will follow the same rules during the historical and the forecasted period. The attained forecasts are a simple consequence of the trend extrapolation. The advantages of forecasts based on trend modelling compared to other forecasting methods are: a) the preliminary evaluation of the forecast stochastic error by trend models and b) the determination of the confidence intervals. As a result, the critical limits, on which the real values of the studied event fluctuate during the forecasted period, can be evaluated.

L. Iliadis et al. (Eds.): AIAI 2014, IFIP AICT 436, pp. 537–545, 2014.

While modeling the trend of time series for forecasting purposes, the issue concerning the length of the period, within which the trends will be studied, frequently arises. Here there are two tendencies opposing each other. The stochastic error of the forecast decreases in longer historical periods. However, by covering a longer historical period, there is a risk for the model to include the influence of factors that used to function in the past, but later disappeared.

Forecasting time series based on trend modeling is often implemented through using statistical packages such as SPSS, Statistica etc. However, for reapplying trend modeling to different length time series and achieving new forecasted values the same procedure is followed manually. In general, in case of forecasting based on large numbers of time series, automatic forecasting is more efficient [2], [4]. Two automatic forecasting time series algorithms based on state space and ARIMA models have been implemented as R packages for statistical computing [5]. Fuzzy time series models and automatic forecasting techniques are developed to improve forecasting accuracy [6], [7].

This work aims to present an adaptive to time series length algorithm in order to achieve the best models and forecasted values automatically. The idea is to reapply forecasting by decreasing time series length and testing different trend models. As a result, different forecasts are achieved giving the opportunity to choose the most proper one based on objective criterions.

2   Materials and Methods

The algorithm is based on the gradual elimination of data from the start of the historical time series by applying the same trend model to each resultant time series. The confidence intervals are evaluated according to the different types of trend models on the grounds of which the forecast is achieved.

Let's denote:

y_i $(i=1, 2,..., n)$ – time series with length n,

$\hat{y}_{n+j}$ – forecasted values, j = 1, 2, ..., L, L – length of the forecasted period

$$S_y = \sqrt{\frac{1}{n} \sum_{i=1}^{n} (y_i - \hat{y}_i)^2} \quad \text{– standard error,} \tag{1}$$

$\hat{y}_i$ $(i=1,2,..., n)$ – smoothed values by the model

The   linear $y = A_0 + A_1 t$ and   the   polynomial   second   degree $y = A_0 + A_1 t + A_2 t^2$ trend models are applied to varying length time series. The stochastic error of the forecasted value $\hat{y}_{n+j}$ for linear (equation 2) and for polynomial second degree (equation 3) model is [8]:

$$e_{\hat{y}_{n+j}} = \frac{S_y}{\sqrt{n-1}} \cdot \sqrt{1 + \frac{1}{n} + \frac{3(n+2j-1)^2}{n(n^2-1)}} \tag{2}$$

$$e_{\hat{y}_{n+j}} = \frac{S_y}{\sqrt{n-1}} \cdot \sqrt{1 + \frac{1}{\displaystyle\sum_{p=1}^{j} t_p^2} \cdot t_{n+j}^2 + \frac{\displaystyle\sum_{p=1}^{j} t_p^4 - (2\sum_{p=1}^{j} t_p^2) t_{n+j}^2 + n.t_{n+j}^4}{n\displaystyle\sum_{p=1}^{j} t_p^4 - (\sum_{p=1}^{j} t_p^2)^2}} \tag{3}$$

where S_y is the standard error by equation 1.

The confidence interval for the linear and polynomial second degree models is evaluated as follows:

$$\hat{y}_{n+j} - t_{[1-\alpha]} e_{\hat{y}_{n+j}} \le \tilde{y}_{n+j} \le \hat{y}_{n+j} + t_{[1-\alpha]} e_{\hat{y}_{n+j}} \tag{4}$$

$\tilde{y}_{n+j}$ – the attained value of y for the jth year of the forecasted period;

$t_{[1-\alpha]}$ – Student's test at a level of significance α and n-2 degrees of freedom.

From equations 2 and 3 it is noticed that the value of the stochastic error depends directly on the length of time series n to which the trend model is applied and on the length of the forecasted period j. When n increases the value of stochastic error decreases while when j increases the stochastic error also does. Stated in other words, the greater the historical period is, the smaller the stochastic error becomes and on the other hand, the longer the forecasted period is, the greater the stochastic error becomes.

A criterion is created in order to evaluate the forecasts attained from time series with different length. The criterion is related to the stochastic error (equation 2 and 3). During the elimination of data of time series by decreasing each time the value of n by 1, the ratio between the average of the stochastic errors for the forecasted period and the average of real values are computed according to the next equation:

$$C_1(d) = \left(\frac{1}{L} \sum_{j=1}^{L} e_{\hat{y}_{d+j}} \right) \bigg/ \left(\frac{1}{n} \sum_{i=1}^{n} y_i \right), \quad d=n, n-1,\ldots,m \tag{5}$$

m is the minimal value of time series length, $(m < n)$. The best trend model and forecasting are achieved for the minimal value of $C_1(d)$ criterion when varying the length time series.

The created forecasting algorithm, based on trend models applied to varying length time series, is presented here for one time series by the following steps:

1. Implement steps 2-6 for both trend models;
2. Procedure for trend modeling;
3. Procedure for trend model adequacy by F-test. If the model is adequate then go to step 4 else go to step 6;
4. Computation of smoothed values $\hat{y}_i$, i=1, 2, ...n and Standard error S_y;
5. Procedure for estimation and storing of:

- Forecasted values $\hat{y}_{n+1}$, $\hat{y}_{n+2}$, ..., $\hat{y}_{n+L}$;
- Stochastic errors $e_{\hat{y}_{n+1}}$, $e_{\hat{y}_{n+2}}$ ..., $e_{\hat{y}_{n+L}}$;
- Confidence intervals by equation 4 applying Student's test;
- $C_1(n)$ value.

6. Next operation;
7. $n=n-1$;
8. If $n < m$ then go to step 10 else go to step 9;
9. Elimination of the most distant year from time series and go to step 1;
10. Application of the criterion for the best forecasting – MIN $(C_1(n), C_1(n-1), ..., C_1(m))$ for both trend models;
11. Evaluation and printing the best forecasting results based on both trend models.

The algorithm is adaptive to time series length and as a result it gives the ability to achieve automatically forecasts based on trend modelling. The perspective of this algorithm is to apply different weights on the time series members used for trend modeling. The study of this approach will reveal abilities for attaining more precise forecasting values.

3   Results and Discussion

3.1   Software and Simulation Experiment

Component oriented approach and VB programming language have been used for software implementation according to the algorithm. The procedures are placed in two modules:

- The first of them is consisted of procedures for estimating the descriptive statistics (mean value, standard deviation, standard error etc) as well as for evaluating the F-test and Student test;
- The second one contains the procedures for trend modelling, managing the length of time series, stochastic error, confidence intervals, application of the criterion and presentation of the results.

Generally, the procedures in the second module, "call" the procedures from the first one. VB programming code is also created for managing Excel-sheets which are proper for storing tables containing automatically accessed theoretical values of F-test and Student test by the trend modelling and forecasting and forecasting results as well.

A simulation experiment has been carried out to test the forecasting software using multidimensional time series database on sea fishery in Greece. Data in database concerns quantities and values of fish catch by areas, fish species, fishing tools and category as well as months, kinds of fishery and employment for the period 1990-2011 [9], [10]. Each time series extracted from FTS database consists of 22 members. For the purpose of forecasting it is divided into two parts: the first twenty members are used as historical time series and the last two (2010, 2011) are compared with the obtained forecasted values for the same years (proportion 20:2).

3.2   Forecasting Results

The forecasting results presented here are oriented to time series on catch quantity of 71 sea fish species statistically registered in Greece. Table 1 presents the part of the obtained results concerning only 10 fish species. Generally, for forecasting purposes the optimal time series length on the particular fish species is different. Besides the type of trend model used for the forecasts is also different. From 71 fish species there are 6 without adequate trend because of big random factor and as a result no forecasts are proposed by the algorithm. Almost all these cases concern fish species with small amount of catches – mean value for the studied period of time do not exceed 200 metric tons (Sprat, Skipjack etc).

Table 1. Trend model and optimal length time series

Fish code	Fish name	Trend model	Optimal length time series
15	Bogue	Linear	12
23	Goatfish	Polynomial 2 degree	17
25	Red bream	Polynomial 2 degree	10
31	Red mullet	Linear	12
35	Bonito	Polynomial 2 degree	15
36	Sprat	No adequate model	-
41	Skipjack	No adequate model	-
43	Goldline	Linear	13
55	Tune fish	Linear	16
68	Cuttle fish	Polynomial 2 degree	18

Table 2 presents more detailed results of the forecasting algorithm applied on quantity of catch for fish species "Bogue", based on linear trend modeling. For each varying length historical time series (n = 15, 14, 13, 12, 11) the forecasted values for the years 2010 and 2011 as well as the confidence interval (equation 4) and the C_1

criterion (equation 5) are also presented. The relative error shows the percentage deviation between real and forecasted values. The best forecast is obtained for time series length 12 when C_1 criterion has the minimum value (Figure1). The obtained relative error values are considered satisfactory – 10.87% and 2.47% for the years 2010 and 2011 respectively.

Table 2. Forecasting results for fish species Bogue by automatic application of linear trend model to varying length time series

Time series length	Year	Fore-casted values	Confidence interval		Real values	C_1	Relative error %
15	2010	2882.92	2464.92	3300.92	3201.69	14.50	9.96
	2011	2683.34	2255.65	3111.03	3405.40	15.94	21.20
14	2010	3052.99	2645.02	3460.96	3201.69	13.36	4.64
	2011	2885.30	2466.59	3304.00	3405.40	14.51	15.27
13	2010	3311.15	2993.61	3628.69	3201.69	9.59	3.42
	2011	3095.09	2867.97	3522.21	3405.40	10.24	9.12
12	**2010**	**3549.77**	**3363.76**	**3735.78**	**3201.69**	**5.24**	**10.87**
	2011	**3484.84**	**3292.34**	**3677.34**	**3405.40**	**5.52**	**2.33**
11	2010	3579.59	3373.36	3785.82	3201.69	5.76	11.80
	2011	3521.54	3306.89	3736.19	3405.40	6.10	3.41

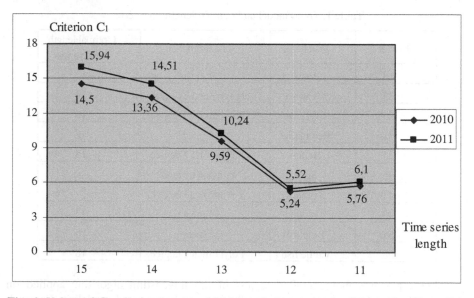

Fig. 1. Values of C_1 criterion between 15-11 length time series for fish species "Bogue" in Greek fishery

Table 3 presents the results of the forecasting subroutine application, based on polynomial second degree trend model, on the "total fish quantity" time series. It is evident that the best forecasting result is achieved for time series length 9, because the value of criterion C_1 is the smallest – 2.96 and 3.75, for years 2010 and 2011, respectively.

Table 3. Forecasting results for total quantity of fish catches in Greece

Time series length	Year	Forecasted values	Confidence interval		Real values	C_1	Relative error %
20	2010	61479.50	50465.45	72493.55	70122.20	17.91	12.33
	2011	55345.39	43955.74	66735.04	62871.50	20.58	11.97
19	2010	69345.30	59130.43	79560.17	70122.20	14.73	1.11
	2011	66207.68	55611.86	76803.50	62871.50	16.00	5.31
18	2010	78486.68	69918.60	87054.77	70122.20	10.92	11.93
	2011	79005.61	70086.60	87924.63	62871.50	11.29	25.66
17	2010	86653.35	79761.13	93545.58	70122.20	7.95	23.57
	2011	90610.88	83407.19	97814.57	62871.50	7.95	44.12
16	2010	92217.97	86049.52	98386.42	70122.20	6.69	31.51
	2011	98648.66	92171.04	105126.28	62871.50	6.57	56.91
15	2010	89754.04	83290.40	96217.68	70122.20	7.20	28.00
	2011	95025.24	88200.23	101850.25	62871.50	7.18	51.14
14	2010	91224.82	84201.31	98248.32	70122.20	7.70	30.09
	2011	97231.40	89767.06	104695.74	62871.50	7.68	54.65
13	2010	89050.90	81438.80	96663.00	70122.20	8.55	26.99
	2011	93898.06	85745.68	102050.44	62871.50	8.68	49.35
12	2010	80442.93	75340.46	85545.40	70122.20	6.34	14.72
	2011	80371.25	74855.87	85886.62	62871.50	6.86	27.83
11	2010	79141.56	73427.51	84855.61	70122.20	7.22	12.86
	2011	78269.03	72022.79	84515.28	62871.50	7.98	24.49
10	2010	71699.59	69295.85	74103.32	70122.20	3.35	2.25
	2011	65865.75	63201.50	68529.99	62871.50	4.04	4.76
9	**2010**	**69385.92**	**67329.52**	**71442.32**	**70122.20**	**2.96**	**1.05**
	2011	**61869.42**	**59550.33**	**64188.50**	**62871.50**	**3.75**	**1.59**
8	2010	68904.13	66429.77	71378.50	70122.20	3.59	1.74
	2011	61002.20	58149.52	63854.88	62871.50	4.68	2.97
7	2010	66870.92	64342.34	69399.50	70122.20	3.78	4.64
	2011	57161.69	54161.67	60161.71	62871.50	5.25	9.08

Figure2 presents the real values for total fish catch quantity for time period 1999-2011 and the forecasted values based on time series length n=11, 9 and 7 (polynomial second degree trend model). The results have the following characteristics:

- Forecast of length 11 is optimistic because trend model uses much bigger values of catch quantity at the beginning of the period 1999–2009 than at the end. For the opposite reason forecast of length 7 spanning the period 2003–2009 is pessimistic.
- The most important result consists of the fact that the algorithm automatically finds the optimal time series length 9 and the polynomial second degree model for achieving the best forecast.

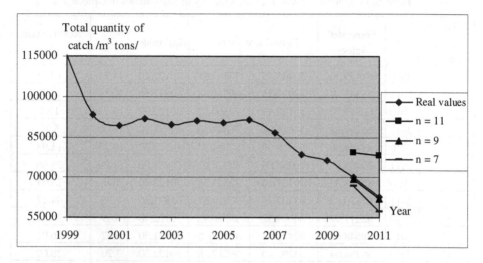

Fig. 2. Forecasting results for varying time series length n

The obtained results show that the algorithm is adaptive to the dynamic of the fish catch quantity process and automatically achieves forecasting results.

4   Conclusion

This work presents an algorithm applied on time series, capable to achieving optimum trend models and forecasting with minimum human intervention. The most significant characteristic of the developed algorithm is the varying length time series used for automatic model fitting. A criterion related to the stochastic error of the forecasted values is constructed for the estimation of the best forecast. When analyzing the set of time series the algorithm behaviour is adaptive to time series length and trend model. As a result, the forecasts are achieved automatically.

A component oriented approach has been used for the implementation of the software. The forecasting algorithm has been tested on time series sets concerning sea fishery in Greece and acceptable forecasting results are attained. The same algorithm can be applied in appropriate time series of other branches giving equivalent results as well. More efficient results are achieved in case forecasting is applied on large numbers of time series because it saves time and efforts.

References

1. Singh, S.: Pattern Modelling in Time-Series Forecasting. An International Journal of Cybernetics and Systems 31(1), 49–66 (2000)
2. Makridakis, S., Wheelwright, S., McGee, V.: Forecasting: Methods and Applications, p. 926. Wiley, New York (1983)
3. Leonard, M.: Large-Scale Automatic Forecasting: Millions of Forecasts, SAS Institute, http://citeseerx.ist.psu.edu/viewdoc/ download?doi=10.1.1.72.9518&rep=rep1&type=pdf
4. Marahaj, E.A., Inder, B.: Forecasting Time Series from Clusters, Working Paper 9/99, Monash University, Australia, http://ideas.repec.org/p/msh/ebswps/1999-9.html
5. Hyndman, R.J., Khandakar, Y.: Automatic Time Series Forecasting: The forecast Package for R. Journal of Statistical Software 27(3), 1–22 (2008)
6. Wong, W.K., Bai, E., Chu, A.W.: Adaptive Time-Variant Models for Fuzzy-Time-Series Forecasting. IEEE Transactions on Systems, Man, and Cybernetics, Part B: Cybernetics 40(6), 1531–1542 (2010)
7. Chen, S.M., Tanuwijaya, K.: Multivariate fuzzy forecasting based on fuzzy time series and automatic clustering techniques. Journal of Expert Systems with Applications 38(8), 10594–10605 (2011)
8. Velickova, N.: Statistical methods for study and forecasting the development of socioeconomic processes. Publishing House "Science & Art", Sofia (1981)
9. Tegos, G.: PC-Information System concerning sea fishery time series in Greece, PhD thesis, Computer science department, Agricultural University, Plovdiv (2005)
10. Onkov, K., Tegos, G.: Management solutions for fish resources in Greece based on analysis of multi-dimensional database. In: Proceedings of the International Conference "Sustainable Landscape Planning and Safe Environment", pp. 95–103 (2012)

Six Degrees of Freedom Implicit Haptic Rendering

Konstantinos Moustakas

Electrical and Computer Engineering Department, University of Patras,
Rio-Patras, Greece
moustakas@upatras.gr

Abstract. This paper introduces a six degrees of freedom haptic rendering scheme based on an implicit support plane mapping representation of the object geometries. The proposed scheme enables, under specific assumptions, the analytical reconstruction of the rigid 3D object's surface, using the equations of the support planes and their respective distance map. As a direct consequence, the problem of calculating the force feedback can be analytically solved using only information about the 3D object's spatial transformation and position of the haptic probe. Several haptic effects are derived by the proposed mesh-free haptic rendering formulation. Experimental evaluation and computational complexity analysis demonstrates that the proposed approach can reduce significantly the computational cost when compared to existing methods.

Keywords: Haptic rendering, implicit representation, six degrees of freedom.

1 Introduction

Human perception combines information of various sensors, including visual, aural, haptic, olfactory, in order to perceive the environment. Virtual reality applications aim to immerse the user into a virtual environment by providing artificial input to its interaction sensors (i.e., eyes, ears, hands). The visual and aural inputs are the most important factors in human-computer interaction (HCI). However, virtual reality applications will remain far from being realistic without providing to the user the sense of touch. The use of haptics augments the standard audiovisual HCI by offering to the user an alternative way of interaction with the virtual environment [1]. However, haptic interaction involves complex and computationally intensive processes, like collision detection or distance calculation [5], that place significant barriers in the generation of accurate and high fidelity force feedback.

Seen from a computational perspective, haptic rendering can be decomposed in two different but heavily interrelated processes, namely collision detection and force calculation. Initially, collisions have to be identified and localized and then the resulting force feedback has to be estimated so as to accurately render the force that will be fed back to the user using specific assumptions on the physical

L. Iliadis et al. (Eds.): AIAI 2014, IFIP AICT 436, pp. 546–555, 2014.
© IFIP International Federation for Information Processing 2014

model involved. Concerning haptic rendering research can be divided into three main categories [2]: Machine Haptics, Human Haptics and Computer Haptics [3]. Machine Haptics is related to the design of haptic devices and interfaces, while Human Haptics is devoted to the study of the human perceptual abilities related to the sense of touch. Computer Haptics, or alternatively haptic rendering, studies the artificial generation and rendering of haptic stimuli for the human user [4], [5]. It should be mentioned that the proposed framework takes into account recent research on human haptics, while it provides mathematical tools targeting mainly the area of computer haptics.

The simplest haptic rendering approaches focus on the interaction with the virtual environment using a single point. Many approaches have been proposed so far both for polygonal, non-polygonal models, or even for the artificial generation of surface effects like stiffness, texture or friction, [6]. The assumption, however, of a single interaction point limits the realism of haptic interaction since it is contradictory to the rendering of more complex effects like torque. On contrary, multipoint, or object based haptic rendering approaches use a particular virtual object to interact with the environment and therefore, besides the position of the object, its orientation becomes critical for the rendering of torques. Apart from techniques for polygonal and non-polygonal models [6], voxel based approaches for haptic rendering including volumetric haptic rendering schemes [7] have lately emerged. Additionally, research has also tackled with partial success the problem of haptic rendering of dynamic systems like deformable models and fluids [8].

In general, with the exception of some approaches related to haptic rendering of distance or force fields, one of the biggest bottlenecks of current schemes is that haptic rendering depends on the fast and accurate resolution of collision queries. The proposed approach aims to widen this bottleneck by providing a free-form implicit haptic rendering scheme based on support plane mappings able to provide a six degrees of freedom haptic feedback. In particular, a 3D object is initially modelled using the associated support plane mappings [9]. Then the distance of the object's surface from the support plane is mapped at discrete samples on the plane and stored at a preprocessing step following the same procedure presented in [10]. During run-time and after collision queries are resolved, estimation of the force feedback can be analytically estimated, while several haptic effects, like friction and texture can be easily derived. This results in constant or linear time complexity haptic rendering based only on the 3D transformation of the associated object and the position of the haptic proxy.

2   Support Plane Mappings

Support planes are a well studied subject of computational geometry and have been employed in algorithms for the separation of convex objects [9,11,12]. From a geometrical perspective, a support plane E of a 3D convex object O is a plane such that O lies entirely on its negative halfspace H_E^-. Support planes have become useful in previous algorithms based on the concept of support mappings. A support mapping is a function that maps a vector $\mathbf{v}$ to the vertex of

548   K. Moustakas

vert(O) of object O that is "most" parallel to $\mathbf{v}$ [9,13]. As a direct consequence, a support plane can be defined as the plane that passes through $s_O(\mathbf{v})$, the support mapping of $\mathbf{v}$, and is parallel to $\mathbf{v}$.

The importance of support planes is intuitively apparent: they provide an explicit way of deciding whether another object could possibly intersect with the one that the support planes refers to. Based on this simple but important feature of support planes, a slightly more generalized formulation has been derived [14] introducing the concept of support plane mappings for collision detection. The approach described in [14] is used in the proposed framework for collision detection.

After collision is detected, the force feedback provided to the user through the haptic device has to be calculated. In the present framework, force feedback is obtained directly from the model adopted for collision detection, thus handling collision detection and haptic rendering in an integrated way, as described in the sequel.

Let the parametric form of the support plane equation $S_{SP}(\eta, \omega)$ be:

$$\mathbf{S_{SP}}(\eta, \omega) = \begin{bmatrix} x_0 + \eta u_1 + \omega v_1 \\ y_0 + \eta u_2 + \omega v_2 \\ z_0 + \eta u_3 + \omega v_3 \end{bmatrix}, \forall \eta, \omega \in \Re \tag{1}$$

where $\mathbf{u}$ and $\mathbf{v}$ constitute an orthonormal basis of the support plane and (x_0, y_0, z_0) its origin.

Assuming now a dense discretization of the η, ω space, we can define a discrete distance map of the support plane SP and the underlying manifold mesh surface S_{mesh}, by calculating the distance of each point of SP from S_{mesh}:

$$D_{SP}(\eta, \omega) = ICD(S_{SP}, S_{mesh}) \tag{2}$$

where ICD calculates the distance of every point sample (η, ω) of the support plane SP, alongside the normal direction at point (η, ω), from the mesh S_{mesh} and assigns the corresponding values to the distance map $D_{SP}(\eta, \omega)$. The distance map is used in the sequel to analytically estimate the force feedback.

It should be mentioned that the above procedure results in scalar distance maps that accurately encode the surface if and only if there is a one to one mapping of all surface parts with at least one support plane. If such a mapping does not exist, then vectorial distance maps can be used that include information about the distance of all sections of the ray cast in the normal direction of the support plane to the object mesh.

3 Point-Based (3DoF) Haptic Rendering

Referring to Figure 1, let point $\mathbf{H}_p$ be the position of the haptic probe and S_{mesh} represent the local surface of the object.

Let also S_{SP} represent the distance of point $\mathbf{H}_p$ from the support plane, which corresponds to point $\mathbf{P}_M$ on the SP.

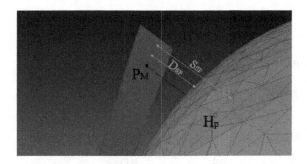

Fig. 1. Distance calculation using distance maps over support planes

If collision is detected, the absolute value of the force fed onto the haptic device is obtained using a spring model as illustrated in Figure 1. In particular:

$$\|\mathbf{F}\| = k \cdot |S_{SP} - D_{SP}(\mathbf{P_M})| \tag{3}$$

where k is the spring constant. $D_{SP}(\mathbf{P_M})$ is the distance of point $\mathbf{P_M}$ from the mesh and is stored in the distance map of the support plane. Notice that the term $|S_{SP} - D_{SP}(\mathbf{P_M})|$ is an approximation of the actual distance of $\mathbf{H}_p$ from the mesh that becomes more accurate if the support plane surface approximates well the mesh.

The direction of the force should in general be perpendicular to the local area, where collision is detected. An obvious solution to the evaluation of the direction of this force would be to detect the surface element (i.e. triangle) where the collision occurred and to provide the feedback perpendicularly to it. This approach is not only computationally intensive, but also results in non-realistic non-continuous forces at the surface element boundaries. In the present framework the analytical approximation of the mesh surface is used utilizing the already obtained SP approximation and the distance map. Based on this approximation the normal to the object's surface can be approximated rapidly with high accuracy. In particular, assuming that (η, ω) form an orthonormal basis of the plane then if $D_{SP}(\eta, \omega)$ is the scalar function of the distance map on the support plane, as previously described, the surface S_{mesh} of the modelled object can be approximated by equation (4) (Figure 1):

$$\mathbf{S}_{mesh}(\eta, \omega) = S_{SP}(\eta, \omega) - D_{SP}(\eta, \omega)\,\mathbf{n}_{SP} \tag{4}$$

where S_{SP} is the surface of the support plane, D_{SP} the associated distance map and $\mathbf{n}_{SP}$ its normal vector that can be easily evaluated through $\mathbf{n}_{SP} = \mathbf{u} \times \mathbf{v}$.

Now the calculation of the force feedback demands the evaluation of the normal vector $\mathbf{n}_S$ on the object's surface. That is obtained through equation (5). In the following the brackets (η, ω) will be omitted for the sake of simplicity.

$$\mathbf{n}_S = \frac{\partial \mathbf{S}_{mesh}}{\partial \eta} \times \frac{\partial \mathbf{S}_{mesh}}{\partial \omega} \tag{5}$$

where

$$\frac{\partial \mathbf{S}_{mesh}}{\partial \eta} = \frac{\partial \mathbf{S}_{SP}}{\partial \eta} - \frac{\partial D_{SP}}{\partial \eta} \mathbf{n}_{SP} - D_{SP} \frac{\partial \mathbf{n}_{SP}}{\partial \eta} \tag{6}$$

Since $\mathbf{n}_{SP}$ is constant over SP, equation (6) becomes:

$$\frac{\partial \mathbf{S}_{mesh}}{\partial \eta} = \mathbf{u} - \frac{\partial D_{SP}}{\partial \eta} \mathbf{n}_{SP} \tag{7}$$

A similar formula can be extracted for $\frac{\partial \mathbf{S}_{mesh}}{\partial \omega}$:

$$\frac{\partial \mathbf{S}_{mesh}}{\partial \omega} = \mathbf{v} - \frac{\partial D_{SP}}{\partial \omega} \mathbf{n}_{SP} \tag{8}$$

All above terms can be computed analytically, except from $\frac{\partial D_{SP}}{\partial \eta}$ and $\frac{\partial D_{SP}}{\partial \omega}$ that are computed numerically.

Substituting now equations (4), (6), (7), (8) in equation (5) the normal direction $\mathbf{n}_S$ can be obtained.

Since, the direction of the normal along the surface of the modelled object is obtained using equation (5), the resulting force feedback is calculated through:

$$\mathbf{F}_h = k \left| S_{SP} - D_{SP}(\mathbf{P}_M) \right| \frac{\mathbf{n}_S}{\|\mathbf{n}_S\|} \tag{9}$$

4   6DoF Haptic Rendering

Let us now assume that the haptic interaction point is actually a haptic interaction object that is also modelled using support plane mappings and distance maps.

Referring to Figure 2 that for simplicity depicts the 2D case, let M_1 and M_2 be the mesh areas of two different objects possibly involved in collision and S_1 and S_2 their respective support planes. Based on the approach presented in [14] it can be decided whether M_1 and M_2 are probably involved in collisions using their SPMs S_1 and S_2. Now the question is: Is it possible to identify the collisions and calculate 6DoF force feedback without entering the computationally intensive narrow-phase [14] of the collision detection algorithm?

This question reduces to the problem of calculating the impact volume (3D case) or impact surface (2D case) S as depicted in Figure 2. Let us consider for sake of simplicity and without loss of generality the 2D case. If $\mathbf{d}$ is a material density function then the mass corresponding to the collision area S can be described by the following formula.

$$V = \int\int_S \mathbf{d} \, dS \tag{10}$$

Now referring again to Figure 2 let W be a sampling point of the support plane S_2 and W_2 the point of mesh M_2 that can be trivially reconstructed by projecting W along side the normal direction n_2 as far as the distance map

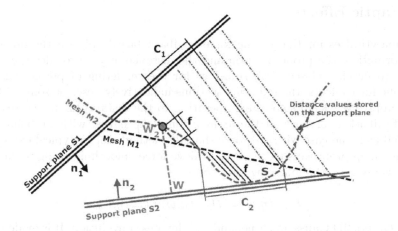

Fig. 2. Distance calculation using distance maps over support planes

$D_{SP}(W)$ dictates. Now, W_2 can be also trivially projected on the support plane S_1 and retrieve the corresponding distance value $D_1(M_1)$ of S_1 from the mesh M_1. Now, omitting the notation (η, ω) in the following for simplicity and assuming the function f is defined as follows:

$$f = D_1(W_2) - D_1(M_1) \tag{11}$$

where $D_1(W_2)$ is the distance of point W_2 from S_1 and $D_1(M_1)$ the corresponding distance of S_1 from mesh M_1. Now, collision can be reported only for cases where function f is positive.

Let now C_1 and C_2 (Figure 2) denote the projection of the colliding volume on the support planes S_1 and S_2 respectively. It is obvious that C_1 is the support set of positive values of function f. Then, equation (10) can be transformed as follows:

$$V = \int_{t_1 \in C_1} \mathbf{d} \cdot f \, dt_1 = \int_{t_2 \in C_2} \mathbf{d} \cdot f \cdot \cos(\mathbf{n}_1 \mathbf{n}_2) \, dt_2 \tag{12}$$

A similar analysis can be followed for the three dimensions and the corresponding formula is:

$$V = \int \int_{\eta, \omega \in C_2} \mathbf{d} \cdot f \cdot \cos(\mathbf{n}_1 \mathbf{n}_2) \, d\eta d\omega \tag{13}$$

Now since the volume of the penetrating object part and its centre of inertia is known, 6DoF haptic feedback can be easily estimated. In the context of the proposed framework a single force is estimated for the penetrating part of the haptic interaction object that is applied on the gravity centre of the colliding volume. Other approaches for approximate 6DoF haptic rendering like the one presented in [15] can be also implemented.

5   Haptic Effects

The analytical estimation of the force feedback based only on the object 3D transformation, the probe position and the distance maps, provides the opportunity to develop closed form solutions for the rendering of physics-based or symbolic force effects; the following sections indicatively describe some of them.

By applying a local smoothing operation to the distance map, the resulting force feedback is smooth in the areas around the edges, without being over-rounded as is the case with the force shading method [16]. A typical example of distance map preprocessing so as to achieve force smoothing using a Gaussian kernel is given by the following equation:

$$D'_{SP}(\eta, \omega) = D_{SP}(\eta, \omega) * G_\sigma(\eta, \omega) \qquad (14)$$

where G_σ is a 2D Gaussian kernel and "$*$" denotes convolution. It is evident that different smoothing operators can be easily applied. A very useful operator that can be implemented is the force smoothing only in areas that are not smooth due to the finite tessellation (sampling) and not in object and surface boundaries, following the popular in computer graphics "crease angle" concept. A haptic "crease angle" rendering can be easily performed by applying anisotropic diffusion or even an edge-preserving smoothing operator on the distance map.

Similarly using the proposed framework for haptic rendering, haptic texture can be also simulated easily by applying appropriate transformations on the acquired distance map. An example for simulating surface roughness is provided below, where Gaussian noise is added on the distance map. No computational cost is added, since the procedures for calculating the force direction are not altered due to the existence of haptic texture. The only difference lies in the evaluation of the magnitude of $\mathbf{F}_{texture}$, which now yields from:

$$\mathbf{F}_{texture} = k\,|S_{SP} - (D_{SP}(\mathbf{P_M}) + n_g)|\,\frac{\mathbf{n}_S}{\|\mathbf{n}_S\|} \qquad (15)$$

where n_g denotes the gaussian noise.

6   Complexity and Experimental Results

In the following an analysis of the computational complexity of the proposed scheme in comparison to the typical state-of-the-art mesh-based haptic rendering scheme is discussed.

Moreover, even if an experimental analysis of the proposed support plane mapping based haptic rendering approach, in terms of timings for simulation benchmarks would not be fair for the state-of-the-art approaches, since it would encode the superiority of SPM based collision detection and would not directly highlight the proposed haptic rendering approach, two experiments are presented where the proposed haptic rendering scheme is compared to the state-of-the-art mesh-based haptic rendering in terms of timings in computationally intensive surface sliding experiments.

After collision is reported, a typical force feedback calculation scheme would need to identify the colliding triangle of the involved 3D object in $O(n)$ time, where n is the number of triangles, or in $O(logn)$ time if bounding volume hierarchies are used. Then the force can be calculated in constant $O(1)$ time. In order to avoid force discontinuities, for example force shading, and if there is no adjacency information then the local neighborhood of the colliding triangle can be found again in $O(n)$ time, where n is the number of triangles, or in $O(logn)$ time if bounding volume hierarchies are used. Finally, the mesh-based haptic rendering scheme has no additional memory requirements per se.

Table 1. Computational complexity comparison

Process	Mesh-based	Free-form
Force	$O(n)$ or $O(logn)$	$O(1)$
6DoF force	$O(mn)$ or $O(mlogn)$	$O(m)$
Smoothing	$O(n)$ or $O(logn)$	$O(1)$
Memory	-	$O(m \cdot s)$

On the other hand, concerning the proposed free-form implicit haptic rendering scheme, after a collision is detected, the resulting force feedback can be calculated in constant time $O(1)$ using equation (9). In order to avoid depth discontinuities the distance map can be smoothed, in an image processing sense, at a preprocessing phase. Even if this step is performed during run-time it would take $O(k)$ time, where k is the local smoothing region or the filtering kernel window. Concerning the 6DoF case, it can be considered requiring to execute m times the 3DoF procedure, where m is the number of support plane sampling points processed. On the other hand the proposed scheme has $O(m \cdot s)$ memory requirements, where m is the number of support planes and s the number of samples per support plane. Taking now into account that the more support planes are used the smaller their size and the less samples are necessary for a specific sampling density we can safely assume that the memory requirements are linear to the total number of samples that depends on the sampling density used.

Table 1 summarizes the computational complexity analysis of the proposed free-form haptic rendering scheme, when compared to the mesh-based approach. Concerning the quantitative results, interaction with two objects was considered, namely the Galeon and Teeth models of 4698 and 23000 triangles respectively. The objects are illustrated in Figure 3 .

Moreover, CHAI-3D was used for interfacing with the haptic devices [17]. Moreover, the force estimation algorithms were applied on a predefined trajectory of the haptic probe, so as to assure fair comparison. In particular, initially the trajectory of the haptic probe in the 3D space has been recorded while being in sliding motion over the objects' surface. Then this trajectory has been used as

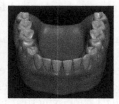

Fig. 3. Galeon model, 4698 triangles. Teeth model, 23000 triangles.

input for both algorithms so as to extract the timings mentioned below. Table 2 presents the mean timings and their standard deviation of the force estimation throughout the simulation using the mesh-based,and the proposed free-form haptic rendering scheme for the case of the Galeon and the Teeth models.

Table 2. Galeon and Teeth models: Interaction timings

| | Galeon | | Teeth | |
Process	Mean time (ms)	σ	Mean time (ms)	σ
Mesh-based	1.2	0.22	4.2	0.82
Free-form	0.028	0.006	0.036	0.005

It should be emphasized that the above timings need to be taken into account under the exact experimental setting. In particular, concerning the proposed approach 1000 support planes were used for the case of the Galeon and 1500 for the case of the Teeth model. Distances are estimated for all support planes and forces are calculated for the closer one. This procedure, could be optimized by partitioning the space in a preprocessing step and knowing beforehand to which support plane, each point in space "belongs to", thus reducing the search from $O(n)$ to $O(logn)$. Moreover, concerning the mesh-based approach force shading has been also implemented.

7   Conclusions

The proposed framework extends the support plane mapping concept and the corresponding collision detection scheme to direct free-form implicit haptic rendering. An analytical scheme to calculate both 3DoF and 6DoF force feedback is proposed. Its most significant property is that there is no need to enter the narrow-phase of the collision detection pipeline so as to identify the colliding triangles and subsequently estimate the reaction forces, on contrary it requires only the distance maps stored on the support planes. It is evident that the proposed scheme reduces significantly the computational cost in the performed simulations. This significant gain comes at an expense of two limitations. Firstly, special care has to be taken at the preprocessing step so that the models are well approximated using the support planes and the distance maps. For example if the objects demonstrate large complex concavities the use of vectorial distance maps is inevitable. Secondly, the proposed scheme cannot be, in its current form, directly applied to deformable models. An extension to piecewise or free-form deformable models, where deformations can be analytically expressed seems possible and remains a direction for future work.

References

1. Burdea, G., Coiffet, P.: Virtual Reality Technology, 2nd edn. Wiley-IEEE Press (2003)
2. Lin, M., Otaduy, M.: Haptic rendering: Foundations, algorithms and applications. A.K.Peters Publishing (2008)
3. Srinivasan, M., Basdogan, C.: Haptics in virtual environments: Taxonomy, research status and challenges. Computers and Graphics, 393–404 (1997)
4. Moustakas, K., Tzovaras, D., Strintzis, M.: Sq-map: Efficient layered collision detection and haptic rendering. IEEE Transactions on Visualization and Computer Graphics 13, 80–93 (2007)
5. Moustakas, K., Nikolakis, G., Kostopoulos, K., Tzovaras, D., Strintzis, M.: Haptic rendering of visual data for the visually impaired. IEEE Multimedia 14, 62–72 (2007)
6. Laycock, S., Day, A.: A survey of haptic rendering techniques. Computer Graphics Forum 26, 50–65 (2007)
7. Palmerius, K., Cooper, M., Ynnerman, A.: Haptic rendering of dynamic volumetric data. IEEE Transactions on Visualization and Computer Graphics 14, 263–276 (2008)
8. Barbic, J., James, D.: Six-dof haptic rendering of contact between geometrically complex reduced deformable models: Haptic demo. In: Proc. of Eurohaptics, pp. 393–394 (2009)
9. van den Bergen, G.: Collision Detection in Interactive 3D Environments. The Morgan Kaufmann Series in Interactive 3D Technology. Morgan Kaufmann (2003)
10. Moustakas, K.: Free-form implicit haptic rendering. In: Joint Virtual Reality Conference, JVRC, pp. 73–76 (March 2005)
11. Dobkin, D.P., Kirkpatrick, D.G.: A linear algorithm for determining the separation of convex polyhedra. Journal of Algorithms 6, 381–392 (1985)
12. Chung, K., Wang, W.: Quick Collision Detection of Polytopes in Virtual Environments. In: ACM Symposium on Virtual Reality Software and Technology 1996, pp. 1–4 (1996)
13. Ericson, C.: Real-Time Collision Detection. The Morgan Kaufmann Series in Interactive 3D Technology. Morgan Kaufmann (2005)
14. Vogiannou, A., Moustakas, K., Tzovaras, D., Strintzis, M.: Enhancing bounding volumes using support plane mappings for collision detection. Computer Graphics Forum 29, 1595–1604 (2010)
15. McNeely, W., Puterbaugh, K., Troy, J.: Six degree-of-freedom haptic rendering using voxel sampling. In: Computer Graphics and Interactive Techniques, pp. 401–408 (1999)
16. Ruspini, D., Kolarov, K., Khatib, O.: The haptic display of complex graphical environments. In: Computer Graphics (SIGGRAPH 1997 Conference Proceedings), pp. 345–352 (1997)
17. Conti, F., Barbagli, K., Morris, D., Sewell, C.: Chai 3d: An open-source library for the rapid development of haptic scenes. In: IEEE World Haptics, Pisa, Italy (March 2005)

Application of Supervised Self Organising Models for Wheat Yield Prediction

Xanthoula Eirini Pantazi[1], Dimitrios Moshou[1], Abdul Mounem Mouazen[2],
Boyan Kuang[2], and Thomas Alexandridis[1]

[1] Aristotle University of Thessaloniki, School of Agriculture, Department of Hydraulics,
Soil Science and Agriculture Engineering, Laboratory of Agricultural Engineering,
P.0 275 54 124, Thessaloniki, Greece
[2] Environmental Technology and Science Department, Cranfield University,
Bedfordshire MK43 0AL, United Kingdom
renepantazi@gmail.com, dmoshou@agro.auth.gr

Abstract. The management of wheat yield behavior in agricultural areas is a very important task because it influences and specifies the wheat yield production. An efficient knowledge-based approach utilizing an efficient Machine Learning algorithm for characterizing wheat yield behavior is presented in this research work. The novelty of the method is based on the use of Supervised Self Organizing Maps to handle existent sensor information by using a supervised learning algorithm so as to assess measurement data and update initial knowledge. The advent of precision farming generates data which, because of their type and complexity, are not efficiently analyzed by traditional methods. The Supervised Self Organizing Maps have been proved from the literature efficient and flexible to analyze sensor information and by using the appropriate learning algorithms can update the initial knowledge. The Self Organizing models that are developed consisted of input nodes representing the main factors in wheat crop production such as biomass indicators, Organic Carbon (OC), pH, Mg, Total N, Ca, Cation Exchange Capacity (CEC), Moisture Content (MC) and the output weights represented the class labels corresponding to the predicted wheat yield.

Keywords: neural networks, data mining, precision agriculture, machine learning.

1 Introduction

A large number of approaches, models, algorithms, and statistical tools have been proposed and used for assessing the yield prediction in agriculture. Many authors used simple linear correlations of yield with soil properties but the results varying from field to field and year to year (Drummond, et al., 1995; Khakural et al., 1999). Many other studies, contain complex linear methods like multiple linear regression, were accomplished with similar results (Drummond et al., 1995; Khakural et al., 1999; Kravchenko & Bullock, 2000). Some authors proposed non-linear statistical methods to investigate the yield response (Adams, et al., 1999; Wendroth, et al,. 1999). Expert systems and artificial intelligent algorithms are a relatively new subset of nonlinear

L. Iliadis et al. (Eds.): AIAI 2014, IFIP AICT 436, pp. 556–565, 2014.
© IFIP International Federation for Information Processing 2014

techniques. They have been proposed in agriculture for decision making and decision support tasks. More specifically, expert systems (Plant & Stave, 1991; Rao, 1992) have been developed and applied in different fields in agriculture to give advice and make management decisions. In this context many studies have been reported using artificial intelligence techniques and a few of them focused on the spatial analysis of produced data in precision agriculture. The most of them use artificial neural networks (ANNs) and machine learning algorithms for setting target yields which is one of the problems in PA (Canteri et al., 2002; Liu, et al., 2001; Miao et al., 2006). Schultz, et al. (2000) summarized the advantages of applying neural networks in agro ecological modeling, including the ability of ANN to handle both quantitative and qualitative data, merge information and combine both linear and non-linear responses. Neural networks have been proposed for identifying important factors influencing corn yield and grain quality variability (Miao et al., 2006), for data analysis (Irmak et al., 2006), for prediction crop yield based on soil properties (Drummond et al., 2003), for setting target corn yields (Liu et al., 2001). Shearer et al. (1999) studied a large number of variables, including fertility, satellite imagery, and soil conductivity, for a relatively small number of observations in one site-year of data.

Self-Organizing Maps (SOMs) are one of the most well-known among the several Artificial Neural Networks architectures proposed in literature (Kohonen, 1988). Their applications have increased during the last decade and they have been applied in several different fields and nowadays they are considered as one of the foremost machine learning tools and an important tool for multivariate statistics (Marini, 2009) Self-Organizing Maps (SOMs) are self-organizing systems able to solve problems in an unsupervised way, without needing target data. In order to cover certain needs, unsupervised models have been extended in order to be able to work in a supervised framework. To this end, methods like counterpropagation Artificial Neural Networks (CP-ANNs), which are very similar to SOMs, since an output layer is added to the SOM layer (Zupan et al., 1995), have been introduced. When dealing with classification issues, CP-ANNs are generally efficient methods for achieving class separation in non-linear boundaries. Recent modifications to CP-ANNs have led to the introduction of new supervised neural network architectures and relevant learning algorithms such as Supervised Kohonen Networks (SKNs) and XY-fused Networks (XY-Fs) (Melssen, 2006).

The aim of the work reported is to present a methodology that can determine wheat yield behavior in precision farming, based on Machine Learning techniques and particularly based on aspects related to cluster visualization. In the current paper several Self Organizing Map models using supervised learning approach and algorithm are used to classify precision agriculture data in order to predict the yield productivity. To achieve this, soil physical and chemical parameters have been fused together with biomass indicators.

2 Materials and Methods

2.1 Crop Parameters Affecting Yield

For the calculation of crop cover the Normalized Different Vegetation Index (NDVI) was used. The NDVI is (NIR-RED)/ (NIR+RED) where NIR is the Near Infrared Radiation (0,725 to 1μm) and RED is the Red Radiation (0, 58 to 0, 68 μm).

The NDVI was calculated based on satellite images that were collected two times on the 2nd May and 3rd June in the spring of 2013 (Fig. 1). These images were provided by DMCii (http://dmcii.tumblr.com/) for the TF1 in the Duck End farm in the UK.

Below is the processing workflow chain for crop NDVI, based on satellite imagery:

1. We received L1R (geo-rectified from DMCii) or L1T (ortho-rectified imagery). With the L1R we had to complete the ortho-rectification using software called 'keystone workstation', otherwise the L1T ortho product should be purchased directly from DMCii (the preferred method),
2. We performed in-band reflectance calibration using ArcGIS,
3. A map of NDVI was created with ArcGIS,
4. The NDVI was calculated for 5mX5m grid resulting 16500 values.

The yield was calculated for 5mX5m grid by a combine harvester that able to measure automatically the yield during harvest resulting in 16500 values.

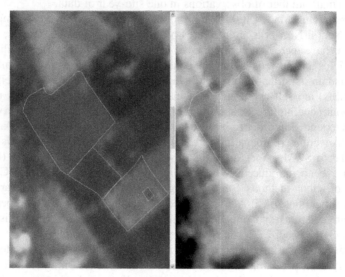

Fig. 1. False colour infrared (left) and NDVI (right) of 3rd Jun 2013 obtained from DMCii UK-DMC2 satellite imagery for TF1 (e.g. Hawnes End field) in the Duck End farm in the UK

2.2   Soil Parameters Affecting Yield

Precision farming requires development of on-line sensors for real-time measurement of soil properties, because these sensors can lead to reducing labor and time cost of soil sampling and analysis. With the emergence of commercial NIR spectrophotometers and multivariate calibration software packages, the vis–NIR spectroscopy has been adopted much widely for soil analysis. Numerous researchers have extended the vis–NIR spectroscopy applications from the measurement of key soil properties (Moisture Content, pH, Soil Organic Matter, Total N, and soil Organic Carbon) with high accuracy to almost all other micro and macro elements with less accuracy.

The analysis of soil with this technique was also extended to soil biological, physical, and engineering properties. Multivariate calibration techniques allowed for simultaneous measurements of several soil properties under consideration. Shibusawa et al. (2001) developed an on-line vis–NIR (400– 1700 nm) sensor to predict Moisture Content, pH, Soil Organic Matter, and NO_3-N. A simpler design to the one of Shibusawa et al. (2001) without sapphire window optical configuration was developed by Mouazen et al. (2005). The system was successfully calibrated for Moisture Content, Soil Organic Matter, Total N, pH, and available P in different soils in Belgium and northern France (Mouazen et al., 2005, 2007, 2009).

On-line measurements of soil were carried out in the selected field of Horn's End. The measurement was carried out by Cranfield University (CU). The raw soil spectra was recorded and stored for time on analysis. Soil samples collected from the field were sent to laboratory analysis with standard methods.

The following soil parameters were estimated from the spectra in a 5mX 5m grid: soil Organic Carbon (OC), pH, Mg, Total N, Ca, Cation Exchange Capacity (CEC) and Moisture Content (MC). This resulted in 16500 vectors of soil parameters for the whole field matching the number of values obtained from the NDVI calculation.

2.3   Counterpropagation Artificial Neural Networks

Counterpropagation Artificial Neural Networks (CP-ANNs) are modeling methods which combine features from both supervised and unsupervised learning (Zupan et al., 1995) CP-ANNs consist of two layers, a Kohonen layer and an output layer, whose neurons have as many weights as the number of classes to be modelled .The class vector is used to define a matrix C, with I rows and G columns, where I is the number of samples and G the total number of classes; each entry cig of C represents the membership of the i-th sample to the g-th class expressed with a binary code (0 or 1). When the sequential training is adopted, the weights of the rth neuron in the output layer (y_r) are updated in a supervised manner on the basis of the winning neuron selected in the Kohonen layer. Considering the class of each sample i, the update is calculated as follows:

$$\Delta y_r = \eta \left(1 - \frac{d_n}{d_{max} + 1} \right) \left(c_i - y_r^{old} \right) \tag{1}$$

where d_{ri} is the topological distance between the considered neuron r and the winning neuron selected in the Kohonen layer; c_i is the ith row of the unfolded class matrix C, that is, a G-dimensional binary vector representing the class membership of the ith sample. At the end of the network training, each neuron of the Kohonen layer can be assigned to a class on the basis of the output weights and all the samples placed in that neuron are automatically assigned to the corresponding class.

2.4   XY-Fused Networks

XY-fused Networks (XY-Fs) (Melssen et al., 2006) are supervised neural networks for building classification models derived from Self- Organizing Maps (SOMs).

In XY-fused Networks, the winning neuron is selected by calculating Euclidean distances between a) sample (x_i) and weights of the Kohonen layer, b) class membership vector (c_i) and weights of the output layer. These two Euclidean distances are then combined together to form a fused similarity, that is used to find the winning neuron. The influence of distances calculated on the Kohonen layer decreases linearly during the training epochs, while the influence of distances calculated on the output layer increases.

2.5   Supervised Kohonen Networks (SKNs)

As in the case for CP-ANNs and XY-Fs, Supervised Kohonen Networks (SKNs) (Melssen et al., 2006) are supervised neural networks derived from Self-Organizing Maps (SOMs) and used to calculate classification models. In Supervised Kohonen Networks, Kohonen and output layers are glued together to give a combined layer that is updated according to the training scheme of Self-Organizing Maps. Each sample (x_i) and its corresponding class vector (c_i) are combined together and act as input for the network. In order to achieve classification models with good predictive performances, x_i and c_i must be scaled properly. Therefore, a scaling coefficient for c_i is introduced for tuning the influence of class vector in the model calculation.

3   Results and Discussion

The values of the eight soil parameters were concatenated with the NDVI values so as to form 16500 feature vectors which correspond to fusion of both soil and crop parameters. Aiming to avoid bias during clustering the fusion vectors were preprocessed so that they had zero mean and start a deviation equal to unity. In order to predict the yield fusion vectors were used as inputs while the yield values were divided in three classes with equal number of samples containing 5500 each in ascending order, thus corresponding to low medium and high yield.

The Supervised Map models Xyf, Skn and Cpann were trained with the fusion vectors as input and the yield classes as output. In order to be able to test the generalization capability of the neural networks cross validation was applied by leaving one out of ten samples randomly so that after training on nine samples the prediction was tested on the tenth. The results of the cross validation are shown in Tables 1, 2, 3 for Xyf, SKn and Cpann. The best overall result is obtained from the Skn network. The Skn gives better results than the other two methods because in the case of Skn the clustering of input layers and output yield is performed using one combined vector and this reduces the possibility of deviating values of yield from affecting the result of classification. In the other two architectures there are values of yield that can affect the clustering in a negative way since it is possible that other non-measured parameters are capable of affecting the yield. In all cases the best prediction is obtained for the low category of yield which is advantageous since the low yield spots in the field require additional fertilization. The SOM clusters of the components of the training vectors are shown in Figures 2, 3, 4. The first subplot corresponds to NDVI component while the second to ninth correspond to Ca, CEC, MC, Mg, OC, P, pH, TN. The last two correspond to target yield and normalized target yield. From this can be seen that the NDVI is the first component and shows partial correlation to the yield map which is shown in the tenth subplot. The other components correspond to soil parameters and provide complementary information to the yield. The Component Maps are useful in interpreting the correlations between the soil factors and NDVI as related to the yield. In the case where tendency of the soil factors is similar to the yield this factor is important for higher yield while in the opposite case, a high value in a soil factor could limit the yield. Generally the NDVI has a positive correlation to the yield.

Table 1. Results of cross validation for Xyf Network

Real Network Estimation	Network Prediction (%)		
	low	medium	high
low	95.62	3.89	0.49
medium	5.45	89.95	4.6
high	5.35	11.29	83.36

Table 2. Results of cross validation for Skn Network

Real Network Estimation	Network Prediction (%)		
	low	medium	high
low	93.11	5.69	1.20
medium	4.09	89.05	6.85
high	2.33	6.64	91.04

Table 3. Results of cross validation for Cpann Network

Real Network Estimation	Network Prediction (%)		
	low	medium	high
low	92.40	4.36	3.24
medium	8.07	81.22	10.71
high	4.18	6.02	89.80

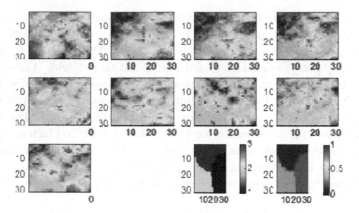

Fig. 2. Skn Component Maps for NDVI and soil parameters

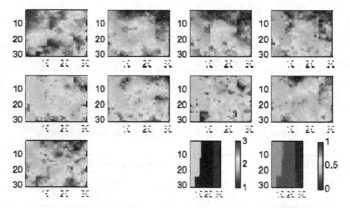

Fig. 3. Xyf Component Maps for NDVI and soil parameters

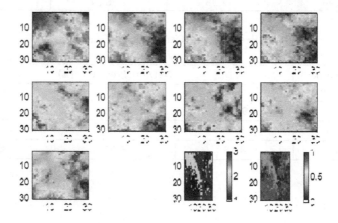

Fig. 4. Cpann Component Maps for NDVI and Soil parameters

The target yield is shown in Fig.5. in three classes labeled as red for high yield, blue for low yield and yellow for medium yield. By looking at Fig.6 one can see that there is a high correlation between the target yield and the predicted yield. The NDVI Shown in Fig.7 demonstrates partial correlation to the yield in the left part of the field, thus explaining the cluster distribution of the Neural Network models. This discrepancy in the right part of the field can be explained by the distribution of the other nutrients which counterbalance the effect on yield. In Figure 8, the spatial variation of total N is shown divided in three classes corresponding to different levels of total N.

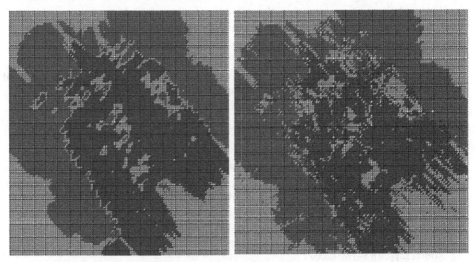

Fig. 5. Target Yield in three classes produced by ArcGIS

Fig. 6. Prediction in three classes produced by ArcGIS

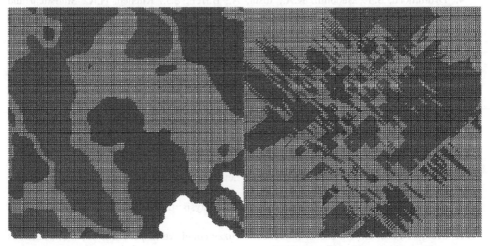

Fig. 7. NDVI Satellite based in three classes produced by ArcGIS

Fig. 8. Total Nitrogen in three classes produced by ArcGIS

4   Conclusions

In this paper, three Self Organizing models were developed that consisted of input nodes representing the main factors in wheat crop production such as biomass indicators, organic carbon (OC), pH, Mg, total N, Ca, Cation Exchange Capacity (CEC), moisture content (MC) and the output weights represented the class labels corresponding to the predicted wheat yield. The results indicate that yield prediction is possible with a very high accuracy reaching 93% and can be used in order to improve fertilizer administration by using the yield prediction models that were presented. Visualization of component maps could reveal extra information that is useful in order to interpret the relations between the soil and crop parameters and the yield.

Acknowledgement. The presented research was carried out in the framework of project FARMFUSE of ICT AGRI 2 ERANET.

References

1. Adams, M.L., Cook, S.E., Caccetta, P.A., Pringle, M.J.: Machine learning methods in site-specific management research: An Australian case study. In: Proceedings of the Fourth International Conference on Precision Agriculture, Madison, USA, pp. 1321–1333 (1999)
2. Canteri, M., Avila, B.C., Dos Santos, E.L., Sanches, M.K., Kovaleschyn, D., Molin, J.P., et al.: Application of data mining in automatic description of yield behavior in agricultural areas. In: Proceedings of the World Congress of Computers in Agriculture and Natural Resources, pp. 183–189 (2002)
3. Drummond, S.T., Sudduth, K.A., Birrell, S.J.: Analysis and correlation methods for spatial data. ASAE Paper No. 95–1335. ASAE, St. Joseph, Michigan (1995)
4. Drummond, S.T., Sudduth, K.A., Joshi, A., Birrell, S.J., Kitchen, N.R.: Statistical and neural methods for site-specific yield prediction. Transactions of the ASAE 46(1), 5–14 (2003)
5. Irmak, A., Jones, J.W., Batchelor, W.D., Irmak, S., Boote, K.J., Paz, J.O.: Artificial neural network model as a data analysis tool in precision farming. Transactions of the ASABE 49(6), 2027–2037 (2006)
6. Khakural, B.R., Robert, P.C., Huggins, D.R.: Variability of corn/soybean yield and soil/landscape properties across a southwestern Minnesota landscape. In: Proceedings of the Fourth International Conference on Precision Agriculture, pp. 573–579 (1999)
7. Kohonen, T.: Self-Organization and Associative Memory. Springer, Berlin (1988)
8. Kravchenko, A.N., Bullock, D.G.: Correlation of corn and soybean grain yield with topography and soil properties. Agronomy Journal 92(1), 75–83 (2000)
9. Liu, J., Goering, C.E., Tian, L.: A neural network for setting target corn yields. Transactions of the ASAE 44(3), 705–713 (2001)
10. Marini, F.: Artificial neural networks in food analysis: trends and perspectives. Analytica Chimica Acta 635, 121–131 (2009)
11. Miao, Y., Liu, Z.: On causal inference in fuzzy cognitive maps. IEEE 1034 Transactions on Fuzzy Systems 8, 107–119 (2000)
12. Melssen, W., Wehrens, R., Buydens, L.: Supervised Kohonen networks for classification problems. Chemometrics and Intelligent Laboratory Systems 83, 99–113 (2006)

13. Mouazen, A.M., De Baerdemaeker, J., Ramon, H.: Towards development of on-line soil moisture content sensor using a fibre-type NIR spectrophotometer. Soil Till. Res. 80, 171–183 (2005)
14. Mouazen, A.M., Maleki, M.R., De Baerdemaeker, J., Ramon, H.: On-line measurement of some selected soil properties using a VIS-NIR sensor. Soil Till. Res. 93, 13–27 (2007)
15. Mouazen, A.M.: The future of for on-line measurement of soil properties with sensor fusion. Landwards, pp. 14–16. Nelson Publishing, Natural Resources Department, Cranfield University, MK43 OAL, United Kingdom (2009)
16. Shearer, S.A., Thomasson, J.A., Mueller, T.G., Fulton, J.P., Higgins, S.F., Samson, S.: Yield prediction using a neural network classifier trained using soil landscape features and soil fertility data, St. Joseph, Michigan, USA. ASAE Paper No. 993042 (1999)
17. Plant, R., Stave, N.: Knowledge based systems in agriculture. McGraw-Hill, USA (1991)
18. Rao, J.P.: Expert systems in agriculture (1992),
    http://www.manage.gov.in/managelib/faculty/PanduRanga.htm
19. Shibusawa, S., Made Anom, S.W., Sato, H.P., Sasao, A.: Soil mapping using the real-time soil spectrometer. In: Gerenier, G., Blackmore, S. (eds.) ECPA 2001, agro Montpellier, vol. 2, pp. 485–490. Montpellier, France (2001)
20. Zupan, J., Novic, M., Gasteiger, J.: Neural networks with counter-propagation learning strategy used for modelling. Chemometrics and Intelligent Laboratory Systems 27, 175–187 (1995)
21. Wendroth, O., Jurschik, P., Nielsen, D.R.: Spatial crop yield prediction from soil and land surface state variables using an autoregressive state-space approach. In: Stafford, J.V. (ed.) Precision Agriculture 1999, Sheffield, U.K, pp. 419–428 (1999)

Prediction of 30-Day Mortality after a Hip Fracture Surgery Using Neural and Bayesian Networks

Dimitrios Galiatsatos[1], George C. Anastassopoulos[1], Georgios Drosos[2], Athanasios Ververidis[2], Konstantinos Tilkeridis, and Konstantinos Kazakos[2]

[1] Medical Informatics Laboratory, Medical School,
Democritus University of Thrace, Greece
dgaliatsatos2010@hotmail.com, anasta@med.duth.gr
[2] Department of Orthopedics, University Hospital of Alexandroupolis, Medical School,
Democritus University of Thrace, GR-68100
drosos@otenet.gr, athanasios@ververidis.net,
tilkerorth@yahoo.com, kazakosk@yahoo.gr

Abstract. Osteoporotic hip fractures have a significant morbidity and excess mortality among the elderly and have imposed huge health and economic burdens on societies worldwide. A medical database of 349 patients that have been operated for hip fracture has been analyzed. Two models of data were used in Multi-Layer Perceptrons, Radial Basis Function and Naïve Bayes networks, in order to predict the 30-day mortality after a hip fracture surgery and also to investigate which is the most appropriate risk factor between the New Mobility Score and Institution factor for the Greek population. The proposed method may be used as a screening tool that will assist orthopedics in the surgery of the hip fracture according to each different patient.

Keywords: hip fracture, artificial neural networks, Bayesian networks, 30-day mortality.

1 Introduction

Hip fractures are common in elderly people although their incidence varies among different countries and populations [1-3]. It has been estimated that according to the epidemiologic projections, the worldwide annual number of hip fractures will rise from 1.66 million in 1990 to 6.26 million by the year 2050 [4]. The mortality, morbidity social-economic costs after a hip fracture are significant [5-7]. Mortality in particular remains significant for decades although there is a geographical and Race/Ethnicity Differences [8-10]. Mortality rates are greatest within the first month [11-14]. The 30-day mortality has been used in several studies because the mortality during this time is directly related to the fracture, its therapy and complications, while the mortality rate beyond this time it may be related to other unrelated causes.

The ability to recognize patients at high risk of poor outcomes before operation would be an important clinical advance. Several preoperative risk factors for the high

L. Iliadis et al. (Eds.): AIAI 2014, IFIP AICT 436, pp. 566–575, 2014.
© IFIP International Federation for Information Processing 2014

mortality have been identified but the most predominant did not, and whether other factors could be considered as predictors are still unknown [15-17].

There are several reports concerning the pre-operative risk factors or scores like the Abbreviated Mental Test Score (AMTS) or the New Mobility Score that have been used as predictors for the 30-day mortality in patients with hip fractures. One of these scores is the Nottingham Hip Fracture Score (NHFS) [13, 18-19]. This score is using the following factors: age, sex, hemoglobin in admission, Abbreviated Mental Score (MMTS), and living or not in an institution, number of co-morbidities (other illness) and suffering from malignancy.

Patients that live in an institute are those that cannot live alone and the factor "living in an institute" is related to the "mobility status" of the patients. In some countries –like our country- elderly people usually live with relatives (in their own homes or in their relatives' homes). Therefore, the factor "living in an institute" is not a reliable factor. The New Mobility Score is a score related with the "mobility status" of these patients and has been found dependent with morbidity [20]. Patients with New Mobility Score 6 and above can walk out of the house and go for shopping even if they use a stick, while patients with score below 6 need help from another person. This score was used in this study in order to identify whether replacing the factor "living in an institute" with this score the results are more reliable or not.

The Artificial Neural Network (ANN) simulating high-level human brain functions, is a computational modeling tool that has become widely accepted for modeling complex real-world problems. Although it has been explored in many areas of medicine, such as nephrology, microbiology, radiology, neurology, cardiology, etc, its use in the orthopedic field is still rare [21].

The aim of this work was to apply three different ANN - based methods (Multi-Layer Perceptrons (MLPs), Radial Basis Function (RBF) Networks and Bayesian Networks) to evaluate their prediction of mortality using the factors of NHFS. In more details, the 3 methods were tested in a given number of patients with known characteristics (factors) and a known 30-day mortality. More specifically a decision support tool has been developed to help clinicians identify which people are at increased risk after the hip fracture surgery. This application area is considered as extremely important since it is associated with increased morbidity and mortality and high socio-economic costs. This paper was tried to focus on the three proposed methods and their performance in clinical data.

The paper is structured as follows: In the next section the hip fracture, as well as the risk factors that affect the 30-day mortality after the operation of the hip fracture are presented. Also, two models of data are described and three methods are illustrated. Section 3 analyzes the results, while in section 4 the conclusions are discussed.

2   Materials and Methods

2.1   Hip Fracture

This study assessed the predictive capability of Neural and Bayesian networks for 30-day mortality after surgery for hip fracture in the Greek population.

Data were collected from 349 patients that have been operated for hip fracture in the Orthopedic department of the University Hospital of Alexandroupolis. All these patients were followed-up and the 30-day mortality was recorded. The number of patients who died after 30-day of monitoring was 33. Apart from patients demographics, other factors that may be related to patients' mortality were also recorded and categorized as follows; Age (between 65 and 85 and above 85 years), Sex (male or female), Hemoglobin (Hb) in admission (below 10 g/dl and above 10 g/dl), Mini Metal Score (MMTS) (below 6 or above 6), Living or not in an Institute, Number of co-morbidities (other illness) (less than 2 or more than 2), suffering from Malignancy or not, and the New Mobility Score (score 6 and below or above 6).

Two models of data set have been used as follows: The first model (Model 1) depends on the factors: Age, Sex, Hb, MMTS, New Mobility Score, Number of co-morbidities and Malignancy. In the second model the New Mobility Score had been replaced by the Institute factor (Model 2). This model is using the factors that are being used in the Nottingham Hip Fracture Score. Both models consisted of seven almost identical attributes (independent variables), while the death was the dependent variable.

2.2 Neural and Bayesian Networks

Trying to find the best model for 30-day mortality after surgery for hip fracture, two most common algorithms of Artificial Neural Networks have been used, the Multilayer Perceptron and the Radial Basis Function. In addition to these ANN algorithms, the simplest Bayesian algorithm (Naïve Bayes) has been utilized. As a validation measurement of the data the k-fold cross validation has been utilized in all of the above methods.

All data were analyzed and processed with the assistance of Weka, which is a machine learning software written in Java and is an open source application that is freely available. Weka contains tools for data processing, classification, clustering, feature selection and visualization. The greatest capability of this machine learning is to perform useful information after learning from training data. [22].

Multilayer Perceptron is a feed-forward algorithm with one or more hidden layers between the input and the output layer. It utilizes a supervised learning technique called back propagation for training the network. In the medical field, MLP algorithm is widely used and gives great classification accuracy for some diseases such as breast cancer [23-24].

Radial Basis Function networks are feed-forward networks trained using a supervised training algorithm. They are similar to back propagation networks in many aspects although RBF have a few advantages. The main advantage in the biomedicine applications is that they train much faster and more accurate than back propagation networks [25-26]. In contrast with the time, the models are less robust than other methods (for example logistic regression) [21, 27].

Bayesian Networks are very efficient classifiers in many health related datasets (Natural Language Processing, Computer Vision, Medical Diagnosis, Bioinformatics, etc) [28]. Naïve Bayes (NB) Classifier is the simplest Bayesian approach algorithm assuming conditional independence between the variables in the models [29]. This simplicity makes naïve bayes techniques attractive and suitable in many domains. The connection in this classifier appears only between the variables and the main node in each model (output/dependent variable). The main advantages are its easy construction (known a priori) and its effective classification process. However there is a strong disadvantage that makes it unsuitable for every set of data. The conditional independence assumptions between nodes rarely are true in the most real world applications. Despite although strong dependencies between nodes, researchers had shown that naïve bayes performs good classification [30]. An updated form of the naïve bayes algorithm such as the augmented naïve bayes algorithm overcomes the conditional independence and it is an innovative algorithm based on the original one. In recent years many researchers had tried to evolutes naive bayes classifiers as they select feature subset and relax independence assumptions [31-32].

In k-fold cross validation the data set separated in k equal size subsets. From the k subsets, a single subset preserved as a validation data and the remaining k-1 samples are utilized as training data. The cross validation process repeated k times with each of the k subsets used exactly once. Then the k results are averaged and came of the estimation. The advantage of this method is that each observation is used in training and validation process. In this work, 3, 10 and 20 fold cross validation has been performed in each technique. In experiments, such as in Hyperthyroid disease, a 6-fold cross validation performed maximum accuracy [33].

3   Results

For each one of the two models, the entire set of 349 data records were used, in order to evaluate the best classification method between multilayer perceptron, radial basis function and naïve bayes networks. For the validation of the data, 3, 10 and 20-fold cross validation for each classified algorithms has been used. For each neural network (MLP and RBF) many experiments have been made using a lot of different topologies in terms of hidden layer etc. The results presented below correspond to the best topology of each method which had the best performance.

In figure 1a, MLP topology of the second model is presented. Seven factors are presented in the input layer, whereas four nodes consist the hidden layer. The same topology had been used in the RBF network.

For the evaluation of the models except from Artificial Neural Networks (MLP and RBF), Bayesian neural networks (naïve Bayes) had utilized in order to find the best classified model.

In figure 1b the direct dependencies between the dependent variable (Death) and the independent variables of the first model in the naïve bayes networks are presented. As it can be observed this specified algorithm has given no direct dependencies between the independent variables. For the better explanation of the Bayesian network, probabilities were concluded in each confusion matrix between variables given the output. The same topology with figure 1a has the naïve bayes form of the second model. This is due to the assumption of non-existence of direct dependencies between the independent variables. Even though the similarity of the graphs, there are strong differences between probabilities in the factors. The most important in the first model which is illustrated below are the probabilities of New Mobility Score which are above or equal to six given that the patient is dead (Pr=0.809) and the number of diseases which are over or equal to three given death (Pr=0.962). In the second model the highest scoring probability corresponds to Number of co-morbidities factor given death (Pr=0.926). The lowest scoring in both models is the probability between institute given death (Pr=0.126). The probability of having a New Mobility Score below 6 given Death has a huge difference than the probability living in an Institution given death. The real differentiation between the two pre-mentioned results reflected into the patients with hip fracture in the Greek society. A probability result of 0.809 in New Mobility score translated that a patient with score below 6 has over 80% possibility to die in contrast with the patient who was living in an Institute who has approximately 12%. Due to this reason, the institution factor in the Greek population is unsuitable for study although the classification of the Bayesian network are higher in the model which included.

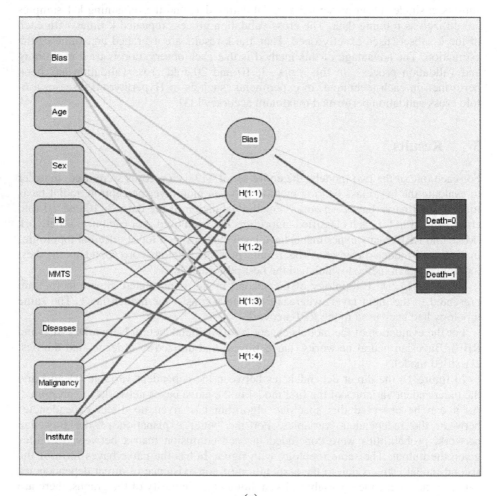

(a)

Fig. 1a,b. Neural and Naïve Bayes Networks

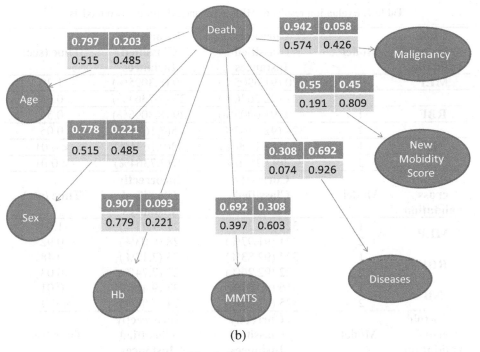

(b)

Fig. 1. (*Continued.*)

Table 1 presents the comparisons between multilayer perceptron, radial basis function and naïve bayes networks between model 1 and model 2 for different cross validation folds. Results had shown that in hip fracture models, the 10–fold cross validation was the most appropriate for the most of the cases. In the first and second columns of the table 1, the experiments of the neural and Bayesian networks are presented for each model, respectively. For the accuracy of the networks a correctly classified and an incorrectly classified index was used. The first index sums the true positive and the false negatives values of the confusion matrix and shows the proportion of the total number of predictors that were correctly identified, whereas the second index sums false positives and true negatives values and indicates the predictors that were incorrectly recognized. The time is the processing time value that is required to build the model, measured in seconds.

The processing time of the MLP algorithm is much higher than that of RBF and naïve bayes network algorithms. The same results are illustrated graphically in figure 2 for the 10-fold cross validation. Although both the neural networks classifiers had resulted that the first model are the most appropriate to predict the 30-day mortality after surgery for hip fracture, the results from naïve Bayes classifier had given the impression that the second model is the most suitable.

Table 1. Evaluation results of MLP, RBF and NB for the two models

3-fold cross validation	Model	Correctly Classified Instances	Incorrectly Classified Instances	Time (sec)
MLP	1	320 (91.6905%)	29 (8.3095%)	1.28
	2	322 (92.2636%)	27 (7.7364%)	0.95
RBF	1	320 (91.6905%)	29 (8.3095%)	0.59
	2	324 (92.8367%)	25(7.1633%)	0.05
NB	1	317(90.8309%)	32(9.1691%)	< 0.01
	2	322(92.2636%)	27(7.7364%)	< 0.01
10-fold cross validation	**Model**	**Correctly Classified Instances**	**Incorrectly Classified Instances**	**Time (sec)**
MLP	1	322 (92.26%)	27 (7.74%)	1.09
	2	321 (91.97%)	28 (8.03%)	0.92
RBF	1	324 (92.83%)	25 (7.17%)	0.48
	2	322 (92.26%)	27 (7.74%)	0.03
NB	1	316 (90.54%)	33 (9.46%)	0.01
	2	325 (93.12%)	24 (6.88%)	< 0.01
20-fold cross validation	**Model**	**Correctly Classified Instances**	**Incorrectly Classified Instances**	**Time (sec)**
MLP	1	320 (91.6905%)	29 (8.3095%)	0.87
	2	327 (93.6963%)	22 (6.3037%)	1.13
RBF	1	319 (91.404%)	30 (8.596%)	0.03
	2	323 (92.5501%)	26 (7.4499%)	0.03
NB	1	314 (89.9713%)	35 (10.0287%)	< 0.01
	2	325 (93.1232%)	24 (6.8768%)	< 0.01

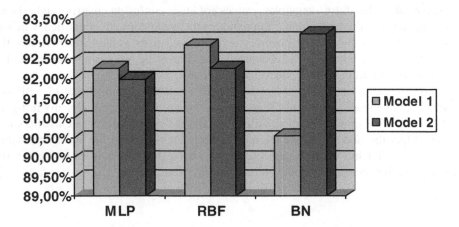

Fig. 2. Percentage of Correctly Classified Instances

From table 1 and figure 2 it is denoted that the instances in the first model were classified in a better way than in the second model for MLP (92.26% in the first, whereas 91.97% in the second model) and RBF (92.83% in the first, whereas 92.26% in the second model) algorithms. The naïve Bayes classifier evaluated significantly good classification for the second model (93.12%) in hip fracture data (all the classification results are above 90%). Also, the processing time for the classification is quite smaller in NB networks. The first model which included the New Mobility Score matched better in the citizens of Greece than the model included institution factor.

4   Conclusion

The Artificial Neural Networks and the Bayesian Networks are subfields of the computational intelligence. Although they are widely used in most research areas, their use in the field of medicine is still limited, especially the use of Bayesian Networks is almost nonexistent in the orthopedic field.

In this paper, the application of MLP, RBF and NB networks that have been applied to medical data, in order to predict the 30-day mortality after surgery for hip fracture in the Greek population is presented. In addition, two models of the medical data have been used, in order to investigate who is the most appropriate risk factor for the Greek population, between the New Mobility Score and the Institution factor, as it is described in the NHFS.

All the methods performed prediction of 30-day mortality over 90%. That means that all these methods are suitable for designing and developing a decision support tool to help clinicians identify which people are at increased risk after the hip fracture surgery. Moreover, the neural networks had resulted that the first model are the most appropriate to predict the 30-day mortality after surgery for hip fracture, but the results from naïve Bayes network had given the impression that the second model is the most suitable.

In the future, more computational intelligence methods have to be investigated (especially Bayesian Networks) in more clinical records from the Greek population, in order to find the best method that could predict the 30-day mortality after hip fracture surgery with maximum reliability. In addition, the institution factor in accordance with the New Mobility Score must be further investigated in order to decide which one of these factors is the most suitable for the Greek population.

References

1. Chang, K.P., Center, J.R., Nguyen, T.V., Eisman, J.: Incidence of hip and other osteoporotic fractures in elderly men and women: Dubbo Osteoporosis Epidemiology Study. J. Bone Miner. Res. 19, 532–536 (2004)
2. Court-Brown, C.M., Caesar, B.: Epidemiology of adult fractures: a review. Injury 37, 691–697 (2006)
3. Dhanwal, D.K., Dennison, E.M., Harvey, N.C., Cooper, C.: Epidemiology of hip fracture: Worldwide geographic variation. Indian J. Orthop. 45, 15–22 (2011)

4. Kannus, P., Parkkari, J., Sievänen, H., Heinonen, A., Vuori, I., Järvinen, M.: Epidemiology of hip fractures. Bone 18(suppl. 1), 57S–63S (1996)
5. Keene, G.S., Parker, M.J., Pryor, G.: Mortality and morbidity after hip fractures. BMJ 307(6914), 1248–1250 (1993)
6. Braithwaite, R.S., Col, N.F., Wong, J.: Estimating hip fracture morbidity, mortality and costs. J. Am. Geriatr. Soc. 51, 364–370 (2003)
7. Abrahamsen, B., van Staa, T., Ariely, R., et al.: Excess mortality following hip fracture: a systematic epidemiological review. Osteoporos Int. 20, 1633–1650 (2009)
8. Giversen, I.: Time trends of mortality after first hip fractures. Osteoporos Int. 18, 721–732 (2007)
9. Haleem, S., Lutchman, L., Mayahi, R., Grice, J.E., Parker, M.: Mortality following hip fracture: trends and geographical variations over the last 40 years. Injury 39, 1157–1163 (2008)
10. Sterling, R.: Gender and race/ethnicity differences in hip fracture incidence, morbidity, mortality, and function. Clin. Orthop. Relat. Res. 469(7), 1913–1918 (2011)
11. Bass, E., French, D.D., Bradham, D.D., Rubenstein, L.: Risk-adjusted mortality rates of elderly veterans with hip fractures. Ann. Epidemiol. 17, 514–519 (2007)
12. Giversen, I.: Time trends of mortality after first hip fractures. Osteoporos Int. 18, 721–732 (2007)
13. Maxwell, M.J., Moran, C.G., Moppett, I.: Development and validation of a preoperative scoring system to predict 30-day mortality in patients undergoing hip fracture surgery. Br. J. Anaesthesia 101, 511–517 (2008)
14. Hu, F., Jiang, C., Shen, J., Tang, P., Wang, Y.: Preoperative predictors for mortality following hip fracture surgery: a systematic review and meta-analysis. Injury 43(6), 676–685 (2012)
15. Hannan, E.L., Magaziner, J., Wang, J.J., et al.: Mortality and locomotion 6 months after hospitalization for hip fracture: risk factors and risk-adjusted hospital outcomes. JAMA 285, 2736–2742 (2001)
16. Johansen, A., Mansor, M., Beck, S., et al.: Outcome following hip fracture: post-discharge residence and long-term mortality. Age Ageing 39, 653–656 (2010)
17. Ozturk, A., Ozkan, Y., Akgoz, S., et al.: The risk factors for mortality in elderly patients with hip fractures: postoperative one-year results. Singapore Med. J. 51, 137–143 (2010)
18. Moppett, I.K., Parker, M., Griffiths, R., Bowers, T., White, S.M., Moran, C.: Nottingham Hip Fracture Score: longitudinal and multi-assessment. Br. J. Anaesth. 109(4), 546–550 (2012)
19. Gunasekera, N., Boulton, C., Morris, C., Moran, C.: Hip fracture audit: the Nottingham experience. Osteoporos Int. 21(suppl. 4), S647–S653 (2010)
20. Parker, M.J., Palmer, C.: A new mobility score for predicting mortality after hip fracture. J. Bone Joint Surg. Br. 75(5), 797–798 (1993)
21. Tseng, W.-J., Hung, L.-W., Shieh, J.-S., Abbod, M.F., Lin, J.: Hip fracture risk assessment: artificial neural network outperforms conditional logistic regression in an age- and sex-matched case control study. BMC Musculoskeletal Disorders 14, 207 (2013), doi:10.1186/1471-2474-14-207
22. http://www.cs.waikato.ac.nz/ml/weka/ (last accesed April 21, 2014)
23. Breast Cancer Diagnosis on Three Different Datasets Using Multi-Classifiers. International Journal of Computer and Information Technology
24. Othman, M.F.B., Yau, T.M.S.: Comparison of Different Classification Techniques. Using WEKA for Breast Cancer

25. Venkatesan, P., Anitha, S.: Application of a radial basis function neural network for diagnosis of diabetes mellitus. Tuberculosis Research Centre, ICMR, Chennai 600 031, India
26. Del_can, Y., Özyilmaz, L., Yildirim, T.: Evolutionary Algorithms Based RBF Neural Networks For Parkinson's Disease Diagnosis. In: ELECO 2011 7th International Conference on Electrical and Electronics Engineering, Bursa, Turkey, December 1-4 (2011)
27. Escaño, L.M.E., Saiz, G.S., Lorente, F.J.L., Fernando, Á.B., Ugarriza, J.M.V.: Logistic Regression Versus Neural Networks For Medical Data. Monografías del Seminario Matemático García de Galdeano 33, 245–252 (2006)
28. Comparison of Different Classification Methods Reihaneh Rabbany Khorasgani, Department of Computing Science, University of Alberta, Edmonton, Canada
29. Tomar, D., Agarwal, S.: A survey on data mining approaches for healthcare. International Journal of BioScience and Biotechnology 5(5), 241–266 (2013)
30. Zhang, H.: The Optimality of Naïve Bayes, Faculty of Computer Science, University of New Brunswick, Fredericton, New Brunswick, Canada
31. Langseth, H., Nielsen, T.: "Classification using Hierarchical Naïve Bayes models". Machine Learning 63(2), 135–159 (2006)
32. Zhang, N.: Hierarchical latent class models for cluster analysis. In: 18th National Conference on Artificial Intelligence, pp. 230–237 (2002)
33. Diagnosis And Classification Of Hypothyroid Disease Using Data Mining, Techniques. International Journal of Engineering Research & Technology (IJERT)

Application of Artificial Neural Network to Predict Static Loads on an Aircraft Rib

Ramin Amali[1], Samson Cooper[1], and Siamak Noroozi[2]

[1] University of the West of England, Bristol, UK
ramin2.amali@uwe.ac.uk, samson2.cooper@live.uwe.ac.uk
[2] Bournemouth University, UK
snoroozi@bournemouth.ac.uk

Abstract. Aircraft wing structures are subjected to different types of loads such as static and dynamic loads throughout their life span. A methodology was developed to predict the static load applied on a wing rib without load cells using Artificial Neural Network (ANN). In conjunction with the finite element modelling of the rib, a classic two layer feed-forward networks were created and trained on MATLAB using the back-propagation algorithm. The strain values obtained from the static loading experiment was used as the input data for the network training and the applied load was set as the output. The results obtained from the ANN showed that this method can be used to predict the static load applied on the wing rib to an accuracy of 92%.

Keywords: Static load, Finite Element Analysis, Artificial Neural Network, Aircraft Rib, MATLAB.

1 Introduction

Research into ANN and its application to structural engineering problems has grown significantly from increased interests over time. However, in the search for an optimum design, it is essential for engineers to evaluate the structural responses quickly at any design stage. Thus, the use of continuum models for the analysis of complex aircraft structures is an attractive idea in modern aerospace engineering especially at the conceptual and preliminary design stages [1]. Extensive research has been carried out using ANN to model operational loads experienced by a fixed-wing aircraft structure [2]. Flight loads on a fixed-wing aircraft can generally be separated into gust and manoeuvre dominated loads, the majority of which tend to occur at frequencies of less than a few Hertz (Hz) and in the case of a rotary-wing aircraft, the loading spectrum experienced by the airframe structure is significantly more complex. There have been a number of attempts in the last few decades at estimating these loads on the helicopter indirectly from flight state parameters or fixed points on the airframe with varying success [3].

In this work, several sets of data obtained from the static test conducted on a wing rib were analysed. This paper presents the implementation details of how accurate an artificial neural network has been used to predict the static load applied on the wing rib, the data sets are standard strain values recorded from the experimental test.

L. Iliadis et al. (Eds.): AIAI 2014, IFIP AICT 436, pp. 576–584, 2014.
© IFIP International Federation for Information Processing 2014

2   Experiment

The main aim of this research section was to measure the strain values which will be used as the inputs for the neural network training on MATLAB. The strain values are measured by subjecting the rib to a range of static loads using a powered hydraulic system.

2.1   Experiment Design

The rib is to be tested in two different positions due to the number of strain gauges and the location that they are attached on the rib. The testing equipment was set up as shown in Figure 1; the rib is supported in the frame with the aid of four steel rods bolted from the top surface of the shear tie to the top frame and another four steel rods connecting the bottom surface of the cap to the ground.

The fifteen strain gauges attached to the rib are channelled to the smart strain scanner which is in turn connected to a computer to run and record the strain values during the test process.

The cylinder is attached to the top part of the frame with a steel block bolted to the frame as shown in the experiment set-up; the cylinder is then connected to the manual hand pump with two high pressure hydraulic hoses as shown in Figure 1.

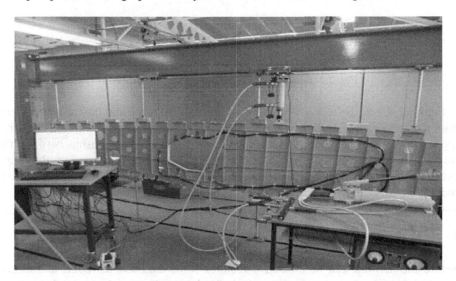

Fig. 1. Static Experiment Set-up

To conduct the test the manual hydraulic pump was used to apply a gradual load on the shear tie of the rib as shown above in Figure 1. The aim was to apply the specified load to the rib as quick as possible without physically damaging any part of the rib, while the smart strain scanner records the strain output of the rib due to the applied load over a time period. The software used for recording the strain output is precise enough to measure the applied load and strain to the nearest one percent.

However due to the fact that the test is a non-destructive test, the maximum load that could be potentially applied to the rib without causing any damage has been set to 3000N.

The strain data was gathered using the smart strain software, the overall test process was repeated three times on several locations on the rib where the cylinder comes in contact with the surface of the shear tie to get as many data and accurate results from the experiment.

2.2   Results

The results obtained were analysed by plotting a graph to show the relationship between the force applied on the rib over a period of time and the output strain against time.

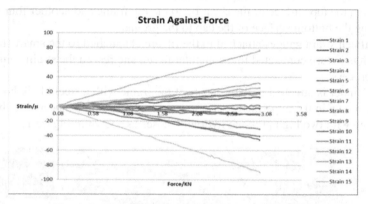

Fig. 2. Strain against Force Graph-Experiment Result

Figure 2 shows the relationship between the applied force and the measured strain from the experiment, it can be observed that the strain increases linearly as the force applied increases. Therefore it can be concluded that the relationship between the output and the input which would be used later in the neural network training are linearly related, this result also ascertain the assumption which was made before conducting the experiment.

3   Data Analysis

Before the network can be created it requires some data sets as inputs and targets obtained from the experiment, the data obtained from the experiment are in ordinary form of forces and strains. For the network to be trained it requires data from which it can learn the appropriate weight and biases that will help the network to produce the desired result. The data is obtained from the linear relationship between the force and strain graphs obtained from the experiment.

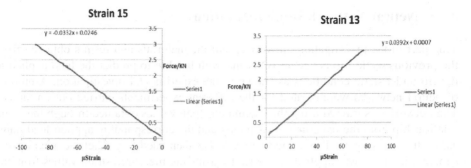

Fig. 3. Force against Strain- Strain 13 and 15

To calculate the strain values which are used as the set of inputs for the network training, a line of best fit was plotted on each strain line in Figure 2 to obtain the linear relationship between the force and the strain. Figure 3 shows some examples of the line of best fit on the graph which also generates the equation that relates the parameters together.

The linear equation was attained for the fifteen strain lines present in each experiment which means that for each experiment there are fifteen linear equations relating the force and the strain which have two constant values.

In general the linear equation can be written in broad terms of the applied force and the strain as shown in the equation (1)

$$F = a_n \varepsilon_n + b_n, \qquad n = 1 - 15 \tag{1}$$

Where F is the force which indicates y in the equations in figure 3,

ε_n is the strain indicating x on the graph,

a and b are the intercept and gradient values in mathematical terms.

By repeating the same step for the experimental test that was conducted, a number of equations are generated in generic terms as shown in equation (2)

$$F_1 = a_n \varepsilon_n + b_n, F_2 = c_n \varepsilon_n + d_n \tag{2}$$

Where F_1 and F_2 are the forces applied for two different experiments ε_n is the corresponding strains and n represents strains 1-15 and

a_n, b_n, c_n, and d_n, etc are constants for each corresponding strain line. Equation 2 represents a generic format of how the linear relationship between the force and the strain obtained from each experiment can be written.

By adding up F_1 and F_2 in the previous equation and rearranging to make ε_n the subject, a set of strain values can be calculated from the equation which is used as the inputs for the training data. The generic form of the equation is written below:

$$2\varepsilon_n = \left(\frac{F_1 - b_n}{a_n}\right) + \left(\frac{F_2 - d_n}{c_n}\right) \tag{3}$$

Finally, for the network to be trained accurately it requires a large number of data for the inputs and target. To achieve this, a set of random numbers ranging between zero and three were generated and then used to represent the targets which indicates the force in equation (3). Using the relationship constant obtained from each strain line graph of the experiment, the corresponding ε_n values were calculated.

4   Neural Network Implementation

This part of the investigation aims to use all the materials and results obtained from the previous section to create an ANN that will be able to predict the load applied to the rib under static conditions as shown in the experimental investigation. A number of neural networks were created for the different experiments carried out on the rib. The network was trained using the calculated strain values obtained through the strain relationship from the experiments as inputs and the corresponding applied load range as the training targets. The trained network was then used to predict the actual experimental load that was applied on the rib by using the measured strain values from the experiments as inputs, the results from the network were also compared with the experimental result to define how well the network was able to predict the applied load.

4.1   Training Data

The training data comprises of two sets which are the input sets and the targets sets, as previously mention the inputs are the calculated strains also referred to as ε_n and the targets are the random numbers ranging from zero to three.

For the network training the inputs was set to a 100x15 matrix configuration and the corresponding set targets had a 100x2 matrix configuration, the 100x2 matrix represents the random forces ranging from zero to three and the 100x15 represents the corresponding calculated strain.

4.2   Neural Network Creation

The simulation and analysis of the network was executed on MATLAB using the neural network toolbox function. The first designed network is a two layer feed forward network with sigmoid functions, ten hidden neurons and two output functions feed forward network is mainly used for fitting multi-dimensional mapping problems when given consistent data. Figure 4 shows an illustration of the network configuration from MATLAB. This type of network is mainly used for input and output curve fitting which is required for this analysis. The network is trained with Levenberg-Marquardt back propagation algorithm.

With the number of hidden neurons chosen, the network must allocate the data sets into three different parts which are training, validation and testing. The training data sets are presented to the network during training and the network is adjusted according to its error. The validation sets are used to measure network generalisation and to end training when generalisation stops improving. While the testing data sets which has no effect on the training is used to provide an independent measure of the network performance during and after training. For this neural network development, the default settings was selected on MATLAB which allocates 70% of the given data for training and 15% each for validation and testing.

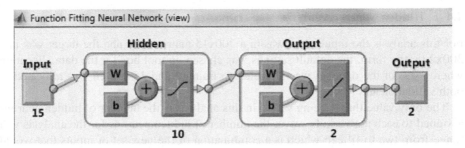

Fig. 4. Nueral Network Configuration

4.3   Network Validation

The network created above uses a data set that its previously seen, therefore it is important to access the performance and capability of the trained network and how closely it would match other outputs values when a new set of inputs are introduced to the network. To validate the network a new set data calculated from the measured strains was set as the input in a 15x2 matrix format to obtain a predicted load as the new output using the trained network. The output from the network is the predicted force or load which is then compared with the actual load that was applied to the rib during the experiment. Figure 5 shows an illustration of a bar chart comparing the predicted load obtained from the network with the experimental load.

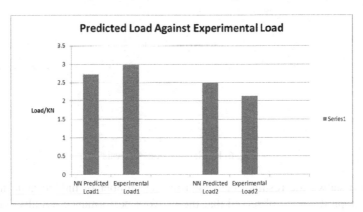

Fig. 5. Neural Network Result comparison

The bar chart shows how well the network has been able to predict the load from the set of new strain data which was set as the new inputs. From the chart it can be observed that the error between the first predicted load and the experimental load is as low as 0.25, which in percentage terms is approximately 8% and the second predicted load is 0.36. The calculated error shows that the initial setting that was designed in the network before training the data and the selection of ten hidden neurons produces an appropriate result from the validation aspect of the network. To obtain a better accurate prediction an analysis was carried out on how the variation of the number of hidden neurons would affect the predicted load which is the output of a network.

4.4   Hidden Neuron Analysis

For this analysis the inputs data was in a 300x15 matrix form and the target was in a 300x2 matrix form, the default settings was chosen on tool box for the data allocation where 70% of the data set was allocated to training and 15% each was allocated to both validation and testing.

The only value that is being varied in this analysis is the number of hidden neurons assigned to each network; however the number of hidden neurons for the analysis will range from two to fifteen which is a combination of the new set of inputs that would be used to validate each network created.

For all the networks trained for this analysis, a new set of data points was introduced to each network to obtain a new set of output which represents the predicted load, a graph was then plotted to show the relationship between the number of hidden neurons and the predicted load as shown in the graph plotted in figure 6. It is expected that an increase in the number of neurons in the hidden layer should initially improve the performance of the network, however there is a downside to increasing the hidden neurons. This causes the network to over fit the training data which will therefore reduce the effectiveness of the network when a new set of data is introduced to the network.

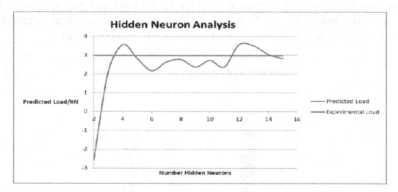

Fig. 6. Hidden Neuron Analysis

Figure 6 shows the results obtained from the number of hidden neuron analysis, from the graph it can be observed that the network performance initially improves as the number of hidden neurons increases. The graph also shows that the network performs best in the range of 6 to 10 hidden neurons as the predicted load obtained from those set of networks lie within the experiment threshold.

The results obtained from this analysis also indicates that there is a risk with increasing the number of hidden neurons where the network can begin to over fit the training data as shown in figure 6. From eleven hidden neurons the predicted load starts to increases again, however to counter this effect it is possible to increase the number of data samples that is being introduced to the network for training.

4.5   Neural Network Result Disscussion

In conclusion the results for the entire neural network tests were positive as indicated from the range of several networks that was created. The results show that it is possible to predict or estimate the static load applied on the rib to an accuracy of 92%.

However there is a requirement that must be met when training a network, this requirement is related to the number of hidden neurons used in training the samples and also the amount of data used to train and validate the network. To improve the accuracy of a neural network it is important to use as many data sets as possible, this will give the network more data samples to allocate for training, validation and testing which should lead to a better result.

Following the success of the prediction part of the networks, attempts were made to produce a method by which the networks would predict lower force values, this was achieved by changing the targets output and the targets of the network from the usually matrix format of [500x2] to a [500x1]. This new matrix will allow a new set of inputs in the matrix form of [15x1] to be introduced to the network to validate the performance of the network.

Also the new set of inputs values that was used for the training of this network has a range of strain values from correspond load applied between 0 and 1.1KN, it is expected that the network should be able to predict the load as it different from the initial loads that have been predicted. The result obtained from this final validation shows that the network is able to predict any static load applied to the rib to an accuracy of 95% provided a consistent set of strain inputs are introduced to the network.

5   Conclusion

This paper presents the use of an ANN to predict the static load applied on a wing rib. It can be stated that this research has shown that it is possible to design a set of neural networks that will correctly predict or estimate the load applied on the rib provided a consistent set of strain inputs values are introduced to the network.

Although a large portion of the analysis has been carried out on the location of the rib where strain gauge rosettes have been attached, it is also possible to use the same technique to analyse other areas or the overall rib by attaching more strain gauges to the rib to capture accurate data when conducting similar static test.

The percentage error between the neural network prediction and the experiment can be classified as the error obtained from using some noisy data to train the network. This however does not affect the performance of the network or how well the network is able to predict other force values when a corresponding set of strain inputs are introduced to the network.

It should be stated that during the data analysis stage of the research, not all the fifteen strain gauges had a perfect linear relationship between the applied force and the measured strain, which could have also some error in the new data set that was used as inputs. Further improvement in the experimental procedure and more allocation of strain gauges to the rib will allow the measuring equipment to capture strain values more accurately which would lead to a better neural network creation and load prediction in the future.

584   R. Amali, S. Cooper, and S. Noroozi

References

1. Liu, S., Tong, F.: Fuzzy pattern recognition of impact acoustic signals for non-destructive. Sensors and Actuators: Physical 167, 588–593 (2011)
2. Reed, S., Cole, D.: Development of a Parametric Aircraft Fatigue Monitoring System using Artificial Neural Networks. In: Proceedings of the 22nd Symposium of the International Committee on Aeronautical Fatigue, Lucerne, Switzerland (March 2003)
3. Polanco, F.: Estimation of Structural Component Loads in Helicopters: A Review of Current Methodologies, DSTO-TN-0239 (December 1999)
4. Mitchell, T.M.: Machine Learning, pp. 85–100. McGraw-Hill, New York (1997)
5. Bishop, C.M.: Pattern recognition and machine learning, pp. 226–290. Springer, New York (2006)
6. Mathworks, Neural Network Toolbox User Guide. s.l.:s.n (2012)

Android Based Electronic Travel Aid System for Blind People

Paraskevas Diamantatos and Ergina Kavallieratou

Dept. Information and Communication Systems Engineering
University of the Aegean
Samos, Greece
{Diamantatos,kavallieratou}@aegean.gr

Abstract. Blindness is the condition of lacking visual perception due to physiological or neurological factors. Blind people do not have the full perception of the surrounding environment, though navigating, in an unknown environment or/and with obstacles on route, can be a very difficult task. In this paper, an information mobile system is presented, that acts as an electronic travel aid, and can guide a blind person through a route, inform him about imminent obstacles in his path and help him orientate himself. The current prototype consists of a mobile phone, and the developed application.

Keywords: Blind, navigation, mobile phone.

1 Introduction

It is absolutely crucial in our time to be able to help blind people move in urban environments effectively, with a simple, easy accessible, low cost system. According to existed research there are almost 40 million blind people [1] that naturally use a simple cane, or a guide dog to help them move around. Both ways have disadvantages: a cane cannot detect obstacles further away than its height, and the guide dogs while more expensive, are not enough for all the blind people.

The majority of the suggested systems [2-9] are often too expensive and unpractical in use, while most of the available navigation systems, cannot detect obstacles in a specific route, they simply give directions to the user for a point A to a point B [2]. The proposed system helps the user to move in the environment, detecting obstacles and finding the best path, in order to avoid them. It can give vocal instructions to the user, or use the mobiles vibration in order to give navigation instructions. It is also able to orientate the user, using military vocal orientations e.g. north 12 o'clock.

In the following section 1.1, a short description of previous work is given, while in section 2, the proposed system is described in detail. Finally, some comparison results and conclusion are mentioned in sections 3 and 4, respectively.

1.1 Literature Review

There are two major elements of the mobility problem that a Blind person has to deal with. First, the blind must sense obstacles in his environment and avoid them. The second problem, equally serious as the first, is the person navigation [3].

L. Iliadis et al. (Eds.): AIAI 2014, IFIP AICT 436, pp. 585–592, 2014.

In the recent years a number of navigation system products became available to the public. Systems, like Google maps [2] , that offer navigation from a point A to a point B, and can also support voice interaction with the user, braille interface, and screen reader interface especially for blind people [2].While the navigation problem, seems to be a solved problem for outdoor navigation, mostly, due to the quality of the offered services and products to the public, the obstacle detection problem remains.

ETAs, Electronic Travel Aids, try to deal with this problem. ETAs development started at the mid 60's. Some of the most known applications include the LaserCane [4], which is a regular long cane with a built-in laser ranging system. The Mowat Sensor is an example of a pocket-sized device containing an ultrasonic air sonar system [5]. When it detects an obstacle, the device vibrates, thereby signaling to the user. Similar technology is used on NavBelt [6], a belt that utilizes an ultrasonic sonar system, and a guide cane, a cane that has a robot attached that also utilizes ultrasonic sonar system in order to guide the user.

While most of the ETAs try to address the obstacle detection problem, some ETAs try to provide navigation for indoor uncharted buildings. More specifically, Navatar [7] utilizes several sensors that can be found on modern smartphones like accelerometer and compass to determine the movement of the user in the space. It also takes a user input into account, e.g. if a user finds a door he can save it on the system. By this information, Navatar can determine the user's location inside a building [7]. Another system for indoor navigation is a case study from the school of engineering of University of California where a system was developed that utilizes mobile phone cameras and searches for specific markers on the walls [8].

A different approach for developing an ETA was used on tyflos system [9]. This system uses stereoscopic cameras in order to detect the depth of obstacles in the user's route and notify the user through a vibration belt. The research problem of designing a better ETA is a tough one. Despite 50 years of effort, no one has been able to design an electronic device that can replace the long cane.

2   The Proposed System

An application was developed that makes use of an android mobile phone with cameras along with sensors that can be found on every android smartphone, like accelerometer and compass. By analyzing the image taken by the camera and creating a depth map it is possible to detect obstacles in the user's path and redirect him by a different route.

In order to achieve that, two consecutively photos are considered and their common points are located. Then the optical flow between them is established, allowing extracting the homography and the fundamental matrices. This step is important in order to rectify the images and transform them to stereo images. This step allows the creation of a depth map that contains any objects found, along with their relative position in space. Next the routing subsystems computes a route with no obstacles and notifies the user. In this section, all the mentioned modules of the proposed system (fig.1) will be further analyzed.

The proposed system is able to help blind people to be orientated in their environment by informing them orally on the location they are staring: north, south, east, west and their combinations. It consists of a total of 12 subsystems, as it is presented in figure 1. The system was developed in the Android mobile platform, and it also requires the Opencv for android library.

The Camera subsystem is responsible to capture two sequential frames using the user's mobile phone camera. The output is two bmp images with resolution 352*288. Low resolution images are used in order to achieve lower computational time. Thus, the system can analyze the image in real time. In the prototype version, the two frames are shown side by side on the user's mobile phone screen, just for testing purposes.

The Points subsystem is responsible to identify common points between the two images acquired by the Camera subsystem, using a detector. Various detectors were tried, offered by opencv library for this task, like SURF [10]: a robust local feature detector that is based on sums of 2D Haar wavelet responses, ORB [11]: a very fast binary descriptor based on BRIEF, and GFTT [12]: a detector that finds the most prominent corners in the image. The results prooved that GFTT (good features to track) detector was ideal for the task, because of its very low computational time.

The Optical Flow subsystem is responsible to calculate the optical flow, i.e. the distance of the corresponding points (shift) between the first and the second frame, detected in the previous subsystem. It is absolutely crucial for the entire system that the points and optical flow subsystems return exact results, since all the next steps depend on those points and their optical flow.

The Fundamental subsystem, along with the Homography subsystem are responsible to calculate the fundamental [13] and the homography [14] matrices, respectively. The fundamental matrix is a 3×3 matrix which relates the corresponding points in stereo images. While homography relates the transformation from a projective space to itself that maps straight lines to straight lines. Homography has many practical applications, such as image rectification, which our system performs on the Rectify subsystem. It is also used for image registration, or computation of camera motion (rotation and translation) between two images. Once camera rotation and translation have been extracted from an estimated homography matrix, this information may be used for navigation, or to insert models of 3D objects into an image or video, so that they are rendered with the correct perspective and appear to be part of the original scene.

The rectify subsystem is responsible to transform the two bmp images in stereoscopic images [15], images taken by a stereoscopic camera. Thus, their common points are located on the same axis. Opencv provides the tools for the image rectification process [16]. It is suggested that the rectification is made by using the camera characteristics along with the homography matrix. The problem here is that the camera characteristics are not available for every camera that someone can use in a mobile phone. However, the rectification is also possible by using only the homography matrix.

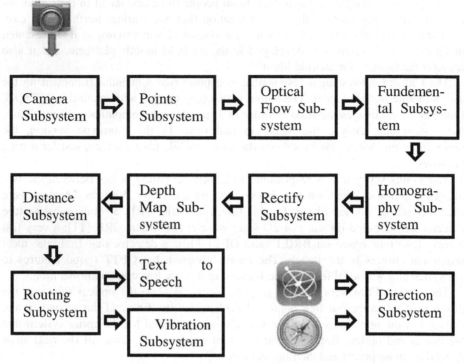

Fig. 1. Overview of the proposed system

The Depth Map subsystem is responsible to calculate the disparity depth map of the two stereo images of the rectify subsystem. A disparity depth map [16] is a 2D image where the color of each pixel represents the distance of that point from the camera. In more detail, light pixels are near and dark pixels are far. The disparity depth map will be used in order to identify obstacles near the user, for the calculation of the path that the user should follow. Block matching algorithms [16] offered by openCV were used, in order to produce the depth map.

The Distance subsystem is responsible to roughly calculate the distances of all the objects that can be found on the depth map. This was implemented by dividing the densities of the pixels from the depth map in 5 sub images. For objects very near to the camera (figure 2), are considered those with densities in the range of 220 – 255. Respectively a rough estimation (figure 3) is done for all the other objects (near, far, very far, and extremely far).

The Routing subsystem is responsible to calculate the area in the image (left, right or center) that includes the freest space. This is done by calculating the horizontal histograms on each of the images extracted by the Distance subsystem. After an iterative process, the area in the image with the freest space is localized. For example on figures 2 and 3, on the right side, the depth map of a scene is presented, while on the left only specific pixel densities are selected to be on. By considering the objects located very near (fig. 2, left image), and calculating the horizontal histogram of the left image, it is computed that on the left side of the scene, the user has free space to move, while on the right side of the scene, the user has no free space.

Fig. 2. Objects located very near. Depth Map on right, Left only pixel densities between 220-255 are on.

Fig. 3. Objects located extremely far. Depth Map on right, Left only pixel densities between 0-80 are on.

The Text-to-speech subsystem is responsible to notify the user with voice commands, while the Vibration subsystem, if selected, is responsible to notify the user with vibration codes. One vibration stands for left, two vibrations for right.

A Direction subsystem was also implemented that is responsible to notify the user of the direction he is headed, orally, using military codes. For example, if the user is headed north, the system will notify the user that he is headed at 12 O' clock.

3   Comparison Results

It is hard to actually measure the results, mostly because there is no standard evaluation technique. However, it is possible to check, if we get the expected results from the subsystems.

For the Optical Flow subsystem, that is based on the Points subsystem, it was possible to measure the detected distances of the points, manually. The followed procedure should save the 2 sequential frames, along with the points detected, and their optical flow. The extracted information for 20 different sets of frames was reviewed. The success rate of this subsystem was 85%, due to some misplaced points, that is a normal behavior for this kind of subsystem.

The Rectify subsystem that is based on the Fundemental and Homography subsystems is the one responsible for the bad output of our algorithm. The accuracy of this system was also measured manually. 50 rectified images were kept, and later reviewed. Only the rectified images onto a common image plane in such a way that the corresponding points have the same row coordinates, where considered to be the ones expected. Specifically only the 4 % of the frames work as expected, making things worse.

The Depth Map subsystem works as expected. Known datasets were used to test the performance of this subsystem, along with their ground truth data. By tuning the characteristics of this subsystem an identical image to the ground truth was achieved. A total of 20 sets of stereo images where used, to output their depth maps. Overall the success rate of this subsystem found to be 95%.

The results revealed that the depth map creation is exceptional. On figure 2 we can see a depth map created from our system on a monocamera device, while on figure 3 a depth map created from images taken from a stereoscopic camera device is presented.

Fig. 4. Depth map from monocamera device

Fig. 5. Depth map from stereoscopic camera device

For the Routing subsystem that is based on the Distance subsystem several depth maps were used, and on each one of the depth maps, a decision was made over which direction the user should go. The success rate of our algorithm was 70%.

4   Conclusion

Although our system works in some frames, it is not the case for all of them. Thus, it cannot be used in a real time application. Responsible for that is the camera characteristics that are not available on most of the mobile phones. Usually mobile phone manufacturers do not care to have available any of the camera characteristics per device. They just care for the camera to work.

This is making the development of such application, which utilizes only one camera, difficult, because it also requires a subsystem for camera calibration. Implementing one in our tests, different camera characteristics were extracted every time. Without camera characteristics, another function for camera calibration (stereoBM) was used that did not give so good results.

Although the final results are not perfectly good, and this system cannot be used by blind people yet, for safety reasons, we strongly believe that this kind of application is possible to be implemented in 3D mobile phones with 3D stereo cameras.

References

1. World Health Organization 2012, Global Data on Visual Impairment (2010)
2. Official Blog: Speech-friendly textual directions from Google Maps (2006), http://googleblog.blogspot.com/2006/12/speech-friendly-textual-directions.html (March 4, 2013)
3. National Research Council (U.S.). Committee on Vision. Working Group on Mobility Aids for the Visually Impaired and Blind, Electronic travelaids: new directions for research (1986)
4. Benjamine, J.M., Ali, N.A., Schepis, A.F.: A laser cane for the blind. In: Proceeding of the San Diego Biomedical Symposium, vol. 12, pp. 53–57 (1973)
5. Nagarajan, R., Yaacob, S., Sainarayanan, G.: Role of Object Identification in Sonification System for Visually Impaired. In: IEEE Tencon (IEEE Region 10 Conference on Convergent Technologies for the AsiaPacific), Bangalore, India, October 15-17 (2003)
6. Shoval, S., Ulrich, I., Borenstein, J.: NavBelt and the Guide-Cane [obstacle-avoidance systems for the blind and visually impaired]. IEEE Robotics & Automation Magazine 10(1), 9–20 (2003)
7. Apostolopoulos, I.: Integrating Minimalistic Localization and Navigation for People with Visual Impairments (May 2011); Warren, D.H., Strelow, E.R.: Electronic spatial sensing for the blind (1985)
8. Manduchi, R., Kurniawan, S., Bagherinia, H.: Blind guidance using mobile computer vision: a usability study. In: Proceedings of the 12th International ACM SIGACCESS Conference on Computers and Accessibility, October 25, pp. 241–242 (October 25, 2010)
9. Bourbakis, N.: Sensing surrounding 3-D space for navigation of the blind. IEEE Engineering in Medicine and Biology Magazine 27(1), 49–55 (2008)
10. Bay, H., Ess, A., Tuytelaars, T., Van Gool, L.: SURF: Speeded Up Robust Features. Computer Vision and Image Understanding (CVIU) 110(3), 346–359 (2008)
11. Rublee, E., et al.: ORB: an efficient alternative to SIFT or SURF. In: 2011 IEEE International Conference on Computer Vision (ICCV), pp. 2564–2571 (November 6, 2011)
12. Shi, J., Tomasi, C.: Good Features to Track. In: Proceedings of the IEEE Conference on Computer Vision and Pattern Recognition, pp. 593–600 (June 1994)

13. Luong, Q.T., Faugeras, O.D.: The Fundamental Matrix: Theory, Algorithms, and Stability Analysis. International Journal of Computer Vision (1996)
14. Hartley, R., Zisserman, A.: Multiple view geometry in computer vision, Cambridge (2000)
15. N. Holliman, "3D Display Systems." to appear; Handbook of Opto-electronics, IOP Press, Spring 2004, ISBN 0-7503-0646-7.
16. Dröppelmann, S., et al.: Stereo Vision using the OpenCV library (2010)

Shoreline Extraction from Coastal Images
Using Chebyshev Polynomials and RBF Neural Networks

Anastasios Rigos[1], Olympos P. Andreadis[2], Manousakis Andreas[1],
Michalis I. Vousdoukas[3], George E. Tsekouras[1], and Adonis Velegrakis[2]

[1] Department of Cultural Technology and Communication, University of the Aegean, Greece
[2] Forschungszentrum Küste Institution, Hannover, Germany
[3] Department of Marine Sciences, University of the Aegean, Greece
a.rigos@aegean.gr, Olympros@marine.aegean.gr,
vousdoukas@fzk.uni-hannover.de,
{ct10087,gtsek}@ct.aegean.gr,
Adonis.velegrakis@gmail.com

Abstract. In this study, we use a specialized coastal monitoring system for the test case of Faro beach (Portugal), and generate a database consisted of variance coastal images. The images are elaborated in terms of an empirical image thresholding procedure and the Chebyshev polynomials. The resulting polynomial coefficients constitute the input data, while the resulting thresholds the output data. We, then, use the above data set to train a radial basis function network structure with the aid of input-output fuzzy clustering and a steepest descent approach. The implementation of the RBF network leads to an effective detection and extraction of the shoreline of the beach under consideration.

Keywords: Coastal morphodynamics, remote sensing, RBF neural networks, fuzzy clustering, steepest descent.

1 Introduction

Monitoring of the shoreline position has become an issue of urgency given the high socio-economic value and population density of the coastal zone [1, 2], the increasing erosion as well as the projected sea-level rise. Beach morphology is known to change in different spatial and temporal scales, a fact that requires intensive monitoring schemes while the energetic conditions make non-intrusive techniques very attractive [3, 4]. As a result, the application of coastal video monitoring has been increased during the last three decades [5, 6] allowing non-intrusive, continuous measurements at temporal and spatial scales and resolutions for which in situ data collection would demand much greater than acceptable inputs of personnel, equipment, and cost. However, despite the application of coastal video monitoring systems for more than two decades still, developing a universal and robust automatic shoreline detection procedure remains a challenge, due to the variety of intra-annual environmental, hydrodynamic and morphological conditions at the coastal zone [6]. Against the foregoing background [7, 8], the present contribution aims to build a systematic

L. Iliadis et al. (Eds.): AIAI 2014, IFIP AICT 436, pp. 593–603, 2014.

methodology for coastal shoreline detection using radial basis function (RBF) artificial neural networks.

RBF networks have been exercised as efficient black-box techniques that have been implemented in a wide range of applications [9-12]. The training process is based on estimating three kinds of parameters: the neuron centers and widths, and the connecting weights between neurons. The estimation of the centers is carried out through cluster analysis; the corresponding widths are evaluated in terms of the above centers, and finally the connection weights using least squares or gradient descent.

In this paper, we present an automated methodology for the shoreline extraction from coastal grayscale variance images. The proposed methodology encompasses three modules. The first module performs an empirical image thresholding process. The second module employs the Chebyshev polynomials [13-14] to approximate the histograms of the resulting images. Finally, the third module applies an RBF network, which is trained in terms of input-output fuzzy clustering and a steepest descent approach based on Armijo's rule [15]. The result is an integrated system able to accurately detect and extract the shorelines.

The rest of the paper is synthesized as follows: Section 2 describes the location of interest and the monitoring system. Section 3 presents the proposed methodology in details. The experimental results are provided in Section 4. Finally, the paper concludes in Section 5.

2   Study Area and Monitoring System

The study area is the Faro Beach (Praia de Faro) located along the central and eastern parts of the Ancão Peninsula, in the westernmost sector of the Ria Formosa barrier island system in Portugal.

Tides in the area are semi-diurnal, with average ranges of 2.8 meters (m) for spring tides and 1.3 m for neap tides although a maximum range of 3.5 m can be reached [16]. Faro Beach is a 'reflective' beach [4] with beach-face slopes typically over 10% and varying from 6% to 15%, and a tendency to decrease eastwards, where a 'low tide terrace' beach state is found [16]. Beach sediments are medium to very coarse sands with d50~0.5 mm and d90~2 mm.

Coastal imagery was provided by a coastal video [4] consisting of two Mobotix M22, 3.1 megapixel (2048 x 1536 resolution), Internet Protocol (IP) cameras, installed on a metallic structure, placed on the roof of a restaurant in Faro Beach, and connected to a PC. The elevation of the centre of view (COV) is around 20 m above mean sea level (MSL).

The image acquisition took place at 1 Hz, at hourly 10 min bursts, during daylight. After each 10 minutes image acquisition set, the system was scheduled to run processing scripts which generate the 'primary products', i.e. snapshot images, time averaged (TIMEX) images, variance images, and timestack images [5].

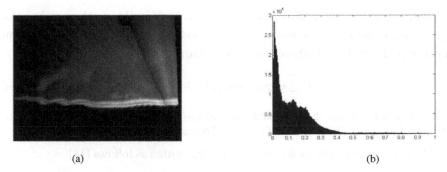

(a)   (b)

Fig. 1. (a) Typical SIGMA image and (b) the corresponding histogram

3   The Proposed Methodology

In this study, we obtain grayscale variance images of the coastal line, commonly called SIGMA images [4, 16]. These images represent the sum of the absolute pixel intensity differences between consecutive images and can be considered as 'accumulated motion images'. Using the above monitoring system, we generated 1600 SIGMA images for various angle view positions of the camera. Fig. 1(a) depicts a typical SIGMA image, where high intensity values are related to high wave-breaking and swash activity. The corresponding normalized histogram is given in Fig. 1(b).

The proposed method consists of three steps aiming towards developing a fully automated methodology able to exhibit accurate shoreline detection and extraction. The first two steps generate the input data and the output data, and the third one uses the above data to train the RBF network.

3.1   Histogram Approximation Using Chebyshev Polynomials

In this step we generate the input data used by the RBF network. Specifically, we attempt to approximate the histogram of each SIGMA image using Chebyshev polynomials, where each polynomial coefficient defines one and only one dimension of the input space of the RBF network. Thus, the number of polynomial coefficients defines the dimension of the network's input space. This fact justifies our choice to use the Chebyshev polynomials because they constitute orthogonal polynomials and they possess and inherent ability to exhibit high approximation accuracy by utilizing a small number of coefficients. Eventually, the input space dimension is kept within reasonable levels. According to Weierstrass's theorem [14] every continuous function $f(x)$ on a closed interval can be approximated as closely as desired by a polynomial function

$$\tilde{f}(x) = \sum_{p=0}^{r} c_p T_p(x) \tag{1}$$

where $\tilde{f}(x)$ is the approximation of $f(x)$, r is the polynomial's order, c_p $(1 \le p \le r)$ are the polynomial coefficients, and $T_p(x)$ is a continuous polynomial of order p. The order r is chosen as to approximate the following error function,

$$J_{error} = \min\left\{\max\left|f(x) - \tilde{f}(x)\right|\right\} \tag{2}$$

In what follows, we shall consider one-dimensional space, which is also our case since the histogram distributions of the SIGMA images are one-dimensional.

The Chebyshev polynomial of order r can be written as follows [14],

$$T_r(x) = \cos r\theta = \sum_{p=0}^{[r/2]}\left((-1)^p \sum_{j=p}^{[r/2]}\binom{r}{2j}\binom{j}{k}\right)x^{r-2k} \tag{3}$$

with $x = \cos\theta$. The Chebyshev polynomials are orthogonal in $[-1, 1]$ with respect to the subsequent weighting function,

$$h(x) = \frac{1}{\sqrt{1-x^2}} \tag{4}$$

The orthogonality is defined according the to the following relationship,

$$\langle T_p(x), T_q(x) \rangle = \int_{-1}^{1} h(x)T_p(x)T_q(x)\,dx = \begin{cases} 0, & p \neq q \\ \lambda_p, & p = q \end{cases} \tag{5}$$

where $\langle \cdot \rangle$ stands for the inner product and,

$$\lambda_p = \langle T_p(x), T_q(x) \rangle = \int_{-1}^{1} h(x)\left(T_p(x)\right)^2 dx = \begin{cases} \pi, & p = 0 \\ \pi/2, & p = 1,2,...,r \end{cases} \tag{6}$$

The polynomial coefficients of the function expansion in (1) can be written as,

$$c_p = \frac{\langle f, T_p \rangle}{\langle T_p, T_p \rangle} = \frac{1}{\lambda_p}\int_{-1}^{1} h(x)f(x)T_p(x)\,dx \tag{7}$$

with $p = 1,2,...,r$. The orthogonality implies that the polynomial functions do not overlap with each other. It thus appears that each coefficient c_k can be adjusted without causing any side effects to the rest of coefficients, meaning that they are independent each other. This is the property we wanted to obtain in the first place because, in our approach, the polynomial coefficients define the input space dimensions.

Since our data are discrete, the discretization of the above inner product relationship can be obtained by implementing the Forsythe's algorithm [13]. This method assumes a set $\{x_k; y_k\}\big|_{k=1}^{N}$ of input-output data with $x_k \in [-1,\ 1]$ and minimizes the error function in eq. (2) by using a monic polynomial of the form,

$$t_0(x_k) = 1, \qquad t_1(x_k) = x_k, \qquad t_2(x_k) = x_k\, t_1(x_k) - \frac{1}{2}t_0(x_k) \qquad (8)$$

$$t_r(x_k) = x_k\, t_{r-1}(x_k) - \frac{1}{4}t_{r-2}(x_k) \qquad (r > 2) \qquad (9)$$

To this end, the Forsythe's algorithm modifies the equation (7) and states that the best fitting of the available discrete data set is obtained if the polynomial coefficients are calculated according to the subsequent formula,

$$c_p = \frac{\sum\limits_{k=1}^{N} h(x_k)\ y_k\ t_p(x_k)}{\sum\limits_{k=1}^{N} h(x_k)\ \bigl(t_p(x_k)\bigr)^2} \qquad (10)$$

3.2   Image Thresholding

In this step, we generate the output data. To accomplish this task, we employ the thresholding process developed in [4]. The basic issue of this approach is to normalize the pixel intensities as follows:

$$\hat{I}_{i,j} = \frac{I_{i,j}}{I_j^{\max}} \qquad (11)$$

where $I_{i,j}$ is the pixel original intensity, where $1 \le i \le M$ and $1 \le j \le N$ indicate horizontal and vertical pixel dimensions, respectively, of an image of size $M \times N$ pixels. The quantity $I_j^{\max}$ corresponds to the smoothed alongshore pixel intensity maxima vector. Given that a standard region of interest was considered, the only input parameter for the shoreline detection model was the pixel intensity threshold $\hat{I}_{thr}$, which has been shown to be related to the pixel intensity histograms. The interested reader can find a detailed presentation of the thresholding approach in [4].

3.3   Training Process of the RBF Network

Based on the above process, the data set to train the network is formulated as follows:

$$S_{IO} = \Bigl\{ (x_k, y_k): x_k = \bigl[c_{k1}, c_{k2}, ..., c_{kr}\bigr]^T, y_k = I_{thr}^k, 1 \le k \le Q \Bigr\} \qquad (12)$$

where Q is the number of the training data, r is the order of the Chebyshev polynomial which coincides with the input space dimension and is the same for all images, and c_{kj} $(1 \le j \le r)$ are the polynomial coefficients. Notice that $Q < 1600$. In order to train the RBF network we must determine effective input-output relationships, which are provided by an optimal set of values of the network's parameters. The most crucial issue is the determination of the radial basis function centers. Fuzzy cluster analysis has been intensively involved in deciding appropriate values for those parameters. In this paper, we employ the algorithm developed by Pedrycz in [17], which is based on the implementation of the fuzzy c-means. The algorithm performs a separate fuzzy clustering in the output space. The number of clusters defines the number of contexts in the input space. The input data belonging to the same context correspond to the output data that belong to the same cluster in the output space. Then, a separate conditional cluster analysis is implemented with respect to each context, where all contexts are partitioned into the same number of clusters. In both clustering schemes the fuzziness parameter was selected to be equal to 2. To this end, the form of the radial basis function is,

$$g_i(x_k) = 1 \bigg/ \sum_{j=1}^{n} \left(\frac{\|x_k - v_i\|}{\|x_k - v_j\|} \right)^2 \tag{13}$$

where n is the number of hidden nodes (i.e. radial basis functions), and $v_i \in \Re^r$ the center of the i-th basis function. The interested reader can find a detailed presentation of this algorithm in [17]. The estimated network's output is,

$$\tilde{y}_k = \sum_{i=1}^{n} w_i \, g_i(x_k) \qquad (1 \le k \le Q) \tag{14}$$

with w_i being the connection weight of the i-th hidden node. Notice that the above basis function does not use any width parameter, since this parameter has been absorbed by the radial basis function form.

The connection weights are calculated by a steepest descent approach, which is based on Armijo's rule [15]. The objective is to minimize the network's square error,

$$J_{SE} = \sum_{k=1}^{Q} \|y_k - \tilde{y}_k\|^2 \tag{15}$$

Then, for the $t+1$ iteration,

$$w(t+1) = w(t) - \alpha(t) \nabla J_{SE}(w(t)) \tag{16}$$

where $w = \begin{bmatrix} w_1 & w_2 & ... & w_n \end{bmatrix}^T$ and,

$$\nabla J_{SE}(w) = \left[\frac{\partial J_{SE}}{\partial w_1}, \frac{\partial J_{SE}}{\partial w_2}, ..., \frac{\partial J_{SE}}{\partial w_n} \right]^T \tag{17}$$

The partial derivatives in (21) are,

$$\frac{\partial J_{SE}}{\partial w_i} = (-2) \sum_{k=1}^{n} (y_k - \hat{y}_k) f_i(x_k) \tag{18}$$

The parameter $\alpha(t)$ in (16) is: $\qquad \alpha(t) = \beta^\mu \tag{19}$

where $\beta \in (0,1)$. The parameter μ is the smallest positive integer such that,

$$J_{SE}\left(w(t)-a(t)\nabla J_{SE}(w(t))\right)-J_{SE}\left(w(t)\right)<-\varepsilon\,a(t)\left\|\nabla J_{SE}(w(t))\right\|^{2} \tag{20}$$

with $\varepsilon \in (0,1)$.

4   Experimental Study

The data set consists of 1600 SIGMA images and therefore the available data set includes $L = 1600$ data. The first experiment concerns the approximation of the image histograms by the Chebyshev polynomials.

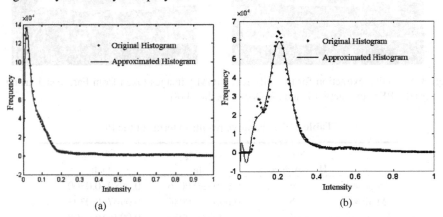

(a)   (b)

Fig. 2. Histogram approximations using Chebyshev polynomials with degrees equal to: (a) $r = 20$ and (b) $r = 6$

Table 1. Mean RMSE values and the corresponding standard deviations of the histogram approximations for various polynomial degrees

Polynomial Degree	RMSE±Standard Deviation
6	0.4354± 0.1072
10	0.4035±0.1138
14	0.3899±0.0933
18	0.3746±0.0866

To perform the experiment we used $N = 200$ discrete intensity levels. Thus, for each image, the Chebyshev polynomials were used to fit 200 image data. The approximation accuracy was evaluated in terms of the root mean square error,

$$RMSE = \sqrt{\sum_{l=1}^{N}\left\|fr(\hat{I}_{l})-\widetilde{fr}(\hat{I}_{l})\right\|^{2}\Big/N} \tag{21}$$

where $fr(\hat{I}_{l})$ is the original histogram frequency function and $\widetilde{fr}(\hat{I}_{l})$ the approximated one, which is provided by the expansion in eq. (1). Fig. 2 depicts the approximations of two different images with two different polynomial degrees. Notice that as the number of coefficients (i.e. the polynomial degree) increases so does the approximation accuracy. This fact is strongly supported by Table 1, where the mean values of the RMSE along with the respective standard deviations for all the 1600 SIGMA images are reported.

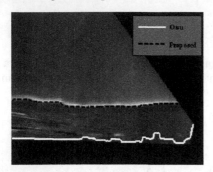 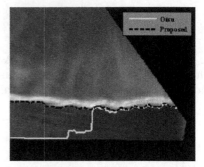

Fig. 3. Shoreline extraction for two different SIGMA images taken from Faro beach using the proposed RBF neural network and the Otsu's method [18]

Table 2. Comparative results in terms of the PI

	No. of Hidden Nodes	Training data	Testing Data
Proposed	6	0.0993±0.0036	0.0989±0.0300
Method	8	0.0963±0.0036	0.0967±0.0303
	10	0.0907±0.0036	0.0946±0.0308
	12	0.0863±0.0035	0.0914±0.0312
Otsu's Model		0.1631±0.01353	

However, in Table 1, the differences are not too large. Based on the last remark, and since the number of polynomial coefficients define the dimensionality of the network's, which must be kept within reasonable levels, we choose to use approximations obtained by the Chebyshev polynomials of degree equal to $r = 6$. Therefore, in this experimental case, the input space dimension of the RBF network is equal to 7. The performance index (PI) used to evaluate the network is,

$$PI = \sqrt{J_{SE}/Q} \qquad (22)$$

To train the network we divided the available 1600 data into $Q = 960$ training data (i.e. 60%), while the rest 640 (i.e. 40%) as testing data. In addition, the proposed method is compared to the well-known Otsu's image thresholding algorithm [18]. Fig. 3 illustrates the shoreline extractions of two different SIGMA images obtained by the proposed method and Otsu's algorithm. To obtain this figure we used a RBF network with $n = 8$ hidden nodes. According to this figure our approach significantly outperformed the other method. This result is strongly evident in Table 2, which presents a simulation comparison in terms of the mean values of the PI and the corresponding standard deviations.

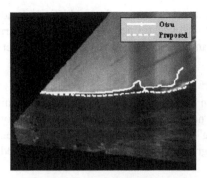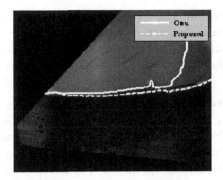

Fig. 4. Shoreline extraction for two different SIGMA images taken from Ammoudara beach using the proposed RBF neural network and the Otsu's method

Table 3. Comparative results in terms of the PI

	No. of Hidden Nodes	Training data	Testing Data
Proposed	6	0.1104±0.0058	0.1298±0.0444
Method	8	0.1081± 0.0122	0.1277±0.0388
	10	0.1004±0.0091	0.1248±0.0202
	12	0.0986±0.0066	0.1197±0.0456
Otsu's Model		0.1831±0.00998	

To further evaluate our method, we used a data set coming from Ammoudara beach, which is located to the city of Heraklion at the northern coast of Crete (Greece). We used exactly the same RBF network that was trained by the data set from Faro beach. That is to say, the data set form Ammoudara beach are used all as testing data and the network had exactly the same parameter values as in the case of Faro beach. Again we compared our network with Otsu's method [18]. The results of this experimental case are reported in Fig. (4) and Table 3. According to those results, our network performed remarkably well, although the data set was "unknown" to it. Finally, similarly to the previous case, we can easily verify the superior behavior of the proposed method.

5   Conclusions

In this paper we have investigated the efforts to assess and enhance the automatic shoreline detection from images representing the sum of the absolute pixel intensity differences between consecutive images, obtained by a specialized monitoring system. The approach was exclusively carried out in terms of a sophisticated radial basis function neural network structure, which was trained by a fuzzy clustering scheme. The experimental findings showed an effective performance with respect to the shoreline detection and extraction.

Acknowledgments. This research has been co-financed by the European Union (European Social Fund – ESF) and Greek national funds through the Operational Program "Education and Lifelong Learning" of the National Strategic Reference Framework (NSRF) - Research Funding Program: THALES. Investing in knowledge society through the European Social Fund.

References

1. Goncalves, R.M., Awange, J.L., Krueger, C.P., Heck, B., dos Santos Coelho, L.: A comparison between three short-term shoreline prediction models. Ocean & Coastal Management 69, 102–110 (2012)
2. Huisman, C.E., Bryan, C.E., Coco, G., Ruessink, B.G.: The use of video imagery to analyse groundwater and shoreline dynamics on a dissipative beach. Continental Shelf Research 31(16), 1728–1738 (2011)
3. Pearre, N.S., Puelo, J.A.: Quantifying seasonal shoreline variability at Rehoboth beach, Delaware, using automated imaging techniques. Journal of Coastal Research 25(4), 900–914 (2009)
4. Vousdoukas, M.I., Ferreira, P.M., Almeida, L.P., Dodet, G., Andriolo, U., Psaros, F., Taborda, R., Silva, A.N., Ruano, A.E., Ferreira, O.: Performance of intertidal topography video monitoring of a meso-tidal reflective beach in South Portugal. Ocean Dynamics 61, 1521–1540 (2011)
5. Holman, R.A., Stanley, J.: The history and technical capabilities of Argus. Coastal Engineering 54(5-6), 477–491 (2007)
6. Plant, N.G., Aarninkhof, S.G.J., Turner, I.L., Kingston, K.S.: The performance of shoreline detection model applied to video imaginary. Journal of Coastal Research 23(3), 658–670 (2007)
7. Madsen, A.J., Plant, N.G.: Intertidal beach slope predictions compared to field data. Marine Geology 173(1-4), 121–139 (2001)
8. Uunk, L., Wijnberg, K.M., Morelissen, R.: Automated mapping of the intertidal beach bathymetry from video images. Coastal Engineering 57, 461–469 (2010)
9. Filho, O.M.M., Ebecken, N.F.F.: Visual RBF network design based on star coordinates. Advances in Engineering Software 40(9), 913–919 (2009)
10. Er, M.J., Chen, W., Wu, S.: High-speed face recognition based on discrete cosine transform and RBF neural networks. IEEE Transactions on Neural Networks 16(3), 679–6915 (2005)
11. Pedrycz, W., Park, H.S., Oh, S.K.: Granular-oriented development of functional radial basis function neural networks. Neurocomputing 72, 420–435 (2008)
12. Roh, S.B., Ahn, T.C., Pedrycz, W.: The design methodology of radial basis function neural networks based on fuzzy K-nearest neighbors approach. Fuzzy Sets and Systems 161(13), 1803–1822 (2010)

13. Mason, J.C., Handscomb, D.C.: Chebyshev Polynomials. Chapman & Hall/CRC, NY (2003)
14. Rivlin, T.J.: The Chebyshev Polynomials. John Wiley & Sons, NY (1974)
15. Armijo, L.: Minimization of functions having Lipschitz continuous first partial derivatives. Pacific Journal of Mathematics 16(1), 1–3 (1966)
16. Vousdoukas, M.I., Wziatek, D., Almeida, L.P.: Coastal vulnerability assessment based on video wave run-up observations at a meso-tidal, reflective beach. Ocean Dynamics 62, 123–137 (2012)
17. Pedrycz, W.: Conditional fuzzy clustering in the design of radial basis function neural networks. IEEE Transactions on Neural Networks 9(4), 601–612 (1998)
18. Otsu, N: A threshold selection method from gray-level histograms. IEEE Transactions on Systems, Man and Cybernetics 9, 62–66 (1979)

Automation System Development
for Micrograph Recognition for Mineral
Ore Composition Evaluation in Mining Industry

Olga E. Baklanova, Olga Shvets, and Zhanbai Uzdenbaev

D.Serikbayev East-Kazakhstan State Technical University,
Ust-Kamenogorsk, The Republic Kazakhstan
O_E_Baklanova@mail.ru, {OShvets,ZhUzdenbaev}@ektu.kz

Abstract. The article deals with the problem of micrographs automated obtaining and micrographs automated processing ore samples of ore dressing processes in ferrous metallurgy. It is described the task of interpreting the results and interaction subsystem automated obtaining and automated processing of samples of ore mining micrographs and beneficiation processes ferrous metals, used to analyze the quality of mineral rocks with other automation systems. It is defined specifies requirements for the subsystem micrographs analysis.

Keywords: Petrographic analysis, Digital microscopy system, Image recognition, Computer vision, SCADA system.

1   Introduction

Petrography is the science that studies the material composition of the rocks. Unlike minerals, rocks are aggregates composed of different minerals [1]. Minerals are homogeneous in composition and structure of the rocks and ores. They are natural chemical compounds resulting from various geological processes. Historically minerals initially determined by color and shape.

Some mining and processing enterprise mixed ore from several fields. In this case it is important to determine the mixing ratio. It is necessary to find a mineral composition of the ore, which can be determined by optical microscopic analysis. In some enterprises, it is implemented manually by technologist-mineralogist.

Nowadays, some complexes micrographs analyzed manually locally in the enterprise by technologist - mineralogist, or in the center of the head group of companies with the transition to the Internet or other communication channels. Manual analysis has drawbacks such as dependency from psychophysiological properties of laboratory specialist (human factor) and the long duration of treatment of one of the micrograph. But the process of analyzing ore micrographs can be automated and implement it locally in the enterprise, due to the fact that subsystem micrographs analysis oriented to samples of a particular ore easier to create than a universal program for quantitative analysis of samples of any material [2].

L. Iliadis et al. (Eds.): AIAI 2014, IFIP AICT 436, pp. 604–613, 2014.

There are the following automation tasks from all at mining complexes:

— Automated obtaining of micrographs;
— Quantitative analysis of mineral ore samples by micrographs.

2   Materials and Methods

2.1   Methods of Automated Acquisition of Mineral Rocks Images

It is required for automated obtaining of micrographs automated microscope containing a motorized specimen stage, a mechanism to change the filters, focusing mechanism, a revolver change lenses [3].

Typically, in this case it is supposed to use the manufacturer's software, which is responsible for the control for motorized units of microscope. Simplified classification of microscopy cameras is shown in Figure 1. Preferably use trinocular microscope with a camera that does not require an optical adapter with a digital interface USB, controlled by a computer and with the TWAIN support.

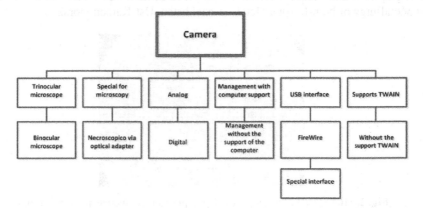

Fig. 1. Classification of microscopy cameras

Thus, the process of obtaining images can pass without operator as in the case of a fully automated microscope and with the participation of the operator. In the case of automated microscope software fully controls the microscope, after getting photo in the memory of a computer program for image analysis starts. It is required from analysis program that the results were collected to a database accessible over the LAN. From database, these results will be perceived by SCADA- system, and the system makes decisions.

In the case of partially automated or non-automated microscope operator is responsible for changing the samples, the selection of the microscope objective, take a photo on your computer, run the program and image analysis. As in the previous case, the analysis program is required to enter the results into a database, where SCADA-system can use it, and the system makes decisions.

This paper describes the problem of optical microscopy analysis on the mining complex. It is determined the requirements for obtaining and analyzing subsystem micrographs. The most important of them is the requirement to the camera, the serial processing micrographs of various samples of ore analysis program should each segment micrograph segmentation method has been proposed, then point of interest must be identified as certain minerals, the quality of the segmentation is more significant than the processing of a single image, the program should transmit the results of its work to the database [4].

Theoretically possibility to determine the mineral ore targets on the microscopic image substantiated by author M.P. Isayenko [5].

2.2   Methods of Automated Processing of Mineral Rocks Images

Very often different minerals on the micrographs correspond to objects of different types of shapes and colors. This allows the identification of various minerals in the form and color of objects. In some cases, also take into account the polarization of certain minerals sample [6, 7].

In this case, it is necessary to make several pictures, accompanying it by turning the sample. Consider a sample of slag copper anode as an example (Figure 2). Micrographs of this sample were kindly provided Eastern Research Institute of Mining and Metallurgy of Non-ferrous Metals (Kazakhstan, Ust-Kamenogorsk).

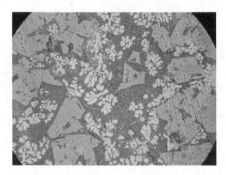

Fig. 2. Micrograph of a sample of slag copper anode, increasing in 500 times

According to experts on microscopy of minerals from Eastern Research Institute of Mining and Metallurgy of Non-ferrous Metals at this picture there is no minerals having dependent on the direction of the plane of polarization of light. In this picture you can detect metallic copper and the following minerals: cuprite Cu_2O, magnetite Fe_3O_4, Delafosse $CuFeO_2$, silicate glass.

Cuprite Cu_2O can be identified as follows: it is characterized by the shape of a round shape, color - it is light gray (sometimes with a slight bluish tint). Figure 3 shows the graphical representation of cuprite.

Fig. 3. Cuprite on micrographs

Fe3O4 magnetite on micrographs may also be detected by color and shape. Color of magnetite on micrographs is dark gray. Shape is angular, as expressed by technologists, "octahedral". Figure 4 shows magnetite apart from other minerals picture.

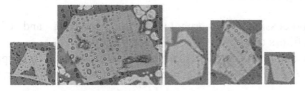

Fig. 4. Magnetite on micrographs

Delafossite CuFeO2 micrographs can allocate to the needle shape and gray (with a brownish tint) color. On Figure 5 it can be seen delafossite on the micrographs

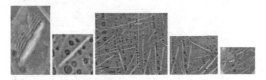

Fig. 5. Delafossite on micrographs

Metallic copper on the micrographs can be found on the following criteria: color - yellow, shape - round, without flat faces. Figure 6 represents a micrograph metallic copper

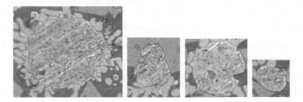

Fig. 6. Metallic copper on micrographs

Silicate glass - is a dark gray mass fills the rest of the space that is left of the other minerals.

These data indicate that for real micrographs slag samples (and some other minerals) it is possible to use automated qualitative assessment of the mineral composition.

After receiving the full image it is often needed to treat it, mainly to simplify further analysis. Available processing capabilities are as follows:

- color correction / color control;
- sharper image display of objects;
- contrast enhancement;

- contrast or inversion;
- alignment of images;
- mosaics;
- correction of spherical aberration;
- morphological filters.

Figure 2 processing by morphological dilation filter and effects of some morphological filters can be viewed on Figure 7 [9].

Basically, image processing is required because of the need to simplify the process of analysis and / or to improve visual perception of the operator-mineralogist [10].

Fig. 7. Micrograph of slag copper anode after 6 times of the use of morphological dilation

Any micrograph can be described in different color spaces. Usually, transformation into another color space is used to segment the micrograph particular method. Known space RGB (red, green, blue), YCrCb (brightness, hue, and saturation), YIQ (brightness, hue, saturation), L*a*b* CIE 1976 (takes into account the human perception of color), L*C*H* (psychometric brightness, hue and saturation), HSI (hue, saturation , brightness), HSL (hue, saturation , lightness), L*u*v* CIE 1973 (equal contrast color space) [4].

Micrographs analysis suggests image segmentation, classification of found objects and calculating the percentage ratio of each type of object. It can provide such a method of segmentation as color segmentation. The user determines what colors correspond to certain types of objects, which are created on the basis of the color clusters. Micrograph segment to be allocated may not contain all the colors of the cluster, but includes most of these colors. Micrograph is divided on contours according to the information of the color clusters.

As a criterion for shape analysis it can be suggested the following. Selected contour is transferred to the Cartesian coordinate system. Then, from the bottom left point after some distance is drawn straight to the point of standing up and to this line is constructed perpendicular and calculated angle perpendicular. It would be considered the first perpendicular. Next contour is gone by the same way clockwise. Next perpendicular constructed similarly. The absolute value of the difference between the angle of each of the current and preceding perpendicular will be remembered by the program.

When all of the perpendiculars will be built program adds all absolute difference values of angles and divides it to the length of their boundary contour formed by segments, which was perpendicular. Cuprite calculated value for the relative rotation angle correlated to the length of the contour composed segments, which was perpendicular, is more than for magnetite. Method can be very sensitive to random points of the real contour, so you should carefully choose the length of the segments, which are constructed perpendiculars. For each type of object it is required to make the range sum of relative angles values correlated to the length of the contour.

Segmentation is also possible by determining the differences of brightness, color gradients, calculating the area of contours [5].

For analysis micrograph program more qualitative segmentation is override the minimum analysis time of one micrograph. It also requires a serial processing of multiple images ore samples.

The analysis result of a micrograph is a table in the same column indicates the name of a mineral in the other relative area occupied by them in the micrograph. Also there are specifies the identity of the image, such as the name of the field, the number of test samples, the number of files in a batch of samples.

3   Results and Discussion

Feature optical microscopic analysis is the percentage of minerals detected on the surface of the sample may be different from the ratio, identified by chemical analysis. This is due to the fact that a picture of the sample surface cannot be analyzed, but it can be different from that on the surface. Therefore it is necessary to analyze more than one ore sample and calculate the percentage of minerals in several pictures. Next SCADA- system calculates the corrected value ratios of minerals in the ore [11].

Interpretation of the results is required due to the fact that the quality measures adopted by the mining industry cannot be directly obtained as a result of analysis techniques. I.e. there are input data as a percentage of a certain type of objects. It is required to interpret these data to obtain quality indicators adopted in the enterprise. Since the quality criteria against mining products in the enterprise depend on the ore and technology enterprise, the system should provide the ability to customize rules interpretation.

Interpretation rules include some mathematical and logical operations. Speed is important. It is required to accept the interpretation of the rules on the basis of the SCADA-system applicability, which has become an integral part of APCS. Image analysis results also include identification data, carrying information about the name of the field and the party ore. The scheme of the flow of information in the automated system of processing of the image to evaluate the qualitative composition of mineral rocks is shown on Figure 8.

Let us consider implementing storage of analysis results. Imaging results of microscopic images are stored in databases. Each field has its own table in the database. This is due to the fact that a single table for the different fields can be unjustified, as the number of allocated objects on microscopic images for a variety of fields may be different. Results interpretation system would be a database, and a shell, leading these databases.

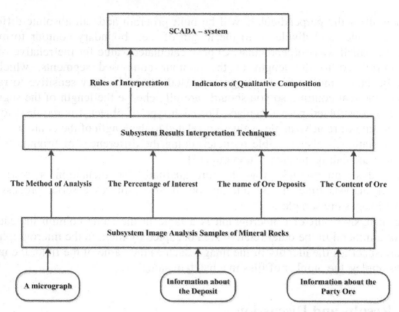

Fig. 8. The scheme of the flow of information in the automated system of processing of the image to evaluate the qualitative composition of mineral rocks

It can be used in multiple tables within a single database. Each source image corresponds with own table. Each table contains information such as the maintenance of a certain color and morphology, identification of mineral rock party. Example of a table for one field is shown on Figure 9.

Identification data of party	Object 1	Object 2	Object 3	...	Object N
AMIRKN11212 2010112	6%	3%	8%	...	Z %
...					

Fig. 9. Identification of mineral rock party

The most obvious solution to the problem of interpretation of the rules will be use to build support scripting subsystem in interpreting the results. For this purpose scripting languages were considered for claims processing speed, ease of syntax, prevalence, the ability to integrate into the overall system.

Input interpretation subsystem in this case composed of the main attributes such as:

— type of treatment - planimetric method or Delesse-Rozival;
— internal identifying name of the object and percentage;
— information on the name field , she entered during the setup phase of the source image;
— identification number of the party ore, it is removed from the file names.

Also it is possible to change the interpretation or clarification of the rules by transferring them from the SCADA-system.

It was analyzed the most common languages Python, Ruby, Perl. However, these scripting languages have not satisfied the required specification [11-13]. Next, it was considered possible solutions to the problem, based on scripting languages by going to rule language of embedded scripting languages SCADA-systems.

Using a scripting language built into SCADA-system, it is convenient that the technical staff of enterprises has gained considerable experience in dealing with them, which means less time debugging rules interpretation.

Table 1 shows a comparison of SCADA-systems according the presence of the scripting language. For comparison, we have used some common SCADA-system.

Table 1. Comparison of SCADA-systems according the presence of a scripting language

Name of SCADA-system	Presence of built scripting language	Name of built scripting language
Genesis 32	yes	Visual Basic for Applications, JScript, VBScript
InTouch	yes	QuickScripts
Citect	yes	CiCode, CitectVBA.
TraceMode	yes	Control algorithms based on language of IEC 61131-3 standard
MasterSCADA	yes	Propriety on 150 functions
iFix	yes	Visual Basic for Applications
Advantech Studio	yes	Propriety on 100 functions
GeniDAQ	yes	BasicScript, identical VBA
WinCC	yes	Visual Basic for Applications, ANSI-C

Most SCADA-systems support the use of Visual Basic for Applications 6.x. Regarding the solution of the problem it is appropriate to use Visual Basic for Applications, as it is available in the script editor of leading SCADA-systems, which is the reason that the technical staff are well acquainted with the language.

An important requirement is to ensure the transfer of the results of image processing micrographs mineral rocks in SCADA-system with the speed with which the act interpreted results of treatment.

It was considered existing technologies and exchange of data between different applications used in process automation OLE for Process Control (OPC) [10], [17], [18]. It was analyzed OPC DA, OPC AE, OPC Batch, OPC DX, OPC HDA, OPC Security, OPC XML-DA, OPC UA. In all of these (Table 1) SCADA-systems standards implemented OPC DA and the vast majority of systems OPC HDA.

Exchange of information with superior enterprise systems can be implemented by standard OPC DA - it should be considered an automated image processing system as several pseudo sensors, the number of which is determined by the number of image sources. Scheme of interaction between source and receiver data is shown on Figure 10.

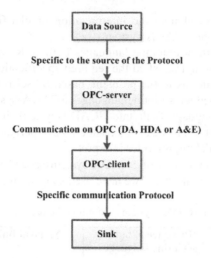

Fig. 10. Scheme of interaction between source and receiver data

Standard OPC DA implements the transfer of three types of attributes; it is a variable, the value and time of receipt of the data.

Based on the aforementioned arguments, as the basis of information exchange it was recommended OPC DA technology.

4   Conclusion

It is described the problem of optical microscopy analysis on the mining complex. It is determined the requirements for obtaining and analyzing subsystem micrographs. The most important of them is the requirement to the camera, the serial processing micrographs of various samples of ore analysis program should each segment micrograph segmentation method has been proposed, then point of interest must be identified as certain minerals, the quality of the segmentation is more significant than the processing of a single image, the program should transmit the results of its work to the database.

It is considered tasks of the interpretation results of the analysis to convert the results of the image processing in accordance with some specified quality requirements. It is analyzed for the interpretation of languages, such as, Python, PERL and Ruby. It is considered a variant of the implementation language interpretation rules based on the syntax of Visual Basic for Applications as a scripting language with a rather simple syntax. Among the possible ways to implement the technology options were considered OLE for Process Control (OPC) and CORBA.

Further research should be focused on the development of methods and algorithms by which can be automated solving phase analysis of samples of mineral rocks. In order to define a class of algorithms and techniques required to make a list of functions "universal" system, which allows automates key tasks of microscopy. Also, the system should be implemented data transmission module to higher systems.

References

1. Harvey, B., Tracy, R.J.: Petrology: Igneous, Sedimentary, and Metamorphic, 2nd edn. W.H. Freeman, New York (1995)
2. Clarke, A.R., Eberhardt, C.N.: Microscopy Techniques for Materials, p. 459. Science Woodhead Publishing, CRC Press (2002)
3. Panteleev, C., Egorova, O., Klykova, E.: Computer microscopy. Technosphere, p. 304 (2005)
4. Baklanova, O.E.: Development of algorithms for image recognition needed to assess the quality of the mineral species in the mining industry. In: Abstracts of International Conference on Mathematical and Informational Technologies, MIT 2013, VrnjackaBanja, Budva, September 5-14, pp. 63–64 (2013)
5. Isayenko, M.P., Borishanskaya, S.S., Afanasyev, E.L.: Keys to the main ore minerals in the reflected light, p. 382. Nedra, Moscow (1986)
6. Farndon, J.: The practical encyclopedia of rocks and minerals. How to Find, Identify, Collect and Maintain the World's best Specimens, with over 1000 Photographs and Artworks. Lorenz Books, London (2006)
7. Chris, P.: Rocks and Minerals. Smithsonian Handbooks. Dorling Kindersley, New York (2002)
8. Shaffer, P.R., Herbert, S.Z., Raymond, P.: Rocks, Gems and Minerals, Rev. edn. St. Martin's Press, New York (2001)
9. Gonsales, R.C., Woods, R.E.: Digital image processing, 3rd edn., p. 976. Pearson Education (2011)
10. Baklanova, O.E., Uzdenbaev, Z.S.: Development of methodology for analysis of mineral rocks in the mining industry. In: Joint Issue of the Bulletin of the East Kazakhstan State Technical University and Computer Technology of Institute of Computational Technologies, Siberian Branch of the Russian Academy of Sciences, Part 1, pp. 60–66 (September 2013)
11. Python vs C++ comparison. - [Electronic resource]. - Mode of access,
    `http://shootout.alioth.debian.org/u32/`
    `benchmark.php?test=all&lang=python&lang2=gpp&box=1`
12. Ruby vs C++ comparison. [Electronic resource]. - Mode of access,
    `http://shootout.alioth.debian.org/u32/`
    `benchmark.php?test=all&lang=yarv&lang2=gpp`
13. PERL vs C++ comparison. - [Electronic resource]. - Mode of access,
    `http://shootout.alioth.debian.org/u32/`
    `benchmark.php?test=all&lang=perl&lang2=gpp`
14. Moskalev, A.A.: CORBA in industrial applications. In: World com-computer automation (On-line), N5, 2001.- [Electronic resource]. - Mode of access (2001),
    `http://www.mka.ru/?p=42049`
15. Object Management Group: "The Common Object Request Broker: Architecture and Specification", Version 2.5 September 2001 (2001), `http://www.omg.org`
16. Schmidt, D.C., Kuhns, F.: An Overview of the Real-Time CORBA Specification. IEEE Computer, 56–63 (June 2000)
17. Grigoriev, A.: Standard OPC path to integrate disparate systems, part 1. In: PC Week/RE, N 32, 2001.- [Electronic resource]. - Mode of access (2001),
    `http://www.pcweek.ru/industrial/article/detail.php?ID=58992`
18. Grigoriev, A.: Standard OPC path to integrate disparate systems, part 2, In: PC Week/RE, N 33, 2001. - [Electronic resource]. - Mode of access (2001),
    `http://www.pcweek.ru/industrial/article/detail.php?ID=59064`

Similarity Based Cross-Section Segmentation in Rough Log End Images

Rudolf Schraml and Andreas Uhl

University of Salzburg, Jakob Haringer Str. 2,
5020 Salzburg, Austria

Abstract. This work treats cross-section (CS) segmentation in digital images of rough wood log ends. Existing CS segmentation approaches are focused on computed tomography CS images of logs and no approach and experimental evaluation for digital images has been presented so far. Segmentation of cross-sections in rough log end images is a prerequisite for the development of novel log end analysis applications (e.g. biometric log recognition or automated log grading). We propose a simple and fast computable similarity-based region growing algorithm for CS segmentation. In our experiments we evaluate different texture features (Local binary patterns & Intensity histograms) and histogram distances. Results show that the algorithm achieves the most accurate results in combination with intensity histograms and the earth movers distance. Generally, we conclude that for certain applications simple texture features and a matured distance metric can outperform higher-order texture features and basic distance metrics.

Keywords: Wood imaging, Cross-section analysis, Cross-section segmentation, Rough log end images.

1 Introduction

In case of digital images of log ends there exist just a few publications which deal with cross-section (CS) analysis. The works of [1],[2],[3],[4],[5] present approaches for pith estimation and in [2],[6],[7],[8] annual ring analysis approaches are proposed. Present CS analysis literature focuses on computed tomography (CT) cross-section images. Compared to conventional log end images, CT cross-section images do not exhibit disturbances due to cutting or dirt. Furthermore, the wood properties are clearly visible and distinguishable. This enables segmentation of the entire CS and of wood properties/ defects using basic thresholding techniques (literature review see [9]). To our knowledge, no publications related to cross-section segmentation in digital images of rough log ends were published. Digital images of rough log ends can be taken at little cost and at almost every stage in the wood processing chain. The location, size and shape of the CS are crucial to perform sophisticated image analysis tasks. Thus, segmentation is a prerequisite for the development of novel applications (e.g. automated log grading or biometric traceability of logs - see [10]) using images of rough log

L. Iliadis et al. (Eds.): AIAI 2014, IFIP AICT 436, pp. 614–623, 2014.

end faces. Beneath a high segmentation accuracy, the timing performance of any segmentation approach is crucial for any real world application.

Basically, a cross-section is built up of an annual ring texture which varies strongly locally. Variations are caused by the the circular alignment, different widths, colour features and disturbances of annual rings. Hence, CS segmentation is a typical task for region-based segmentation approaches which are able to segment textured regions. State-of-the-art region-based segmentation approaches are based on graph cuts ([11],[12]), active contours ([13],[14]) or mean shift ([15]) algorithms. In the past two decades manifold publications showed up high level segmentation algorithms combined with well matured texture features like Markov-random-field statistics ([16]), Wavelet features ([17],[18]), Gabor filters ([19]) and Fractal features ([20]). Furthermore, there exist many more textural spectral-temporal features (literature review see [21]) and various new texture features and extensions were published (e.g. [22], [23], [24], [25]).

Recently, in [26],[27] the region-based active contour approach proposed by [14] is extended to deal with similarity of adjacent patches which enables segmentation of inhomogeneous textured regions. This approach is applicable to CS segmentation, but requires a high computational effort. For this first work dealing with CS segmentation we pay attention to the timing performance. Instead of using active contours we utilize an ease and fast computable region-growing approach which is based on similarity of adjacent patches. Similarity is assessed by comparing the histogram distances of texture features. For this purpose two fast computable features are evaluated as texture features: Intensity- and local binary pattern (LBP) histograms. The utilized region-growing approach can be combined with any comparable texture feature.

This work contributes to the development of new CS analysis applications suggesting a simple, accurate and fast computable approach for CS segmentation in rough log end images. A testset consisting of 108 different rough log end images is utilized to perform experiments. Results show that for the utilized algorithm simple intensity histograms as texture features and the earth movers distance (EMD) outperform a well-matured texture descriptor - the LBP operator. Additionally, the experiments examine different settings and parameters required for the segmentation algorithm.

At first, Section 2 introduces the utilized texture analysis methods. In Section 3 the region-growing based CS segmentation algorithm is outlined. The experimental setup and results are presented in Section 4 followed by the conclusion in Section 5.

2 Texture Analysis Background

This section summarizes two different texture features and a set of histogram distances utilized for the CS segmentation algorithm and the experiments.

Intensity Histograms do not extract information about the topology of the pixels. Generally, histograms of images or image sections are probability distributions describing the frequency of each color or gray value.

Local Binary Patterns (LBPs) were introduced by [28]. In counting the frequency of micro texels in an image or image section, LBPs unify statistical and structural principles of texture analysis. For each pixel in an image or image section the local binary pattern is computed by analysing N neighbours in a circular neighbourhood. The occurrence of each of the 2^N possible patterns is stored into a feature histogram, where each bin represents a single pattern and its frequency. Additionally to the generic formulation of the LBP operator, the authors of [29] and [30] introduced uniform, rotation invariant and multi-scale LBP variations.

Histogram Distances are used as similarity measures between feature histograms of adjacent image sections. Four bin-by-bin distances (L1-Norm - L_1, L2-Norm - L_2, Chi-quadrat distance - X^2, Hellinger distance - H) and one cross-bin distance (EMD) are examined. The EMD [31] computes the minimal cost required to transform one histogram (P) into another (Q). In case of two one-dimensional probability distribution functions P and Q the EMD is simply given by the L_1 norm between their cumulative distribution functions.

3   Cross-Section Segmentation Algorithm

Our algorithm is inspired by the EMD-region-based level set formulation in [14],[27] and is adopted to a region growing procedure. The procedure is based on similarity of adjacent image sections. For this purpose, the image is subdivided into blocks. The block size and the overlapping factor between the blocks are crucial parameters which strongly influence the segmentation accuracy and timing performance. Our algorithm is subdivided into three consecutive stages which are described subsequently.

3.1   Cluster Initialisation

For the selection of seed blocks and the subsequent initialisation of clusters the position of the pith is utilized. We utilize the suggested algorithm and configuration presented in [5]. This pith estimation algorithm computes local orientations in the Fourier Spectra of image sections. The pith position is determined by intersection of the local orientations.

A predefined number of blocks (e.g. 4) which are equally distributed close around the pith position are selected. Subsequently, for each single seed block the adjacent neighbours in a four-neighbourhood are added and then initialised as a single cluster.

Each cluster is initialised by computing three features which describe the contained texture: mean gray value/variance, mean entropy/variance and mean intensity histogram or LBP histogram distances between the blocks of the cluster and the corresponding variance.

3.2   Growing Procedure

The cluster growing procedure starts with the selection of one of the initialised clusters. Then the four-neighbourhood of each border block is analysed. Only neighbours which are not allocated to a cluster are used as candidate blocks.Each candidate block is compared to the cluster which leads to a decision whether or not the block is allocated to the cluster.

In Fig. 1 the blocks of a cluster and the decision making procedure are illustrated. For the block labelled B3 three candidate blocks (nb1, nb2, nb3) are available. The final decision whether one of the candidates (e.g. nb2) is allocated requires that each of the three comparisons satisfies predefined conditions. For example, the mean gray value of the block must range between the cluster mean gray value $+/-$ its variance. The feature-variances are used to regulate the restrictiveness of the decision making procedure. For this purpose, the variance is multiplied by a factor ranging from [-2,2]. Preliminary tests showed that other factors are either too restrictive or too tolerant for the growing procedure. If the block is allocated to the cluster, the cluster features are updated. The procedure continues with the next border block

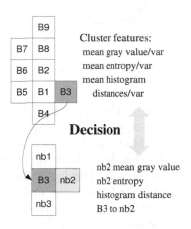

Fig. 1. Illustration of the cluster growing procedure

until no border block has further candidate blocks that fit to the cluster. Then the algorithm continues with the next cluster. The cluster-growing algorithm finishes after all clusters are processed. Finally, the clusters are merged and it is assumed that the merged cluster represents the area of the CS.

An exemplary result is shown in Fig. 2 . It is clearly visible that the resulting cluster covers the entire CS. Unfortunately, similar textured CSs in the background are difficult to distinguish from the main CS. This problem is visible in the bottom right of Fig. 2 .

In three consecutive steps the final estimate of the CS boundary is computed. First, all cluster blocks which have at least one neighbour that is not in the cluster are selected (see Fig. 3a). This step reduces the amount of blocks and all of the blocks representing parts of the CS boundary still remain. In the second step, it is assumed that CSs are approximately elliptical. First, a circle is fitted into

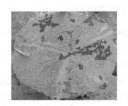

Fig. 2. Cluster initialisation/growing/merging result using 16x16 half-overlapping pixels blocks

the point cloud of the remaining blocks. Blocks within the circle are removed. This improves the accuracy of the ellipse fitting procedure. After ellipse fitting all blocks located outside of the ellipse are removed (see Fig. 3b). In the best case and especially for elliptical CSs this procedure decreases the probability of including incorrectly segmented blocks (e.g. caused by neighboured CSs in the background).

(a) Select border blocks   (b) Fit circle/ ellipse   (c) Alpha shape (α =50)

Fig. 3. Cross-section (CS) boundary estimation

Finally, the alpha shape [32] of all remaining blocks is computed. Alpha shapes are an approach to compute concave hulls and are a generalisation of convex hulls. They can be tuned by the α parameter. The higher the α value is, the better the alpha shape approximates the convex hull and details of the boundary are neglected. The lower the value of α is, the more details are considered. In contrast, a lower value makes the alpha shape computation vulnerable to large gaps between the outermost blocks. Finally, the alpha shape (see Fig. 3c) is used as a final estimate of the CS boundary.

4   Experiments

Three experiments with different configurations are used to assess the general performance as well as the impact of: Different block sizes and half-overlapping/ non-overlapping blocks, using intensity or LBP histograms as texture features, different local histogram distances and different variance factors (V_f) and the circle/ ellipse fitting step. Additionally, one experiment evaluates the timing performance.

Testset: 108 (1024x768) circular saw-cut spruce log ends were captured with a digital camera (Samsung WB2000), without flash and at approximately the same distance from the log end surface to the camera. The images were converted to grayscale using equal weighting factors. The accuracy for a certain configuration is evaluated computing the mean segmentation accuracy (Mean) and the deviation (StDev) across all of the 108 images. R describes the span between the minimum and the maximum of the computed segmentation accuracies. Additionally, the computation time in milliseconds [ms] provides information on the timing performance. The percentage of difference between the pixels of the

(a) Original image   (b) Ground truth mask   (c) Computed CS mask

Fig. 4. Segmentation accuracy computation masks

manually determined ground-truth mask (Fig. 4b) and the computed CS mask (Fig. 4c) is defined as the segmentation accuracy.

All steps of the CS segmentation algorithm have been implemented in Java and all experiments have been performed on an Intel Core i7-2620M processor with 2.7 GHZ and 8 GB RAM, JRE 1.7.

4.1 Experiment #1 - Intensity Histograms

First, the applicability of intensity histograms as texture features for our algorithm is evaluated. Furthermore, block sizes of 16x16 and 32x32, half- and non-overlapping and all histogram distances noticed in Section 2 are examined.

For the cluster growing procedure each histogram distance is tested with variance factors (V_f) in a range between [-2.0,2.0]. For the entropy and mean gray value comparisons a constant $V_f = 1$ is used. The circle/ellipse fit step is not performed. The best results for all tested configurations are presented in Table 1. For three of four configurations the EMD achieves the best segmentation accuracy. The most accurate results are reached with EMD and 16x16 half-overlapping pixels blocks (Mean: 5.55, StDev: 3.3). It can be stated that the EMD is a robust and appropriate histogram distance for all tested configurations.

Table 1. Experiment #1 - Statistical analyisis

H_d	V_f	Mean	StDev	R	[ms]		H_d	V_f	Mean	StDev	R	[ms]
Blocksize 16x16							Blocksize 32x32					
non-overlapping							non-overlapping					
L_1	0.9	7.29	3.74	17.0	312		L_1	1.3	10.05	3.46	18.0	258
L_2	1.0	7.97	3.4	15.0	310		L_2	1.7	11.13	3.87	17.0	261
H	0.9	7.72	4.16	17.0	312		H	1.0	9.45	3.81	18.0	267
X^2	0.6	8.55	4.27	17.0	310		X^2	0.6	10.88	4.11	19.0	260
EMD	0.5	6.45	3.41	17.0	304		EMD	1.2	9.34	3.49	16.0	262
half-overlapping							half-overlapping					
L_1	−0.5	10.04	5.64	20.0	2135		L_1	−0.3	8.03	3.03	14.0	1103
L_2	−0.4	10.52	4.68	20	2175		L_2	−0.2	8.44	3.12	15.0	1090
H	−0.3	11.11	5.99	21.0	2278		H	−0.3	7.13	3.16	14.0	1078
X^2	−0.9	6.71	4.14	19.0	2000		X^2	−0.6	8.12	3.05	15.0	1064
EMD	−0.1	5.53	3.3	14.0	1867		EMD	0.4	8.35	3.57	17.0	1043

Results show that smaller block sizes and overlapping blocks increase the segmentation accuracy. Especially for shape estimation, a smaller distance between the block-center points leads to a refinement and increases the accuracy of the CS boundary estimation. On the other hand, the amount of blocks is a crucial factor for the timing performance. With an increasing amount of blocks the timing performance decreases significantly. For example, compare the non- and half-overlapping timing measurements in Table 1 .

Furthermore, the results show that for non-overlapping blocks lower variance factors (V_f) and for half-overlapping blocks higher variance factors are required. Overlapping and smaller blocks are more similar than larger sized ones which are not overlapping. Thus, the variance factors for larger sized blocks need to be higher and consequently less restrictive. Smaller block sizes and overlapping blocks require lower variance factors and enable a more accurate segmentation.

4.2   Experiment #2 - LBP Histograms

Second, LBPs are examined as texture features. For this purpose, equal settings as in Experiment #1 are utilized. Due to an improper ordering of the histogram bins the EMD is not evaluated. Four different LBP variations (3x3 neighbourhood) are evaluated: basic LBPs, uniform LBPs, multiscale LBPs (three scales - see [29]) and multiscale & uniform LBPs. In Table 2 the best results for certain block settings and LBP variations are shown.

Nearly all of the most accurate results for a certain configuration are computed using the L_2 norm. Contrary to the first experiment, smaller blocksizes and overlapping blocks do not increase the seg-

Table 2. Experiment #2 Statistical Analysis (no = non-overlapping, ho = half-overlapping)

LBP histograms						
Config.	H_d	V_f	Mean	StDev	R	[ms]
3x3 LBPs						
16x16	no	L_2	0.9	9.85	4.75	20.0	772
32x32	ho	L_2	−0.2	8.88	3.83	25.0	3273
3x3 uniform LBPs						
16x16	ho	X^2	−0.4	8.57	2.91	14.0	6636
32x32	no	L_2	0.5	9.66	3.64	14.0	711
3x3 multiscale LBP						
16x16	no	L_2	0.0	9.37	7.31	73	1294
32x32	ho	L_2	0.3	7.53	3.45	14.0	5084
3x3 multiscale & uniform LBPs						
16x16	no	L_1	0.9	8.89	4.29	16.0	1250
32x32	ho	L_2	0.1	7.5	3.41	15.0	5206

mentation accuracy. Due to strongly varying accuracies over all results it is not possible to draw further conclusions about the impact of the blocksize and half-/ non-overlapping blocks on the accuracy.

The most accurate result is achieved using the L_1 norm and multiscale & uniform LBPs (L_1/ 32x32/ half-overlapping – Mean: 7.53, StDev: 3.41). It is difficult to conclude that using multiscale LBPs or uniform LBPs instead of normal 3x3 LBPs improves the segmentation accuracy. However, the two most accurate results are reached with the LBP extensions and especially with multiscale & uniform LBPs.

Results show that the two LBP extensions influence the timing performance differently. While for uniform LBPs the number of bins in the LBP histogram decreases, for multiscale LBPs the number of bins increases with each scale. Consequently, uniform LBPs increase the timing performance and multiscale LBPs cause a significant deterioration of the timing performance.

The most important conclusion is that the results of the first experiment are more robust over all configurations and histogram distances. The best results of the first experiment using intensity histograms remarkable outperform the segmentation accuracies and timing measurements of the second experiment using LBP histograms.

4.3   Experiment #3 - Circle/Ellipse Fitting

To assess the impact of the circle/ellipse fitting step the best configurations (Intensity histograms/ EMD/ 16x16 pixels blocks) are recomputed including circle/ellipse fitting (Table 3). Compared to the corresponding results in Table 1 the results in Table 3 show that the circle/ellipse fit step is increasing the segmentation accuracy. Unfortunately, the circle-/ellipse fit approach is very time consuming.

Table 3. Experiment #3 Circle/ellipse fitting improvement

16x16	H_d	V_f	Intensity histograms Mean	StDev	R	[ms]
no	EMD	0.4	5.99	2.83	12.0	610
ho	EMD	−0.1	5.1	2.85	13.0	2597

4.4 Timing Performance Evaluation

For evaluation of the timing performance, the testset images were scaled in a range from $\{0.1, 0.2, ..., 1, ..., 2\}$ according to the amount of pixels. For each scale, the segmentation accuracies and timings are computed as described in Experiment #1 using EMD and half-overlapping blocks. The most accurate results for each scale and blocksize are used to illustrate the timing performance in Fig. 5.

Compared to 32x32 blocks, with 16x16 blocks the amount of analysed

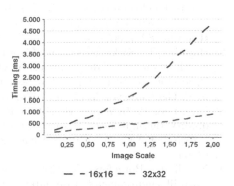

Fig. 5. Timing performance

blocks increases approximately by a factor of 4. Results show, that the timings for 32x32 blocks increase roughly linearly. The different increase of the 16x16 timings is caused by the Alpha shape computation step which has a complexity of $O(nlogn)$. In case of 16x16 blocks and larger scales, the number of points used for Alpha shape computation increases rapidly. Thus, the impact on the timing performance becomes clearly visible. Finally, it can be stated that an appropriate selection of the blocksize is crucial for the timing performance and the accuracy of the CS segmentation algorithm.

5 Conclusion

We proposed a similarity-based region growing algorithm which is suitable for segmentation of CS in images of rough log end faces. In combination with intensity histograms as texture features and the EMD as histogram distance our algorithm is fast, accurate and robust.

Surprisingly, LBP histograms as texture features achieve less accurate and very irregular results. Additionally, the timing performance decreases remarkable. These observations lead to an interesting conclusion. For particular segmentation applications, simple texture features and a more sophisticated similarity measure can outperform matured texture features and a simple similarity measure.

The experimental evaluation of our work forms a solid basis for further research on CS segmentation. In our future research further segmentation techniques will be evaluated based on their accuracy and timing performance. Apart from the computational effort, active contour approaches as presented by [27] are of interest because they are probably more robust to similar textured cross-sections in the background.

References

1. Wu, J., Liew, D.: A computer vision method for detection of external log cracks and pith in log cross-section images. In: Procs. of the World Automation Congress: International Symposium on Intelligent Automation and Control (ISIAC 2000), Hawaii, USA (2000)
2. Hanning, T., Kickingereder, R., Casasent, D.: Determining the average annual ring width on the front side of lumber. In: Osten, W., Kujawinska, M., Creath, K. (eds.) Proceedings of SPIE: Optical Measurement Systems for Industrial Inspection, Munich, Germany, vol. 5144, pp. 707–716 (2003)
3. Österberg, P., Ihalainen, H., Ritala, R.: Method for analyzing and classifying wood quality through local 2d spectrum of digital log end images. In: Proceedings of International Conference on Advanced Optical Diagnostics in Fluids, Tokyo, JP (2004)
4. Norell, K., Borgefors, G.: Estimation of pith position in untreated log ends in sawmill environments. Computers and Electronics in Agriculture 63, 155–167 (2008)
5. Schraml, R., Uhl, A.: Pith estimation on rough log end images using local fourier spectrum analysis. In: Proceedings of the 14th Conference on Computer Graphics and Imaging (CGIM 2013), Innsbruck, AUT (2013)
6. Cerda, M., Hitschfeld-Kahler, N., Mery, D.: Robust Tree-Ring Detection. In: Mery, D., Rueda, L. (eds.) PSIVT 2007. LNCS, vol. 4872, pp. 575–585. Springer, Heidelberg (2007)
7. Norell, K.: An automatic method for counting annual rings in noisy sawmill images. In: Foggia, P., Sansone, C., Vento, M. (eds.) ICIAP 2009. LNCS, vol. 5716, pp. 307–316. Springer, Heidelberg (2009)
8. Norell, K.: Counting annual rings on pinus sylvestris end faces in sawmill industry. Computers and Electronics in Agriculture 75, 231–237 (2010)
9. Longuetaud, F., Mothe, F., Kerautret, B., Krähenbühl, A., Hory, L., Leban, J.M., Debled-Rennesson, I.: Automatic knot detection and measurements from x-ray ct images of wood: A review and validation of an improved algorithm on softwood samples. Computers and Electronics in Agriculture 85, 77–89 (2012)
10. Schraml, R., Uhl, A.: Temporal and longitudinal variances in wood log cross-section image analysis. In: IEEE International Conference on Image Processing 2014 (ICIP 2014), Paris, FR (2014)
11. Boykov, Y., Jolly, M.P.: Interactive graph cuts for optimal boundary & region segmentation of objects in n-d images. In: Proceedings of the 8th IEEE International Conference on Computer Vision (ICCV 2001), vol. 1, pp. 105–112 (2001)
12. Boykov, Y., Funka-Lea, G.: Graph cuts and efficient n-d image segmentation. International Journal of Computer Vision 70, 109–131 (2006)
13. Chan, T., Vese, L.: Active contours without edges. IEEE Transactions on Image Processing 10, 266–277 (2001)

14. Chan, T., Esedoglu, S., Ni, K.: Histogram based segmentation using Wasserstein distances. In: Sgallari, F., Murli, A., Paragios, N. (eds.) SSVM 2007. LNCS, vol. 4485, pp. 697–708. Springer, Heidelberg (2007)
15. Comaniciu, D., Meer, P.: Mean shift: a robust approach toward feature space analysis. IEEE Transactions on Pattern Analysis and Machine Intelligence 24, 603–619 (2002)
16. Haindl, M., Mikeš, S.: Unsupervised texture segmentation using multiple segmenters strategy. In: Haindl, M., Kittler, J., Roli, F. (eds.) MCS 2007. LNCS, vol. 4472, pp. 210–219. Springer, Heidelberg (2007)
17. Unser, M.: Texture Classification and segmentation using wavelet frames. IEEE Transactions on Image Processing 4, 1549–1600 (1995)
18. Wang, B., Zhang, L.: Supervised texture segmentation using wavelet transform. Proceedings of the International Conference on Neural Networks and Signal Processing 2, 1078–1082 (2003)
19. Weldon, T.P., Higgins, W.E., Dunn, D.F.: Efficient gabor filter design for texture segmentation. Pattern Recognition 29, 2005–2015 (1996)
20. Eiterer, L., Facon, J., Menoti, D.: Postal envelope address block location by fractalbased approach. In: Proceedings of the 17th Brazilian Symposium on Computer Graphics and Image Processing, pp. 90–97 (2004)
21. Materka, A., Strzelecki, M.: Texture analysis methods - a review. Technical report, Institute of Electronics, Technical University of Lodz (1998)
22. de Oliveira Nunes, E., Conci, A.: Texture segmentation considering multiband, multiresolution and affine invariant roughness. In: XVI Brazilian Symposium on Computer Graphics and Image Processing, pp. 254–261 (2003)
23. Stitou, Y., Turcu, F., Berthoumieu, Y., Najim, M.: Three-dimensional textured image blocks model based on wold decomposition. IEEE Transactions on Signal Processing 55, 3247–3261 (2007)
24. Eckley, I.A., Nason, G.P., Treloar, R.L.: Locally stationary wavelet fields with application to the modelling and analysis of image texture. Journal of the Royal Statistical Society: Series C (Applied Statistics) 59, 595–616 (2010)
25. Atto, A., Berthoumieu, Y., Bolon, P.: 2-d wavelet packet spectrum for texture analysis. IEEE Transactions on Image Processing 22, 2495–2500 (2013)
26. Jung, M., Peyré, G., Cohen, L.D.: Texture segmentation via non-local nonparametric active contours. In: Boykov, Y., Kahl, F., Lempitsky, V., Schmidt, F.R. (eds.) EMMCVPR 2011. LNCS, vol. 6819, pp. 74–88. Springer, Heidelberg (2011)
27. Jung, M., Peyré, G., Cohen, L.D.: Nonlocal active contours. SIAM Journal on Imaging Sciences 5, 1022–1054 (2012)
28. Ojala, T., Pietikäinen, M., Harwood, D.: A comparative study of texture measures with classification based on featured distributions. Pattern Recognition 29, 51–59 (1996)
29. Ojala, T., Pietikainen, M., Maenpaa, T.: Multiresolution gray-scale and rotation invariant texture classification with local binary patterns. IEEE Transactions on Pattern Analysis and Machine Intelligence 24, 971–987 (2002)
30. Mäenpää, T.: The Local Binary Pattern Approach to Texture Analysis – Extensions and Applications. PhD thesis, University of Oulu (2003)
31. Rubner, Y., Tomasi, C., Guibas, L.: A metric for distributions with applications to image databases. In: Sixth International Conference on Computer Vision, pp. 59–66 (1998)
32. Edelsbrunner, H., Kirkpatrick, D., Seidel, R.: On the shape of a set of points in the plane. IEEE Transactions on Information Theory 29, 551–559 (1983)

Visual Security Evaluation Based on SIFT Object Recognition

Stefan Jenisch and Andreas Uhl

Multimedia Signal Processing and Security Lab (WaveLab)
Department of Computer Sciences, University of Salzburg
andreas.uhl@sbg.ac.at

Abstract. The paper presents a metric for visual security evaluation of encrypted images based on object recognition using the Scale Invariant Feature Transform (SIFT). The metrics' behavior is demonstrated using three different encryption methods and its performance is compared to that of the PSNR, SSIM and Local Feature Based Visual Security Metric (LFBVSM). Superior correspondance to human perception and better responsiveness to subtle changes in visual security are observed for the new metric.

1 Introduction

Today a number of (format compliant) image encryption techniques exist which allow the encrypted content to be decoded and viewed. To determine the level of security offered by these techniques it is not enough to simply evaluate the cryptographic strength of the encryption cipher used.

For some methods the decoded encrypted image is a low quality version of the original image and certain image features can still be recognised. So beside evaluating the encryption cipher also the *visual security* of the result has to be assessed. In this context we need to deal with the remaining image quality left behind by the encryption process and the recognizability and intelligibility of the encrypted image content.

In order to be able to discuss the exact notion of visual security, we need to distinguish distinct application scenarios of media encryption schemes [1]:

Confidentiality Encryption: Means MP security (mes- sage privacy). The formal notion is that if a system is MP- secure an attacker cannot efficiently compute any property of the plain text from the cipher text. This can only be achieved by the conventional encryption approach, i.e. applying a cryptographically strong cipher to compressed (redundancy-free) image data.

Content Confidentiality: Is a relaxation of confidential encryption. Side channel information may be reconstructed or left in plaintext, e.g. header information, packet length, but the actual visual content must be secure in the sense that the image content must not be intelligible / discernible.

Sufficient Encryption: Means we do not require full security, just enough security to prevent abuse of the data. The content must not be consumable due

L. Iliadis et al. (Eds.): AIAI 2014, IFIP AICT 436, pp. 624–633, 2014.

to high distortion (e.g. for DRM systems) by destroying visual quality to a degree which prevents a pleasant viewing experience or destroys the commercial value. This implicitly refers to message quality security (MQ), which requires that an adversary cannot reconstruct a higher quality version of the encrypted material than specified for the application scenario.

Given these different application scenarios it is clear that depending on the goal, a security metric has to fulfill dif- ferent roles. For example, under the assumption of sufficient encryption a given security metric would have to evaluate which quality is low enough to prevent a pleasant viewing experience.

When it comes to content confidentiality the question of quality is no longer applicable. Content confidentiality requires that image content must not be identified by human or automated recongnition. This requirement also has to be maintained for any part of the image. Image metrics, in general, do not deal with such questions but rate the overall image quality, the question of intelligibility is usually not covered at all. Thus, it seems to be clear that a general purpose metric covering all application scenarios is probably very hard or impossible to design.

Additionally we have to face the fact that different encryption methods introduce different kind of distortions. While some methods shift and morph the images (i.e. chaotic encryption which is mainly based on permutations) others introduce noise and noise like patterns. An ideal metric for assessment of visual security has to be able to deal with those different kind of distortions.

To evaluate the visual security of an encrypted image in an objective manner, often the Peak Signal to Noise Ratio (PSNR) or the Structural Similarity (SSIM) Index are used. Despite the fact that both originally have been developed for image quality assessment, they have also been used for the assessment of encrypted images [2,3,4,5].

Also several attempts have been made to develop a metric specifically for the task of visual security assessment. One popular example is the Local Feature Based Visual Security Metric (LFBVSM [6]), which compares corresponding image regions of the cipher and plain images by their luminance and contour information to evaluate the visual security of an encrypted image. Also further dedicated metrics for visual security evaluation have been proposed (e.g. [7,8,9]).

While there exist particular image encryption techniques for which PSNR, SSIM, and LFBVSM do a reasonable job to rate the visual security of a ciphered image, for many encryption methods these metrics tend to have troubles in the correct assessment of visual security in correspondence to visual perception (as we shall see in the experiments).

Since most of these metrics compare the plain and the cipher images pixel by pixel or region by region (fundamental principles of the Human Vision System (HVS) in terms of luminance and edge perception are considered) a warped image may still be recognisable while the metric rates the image as secure due to large dissimilarities in terms of pixel or local region differences. Also, noise patterns tend to decrease the score rater quickly but leave the content of the image still intelligible. Thus, answering the question if an encryption of this type results in a *content confidential* image, i.e. an image without any intelligible content, can become quite challenging with those metrics.

In this paper, we aim to apply object recognition methods to design a metric for visual security assessment in order to tackle the issue of content recognition and intelligibility in a more appropriate manner. The basic idea of the metric presented is to compare the recognizability of objects found in a reference and a cipher image instead of measuring image quality as such.

In particular, we propose to employ the Scale Invariant Feature Transform (SIFT) [10] for object recognition in ciphered images. Therefore, the metric is termed "Scale Invariant Feature Transform Similarity Score" (SIFTSS).

The paper is divided into four parts. First a description of the SIFTSS is given, followed by a description of the encryption methods used to test the metrics performance. Then the results of the experiments are presented and finally they are discussed in the last section of the paper.

2   SIFT Similarity Score

The SIFT algorithm derives a set of key-points for each image. Each key-point is associated with a descriptor vector (edge histograms). The images are compared using these key-points, i.e. all key-points of the target (ciphered) image are compared to the key-points found in the reference (original) image. The matching process compares the Euclidian distances between descriptor vectors of the reference key-points and the target key-points. A search for the minimum Euclidian distance between their descriptor vectors is carried out.

To improve matching performance, a validity check is performed. The minimum distance found is multiplied by the value of 1.5 and again compared to the set of distances. If the multiplied distance is still smaller than all other distances, the key-point pair involved is considered a match. The check ensures that the match of the key-points is distinctive or dominant in comparison to all other possible matches.

For the implementation of the metric the VLFeat.org [1] implementation of the SIFT algorithm was used. The matching process returns an array of matching image key-points along with their corresponding Euclidian distances, measured between their descriptor vectors. The number of matched key-points as well as the average Euclidian distance of the edge histograms is used to derive a matching score.

In the formula

$$\text{SIFTSS}(A, B) = \left(\frac{\dim(m)}{\min(n_A, n_B)} \right)^{\frac{\mu_m}{|m|_2}} \tag{1}$$

the calculation of the SIFTSS between the images A and B is shown, where m is a vector containing a list of the Euclidian distances of the matched key-points between A and B, and n_A, n_B is the total number of key-points found in each

---

[1] http://www.vlfeat.org

of the two images. The number of matched key-points $dim(m)$ is divided by the maximum number of possible matches, and thus mapped into the interval $[0:1]$. The term is then taken to the power of the average Euclidian distance μ_m divided by the L_2-Norm of m. This also maps the exponent into the interval $[0:1]$. Consequently, while an increasing number of matched key-points increases the score, large Euclidian distances decrease it. Since the SIFT matching process is not commutative and returns different values when matching `imageA` to `imageB` than when matching `imageB` to `imageA` the algorithm calculates both directions and averages the result to restore symmetry.

In Listing 1.1 a pseudo code for calculating the SIFTSS is shown.

Listing 1.1. The calculation of the SIFTSS

```
1   [matchScoresA] = SIFTmatch(imageA, imageB);
2   [matchScoresB] = SIFTmatch(imageB, imageA);
3
4   matchCountA = length(matchScoresA);
5   matchCountB = length(matchScoresB);
6
7   if (matchCountA == 0 || matchCountB == 0)
8       return 0
9   else
10      normA = Norm(matchScoresA);
11      normB = Norm(matchScoresB);
12      if (normA == 0 || normB == 0) then
13          return 0
14      else
15          normA = Norm(matchScoresA);
16          normB = Norm(matchScoresB);
17          if [normA == 0 || normB == 0] then;
18              return 1;
19          else
20              meanEuclidianDistanceA = mean(matchScoresA/normA);
21              meanEuclidianDistanceB = mean(matchScoresB/normB);
22
23              keyPointsOfA = getNumberOfKeypoints(imageA);
24              keyPointsOfB = getNumberOfKeypoints(imageB);
25
26              matchScoreA = matchCountA / min(keyPointsOfA, keyPointsOfB);
27              matchScoreA = power(matchScoreA, meanEuclidianDistanceA);
28
29              matchScoreB = matchCountB / min(keyPointsOfA, keyPointsOfB);
30              matchScoreB = power(matchScoreB, meanEuclidianDistanceB);
31
32              matchScore = (matchScoreA + matchScoreB)/2
33
34              return matchScore;
35
36          end
37      end
38  end
```

The range of SIFTSS values is between 0 and 1 where scores close to 0 signify a better visual security and 1 indicates identical images.

3   Experimental Settings

To give an overview of the metrics performance three case studies using different encryption methods have been selected and will be briefly described below. To establish a standard of comparison these cases are also evaluated using the PSNR, SSIM and LFBVSM. For the experiments the images of the *Kodak Lossless True Color Image Suite*[2] where cropped into a square format (which is required for Arnolds's Cat Map encryption), and scaled down to 150×150 pixels. Metrics results are averaged for this data set, all images have been encrypted using individual random encryption keys.

---

[2] http://r0k.us/graphics/kodak/

The first encryption approach uses Arnold's Cat Map, a chaotic map, for image encryption [11,12]. This type of encryption uses warp and shift operations for rendering the image unintelligible. The map works on an image of size NxN and is defined by the formula

$$\begin{pmatrix} x_{k+1} \\ y_{k+1} \end{pmatrix} = C(x_k, y_k) = \begin{pmatrix} 1 & b \\ a & ab+1 \end{pmatrix} * \begin{pmatrix} x_k \\ y_k \end{pmatrix} \bmod N \qquad (2)$$

where (x_k, y_k) satisfying $0 \leq x_k, y_k < N$ are the positions of the pixels in the original (square) image area while (x_{k+1}, y_{k+1}) is the position of the pixel in the target image, k is the number of the current iteration and $0 \leq a, b < N$, $a, b \in \mathbb{N}$ are the control parameters of the function used as key.

In each iteration the image is warped, cut and transformed back into its squared shape rendering the image more and more random. The operation has the property of a torus restoring the image after a discrete number of iterations. A visual explanation of a single iteration step is shown in Figure 1.

Fig. 1. Arnold's cat map in pictures

The iteration stages evaluated in the experiment were 0, 1, 3, 132, 155, 157, 200, 211, 250, 275, 299 and 300. In Figure 2 one of the encrypted images can be seen in all investigated iteration stages.

Fig. 2. Image transformed with Arnold's cat map (iterations ordered from left to right and top to bottom)

The second encryption method is integrated into the JPEG XR compression standard [13]. It is suggested to encrypt the DC coefficients using Random Level

Shifts (RLS), i.e. to alter the value of the DC coefficients by adding or subtracting random numbers which are derived from a key. To increase the impact of the encryption method on visual security we have recently suggested to apply RLS to all coefficients of a transform block, not only to its DC coefficient [1].

This encryption method introduces noise into the image and the impact on image perception can be seamlessly adjusted from low to high by setting the maximum allowed shift value accordingly. In Figure 3 sample images for increasing maximum shift values can be seen.

Fig. 3. RLS encryption using increasing maximum shift values (values from left to right: 80, 160, 280, 480, 800, 2000)

The third encryption method used for SIFTSS evaluation is the permutation of the coefficients' scan order in a JPEG XR code stream. This method has first been discussed for the JPEG standard [14] and was later proposed to be used for encrypting the LP frequency band of JPEG XR encoded images [15].

In JPEG XR the coefficients are grouped into three frequency bands, the DC-, the Lowpass- (LP) and Highpass (HP) band. In the experiment only the coefficients of the LP and HP band are subject to the permutation process [1]. Swapping of coefficients across frequency bands is not carried out.

As a result there are six possible encryption settings. For each of the two storage modes (spatial and frequency storage mode), the encryption of the LP, of the HP and of both frequency bands can be selected. Sample images for each encryption mode can be found in Figure 4.

To establish a subjective order in terms of visual security the images in Figure 4 are ordered from left (low security) to right (high security). Corresponding settings are: Spatial store mode and HP band encryption, Frequency store mode and HP band encryption, spatial store mode and LP band encryption, spatial store mode LP+HP band encryption, frequency store mode LP band encryption and frequency store mode LP+HP band encryption.

Fig. 4. Lena image scrambling using coefficient scan order permutation

4   Experimental Results

The objective image metrics return values for the first experiment using Arnold's Cat Map can be seen in Figure 5.

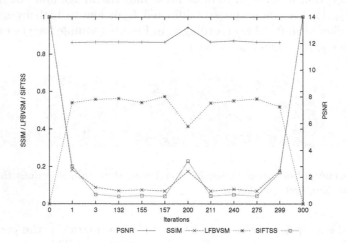

Fig. 5. Averaged return values for Arnold's Cat Map encrypted Kodak Image Database images

The desired behavior of a metric evaluating this set of images would be an indication of content intelligibility for iteration stages 1, 200 and 299 and a contraindication for all others (stages 0 and 300 set aside). Subjectively, these iteration stages do show some resemblance of the original image (see Figure 2).

The SIFTSS indeed shows the desired behaviour for the suspected iteration stages 1, 200 and 299. Interestingly, also the SSIM indicates intelligibility for these stages. The PSNR and the LFBVSM do not show exactly the desired behavior. Notice that no values for iteration stage 0 and 300 are plotted in Figure 5 since the metric returns ∞ for identical images. Both metrics indicate some intelligibility for iteration stage 200 when comparing the return values for this stage with iteration stages 3 to 275. However, they fail to indicate it for iteration stage 1 and 299 which show similar values as for all other iteration stages.

The reason why iteration stage 1 and 299 is handled well by the SSIM may be explained with the fact that most pixels are still neighbouring each other in these stages. Certain image regions are not moved far from their original position (bottom left and top right corner of the image) and additionally certain regions show the same structure after the operation (e.g. an area showing clear blue sky is replaced with clear blue sky during warping). The SSIM metric uses a sliding window approach assigning each pixel of the image (the pixel aligned to the center pixel of the window) a similarity score derived from the area covered from the window. Also the metric uses mainly mean luminance and variance to describe the local structure information but ignores edge orientation and magnitude information which becomes disturbed during the warp operation. The LFBVSM metric however, which also compares image regions, uses edge orientation and magnitiude information contrasting to SSIM. This is probably the reason why the LFBVSM does not indicate intelligibility for those two iteration stages and SSIM does.

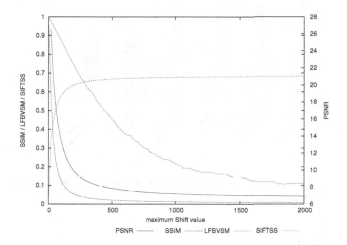

Fig. 6. Averaged return values for RLS encrypted Kodak Image Database images

The objective image metrics averaged return values for the RLS encrypted Kodak Image Database images can be seen in Figure 6. The PSNR, SSIM and LFBVSM values exhibit a steep drop when increasing the maximum shift level from 0 to 300. After reaching a maximum shift value of 300 their slopes flatten out. As can be seen in Figure 3, this behaviour does not correspond well with visual perception where a gradual worsening of image quality and content intelligibility is observed across the entire range of considered shift values.

The SIFTSS on the other hand shows a much less steep drop in its return values when increasing the maximum shift level and flattens out at a much later stage as compared to the other metrics. In the area of a maximum shift value from 500 to 2000 the SIFTSS is still tributing the changes in visual security with much more distinct return values than the other metrics do.

Finally, the objective image metrics' averaged return values for the Coefficient Scan Order Permutation encrypted Kodak Image Database images are shown in Figure 7. The SIFTSS values suggest an ordering with respect to visual security which corresponds to the subjective one shown in Figure 4. Contrasting to that, the return values of the PSRN, SSIM, and LFBVSM do not at all correspond to this ordering of the images. These metrics attest HP encrypted images in spatial store mode a better visual security than images which have an encrypted LP band (also in spatial store mode) which is found in the middle of the plot. Notice that the LFBVSM shows a valley or dent instead of a peak because the metric uses an inverted scale.

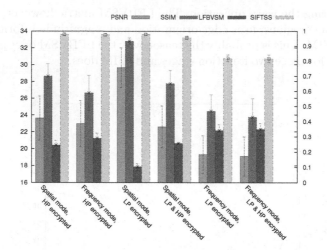

Fig. 7. Averaged return values for Coefficient Scan Order Permutation encrypted Kodak Image Database images

5   Conclusion

We have found that overall, SIFTSS is well suited to assess visual security and its ratings correspond well to subjective perception. In particular, for the three encryption techniques considered, SIFTSS clearly outperforms PSNR, SSIM, and LFBVSM in terms of correspondence to human perception and responsiveness to subtle changes in visual security.

On major drawback of SIFTSS is the computational effort required by the metric. Due to the complexity of the SIFT algorithm itself the metric is *significantly* more demanding as compared to PSNR and SSIM.

The behaviour of SIFTSS might be further improved by extracting additional characteristics from the area around each matched key-point, for example by including the mean luminance and deviation of the subregions around the key-point. These options have not been investigated so far and remain as a suggestion for future work.

References

1. Jenisch, S., Uhl, A.: A detailed evaluation of format-compliant encryption methods for JPEG XR-compressed images. EURASIP Journal on Information Security (6) (2014)
2. Au Yeung, S.K., Zhu, S., Zeng, B.: Quality assessment for a perceptual video encryption system. In: 2010 IEEE International Conference on Wireless Communications, Networking and Information Security (WCNIS), pp. 102–106 (June 2010)

3. Khan, M.I., Jeoti, V., Malik, A.S.: On perceptual encryption: Variants of DCT block scrambling scheme for JPEG compressed images. In: Kim, T.-h., Pal, S.K., Grosky, W.I., Pissinou, N., Shih, T.K., Ślęzak, D. (eds.) SIP/MulGraB 2010. CCIS, vol. 123, pp. 212–223. Springer, Heidelberg (2010)
4. Droogenbroeck, M.V., Benedett, R.: Techniques for a selective encryption of uncompressed and compressed images. In: Proceedings of ACIVS (Advanced Concepts for Intelligent Vision Systems), Ghent University, Belgium, pp. 90–97 (September 2002)
5. Yeung, S.K.A., Zhu, S., Zeng, B.: Partial video encryption based on alternating transforms. IEEE Signal Processing Letters 16(10), 893–896 (2009)
6. Tong, L., Dai, F., Zhang, Y., Li, J.: Visual security evaluation for video encryption. In: Proceedings of the International Conference on Multimedia, MM 2010, pp. 835–838. ACM, New York (2010)
7. Mao, Y., Wu, M.: Security evaluation for communication-friendly encryption of multimedia. In: Proceedings of the IEEE International Conference on Image Processing (ICIP 2004). IEEE Signal Processing Society, Singapore (2004)
8. Sun, J., Xu, Z., Liu, J., Yao, Y.: An objective visual security assessment for cipher-images based on local entropy. Multimedia Tools and Applications 53(1), 75–95 (2011)
9. Yao, Y., Xu, Z., Sun, J.: Visual security assessment for cipher-images based on neighborhood similarity. Informatica 33, 69–76 (2009)
10. Lowe, D.G.: Distinctive image features from scale-invariant keypoints. International Journal of Computer Vision 60(2), 91–110 (2004)
11. Chen, T.Y., Huang, C.H., Chen, T.H., Liu, C.C.: Authentication of lossy compressed video data by semi-fragile watermarking. In: Proceedings of the IEEE International Conference on Image Processing, ICIP 2004. IEEE, Singapore (2004)
12. Huang, L., Zhou, W., Jiang, R., Li, A.: Data quality inspection of watermarked GIS vector map. In: Proceedings of the 18th International Conference on Geoinformatics, Beijing, China (June 2010)
13. Sohn, H., De Neve, W., Ro, Y.M.: Privacy Protection in Video Surveillance Systems: Analysis of Subband-Adaptive Scrambling in JPEG XR. IEEE Transactions on Circuits and Systems for Video Technology 21(2), 170–177 (2011)
14. Tang, L.: Methods for encrypting and decrypting MPEG video data efficiently. In: Proceedings of the ACM Multimedia 1996, Boston, USA, pp. 219–229 (November 1996)
15. Sohn, H., DeNeeve, W., Ro, Y.: Region-of-interest scrambling for scalable surveillance video using JPEG XR. In: ACM Multimedia 2009, Beijing, China, pp. 861–864 (October 2009)

Author Index

636   Author Index

Printed in the United States
By Bookmasters